# CAROTENOIDS AND VITAMIN A IN TRANSLATIONAL MEDICINE

# OXIDATIVE STRESS AND DISEASE

Series Editors

## Lester Packer, PhD
## Enrique Cadenas, MD, PhD

### University of Southern California School of Pharmacy
### Los Angeles, California

# CAROTENOIDS AND VITAMIN A IN TRANSLATIONAL MEDICINE

EDITED BY

## OLAF SOMMERBURG
## WERNER SIEMS
## KLAUS KRAEMER

CRC Press
Taylor & Francis Group
Boca Raton London New York

CRC Press is an imprint of the
Taylor & Francis Group, an **informa** business

CRC Press
Taylor & Francis Group
6000 Broken Sound Parkway NW, Suite 300
Boca Raton, FL 33487-2742

First issued in paperback 2016

© 2013 by Taylor & Francis Group, LLC
CRC Press is an imprint of Taylor & Francis Group, an Informa business

No claim to original U.S. Government works

Version Date: 20140523

ISBN 13: 978-1-138-19947-7 (pbk)
ISBN 13: 978-1-4398-5526-3 (hbk)

### Library of Congress Cataloging-in-Publication Data

Carotenoids and vitamin A in translational medicine / edited by Olaf Sommerburg, Werner Siems, and Klaus Kraemer.
p. ; cm. -- (Oxidative stress and disease ; 33)
Includes bibliographical references and index.
ISBN 978-1-4398-5526-3 (hardcover : alk. paper)
I. Sommerburg, Olaf. II. Siems, Werner. III. Kraemer, Klaus, 1960- IV. Series: Oxidative stress and disease ; 33.
[DNLM: 1. Vitamin A--therapeutic use. 2. Carotenoids--therapeutic use. 3. Translational Medical Research. 4. Vitamin A--supply & distribution. 5. Vitamin A Deficiency--drug therapy. W1 OX626 / QU 167]

615.3'28--dc23                                                                    2012038319

**Visit the Taylor & Francis Web site at**
**http://www.taylorandfrancis.com**

**and the CRC Press Web site at**
**http://www.crcpress.com**

# Contents

## PART I  Carotenoids and Vitamin A in Metabolic Diseases

## PART II  Carotenoids and Vitamin A in Skin, Eye, and Ear Diseases

## PART III   Carotenoids and Vitamin A in Cancer

## PART IV   Carotenoids and Vitamin A in Other
             Diseases

## *PART V  Meeting Dietary Supply of Vitamin A and Carotenoids*

# Preface

This book, *Carotenoids and Vitamin A in Translational Medicine,* not only involves chapters on medical use of carotenoids and vitamin A in metabolic diseases, skin diseases, eye and ear diseases, cancer, lung diseases, and inflammatory bowel disease but also involves chapters on analytics of carotenoids and supply of carotenoids and vitamin A in developing countries.

## CAROTENOIDS: SOME OF THESE USEFUL ANTIOXIDANTS ARE PRECURSORS OF VITAMIN A

Carotenoids are fat-soluble plant pigments. More than 600 carotenoids have been identified in plants, algae, fungi, and bacteria. In 1831, the German scientist Wackenroder coined the term "carotene" while isolating the first of those compounds, beta-carotene, from carrots. Some carotenoids are important to human health. The most common carotenoids in the diet are alpha-carotene, beta-carotene, lutein, zeaxanthin, beta-cryptoxanthin, and lycopene.

Carotenoids are antioxidants that react with free radicals. Structurally, carotenoids consist of 11 (beta-carotene, zeaxanthin, lycopene) or 10 (alpha-carotene, lutein) conjugated double bonds responsible for their antioxidant capability. Antioxidants may protect cells from damage. Claims that carotenoids can protect against diseases such as cancer and cardiovascular diseases are primarily based on their antioxidant properties.

A subgroup of carotenoids including alpha-carotene, beta-carotene, and beta-cryptoxanthine can be converted into vitamin A (retinol). Vitamin A is important for vision, immune response, and growth. It is also important for turning on and off gene expression during cell division and differentiation. Carotenoids and vitamin A need small amounts of fat for absorption from the intestine. Dietary supplements of carotenoids are micronized and some contain oil, which makes them more bioavailable for the body than carotenoids in food.

Carotenoids and retinol (vitamin A) have been studied by their action in the protection of a considerable number of diseases, such as cardiovascular diseases, neurodegenerative diseases such as Alzheimer's disease (AD), certain types of cancer, age-related macular degeneration and other eye diseases, and skin diseases. Experimental evidence suggests that these compounds are important for protecting biological macromolecules against oxidative damage.

# CAROTENOIDS AND VITAMIN A IN VARIOUS FIELDS OF PREVENTION AND THERAPY

## CAROTENOIDS AND CARDIOVASCULAR RISK

Oxidative stress has been considered universally and undeniably implicated in the pathogenesis of all major diseases, including those of the cardiovascular system. Dietary intervention trials suggest that diets rich in fruits and vegetables lead to improvements in coronary risk factors and reduce cardiovascular mortality. The low-density lipoprotein (LDL) oxidation process is one of the most important first steps of atherosclerotic disease and, consequentially, the first pathogenetical step of cerebro- and cardiovascular events like myocardial infarction and stroke, which are the primary causes of death worldwide. Reactive oxygen species (ROS) also seem to be the target of carotenoids' main action, by scavenging singlet oxygen and free radicals. Data in the literature show that ROS increase atherosclerotic individual burden. The scavenging actions by carotenoids could reduce atherosclerosis progression partly or even mainly due to such a decrease in ROS concentrations. Many studies demonstrated such a reduction by analyzing the relationship between carotenoids and intima-media thickness (IMT) of the carotid artery wall, an established marker of reduction of atherosclerosis [1].

## CAROTENOIDS AND RETINOL AND NEURODEGENERATIVE DISEASES SUCH AS ALZHEIMER'S DISEASE

The deposition of amyloid beta-protein in the brain is an invariant feature of AD. Retinal plays a role in maintaining higher functional level in the central nervous system. Plasma or cerebrospinal fluid concentrations of retinol and beta-carotene have been reported to be lower in AD patients. Additionally, these nutrients have been clinically shown to slow the progression of dementia. Retinol, retinal, and retinoic acid as well as beta-carotene have been shown to inhibit the formation, extension, and destabilizing effects of beta-amyloid fibrils. Recently, the inhibition of the oligomerization of amyloid beta has been suggested as a possible therapeutic target for the treatment of AD. The inhibitory effects of vitamin A and beta-carotene on the oligomerization of Aβ40 (amyloid beta 40) and Aβ42 (amyloid beta 42) in vitro have recently been demonstrated. In previous in vivo studies, intraperitoneal injections of retinol decreased brain amyloid beta deposition and tau phosphorylation in transgenic mouse models of AD, attenuated neuronal degeneration, and improved spatial learning and memory. Thus, retinol and beta-carotene could be key molecules for the prevention and therapy of AD [2].

## CANCER CHEMOPREVENTION BY CAROTENOIDS

Dietary intake of carotenoids is inversely associated with the risk of a variety of cancers in different tissues. Preclinical studies have shown that some carotenoids have potent antitumor effects both in vitro and in vivo, suggesting potential preventive and/or therapeutic roles for the compounds. Since chemoprevention is one of the most important strategies in the control of cancer development, molecular mechanism–based cancer chemoprevention using carotenoids seems to be an attractive approach. Various carotenoids, such as beta-carotene, alpha-carotene, lycopene, lutein, zeaxanthin, beta-cryptoxanthin, fucoxanthin, canthaxanthin, and astaxanthin, have been proven to have anticarcinogenic activity in several tissues, although high doses of beta-carotene failed to exhibit chemopreventive activity in clinical trials. The possible mechanisms of anticarcinogenic activity of carotenoids were described intensively [3]. Very recently, studies have shown that patients with cervical intraepithelial neoplasia respond favorably to beta-carotene supplementation [4].

## CAROTENOIDS AGAINST EYE DISEASES AND FOR EYE FUNCTION

In retinitis pigmentosa, patients typically lose night vision in adolescence, side vision in young adulthood, and central vision in later life because of progressive loss of rod and cone photoreceptor cells. Measures of retinal function, such as the electroretinogram, show that photoreceptor function is diminished generally many years before symptomic night blindness, visual-field scotomas, or decreased visual acuity arise. More than 45 genes for retinitis pigmentosa have been identified. Findings of controlled trials indicate that nutritional interventions, including vitamin (retinyl palmitate) and omega-3 fatty acid–rich fish, slow progression of disease in many patients [5].

Lutein is concentrated in the primate retina, where lutein together with zeaxanthin forms the macular pigment. Traditionally, lutein is characterized by its blue light filtering and antioxidant properties. Eliminating lutein from the diet of experimental animals results in early degenerative signs in the retina while patients with an acquired condition of macular pigment loss (*Macular telangiectasia*) show serious visual handicap, indicating the importance of macular pigment. Whether lutein intake reduces the risk of age-related macular degeneration (AMD) or cataract formation is currently an intensively investigated matter of debate. SR-B1 has recently been identified as a lutein-binding protein in the retina, and this same receptor plays a role in the selective uptake in the gut. In the blood, lutein is transported via HDL. Genes controlling SR-B1 and HDL levels predispose to AMD, which supports the involvement of cholesterol/lutein transport pathways. Apart from the beneficial effects of lutein intake on various visual function tests, recent findings show that lutein can affect immune responses and inflammation. Lutein diminishes the expression of ocular inflammation. In vitro studies show that lutein suppresses NF kappa-B activation as well as the expression of iNOS and COX-2. Since AMD has features of a chronic low-grade systemic inflammatory response, attention to the exact role of lutein in this disease has shifted from a local effect in the eye toward a possible systemic anti-inflammatory function [6].

## SUFFICIENT SUPPLY OF RETINOL IS ESSENTIAL FOR
## NORMAL REPRODUCTION AND DEVELOPMENT

The requirement for vitamin A in reproduction was first recognized in the early 1900s, and its importance in the eyes of developing embryos was realized shortly after. A greater understanding of the large number of developmental processes that require retinol emerged first from nutritional deficiency studies in rat embryos and later from genetic studies in mice. It is now generally believed that all-*trans* retinoic acid (RA) is the form of vitamin A that supports both male and female reproduction as well as embryonic development. This conclusion is based on the ability to reverse most reproductive and developmental blocks found in vitamin A deficiency induced either by nutritional or genetic means with RA and the ability to recapitulate the majority of embryonic defects in retinoic acid receptor compound null mutants. The activity of the catabolic CYP26 enzymes in determining what tissues have access to RA has emerged as a key regulatory mechanism and helps explain why exogenous RA can rescue many vitamin A deficiency defects. In severely vitamin A–deficient (VAD) female rats, reproduction fails prior to implantation, whereas in VAD pregnant rats given small amounts of carotene or supported on limiting quantities of RA early in organogenesis, embryos form but show a collection of defects called the vitamin A deficiency syndrome or late vitamin A deficiency. Vitamin A is also essential for maintenance of the male genital tract and spermatogenesis. Recent studies show that vitamin A participates in a signaling mechanism to initiate meiosis in the female gonad during embryogenesis and in the male gonad postnatally. Both nutritional and genetic approaches are being used to elucidate the vitamin A–dependent pathways upon which these processes depend [7].

## ANTIAGING POTENTIAL OF CAROTENOIDS

As an example, effects of astaxanthin should be briefly explained. Astaxanthin, a xanthophyll carotenoid, is a nutrient with diverse clinical benefits. This molecule neutralizes free radicals or other oxidants by either accepting or donating electrons, and without being destroyed or becoming a pro-oxidant in the process. Its linear, polar–nonpolar–polar molecular layout equips it to precisely insert into the membrane and span its entire width. In this position, astaxanthin can intercept reactive molecular species within the membrane's hydrophobic interior and along its hydrophilic boundaries. In double-blind, randomized controlled trials, astaxanthin lowered oxidative stress in overweight and obese subjects and in smokers. It blocked oxidative DNA damage and lowered inflammatory parameters such as C-reactive protein (CRP). Astaxanthin lowered triglyceride concentration and raised HDL-cholesterol concentration in blood plasma in another trial and improved blood flow in an experimental microcirculation model. It improved cognition in a small clinical trial and boosted proliferation and differentiation of cultured nerve stem cells. In several Japanese trials, astaxanthin improved visual acuity and eye accommodation. It improved reproductive performance in men and reflux symptoms in *Helicobacter pylori* patients. In preliminary trials, it showed promise for sports performance (soccers). In cultured cells, astaxanthin protected the mitochondria against endogenous oxygen radicals, conserved their antioxidative capacity, and enhanced their energy production efficiency. The concentrations used in these cells would be attainable in humans by modest dietary intakes. Astaxanthin's clinical success extends beyond protection against oxidative stress and inflammation to demonstrable promise for slowing age-related functional decline [8].

## Access to Evaluate Complexity: A Carotenoid Health Index Based on Plasma Carotenoids and Health Outcomes

In a review of Donaldson [9], 62 studies on total plasma carotenoids and health outcomes, mostly prospective cohort studies or population-based case–control studies, are analyzed together to establish a carotenoid health index. Five cutoff points are established across the percentiles of carotenoid concentrations in populations, from the 10th to the 90th percentile. The cutoff points (mean ± standard error of the mean) are 1.11 ± 0.08, 1.47 ± 0.08, 1.89 ± 0.08, 2.52 ± 0.13, and 3.07 ± 0.20 $\mu M$ total carotenoids. For all-cause mortality, there seems to be a low threshold effect with protection above every cutoff point but the lowest. But for metabolic syndrome and cancer outcomes, there tends to be significant positive health outcomes only above the higher cutoff points, perhaps as a triage effect. Based on these data, a carotenoid health index is proposed with risk categories as follows: very high risk: <1 $\mu M$, high risk: 1–1.5 $\mu M$, moderate risk: 1.5–2.5 $\mu M$, low risk: 2.5–4 $\mu M$, and very low risk: >4 $\mu M$. Over 95% of the U.S. population falls into the moderate- or high-risk category of the carotenoid health index [9].

## Analytics and Biotechnology

High-performance liquid chromatography (HPLC) has become the method of choice for carotenoid analysis. Although a number of normal-phase columns have been reported, reverse-phase columns are the most widely used stationary phases for the analysis of these molecules. C18 and C30 stationary phases have provided good resolution for separations of geometrical isomers with similar polarity. More recently, ultra-high-performance liquid chromatography (UHPLC) has been used. UHPLC has a number of distinct advantages over conventional HPLC. These include faster analyses, narrower peaks giving increased signal-to-noise ratio, and higher sensitivity. High strength silica (HSS) T3 and C18 and ethylene-bridged hybrid (BEH) C18 stationary phases, with sub-2 $\mu m$ particles have been used successfully for UHPLC analysis and separation of carotenoids. A number of spectroscopic and MS techniques have also been used for carotenoid qualitative and quantitative analysis. Matrix-assisted laser desorption ionization time-of-flight mass spectrometry (MALDI/TOF-MS), atmospheric-pressure solids-analysis probe (ASAP), and Raman spectroscopy are used to profile rapidly and qualitative carotenoids present in different extracts. Such detection methods can be used directly for the analysis of samples without the need for sample preparation or chromatographic separation. Consequently, they allow for a fast screen for the detection of multiple samples. Quantitative carotenoid analysis can be carried out using absorbance or mass detectors. Liquid chromatography–tandem mass spectrometry (LC–MS/MS) is efficient for carotenoid identification through the use of transitions for the detection of samples through precursor and daughter ions. This approach is suitable for the identification of carotenoids with the same molecular mass but different fragmentation patterns. Recently, reliable and precise analyses of total carotenoids and vitamin A in food and biological fluids have become possible using portable devices (iCheck™). This has been a significant breakthrough for working in developing countries, as there is no need for shipment of samples out of the country. The latest developments for qualitative and quantitative analysis of carotenoids were reviewed by Rivera and Canela-Garayoa [10].

Carotenoids are isoprenoids with a long polyene chain containing 3–15 conjugated double bonds, which determines their absorption spectrum. Cyclization at one or both ends occurs in hydrocarbon carotene, while xanthophylls are formed by the introduction of oxygen. In addition, modifications involving chain elongation, isomerization, or degradation are also found. The composition of carotenoids in food may vary depending on production practices, postharvest handling, processing, and storage. In higher plants, they are synthesized in the plastid. Both mevalonate-dependent and -independent pathways for the formation of isopentenyl diphosphate are known. Isopentenyl diphosphate undergoes a series of addition and condensation reactions to form phytoene, which gets converted to lycopene. Cyclization of lycopene either leads to the formation of beta-carotene and its derivative xanthophylls, beta-cryptoxanthin, zeaxanthin, antheraxanthin, and violaxanthin or alpha-carotene and lutein. Even though most of the carotenoid biosynthetic genes have been cloned and identified, some aspects of carotenoid formation and manipulation in higher plants especially remain poorly understood. In order to enhance the carotenoid content of crop plants to a level that will be required for the prevention of diseases, there is a need for research in both the basic and the applied aspects [11]. Many studies investigate detailed biological and medicinal effects of carotenoids. For example, Miyashita [12] found antiobesity and antidiabetic effects as specific and novel biofunctions of fucoxanthin. A nutrigenomic study revealed that fucoxanthin induces uncoupling protein 1 expression in white adipose tissue (WAT) mitochondria that leads to oxidation of fatty acids and heat production in WAT. Fucoxanthin improves insulin resistance and decreases blood glucose level, at least in part, through the downregulation of tumor necrosis factor-alpha in WAT of animals. Thus, the specific regulation of fucoxanthin on a particular biomolecule will be responsible for the characteristic chemical structures, which differ depending on the length of the polyene, nature of the end group, and various substituents they contain. The key structure of carotenoids for the expression of antiobesity effect was suggested to be carotenoid end of the polyene chromophore containing an allenic bond and two hydroxyl groups [12].

## CAROTENOID MEETINGS AND COMMUNITY OF CAROTENOID RESEARCHERS: GOING FROM BASIC SCIENCE TO TRANSLATIONAL MEDICINE

Carotenoids have been the focus of research efforts in the fields of plant biology, chemistry, biochemistry, physiology, nutrition, and medicine for over a century. The Gordon Research Conferences on Carotenoids have been held triennially since 1992. The 2010 7th Gordon Research Conference on Carotenoids in Ventura, California (chair Susan T. Mayne, vice chairs Eleanore T. Wurtzel and Xiang-Dong Wang) upheld the tradition of bringing together multidisciplinary research investigators at the forefront of carotenoid science. The conference showcased exciting developments and updates with presentations in genomics/modeling/systems biology; biosynthesis and regulation; photosynthesis; metabolic engineering of provitamin A carotenoids; carotenoid transport and metabolism; biological actions of carotenoids and their metabolites; and carotenoids and chronic disease prevention, including eye, cognition, metabolic syndrome, and cancer. Academic, industrial, and government organizations presented their findings.

Also the 2013 8th Gordon Research Conference on Carotenoids (chairs Eleanore T. Wurtzel and Xiang-Dong Wang, vice chair Johannes von Lintig) which was held in conjunction with the Carotenoids Gordon-Kenan Research Seminar again in Ventura, California, was an outstanding opportunity to promote growth, development, and open communication of frontier developments in carotenoid science. The collegial atmosphere of this conference provided a forum for scientists from different disciplines to brainstorm and promote cross-disciplinary collaborations between carotenoid researchers worldwide.

It should be mentioned that there exists a great tradition in the field of carotenoid research with the International Carotenoid Society and especially with International Symposia on Carotenoids. The 17th International Symposium on Carotenoids will be held in 2014 in Salt Lake City, Utah, and will be hosted and organized by Paul S. Bernstein who is a leading pioneer of carotenoid research. The 16th International Symposium on Carotenoids was held in 2011 in the World Heritage city of Krakow, Poland, and was hosted by Kazimierz Strzalka at the Jagiellonian University.

It is important to thank scientists who are prominent and who have made outstanding contributions in research and in the organization of scientific meetings in the field of carotenoids and vitamin A. The list includes—Norman Krinsky, James Allen Olson, Garry J. Handelman, Hideki Hashimoto, John T. Landrum, Bruce N. Ames, Fiji Yamashita, Elisabeth J. Johnson, Frederick Khachik, Paul S. Bernstein, Richard Bone, Micheline M. Mathews-Roth, Omer Kucuk, Wataru Miki, Takashi Maoka, Susan T. Mayne, George Britton, Elli Wurtzel, Erich Grotewold, Julie Mared, Sherry Tanumihardjo, Harry A. Frank, Cale A. Cooper, Cheryl L. Rock, Wendy S. White, Robert M. Russell, John Erdman, William S. Blaner, A. Catherine Ross, Kalanithi Nesaretnam, John Bertram, Helmut Sies, Georg Lietz, Bruno Roberts, Wolfgang Schalch, Carmen Socaciu, Peter M. Eckl, Kazimierz Strzalka, Johannes von Lintig, Xiang-Dong Wang, Donald S. McLaren, Hans Konrad Biesalski, Alfred Sommer, Keith P. West, Jr., Richard D. Semba, and Wilhelm Stahl.

Special thanks have to be directed to Sight and Life, a nonprofit nutrition think tank of DSM, which conducts important activities in the worldwide fight against vitamin A and other micronutrient deficiencies. It is also important to mention and highlight the former activities of the International Vitamin A Consultative Group (IVACG) and the International Nutritional Anemia Consultative Group (INACG), which were set up more than 30 years ago to assist in the control of specific nutritional deficiencies. IVACG and INACG have been replaced by the all-encompassing Micronutrient Forum (MNF), which started work in 2006 and disbanded after a very successful forum held in Beijing. However, in late 2011, the MNF was revitalized by a group of organizations and individuals with a common interest in improving the nutritional status of populations and in strengthening our ability to prevent micronutrient malnutrition.

## SOME MILESTONES IN THE HISTORY OF CAROTENOIDS AND VITAMIN A

Table 1 presents some of the most important milestones in the history of carotenoid and vitamin A research.

## TABLE 1
## Milestones in the History of Carotenoid and Vitamin A Research— From the Basic Sciences to Translational Medicine

| Year | Discovery |
|---|---|
| 1831 | Isolation of an orange-yellow pigment in carrots which was named carotene (Wackenroder) |
| 1866 | Classification of carotene as a hydrocarbon (Arnaud et al.) |
| 1887 | Widespread presence of carotenes in plants investigated (Arnaud) |
| 1905 | William Fletcher determined that diseases occur if special factors were removed from food; those factors were later discovered and called vitamins |
| 1907 | Consistence of one molecule carotene of 40 C and 56 H atoms was established (Willstatter and Mieg) |
| 1912 | Polish scientist Casimir Funk names the special nutritional parts of food as a "vitamine"/later vitamin after "vita" meaning life and "amine" from compounds found in the thiamine he isolated from rice husks. Together, Hopkins and Funk formulated the vitamin hypothesis of deficiency disease |
| 1912–1914 | First findings of McCollum and Davis and also of Osborne and Mendel on fat soluble substances, e.g., in butter, later known as vitamin A |
| 1914 | First detection of carotene and xanthophylls in human blood plasma (Palmer and Eckles) |
| 1916–1917 | Substances necessary to keep cattle healthy were discovered and called "fat-soluble factor A" (leading to the later name vitamin A) (Hart/McCollum and collaborators [13]; Osborne and Mendel [14]) |
| 1929 | Demonstration of beta-carotene conversion into colorless form of vitamin A in the liver (Moore) |
| 1931 | Determination of structures of beta-carotene and vitamin A (Karrer et al.) |
| 1937 | Paul Karrer receives The Nobel Prize in Chemistry for his investigations on carotenoids, flavins, and vitamins A and B2 |
| 1939 | Conversion of beta-carotene into vitamin A in intestinal mucosa was demonstrated (Wagner et al.) |
| 1946–1950 | Synthesis of biologically active derivative of vitamin A/corresponding methyl ether (Isler et al. [15]), synthesis of the carboxylic acid corresponding to vitamin A (Arens and Dorp [16]), synthesis of beta-carotene was established (1950 Isler and collaborators) |
| 1966 | Beta-carotene was found to be acceptable for use in foods by the Joint FAO/WHO Expert Committee on Food Additives |
| 1979 | Carotene is established as "GRAS," which means that the ingredient is "generally recognized as safe" and can be used as a dietary supplement or in food fortification |
| 1981 | Suggestion that beta-carotene/carotenoid are recognized as important factors (independent of their provitamin A activity) in the potential reduction of the risk of certain cancers (Doll and Peto [17] and Shekelle et al. [18]) and that intake of carotenoid-rich foods is associated with reduced risk of certain cancers |
| 1982 | Investigations of the interaction between oxygen and oxyradicals using carotenoids (Krinsky and Deneke) |

TABLE 1 (continued)

## Milestones in the History of Carotenoid and Vitamin A Research— From the Basic Sciences to Translational Medicine

| Year | Discovery |
|------|-----------|
| 1983 | The U.S. National Cancer Institute (NCI) launches large-scale clinical intervention trials using beta-carotene supplements alone and in combination with other nutrients |
| 1984 | Beta-carotene is demonstrated to be an effective antioxidant in vitro |
| 1984 | Vitamin A deficiency demonstrated to be major cause of child mortality (Sommer) |
| 1988 | Due to the large number of epidemiological studies that demonstrate the potential reduction of cancer incidence with increased consumption of dietary beta-carotene, the U.S. National Cancer Institute (NCI) issues dietary guidelines advising Americans to include a variety of vegetables and fruits in their daily diet |
| 1993 | Availability of results from several large-scale clinical intervention trials using beta-carotene alone or in various other combinations |
| 1997 | Evidence indicates that beta-carotene acts synergistically with vitamins C and E |
| 1999 | The Women's Health Study showed no increased risk of lung cancer for women receiving 50 mg beta-carotene every second day; later, it was found that heavy smokers and asbestos workers have an increased risk of lung cancer if they are supplemented with high dosage of beta-carotene/retinol (ATBC and CARET trials) |
| 2001 | The U.S. Institute of Medicine recommended the retinol activity equivalent (RAE). 1 μg RAE corresponds to 1 μg retinol, 2 μg of beta-carotene in oil, 12 μg of "dietary" beta-carotene, or 24 μg of the three other dietary provitamin-A carotenoids |
| 2001 | Molecular analysis of vitamin A formation (von Lintig and Wyss [19]) |
| 2004 | Results from the French SU.VI.MAX study indicate that a combination of antioxidant vitamins (C, E, and beta-carotene) and minerals lowers total cancer incidence and all-cause mortality in men |
| 2012 | Overview on metabolism of carotenoids and retinoids related to vision (von Lintig [20]) |

## THIS BOOK IS DEDICATED TO HELMUT SIES

Helmut Sies (Figure 1) was born in Goslar in 1942. He is a renowned German physician and biochemist. He has worked and published extensively on micronutrients including carotenoids in relation to biological and medicinal effects of those compounds. Helmut Sies studied medicine at the universities of Tuebingen, Munich, and Paris. He received his MD in 1967 and his habilitation in 1972 from the University of Munich.

**FIGURE 1**   Prof. Helmut Sies in his office in Duesseldorf, Germany.

Helmut Sies had a distinguished career as a scientist and university professor. He served for many years as a professor at Heinrich-Heine University in Duesseldorf, Germany. He worked as a visiting professor at the University of California, Department of Biochemistry; as Burroughs-Wellcome Visiting Professor of Pharmacology at the University of Texas; as Professore Contratto at University degli Studi di Siena; as Miller Visiting Professor at the University of California at Berkeley; as visiting professor at Heart Research Institute in Sydney; as distinguished lecturer at the University of Texas, German-American Academic Council, Austin, Texas; as adjunct professor at the University of Southern California, Department of Pharmacology and Pharmacological Sciences, Los Angeles; and as Burroughs-Wellcome Visiting Professor in Basic Medical Sciences at the University of Southern California, Los Angeles. He is still active as a professor of biology and biochemistry at King Saud University in Riyadh, Saudi Arabia, and as senior scientist at Institut für Umweltmedizinische Forschung (IUF) Duesseldorf, Germany.

Helmut Sies' research interests include biological oxidations, oxidative stress, oxidants and antioxidants, photooxidation, singlet oxygen, biochemical pharmacology and toxicology, nutritional biochemistry, micronutrients such as carotenoids, flavonoids, and selenium, and biochemical processes in liver and skin. He described lycopene as an effective antioxidant and radical scavenger. He has also intensively investigated the protection of the skin against UV radiation by lycopene, other carotenoids, and flavonoids.

Helmut Sies has published more than 500 scientific papers on carotenoids [21–23]. He has been conferred the following awards: with the FEBS Anniversary Prize; with a silver medal from Karolinska Institute Stockholm; with the Ernst Jung Award for Medicine; with the Claudius Galenus Award; with the ISFE Award; and with the Werner Heisenberg Medal of Alexander von Humboldt Foundation. He has also been recognized as an NFCR Fellow (National Foundation of Cancer Research, Bethesda, Maryland) and as Distinguished Foreign Scholar by MASUA (Mid-America State Universities Association). He is Dr. h.c. of University of Buenos Aires. Since 1991 Helmut Sies is Member of Academy of Sciences and Arts of Nordrhein-Westfalen.

Some of the books that Sies has published and coedited include *Carotenoids and Retinoids. Molecular Aspects and Health Issues* [24], *Oxidative Stress and Inflammatory Mechanisms in Obesity, Diabetes, and the Metabolic Syndrome* [25],

*Quinones and Quinone Enzymes* [26], *Antioxidants in Disease Mechanisms and Therapeutic Strategies* [27], *Protein Sensors of Reactive Oxygen Species* [28,29], *Glutathione Transferases and Gamma-Glutamyl Transpeptidases* [30], *Carotenoids and Retinoids* [31], *Oxygen Biology and Hypoxia* [32], *Micronutrients and Brain Health* [33], and *Carotenoids in Health and Disease* [34].

Helmut Sies is very active in research and development in the field of carotenoids. He was chair of a number Gordon Conferences on Carotenoids and served in the Federation of European Biochemical Societies (FEBS), the International Society for Free Radical Research (SFRR), and the Oxygen Club of California. He is editor in chief of *Free Radical Research* and plays an active role in the publication of leading, high-impact journals such as *Archives in Biochemistry and Biophysics* and others. We would like to thank him for being an excellent listener, mentor, speaker, partner in discussion, and friend. We would also like to thank him for his advice, collaboration, and friendship.

**Olaf Sommerburg**
*Heidelberg, Germany*

**Werner Siems**
*Bad Harzburg, Germany*

**Klaus Kraemer**
*Basel, Switzerland*

## REFERENCES

1. Giordano P, Scicchitano P, Locorotondo M, Mandurino C, Ricci G, Carbonara S, Gesualdo M et al. Carotenoids and cardiovascular risk. *Curr Pharm Des* 2012 Jun 28 [Epub ahead of print].
2. Ono K, Yamada M. Vitamin A and Alzheimer's disease. *Geriatr Gerontol Int* 2012 Apr; 12(2): 180–188. doi: 10.1111/j.1447-0594.2011.00786.x [Epub 2011 Dec 23].
3. Tanaka T, Shnimizu M, Moriwaki H. Cancer chemoprevention by carotenoids. *Molecules* 2012; 17(3): 3202–3242.
4. Fuchs-Tarlovsky V, Bejarano-Rosales M, Gutiérrez-Salmeán G, Casillas MA, López-Alvarenga JC, Ceballos-Reyes GM. Effect of antioxidant supplementation over oxidative stress and quality of life in cervical cancer. *Nutr Hosp* 2011; 26(4): 819–826.
5. Hartong DT, Berson EL, Drvia TP. Retinitis pigmentosa. *Lancet* 2006; 368(9549): 1795–1809.
6. Kijlstra A, Tian Y, Kelly ER, Berendschot TT. Lutein: More than just a filter for blue light. *Prog Retin Eye Res* 2012; 31(4): 303–315.
7. Clagett-Dame M, Knutson D. Vitamin A in reproduction and development. *Nutrients* 2011; 3(4): 385–428.
8. Kidd P. Astaxanthin, cell membrane nutrient with diverse clinical benefits and anti-aging potential. *Altern Med Rev* 2011; 16(4): 355–364.
9. Donaldson MS. A carotenoid health index based on plasma carotenoids and health outcomes. *Nutrients* 2011; 3(12): 1003–1022.
10. Rivera SM, Canela-Garayoa R. Analytical tools for the analysis of carotenoids in diverse materials. *J Chromatogr A* 2012 Feb 10; 1224: 1–10 [Epub 2011 Dec 13].

11. Namitha KK, Negi PS. Chemistry and biotechnology of carotenoids. *Crit Rev Food Sci Nutr* 2010; 50(8): 728–760.
12. Miyashita K. Function of marine carotenoids. *Forum Nutr* 2009; 61: 136–146.
13. Hart EB, McCollum EV, Steenbock H, Humphrey GC. Physiological effect on growth and reproduction of rations balanced from restricted sources. *Proc Natl Acad Sci USA* 1917; 3(5): 374–382. With the citation: McCollum EV, Kennedy C. *J Biol Chem* 1916; 24: 491.
14. Osborne TB, Mendel LB. The growth of rats upon diets of isolated food substances. *Biochem J* 1916; 10(4): 534–538.
15. Isler O, Kofler M, Huber W, Ronco A. *Experientia* 1946; 2: 31. See also: Isler, O. (Ed.) *Carotenoids*. Birkhäuser Verlag, Basel, Stuttgart 1971 and Bodnigh J, Camma HR, Collins FD, Morton RA, Gridgeman NT, Isler O, Kofler M, Taylor RJ, Welland AS, Bradbury T. Pure all-trans vitamin A acetate and the assessment of vitamin A potency by spectrophotometry. *Nature* 1951; 168(4275): 598.
16. Arens JF, Dorp DA. Synthesis of some compounds possessing vitamin A activity. *Nature* 1946; 157: 190.
17. Doll R, Peto R. Can dietary beta-carotene materially reduce human cancer rates? *Nature* 1981; 290: 201–208.
18. Shekelle RB, Lepper M, Liu S, Maliza C, Raynor WJ Jr, Rossof AH, Paul O, Shryock AM, Stamler J. Dietary vitamin A and risk of cancer in the Western Electric study. *Lancet* 1981 Nov 28; 2(8257): 1185–1190.
19. von Lintig J, Wyss A. Molecular analysis of vitamin A formation: Cloning and characterization of beta-carotene 15,15′-dioxygenases. *Arch Biochem Biophys* 2001; 385(1): 47–52.
20. von Lintig J. Metabolism of carotenoids and retinoids related to vision. *J Biol Chem* 2012; 287(3): 1627–1634.
21. Stahl W, Sies H. Photoprotection by dietary carotenoids: Concept, mechanisms, evidence and future development. *Mol Nutr Food Res* 2012; 56(2): 287–295.
22. De Spirt S, Sies H, Tronnier H, Heinrich U. An encapsulated fruit and vegetable juice concentrate increases skin microcirculation in healthy women. *Skin Pharmacol Physiol* 2012; 25(1): 2–8.
23. Erhardt A, Stahl W, Sies H, Lirussi F, Donner A, Häussinger D. Plasma levels of vitamin E and carotenoids are decreased in patients with nonalcoholic steatohepatitis (NASH). *Eur J Med Res* 2011; 16(2): 76–78.
24. Packer L, Obermueller-Jevic U, Kraemer K, Sies H (Eds.). *Carotenoids and Retinoids. Molecular Aspects and Health Issues*. AOCS Press, Champaign, IL, 2006.
25. Packer L, Sies H (Eds.). *Oxidative Stress and Inflammatory Mechanisms in Obesity, Diabetes, and the Metabolic Syndrome*. CRC Press Inc., Boca Raton, FL, 2007.
26. Sies H, Packer L (Eds.). *Quinones and Quinone Enzymes. Part B: 382 (Methods in Enzymology)*. Academic Press Inc., San Diego, CA, 2004.
27. Sies H (Ed.). *Antioxidants in Disease Mechanisms and Therapeutic Strategies (Advances in Metabolic Disorders)*. Academic Press Inc., San Diego, CA, 1996.
28. Sies H, Packer L (Eds.). *Protein Sensors of Reactive Oxygen Species. Part A: Selenoproteins and Thioredoxin: 347 (Methods in Enzymology)*. Academic Press Inc., San Diego, CA, 2002.
29. Sies H, Packer L (Eds.). *Protein Sensors of Reactive Oxygen Species. Part B: Thiol Enzymes and Proteins: 248 (Methods in Enzymology)*. Academic Press Inc., San Diego, CA, 2002.
30. Sies H, Packer L (Eds.). *Glutathione Transferases and Gamma-Glutamyl Transpeptidases. Methods in Enzymology*. Academic Press Inc., San Diego, CA, 2005.
31. Packer L, Obermüller-Jevic U, Kraemer K, Sies H (Eds.). *Carotenoids and Retinoids*. AOCS Press, Champaign, IL, 2005.

32. Sies H, Brune B (Eds.). *Oxygen Biology and Hypoxia. Methods in Enzymology.* Academic Press, San Diego, CA, 2007.
33. Packer L, Sies H, Eggersdorfer M (Eds.). *Micronutrients and Brain Health. (Oxidative Stress and Disease: 26).* CRC Press Inc., Boca Raton, FL, 2009.
34. Krinsky NI, Maine ST, Sies H (Eds.). *Carotenoids in Health and Disease. Oxidative Stress and Disease.* Dekker, New York, 2007.

# Series Preface

Through evolution, oxygen—itself a free radical—was chosen as the terminal electron acceptor for respiration, and hence the formation of oxygen-derived free radicals is a consequence of aerobic metabolism. These oxygen-derived radicals are involved in oxidative damage to cell components inherent in several pathophysiological situations. Conversely, cells convene antioxidant mechanisms to counteract the effects of oxidants in either a highly specific manner (e.g., superoxide dismutases) or in a less specific manner (e.g., through small molecules such as glutathione, vitamin E, vitamin C, etc.). Oxidative stress—as classically defined—entails an imbalance between oxidants and antioxidants. However, the same free radicals that are generated during oxidative stress are produced during normal metabolism and, as a corollary, are involved in both human health and disease by virtue of their involvement in the regulation of signal transduction and gene expression, activation of receptors and nuclear transcription factors, antimicrobial and cytotoxic actions of immune system cells, as well as in aging and age-related degenerative diseases.

In recent years, the research disciplines interested in oxidative stress have increased our knowledge of the importance of the cell redox status and the recognition of oxidative stress as a process, with implications for many pathophysiological states. From this multi- and interdisciplinary interest in oxidative stress emerges a concept that attests to the vast consequences of the complex and dynamic interplay of oxidants and antioxidants in cellular and tissue settings. Consequently, our view of oxidative stress is growing in scope and new future directions. Likewise, the term "reactive oxygen species"—adopted at some stage in order to highlight nonradical/radical oxidants—now fails to reflect the rich variety of other species in free radical biology and medicine, encompassing nitrogen-, sulfur-, oxygen-, and carbon-centered radicals. These reactive species are involved in the redox regulation of cell functions and, as a corollary, oxidative stress is increasingly viewed as a major upstream component in cell signaling cascades involved in inflammatory responses, stimulation of cell adhesion molecules, and chemoattractant production and as an early component in age-related neurodegenerative disorders such as Alzheimer's, Parkinson's, and Huntington's diseases and amyotrophic lateral sclerosis. Hydrogen peroxide is probably the most important redox signaling molecule that, among others, can activate NFκB, Nrf2, and other universal transcription factors and is involved in the redox regulation of insulin- and MAPK-signaling. These pleiotropic effects of hydrogen peroxide are largely accounted for by changes in the thiol/disulfide status of the cell, an important determinant of the cell's redox status with clear involvement in adaptation, proliferation, differentiation, apoptosis, and necrosis.

The identification of oxidants in the regulation of redox cell signaling and gene expression was a significant breakthrough in the field of oxidative stress: the classical definition of oxidative stress as an imbalance between the production of oxidants and the occurrence of antioxidant defenses now seems to provide a limited depiction of oxidative stress, but it emphasizes the significance of cell redox status.

Because individual signaling and control events occur through discrete redox pathways rather than through global balances, a new definition of oxidative stress was advanced by Dean P. Jones as a disruption of redox signaling and control that recognizes the occurrence of compartmentalized cellular redox circuits. These concepts are anticipated to serve as platforms for the development of tissue-specific therapeutics tailored to discrete, compartmentalized redox circuits. This, in essence, dictates principles of drug development–guided knowledge of mechanisms of oxidative stress. Hence, successful interventions will take advantage of new knowledge of compartmentalized redox control and free radical scavenging.

Virtually all diseases thus far examined involve free radicals. In most cases, free radicals are secondary to the disease process, but in some instances causality is established by free radicals. Thus, there is a delicate balance between oxidants and antioxidants in health and diseases. Their proper balance is essential for ensuring healthy aging. Compelling support for the involvement of free radicals in disease development originates from epidemiological studies showing that enhanced antioxidant status is associated with reduced risk of several diseases. Of great significance is the role that micronutrients play in modulation of cell signaling: this establishes a strong link between diet and health and disease centered on the abilities of micronutrients to regulate redox cell signaling and modify gene expression.

Oxidative stress is an underlying factor in health and disease. In this series of books, the importance of oxidative stress and diseases associated with organ systems is highlighted by exploring the scientific evidence and clinical applications of this knowledge. This series is intended for researchers in the basic biomedical sciences and clinicians. The potential of such knowledge for healthy aging and disease prevention warrants further knowledge about how oxidants and antioxidants modulate cell and tissue function.

*Carotenoids and Vitamin A in Translational Medicine* is an authoritative treatise that provides the most updated information on carotenoids and vitamin A and their roles in health, disease prevention, and treatment. The book has chapters that cover bioavailability, absorption, and metabolism of carotenoids, their multiple roles on skin, eye, and ear diseases as well as cancer and lung and inflammatory bowel disease. About 600 carotenoids of plant origin are known and many are present in the diet; however, the scarcity of these micronutrients in food sources in large populations around the world, mainly in developing countries, leads to abnormal development and disease. Throughout the book, our current knowledge on carotenoids and vitamin A is highlighted as well as the need for a better understanding of their basic mechanisms of action.

The editors, Olaf Sommerburg, Werner Siems, and Klaus Kraemer, internationally recognized scholars and leaders in the field of carotenoids and vitamin A, are congratulated for assembling this excellent and timely book of the Oxidative Stress and Disease series.

**Lester Packer**
**Enrique Cadenas**

# Editors

**Olaf Sommerburg**, MD, Dr med. received his first degree and doctorate in medicine from the Medical Faculty of the Humboldt University of Berlin, Germany. He currently works as a senior physician in the Division of Pediatric Pulmonology and Allergology at the University Children's Hospital in Heidelberg, Germany. He is also a clinical researcher and is involved as principal investigator in the Translational Lung Research Center Heidelberg (member of the German Lung Research Center, DZL).

Sommerburg started his scientific career 20 years ago with research activities in the fields of oxidative stress and antioxidants. He has over 15 years of experience in research of carotenoids and carotenoid cleavage products. He has always connected basic research with clinical questions and is currently involved in different clinical and translational research projects. In the last decade, he has worked successfully on research projects on ophthalmology, nephrology, and pulmonology. Sommerburg has published more than 60 scientific articles and book chapters.

**Werner Siems**, MD, DrScMed, received his first degree and doctorate in medicine from the Humboldt University of Berlin, Germany. He currently serves as director of KortexMed Institute of Medical Education in Bad Harzburg, Germany (www.kortexmed.de), and of the Research Institute of Physiotherapy and Gerontology at KortexMed.

Siems has over 30 years of experience in clinical work, especially in orthopaedics, in education, and applied biochemical research, as well as in managing educational institutions. He is researcher on the fields of oxidative stress, aging, renal disease, and lymphoedema. He serves in several professional and scientific societies dedicated to lipid peroxidation and cardionephrology. He is a reviewer for a number of scientific journals, has published more than 300 scientific articles and book chapters, and has edited and coedited nine books. Siems is also an active member of the Academy of Sciences, New York.

**Klaus Kraemer**, PhD, received his first degree and doctorate in nutritional sciences from the University of Giessen in Germany. He currently serves as director of Sight and Life (www.sightandlife.org). Sight and Life is a nonprofit nutrition think tank of DSM, which cares about the world's most vulnerable populations and tries to improve their nutritional status. It guides original nutrition research, disseminates its findings, and facilitates dialog to bring about positive change.

Dr. Kraemer has over 25 years of experience in research and advocacy in the field of health, bioavailability, and safety of vitamins, minerals, and carotenoids. He serves as editor of *Sight and Life* magazine, one of the most widely distributed scientific magazines on micronutrients and nutrition in the developing world. He also serves on several professional societies dedicated to nutrition, vitamins, and antioxidants, is a reviewer for a number of scientific journals, has published many peer-reviewed scientific articles, and has coedited nine books. Dr. Kraemer was honored by the Micronutrient Forum and the Oxygen Club of California for his dedication in the fight against micronutrient deficiencies in developing countries. He has recently been appointed as a member of the Flour Fortification Initiative Leaders Group and Steering Committees of the Micronutrient Forum, BOND & INSPIRE (NIH initiatives), and the New York Academy of Sciences' Sackler Institute for Nutrition Science.

# Contributors

**Avdulla Alija**
Department of Biology
University of Prishtina
Prishtina, Kosovo

**Tos T.J.M. Berendschot**
University Eye Clinic Maastricht
Maastricht, the Netherlands

**Dechenla Tshering Bhutia**
Community Medicine
Sikkim-Manipal Institute of Medical
Sciences
and
Central Referral Hospital
Sikkim, India

**Hans-Konrad Biesalski**
Department of Biological Chemistry
and Nutrition
University of Hohenheim
Stuttgart, Germany

**William S. Blaner**
Department of Medicine
College of Physicians and Surgeons
Columbia University
New York, New York

**Christine Boesch-Saadatmandi**
Human Nutrition Research Centre
School of Agriculture, Food and Rural
Development
Newcastle University
Newcastle upon Tyne, United Kingdom

**Ekramije Bojaxhi**
Department of Cell Biology
University of Salzburg
Salzburg, Austria

**Neil Borkar**
Baylor Institute for Immunology
Research
Dallas, Texas

**Nikolaus Bresgen**
Department of Cell Biology
University of Salzburg
Salzburg, Austria

**Charbel Issa Peter**
Department of Ophthalmology
University of Bonn
Bonn, Germany

**Carlo Crifò**
Department of Biochemical Sciences
University of Roma La Sapienza
Rome, Italy

**Joanna Dulińska-Litewka**
Medical College
Jagiellonian University
Cracow, Poland

**Peter M. Eckl**
Department of Cell Biology
University of Salzburg
Salzburg, Austria

**Susan Emmett**
Department of International Health
Center for Human Nutrition
Johns Hopkins Bloomberg School
    of Public Health
and
Department of Otolaryngology Head
    and Neck Surgury
Johns Hopkins School of Medicine
Baltimore, Maryland

**Franziska Ferk**
Institut für Krebsforschung
Inner Medicine I
Medical University of Vienna
Wien, Austria

**Billy R. Hammond, Jr.**
Department of Psychology
University of Georgia
Athens, Georgia

**Masashi Hosokawa**
Faculty of Fisheries Sciences
Hokkaido University
Hokkaido, Japan

**Anita Ratnasari Iskandar**
Jean Mayer/U.S. Department
    of Agriculture
Human Nutrition Research Center
    on Aging
Tufts University
Boston, Massachusetts

**Shanmugam M. Jeyakumar**
Department of Human Nutrition
Ohio State University
Columbus, Ohio

and

Biochemistry Division
National Institute of Nutrition
Hyderabad, India

**Nilesh Kalariya**
Department of Ophthalmology and
    Visual Sciences
and
School of Nursing
The University of Texas Medical
    Branch at Galveston
Galveston, Texas

**Aldona Kieć-Dembińska**
Department of Clinical Biochemistry
Jagiellonian University
    Medical College
Cracow, Poland

**Siegfried Knasmüller**
Institute of Cancer Research
Medical University of Vienna
Vienna, Austria

**Klaus Kraemer**
Sight and Life
Basel, Switzerland

**Frederik J.G.M. van Kuijk**
Department of Ophthalmology
University of Minnesota Medical
    School
Minneapolis, Minnesota

**Piotr Laidler**
Medical College
Jagiellonian University
    Medical College
Cracow, Poland

**Claus-Dieter Langhans**
Paediatric Hospital
University of Heidelberg
Heidelberg, Germany

**Akeem Olawale Lasisi**
Department of Otorhinolaryngology
College of Medicine
University of Ibadan
Ibadan, Nigeria

**Georg Lietz**
Human Nutrition Research Centre
School of Agriculture, Food and Rural
    Development
Newcastle University
Newcastle upon Tyne, United Kingdom

**Giuseppe Martano**
Department of Molecular Biology
University of Salzburg
Salzburg, Austria

**Alan Menter**
Division of Dermatology
Baylor University Medical Center
and
Baylor Psoriasis Research Center
and
University of Texas Southwestern
    Medical Center
Dallas, Texas

**Anthony Oxley**
Human Nutrition Research Centre
School of Agriculture, Food and Rural
    Development
Newcastle University
Newcastle upon Tyne, United Kingdom

**Saskia de Pee**
Policy and Strategy Division
World Food Programme
Rome, Italy

and

Friedman School of Nutrition Science
    and Policy
Tufts University
Boston, Massachusetts

**Kota V. Ramana**
Department of Biochemistry
    & Molecular Biology
The University of Texas Medical Branch
Galveston, Texas

**Barbara Reichert**
Department of Human Nutrition
Ohio State University
Columbus, Ohio

**Costantino Salerno**
Department of Biochemical Sciences
University of Roma La Sapienza
Rome, Italy

**Theodor Sauer**
Duke University School of Medicine
Durham, North Carolina

**Ingolf Schimke**
Pathobiochemie und Medizinische
    Chemie
Charité—Universitätsmedizin Berlin
Berlin, Germany

**Gerd Schmitz**
Department of Clinical Chemistry
    and Laboratory Medicine
University of Regensburg
Regensburg, Germany

**Werner G. Siems**
Kortex Med GmbH
and
Research Institute of Physiotherapy
    and Gerontology
KortexMed Institute of Medical
    Education
Bad Harzburg, Germany

**Olaf Sommerburg**
Division of Pediatric Pulmonology
    & Allergology
University Children's Hospital III
and
Translational Lung Research Centre
Heidelberg
Heidelberg, Germany

**Satish K. Srivastava**
Department of Biochemistry
   & Molecular Biology
The University of Texas Medical Branch
Galveston, Texas

**Hanno Stutz**
Department of Molecular Biology
University of Salzburg
Salzburg, Austria

**Thomas Theelen**
Department of Ophthalmology
Radboud University Nijmegen
   Medical Centre
Nijmegen, the Netherlands

**Barbara Trösch**
DSM NUtritional Products
Basel, Switzerland

**Grace Jui-Ting Tsai**
Bloomberg School of Public Health
Johns Hopkins University
Baltimore, Maryland

**Michael Vogeser**
Institute of Laboratory Medicine
Hospital of the University of Munich
München, Germany

**Cornelia Vogl**
Department of Cell Biology
University of Salzburg
Salzburg, Austria

**Xiang-Dong Wang**
Jean Mayer/U.S. Department of
   Agriculture
Human Nutrition Research Center on
   Aging
Tufts University
Boston, Massachusetts

**Petra Weber**
DSM NUtritional Products
Basel, Switzerland

**Keith P. West, Jr.**
Department of International Health
Center for Human Nutrition
Johns Hopkins Bloomberg School of
   Public Health
and
Johns Hopkins School of Medicine
Baltimore, Maryland

**Ingrid Wiswedel**
Department of Pathological
   Biochemistry
Otto-von-Guericke University
Magdeburg, Germany

**Kryscilla Jian Zhang Yang**
Department of Medicine
College of Physicians and Surgeons
Columbia University
New York, New York

**Rumana Yasmeen**
Department of Human Nutrition
Ohio State University
Columbus, Ohio

**Yumiko Yasui**
Rakuno Gakuen University
Ebetsu, Japan

**Jason J. Yuen**
Department of Medicine
College of Physicians and Surgeons
Columbia University
New York, New York

**Ouliana Ziouzenkova**
Department of Human Nutrition
Ohio State University
Columbus, Ohio

# Part I

## Carotenoids and Vitamin A in Metabolic Diseases

# 1 Hepatic Retinoid Metabolism

## What Is Known and What Is Not

*Jason J. Yuen,\* Kryscilla Jian Zhang Yang,\* and William S. Blaner*

## CONTENTS

## INTRODUCTION/OVERVIEW

It has been estimated that as much as 70% of all retinoid present in the body of a healthy, well-nourished individual is stored in the liver [1]. Consequently, the liver is quantitatively the most important organ site for retinoid storage and metabolism in the body. The goal of this chapter is to review the seminal older literature and to consider newer literature in order to summarize present understanding of hepatic retinoid storage and metabolism. Although there is considerable understanding of the major details regarding how retinoids are metabolized and stored in the liver, there are many specific points about these processes that remain controversial. These, too, will be considered in the chapter.

---

\* The authors Jason J. Yuen and Kryscilla Jian Zhang Yang contributed equally to this chapter.

The two cell types in the liver that are known to have essential roles in hepatic retinoid storage and metabolism are hepatocytes (also referred to as parenchymal cells) and hepatic stellate cells (HSCs) (also referred to as Ito cells, fat-storing cells, or lipocytes) [1–4]. There are four processes that will be considered in this chapter: (a) dietary retinoid uptake into the liver, (b) retinoid processing within hepatocytes and its transfer to HSCs for storage, (c) retinoid storage within the lipid droplets of HSCs, and (d) retinoid mobilization from the liver to meet peripheral tissue retinoid needs. The chapter has been subdivided into sections to consider separately the biochemical events that occur specifically in hepatocytes or specifically in HSCs.

Although retinoic acid is required for maintaining normal transcriptional regulatory activity in both hepatocytes and HSCs, our focus will be on the metabolism of retinol and its storage as retinyl ester, and not on retinoic acid formation. The topic of hepatic retinoic acid metabolism and actions within the liver has been recently reviewed [5]. The reader is referred to this review for more information on these topics.

## HEPATOCYTES

### DIETARY RETINOID UPTAKE

In the postprandial state, the majority of ingested retinoid is associated with chylomicrons in the form of retinyl esters, along with other dietary lipids including cholesterol ester and triglyceride. Approximately a quarter, and up to a third, of the retinyl esters are delivered to extrahepatic tissues, and the rest remain associated with chylomicrons, which are eventually taken up by the liver [6,7]. In the circulation, chylomicrons undergo a series of modifications, which render them into much smaller chylomicron remnants (CRs). These metabolic transformations are characterized first by significant hydrolysis of triglyceride by lipoprotein lipase (LPL), a process whereby hydrolyzed fatty acids are acquired by adipose tissue and muscle. The hydrolytic activity of LPL in the postprandial circulation is necessary for hepatic removal of chylomicrons, as the remnant particles must be made small enough to enter the space of Disse, where they come in direct contact with hepatocytes, the cellular site of uptake into the liver. Indeed, mutations resulting in structural defects and/or deficiency of LPL have been associated with hyperlipidemia characterized by accumulation of CRs in the circulation and, hence, increased plasma levels of triglyceride and cholesterol [8,9]. In addition to triglyceride, a significant amount of retinyl esters associated with chylomicrons is also hydrolyzed in the circulation for uptake by extrahepatic tissues expressing LPL [10,11]. Also required for the maturation of chylomicrons into CRs are changes in protein composition, which include loss of the C apolipoproteins (apoC) and acquisition of apolipoprotein E (apoE) [12–14]. For more information regarding specific details of these essential steps, the reviewer is referred to two excellent and extensive reviews by Cooper et al. [15,16]. In brief, after sufficient hydrolysis of triglyceride, chylomicrons lose apoC, which facilitates the process of triglyceride hydrolysis by serving as an essential cofactor for LPL [17], while displacing and inhibiting the activities of apoE [18–20]. After losing apoC, recruitment of apoE allows for sequestration of the CRs in the space of Disse, where apoE binds to one of several proteins, leading to receptor-mediated endocytosis. This happens via two known, independent but potentially overlapping pathways, involving CR receptors that recognize apoE as a ligand: low-density lipoprotein (LDL) receptor and the LDL receptor-related protein (LRP) [21,22]. These events are summarized in Figure 1.1.

Newly internalized retinyl esters are quickly hydrolyzed to retinol in the liver by one or more members of a group of enzymes collectively referred to as retinyl ester hydrolases (REHs). These REHs consist of known carboxylesterases and lipases that recognize retinyl ester as a substrate. Hydrolysis catalyzed by REHs constitutes a key step in retinol homeostasis, as retinyl ester hydrolysis is necessary in the processes of cellular uptake and translocation of newly absorbed retinoid, as well as in the mobilization of retinyl ester stores from the liver. Most of the early work on hepatic retinyl ester hydrolysis had been focused on the bile salt-activated carboxylester lipase (CEL), which was shown to be a potent REH in vitro [23,24]. However, studies of CEL-deficient and CEL-overexpressing mice established that neither CEL deficiency nor overexpression made a difference in the uptake or hydrolysis of CR-associated retinyl esters [25]. It has not been established whether some retinyl ester hydrolysis can take place in the space of Disse before endocytosis. Nonenzymatic functions have been described for LPL [26,27] and hepatic lipase (HL) [28,29] in the binding and uptake of CRs in the space of Disse, where these enzymes are abundant, but it remains to be demonstrated whether they act as REHs in this location. HL has never been shown to act enzymatically on retinyl esters, but it is well established that LPL hydrolyzes chylomicron-associated retinyl esters in the circulation, albeit only after significant hydrolysis of triglyceride [11]. The liver-specific carboxylesterase ES-2 may also have a role in the hydrolysis of CR-associated retinyl esters. Known as serum esterase, ES-2 lacks the endoplasmic reticulum (ER) retention sequence present in other members of this enzyme family and is secreted by hepatocytes into the space of Disse [30,31]. It is not known, however, whether this enzyme contributes to REH activity in vivo, either in the extracellular space or within hepatocytes.

Much, if not all, of the hydrolysis of CR retinyl esters is thought to take place intracellularly after endocytosis. Studies employing isotopic labeling of CR retinyl esters and subcellular fractionation have shown that, after hepatic uptake, retinyl esters are localized to the plasma membrane and/or endosomal fraction of liver homogenates, which are enriched in both neutral and acidic bile salt-independent REH activities [32–36]. These REH activities are thought to be represented by distinct enzymes present in endosomes that are active in either neutral or acidic conditions, which correspond respectively to the pHs of the interior of early versus late endosomes [34,37], thus allowing for efficient hydrolysis of retinyl esters in the increasingly acidic endosomal compartments. As of yet, the specific enzymes responsible for REH activities observed in the endosomal fractions have not been definitively identified, but several nonspecific carboxylesterases have been isolated from liver microsome fractions, which have been shown to hydrolyze retinyl esters in vitro: ES-2, ES-4, and ES-10 [33,38–41]. The reader should note that overlaps in nomenclature for these carboxylesterases exist in the literature. ES-10, for example, is synonymous with Ces3 and triglyceride hydrolase (TGH). For clarification, the reader is referred to a review by Holmes et al. [42], which proposes a unified nomenclature system for the members of the mammalian carboxylesterase multigene family. ES-2, ES-4, and ES-10 are isoforms of the Ces/CES gene family 1, which are highly expressed in the

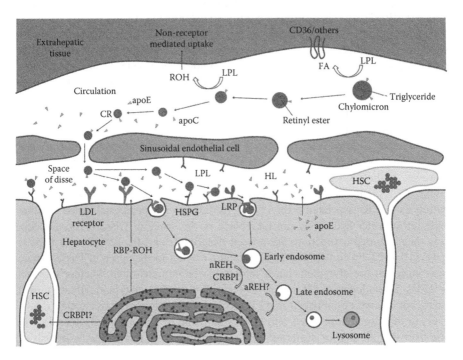

**FIGURE 1.1** Hepatic uptake of dietary retinoid. Newly ingested dietary retinoid enters the circulation as retinyl ester packaged in triglyceride-rich chylomicrons. Up to 70% of this triglyceride is hydrolyzed in the postprandial circulation through the activity of LPL, and the free fatty acids are taken up by extrahepatic tissues via fatty acid transporters, such as CD36. LPL also recognizes retinyl ester as a substrate. Twenty-five to thirty-three percent of chylomicron retinyl ester is hydrolyzed, and the unesterified retinol is taken up by extrahepatic tissues presumably through a receptor-independent mechanism involving "flip-flop" across the plasma membrane. As the chylomicron is progressively reduced in size, it loses apoC and picks up apoE. At this point, it becomes a CR, which can easily traverse the endothelial fenestrae of the liver and enter the space of Disse. Here, apoE on the CR surface can bind with LDL receptor to initiate endocytosis. Alternatively, the CR can acquire additional apoE, which is abundant in the space of Disse, before binding with LRP to be endocytosed. ApoE also binds with heparan sulfate proteoglycan (HSPG), which is expressed on the cell surface of hepatocytes and sinusoidal endothelial cells. This allows for the sequestration of CRs as they enter the space of Disse. HL and LPL also have nonenzymatic roles in the hepatic uptake of CRs, but it is not known whether these enzymes contribute to the hydrolysis of retinyl esters. Once taken up by hepatocyte, the retinyl esters are quickly hydrolyzed in the early endosome by neutral retinyl ester hydrolases (nREHs), and possibly in the late endosome by acidic retinyl ester hydrolases (aREHs). The hydrolysis product retinol is transferred to the ER, probably bound to cellular retinol-binding protein, type I (CRBPI). The metabolic fate of CR retinyl ester is different from that of triglyceride and cholesterol ester, which end up in lysosomes where they are metabolized to unesterified fatty acids and cholesterol. From the ER, unesterified retinol is transferred to HSCs, where it is reesterified and stored in lipid droplets. Alternatively, it can be secreted bound to nascent retinol-binding protein (RBP).

liver [42]. Of these, ES-4 and ES-10 have been estimated to account for up to 94% of total neutral REH activity [40]. ES-10 has also been demonstrated in vitro to exhibit high REH activity in acidic conditions, with little activity toward TG or cholesterol esters [41]. Collectively, these are compelling in vitro data in support of roles for ES-4 and ES-10 in the hydrolysis of newly absorbed CR-associated retinyl esters. However, the physiological significance of these carboxylestcrases remains to be demonstrated in vivo, through either specific inhibition of the enzymes or techniques involving selective gene silencing. These experimental approaches have proven to be difficult given the high degree of sequence identity and similarities in substrate preferences of these carboxylesterases. Recently, a new carboxylesterase, ES-22, has been described to exhibit specific activity against retinyl esters in hepatocytes [43]. Overexpression of ES-22 has been shown to attenuate the accumulation of retinyl esters in COS-7 cells [43]. Further studies will be needed to define the specific context in which the hydrolytic activity of ES-22 has biological relevance.

Other enzymes that may also have significant roles in hepatic metabolism of retinoids are adipose triglyceride lipase (ATGL) and hormone-sensitive lipase (HSL). ATGL does not recognize retinyl ester as a substrate in vitro [44], but mice deficient in ATGL exhibit low plasma levels of retinol-binding protein (RBP), suggesting at least an indirect role for ATGL in retinoid metabolism [45]. HSL, on the other hand, is known to hydrolyze retinyl esters in adipose tissue [46,47], but because of its low levels of expression in the liver, the extent to which it acts as a hepatic REH has not been established.

Unlike CR cholesterol ester and triglyceride, CR-associated retinyl esters do not appear in lysosomes after uptake into the hepatocyte. Instead, the unesterified retinol is transferred to the ER [36], perhaps immediately after hydrolysis in the early endosome, via a process that has yet to be delineated. As shown in Figure 1.1, this divergence in metabolic fate takes place almost immediately after internalization, long before the CR reaches the lysosome. Retinyl ester hydrolysis occurs in the early endosome and may continue into the late endosome stage where acidic REHs are proposed to remain active, but it has not been established whether hydrolysis is required for the transport of retinoid to the ER. Unesterified retinol originally present as retinyl ester in CRs has been found associated with the ER [36], but it is not clear if any of the retinyl esters reach the ER before hydrolysis. In the ER, the metabolic fate of retinol depends on the vitamin A nutritional status of the organism. Early experiments established that during times of vitamin A deficiency, almost all of the hydrolyzed retinol from internalized CRs is bound to RBP and secreted back into the circulation for delivery to extrahepatic tissues [48]. Alternatively, when vitamin A is sufficient, dietary retinol is transferred to HSCs for reesterification and storage [48]. The mechanism by which retinol is transferred to HSCs remains to be elucidated (discussed in more detail later), but it is clear that hepatocytes are the only liver cell type responsible for the uptake of CRs [15].

## ROLE IN RETINOID STORAGE

Early work had proposed that among liver cells, hepatocytes, HSCs, and Kupffer cells could have key roles in retinoid storage. However, studies carried out in the 1980s established convincingly that the HSCs are the major cellular storage site for retinoid in the body [49–51]. It was also established that hepatocytes, but not hepatic endothelial or Kupffer cells, have roles in hepatic retinoid storage. One study by Blomhoff et al. estimated from measurements employing isolated primary rat hepatic cells that as much as 80% or more of total liver retinoid (retinol and retinyl ester) might be stored in HSCs, with the remainder in hepatocytes [49]. Retinoic acid accounts for less than 0.01% of the total retinoid present in a vitamin A-sufficient liver [52,53]. The rats employed for these studies were healthy and maintained on a standard vitamin A-sufficient chow diet. Blomhoff et al. further reported hepatocyte retinoid levels of 0.5–0.8 nmol per $10^6$ cells and much higher HSC levels of 28–34 nmol per $10^6$ cells [49]. A second study by Blaner et al. published at the same time and also employing healthy vitamin A-sufficient chow-fed rats, estimated that hepatocytes account for approximately 9% of the total retinoid in the liver, with HSCs accounting for the remainder [50]. Both groups agreed that 5%–12% of the retinoid present in hepatocytes was in the form of free alcohol retinol with the remaining 88%–95% present as retinyl ester [49,50]. There was also strong agreement between the two studies that 98%–99% of the retinoid present in isolated HSCs is retinyl ester, with the remainder being retinol [49,50].

Careful nutritional studies carried out by Batres and Olson established that the distribution of hepatic retinoid stores between hepatocytes and HSCs is markedly influenced by the vitamin A nutritional status of the animals [51]. For groups of adult Sprague–Dawley rats with mean total retinoid reserves of 1.2, 14.5, and 28.9 μg/g of fresh wet liver, the relative percentage of the total retinoid present within hepatocytes were 83%, 37%, and 18%, respectively, which indicates that the percentage of total hepatic retinoid accounted for by hepatocytes progressively declined as hepatic retinoid levels increased. These authors further reported that HSCs accounted for the remainder of the total retinoid present in the livers of these rats. Interestingly, for the group of rats with the lowest total retinoid reserves, approximately 50% of the total retinoid in the liver was present as the free alcohol retinol. For the other two groups with higher total hepatic retinoid reserves, 87%–91% was reported to be present in the form of retinyl esters.

Blomhoff et al. provided insights into the distribution of several proteins involved in hepatic retinoid metabolism, reporting that hepatocytes account for as much as 75%–80% of the REH and acyl-CoA:retinol acyltransferase (ARAT) activities present in the liver, with HSCs accounting for most of the remainder [49]. Blaner et al. reported, based on radioimmunoassay analyses, that only hepatocytes, and not HSCs, expressed RBP protein [50]. Both groups found upon radioimmunoassay measurements that cellular RBP, type I (CRBPI), is highly expressed in both hepatocytes and HSCs, with approximately 90% of total hepatic CRBPI expressed in hepatocytes [49,50]. It is clear from these reports that both hepatocytes and HSCs have significant metabolic capacities for accumulating and metabolizing retinoids.

The literature establishes that hepatocytes present within the liver of a healthy vitamin A-sufficient animal maintain significant retinoid stores. Although these stores are quantitatively less significant than those of the HSCs, and possibly those of adipocytes, hepatocytes are indeed a major site of retinoid storage in the body.

As noted earlier and discussed in more detail later, hepatic RBP is synthesized and secreted solely by hepatocytes. Thus, hepatocyte retinoid stores are likely very important for meeting routine peripheral tissue needs for retinoid.

## RETINOID MOBILIZATION BY RETINOL-BINDING PROTEIN

Retinol is mobilized from hepatic stores bound to its specific transport protein, RBP, the sole carrier for retinol in the blood [54]. RBP is a 21 kDa protein containing a single site for the noncovalent binding of one molecule of retinol. Although RBP is synthesized and secreted by many tissues including adipose tissue, kidneys, eyes, and lungs, the hepatocyte is the major cell type in the body responsible for RBP synthesis and secretion [49,50,55].

There is substantial evidence to suggest that RBP has a role in ensuring that serum retinol levels do not fall below a minimum of 1.5–2 μM in humans, maintaining a constant supply of retinol to tissues that require it [56]. In mice, serum RBP levels are maintained at 0.7–1.0 μM. During fasts, when dietary retinoid is not available to tissues, the retinol–RBP complex constitutes 99% of all retinoid present in the blood [57]. Studies of RBP-deficient (Rbp$^{-/-}$) mice demonstrate that these mice have impaired vision during their first few months of life, but they later recover within 4 months of age when fed a vitamin A-sufficient diet [58]. When maintained on a vitamin A-deficient diet, however, these mice become completely blind and experience vitamin A deficiency [59]. Total hepatic retinoid levels are also greater in these mice when compared to normal wild-type mice. Reports of humans deficient in RBP due to heterozygous mutations at the Rbp loci show that these patients are developmentally normal, despite low levels of circulating retinol [60]. Besides night blindness and mild retinal dystrophy, the patients did not present with other clinical symptoms. The results from human data are in agreement with those from the studies conducted in mice. Thus, RBP-bound retinol secreted from hepatocytes serves as a crucial and necessary source of retinoid when peripheral tissue demands cannot be satisfied solely by recently ingested dietary retinoid.

Insights into the physiology of RBP synthesized by tissues other than the liver have been obtained from studies of Rbp$^{-/-}$ mice genetically modified to express human RBP (hRBP) in muscle, or hRBP$^{-/-}$ mice. The serum concentration of hRBP in these mice is similar to the serum concentration of RBP in wild-type mice [58]. Interestingly, the hRBP$^{-/-}$ mice have greater muscle retinoid levels than wild-type mice [61]. When placed on a vitamin A-deficient diet, lung retinoid stores of the hRBP$^{-/-}$ mice were depleted more rapidly than that in both wild-type and RBP$^{-/-}$ mice [58]. It was further observed that, on a vitamin A-sufficient diet, hepatic retinoid levels were not different in hRBP$^{-/-}$ mice compared to Rbp$^{-/-}$ mice, but interestingly, these levels were greater than in wild-type mice. Collectively, these observations imply that (i) RBP, irrespective of its tissue origin, can draw upon peripheral tissue retinoid stores to mobilize retinol into the circulation in order to maintain blood retinol levels, and (ii) extrahepatically synthesized RBP is unable to draw upon hepatic retinoid stores. It is tempting to speculate that hepatic RBP may undergo posttranslational modifications to be differentiated from extrahepatic RBP, which is not recognized by hepatocytes. This could explain why extrahepatic RBP may behave differently from hepatic RBP and provide support for the hypothesis that adipose tissue-derived RBP is an adipokine with functions unrelated to retinol, which account for its association with obesity and insulin resistance [62,63].

The blood concentration of the retinol–RBP complex is tightly maintained [56]. Many studies have been conducted to identify the mechanisms through which the body is able to regulate blood retinol levels. It has long been established that the majority of RBP present in the blood is bound to transthyretin (TTR) [54,55]. TTR is a tetrameric 54 kDa protein synthesized and secreted by both the liver and the choroid plexus in the brain. TTR also serves as one of the transport proteins for thyroid hormone. Based on in vitro binding affinity studies, one molecule of tetrameric TTR can bind a maximum of two RBP molecules [64], but in vivo, never more than one molecule of RBP has been found associated with TTR [55]. Plasma concentrations of RBP are high in patients with chronic renal disease, which shows that RBP is catabolized by the kidneys [65]. When given an intravenous injection of hRBP, TTR-deficient (Ttr$^{-/-}$) mice were found to clear the hRBP from their circulation more rapidly than wild-type mice. hRBP was also found to appear more quickly in the kidneys of Ttr$^{-/-}$ mice [66]. These observations are taken to indicate that TTR, due to its size, is able to protect the retinol–RBP complex from renal filtration. Monaco et al. determined that the hydroxyl group of retinol also participates in this interaction with TTR [64]. Noy et al. demonstrated that the holo-to-apo transition of RBP results in a decrease in RBP's affinity for TTR [67]. They further proposed that, because the half-life of the dissociation of apo-RBP from TTR after retinol release is approximately 15 h, RBP is not immediately cleared from the blood. Moreover, if free retinol is present in the serum or in the plasma membrane of cells, apo-RBP can rebind retinol and lengthen its lifetime in the circulation. As such, the interaction between RBP and TTR may be important in the regulation of blood retinol levels.

Other processes that are central for maintaining blood retinol homeostasis are (i) RBP synthesis and secretion from hepatocytes and (ii) peripheral tissue uptake of retinol from retinol–RBP in the circulation. Retinol deficiency has been shown to inhibit the secretion of hepatic RBP, resulting in its accumulation in the liver and causing plasma RBP levels to fall [68]. When retinol is not available in hepatocytes, newly synthesized RBP accumulates in the ER [69]. Upon repletion of retinol to vitamin A-deficient rats, RBP localized in the ER moves rapidly through the Golgi complex and secretory vesicles to the cell surface for secretion into the circulation [65,70]. This suggests that hepatocytes must maintain a minimum store of retinoid or have ready access to HSC retinoid stores in order for retinol to be mobilized bound to RBP. Bellovino et al. reported in vitro studies showing that the RBP–TTR complex can be formed within the ER of hepatocytes [71]. These researchers claimed that RBP and TTR interact within hepatocytes prior to secretion of the retinol–RBP–TTR complex and suggested that this might regulate retinol–RBP mobilization from liver. However, RBP is found in the circulation of Ttr$^{-/-}$ mice [72]. Although hepatic RBP levels are 60% greater in these mice compared to wild-type mice, cultured primary hepatocytes isolated from wild-type and Ttr$^{-/-}$ mice secrete RBP into the culture medium at identical rates [72]. These observations establish that RBP can be secreted from hepatocytes in the absence of TTR and suggest that formation of the retinol–RBP–TTR complex in the ER is not a necessary precondition for RBP secretion, nor is it important for the regulation of this process. Although much has been learned, considerable research is still needed to elucidate the mechanisms underlying hepatic retinol mobilization via RBP and its regulation.

The literature regarding how retinol bound to RBP is transferred to or taken up by cells is also not fully resolved. There have been convincing reports identifying and characterizing a cell surface receptor for RBP, termed Stra6, which binds RBP with high affinity [73–76]. Stra6 was first identified as a gene of unknown function, which was upregulated in WiDr human colon adenocarcinoma cells upon treatment with retinoic acid [75,76], and it was subsequently shown to bind RBP with high affinity [74,77]. Retinol uptake into cells was also demonstrated not to involve internalization of RBP. Moreover, the ability of Stra6 to facilitate cellular uptake of retinol is also dependent on the ability of cells to form retinyl esters from retinol through the activity of lecithin:retinol acyltransferase (LRAT). Other investigators have shown that Stra6 can facilitate retinol efflux from cells and have consequently proposed that Stra6 mediates retinol transport both into and out of the cell [77,78]. Importantly though, the tissue expression pattern of Stra6 does not include all tissues that require retinol. Hence, Stra6 cannot be universally important for retinol uptake by all tissues and cell types. Stra6 is expressed primarily in the eyes, brain, placenta, testes, spleen, and thymus [74–76], but it is not expressed in other tissues like liver and heart, which are centrally involved in retinol metabolism or require retinol [76]. Since Stra6 is not expressed in these tissues, it cannot be involved in mediating retinol uptake by these tissues. Consequently, some investigators believe that retinol may spontaneously dissociate from RBP and "flip-flop" across the plasma membrane of cells expressing LRAT or other proteins involved in the metabolism of retinoids [67]. The diffusion of retinol in this model requires a concentration gradient of retinol across the blood, plasma membrane, and cytosol of the target cell, which is presumably driven by the metabolism of retinol within the cell. This hypothesis is supported by evidence that the dissociation of retinol from the retinol–RBP–TTR complex is orders of magnitude faster than the rate of uptake of retinol by target cells. Furthermore, retinol is solvated after dissociation from RBP and before crossing into the membrane. Still other investigators have reported findings that radiolabeled RBP injected into the circulations of rats can be detected in the liver and that RBP can be endocytosed by hepatocytes and HSCs, further complicating the understanding of how retinol bound to RBP is taken up by cells [79,80]. Two likely mechanisms of retinol uptake are illustrated in Figure 1.2. Additional research is required to clarify and expand our understanding of how cells and tissues take up retinol, and possibly RBP, from the circulation.

## WHAT IS NOT UNDERSTOOD

As recounted earlier, it is solidly established in the literature that dietary retinoid, as retinyl ester associated with CRs, is taken up in the liver specifically by hepatocytes. In times of dietary vitamin A sufficiency, some of this retinoid is retained by hepatocytes and is eventually secreted from the liver into the circulation bound to

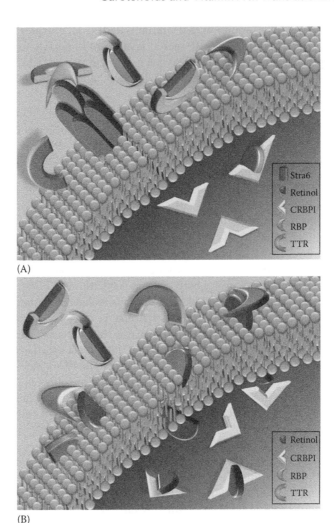

(A)

(B)

**FIGURE 1.2** Mechanisms of cellular uptake of retinol from RBP. Cellular retinol uptake from circulating RBP may occur in different tissues via different processes. For tissues that express Stra6, including the eyes, brain, testes, and spleen, retinol uptake may be facilitated by the Stra6 membrane receptor for RBP (A). Because RBP binds Stra6 with high affinity, the binding of RBP to Stra6 may initiate transfer of retinol into the cell. Note that retinol's hydroxyl group is oriented outward toward the aqueous environment when bound to RBP. As mentioned in the text, this hydroxyl group also interacts with the TTR tetramer to stabilize the RBP–TTR interaction. Thus, there is a decrease in affinity between RBP and TTR once retinol is released from RBP. In healthy humans, the molar ratio of RBP to TTR is approximately 1–3. Once inside the cell, CRBPI binds retinol facilitating retinol transfer within the cytosol. When bound to CRBPI, the hydroxyl group on retinol is buried within the binding protein. Because Stra6 is not expressed in certain tissues, such as the liver and heart, retinol may enter these tissues via a "flip-flop" mechanism, where its hydroxyl group flips from the outer phospholipid layer across the membrane, into the inner phospholipid layer, and then to the cytoplasm (B). Note that this illustration is not to scale.

RBP in order to satisfy peripheral tissue needs for retinoid. The remainder of the internalized dietary retinoid is transferred to HSCs for storage in HSC lipid droplets. A number of very significant questions remain to be answered regarding how this occurs. Possibly the most significant of these is: how is dietary retinoid transferred from hepatocytes to HSCs for storage? What processes and factors are involved in this transfer? The older literature proposed, and it was generally accepted, that RBP is responsible for transferring newly absorbed dietary retinol from hepatocytes to HSCs [1,81–83]. However, investigations of Rbp$^{-/-}$ mice fed a control diet indicate that retinoid stores in HSCs of these mice are the same as, or even slightly greater than, HSC stores of diet-matched wild-type mice [58,61]. Thus, RBP cannot be essentially involved in this transfer process. It has been subsequently suggested that CRBPI, which is expressed in both hepatocytes and HSCs [49,50], may catalyze the transfer process [1,84]. Indeed, CRBPI-deficient (CrbpI$^{-/-}$) mice are reported to have fewer and smaller HSC lipid droplets, suggesting diminished HSC retinoid stores [84]. However, this possibility has not been quantitatively assessed or established. Moreover, the binding of CRBPI to retinol is reported to lessen the hepatic catabolism (loss) of retinol [85], and it is not clear whether the apparent diminished HSC retinoid stores reported for CrbpI$^{-/-}$ mice arises from increased catabolism of hepatic retinol or its impaired transfer to HSC storage sites. There is a clear need for better understanding of these processes, which are important for transferring dietary retinoid from hepatocytes to HSCs for storage.

A second significant question regarding the processing of dietary retinoid by hepatocytes concerns how this retinoid leaves the newly internalized CRs. As noted earlier, the metabolism of CR retinoid diverges very early from the metabolism of CR cholesteryl ester and triglyceride. This divergence in metabolism occurs in the early endosomes [35,36] before the other CR lipids are metabolized in lysosomes. It is not clear how retinoid is transferred to the ER from endosomes, but the process probably involves CRBPI as a carrier protein in the cytoplasm. It has also not been established whether retinyl ester hydrolysis is required for this divergence, but data from studies employing non-hydrolyzable ether analogs of retinyl esters suggest that the subsequent transfer of retinoid to HSCs for storage does require retinyl ester hydrolysis [86]. Yet it is still not known which REHs are physiologically relevant for catalyzing this process within hepatocytes. Although a number of enzymes have been identified that can act as REHs in vitro, it is not clear which are relevant in vivo. Future studies will have to determine which of the proposed REHs are physiologically important for hydrolyzing CR retinyl esters and how this contributes to the removal of the dietary retinoid from the rest of the CR components.

Another very significant and long-standing research question in need of resolution concerns the factors and processes that are important for maintaining and regulating circulating levels of retinol–RBP. It has long been postulated that there are homeostatic mechanisms that regulate blood retinol–RBP levels. Is this truly the case? Or, does the liver simply secrete retinol bound to newly synthesized RBP until hepatic retinoid stores are completely exhausted? What role does receptor-mediated uptake of retinol from RBP by peripheral tissues have in regulating blood levels of retinol–RBP? And how is this related to the secretion of RBP? Does this truly control retinol availability to cells?

## HEPATIC STELLATE CELLS

### DIETARY RETINOID UPTAKE

As discussed earlier, CRs are taken up into the liver by hepatocytes. Although dietary retinoid is quickly removed from newly internalized CRs, allowing for the divergence of retinoid metabolism from that of newly ingested cholesterol and fat [36], there is no evidence that this divergence occurs outside of hepatocytes. Work published in the 1980s convincingly establishes that CR retinoid is taken up first by hepatocytes, prior to transfer to HSCs for storage [48,87]. Since the great majority of hepatic retinoid is stored in HSCs, it would be very tempting to speculate that dietary retinoid can be taken up directly by HSCs, but there is no experimental evidence to support this possibility.

### ROLE IN RETINOID STORAGE

It has been estimated that the liver of a healthy, well-nourished rodent contains approximately 70% of the total retinoid present in the body [1] and that HSCs account for 80% or more of the retinoid present in the liver [49,50]. Thus, HSC stores account for more than half of all retinoid present in the body. This is a striking amount given that HSCs account for approximately 6%–8% of the cells present in the liver and only approximately 1% of hepatic protein [2–4,50]. Retinol and retinyl esters are found at physiologically significant concentrations in the lipid droplets of adipocytes [11,46,55,56,88], and, as noted earlier, are also found in significant concentrations in hepatocytes. Consequently, other lipid-accumulating cell types can be sites for retinoid storage. These observations have led us to wonder what evolutionary pressures would have selected for specific accumulation of such a large percentage of the body's retinoid in a relatively minor hepatic cell type.

The HSC is a specialized cell for retinoid storage. As mentioned earlier, for well-nourished animals, 99% of the retinoid present in HSCs is present as retinyl ester, with the great majority of the remainder present as retinol [49,50]. The predominant retinyl ester is retinyl palmitate, accounting for approximately 75% of the total. The remainder is primarily retinyl stearate, retinyl oleate, and retinyl linoleate. HSCs possess very high concentrations of CRBPI, have high levels of ARAT activity, and express high levels of LRAT protein [49,50]. High levels of REH activity have also been reported to be present in isolated HSCs, but it is not clear which specific enzyme or enzymes are responsible for this activity [49,50]. CRBPI, ARAT, LRAT, and REHs constitute the intracellular factors that are needed for processing, storing, and mobilizing retinoids, and the HSCs possess all of the metabolic machinery needed for undertaking these processes. Like hepatocytes, HSCs do not express Stra6 [76,89], so this membrane receptor for RBP cannot be involved in facilitating retinol uptake into or efflux from HSCs. HSCs express all three of the retinoic acid receptors and all three of the retinoid X receptors and are thus also targets for retinoid action [4,5,89].

Within HSCs, retinoids are stored in lipid droplets, which are a distinguishing morphological feature of these cells [4]. Figure 1.3 provides an image of primary HSCs cultured overnight following their isolation from chow-fed mice. As one can

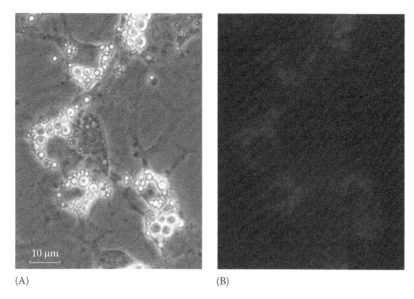

(A)                                                                    (B)

**FIGURE 1.3**   Primary murine HSCs in culture, 24 h after isolation. HSCs are the main storage site for retinoid in the body, accounting for over half of all the retinoid present in the body of a healthy, well-nourished rodent. The large lipid droplets seen on (A) are a distinguishing morphological feature of these specialized cells. Retinoid constitutes 40% of the total lipid content of HSC lipid droplets, and 99% of this retinoid is in the form of retinyl ester. (B) Shows the characteristic autofluorescence of the retinoid present in these cultured HSCs, captured upon excitation at 330 nm.

see from Figure 1.3, lipid droplets are indeed a striking characteristic of these cells. Moriwaki et al. reported for control diet-fed rats that 40% of the lipid present in HSC lipid droplets is retinyl ester, 32% triglyceride, 15% cholesteryl ester, 6% phospholipid, 5% cholesterol, and 2% unesterified fatty acids [90]. Other investigators have reported similar lipid droplet lipid compositions for HSCs isolated from well-nourished animals [91,92]. So there is generally good agreement on HSC lipid droplet composition. Moriwaki et al. further reported that HSC lipid droplet composition was markedly affected by dietary retinoid intake, with relatively greater concentrations of retinyl ester present in HSC lipid droplets isolated from animals maintained on a diet containing high levels of retinol [90]. Animals maintained on a retinol-poor diet were found to possess relatively less retinyl ester in isolated HSC lipid droplets. Dietary triglyceride intake did not influence the lipid composition of HSC lipid droplets but did influence the triglyceride content of the whole liver, and presumably the triglyceride content of hepatocytes [90].

Careful histological investigations of HSCs in situ have shown that two types of lipid droplets can be found in HSCs; one bound by a unit membrane, which does not directly make contact with the lipid, and one lacking a limiting membrane [2–4,93,94]. The possible physiologic or metabolic importance of the two distinct types of lipid droplets has not been established. Straub et al. have reported immunolocalization studies showing that two proteins that are associated with lipid droplets in other cell types, adipocyte differentiation-related protein (ADRP) and TIP47, are also associated with

HSC lipid droplets in mouse, bovine, and human liver [95]. These investigators further showed that perilipin, another common lipid droplet-associated protein, is present in bovine, but not mouse or human, HSC lipid droplets. ADRP, TIP47, and perilipin provide a scaffolding to support the actions of enzymes involved in the synthesis or degradation of lipid droplets or are involved in facilitating the regulation of these processes [96,97]. One would speculate, based on the significant concentration of retinyl ester present in HSC lipid droplets, that both REH activities and LRAT must be localized on this scaffolding, in close proximity to the lipid droplet surface.

## MOBILIZATION OF RETINOID STORES

Maintenance of an animal on a retinoid-deficient diet results in the mobilization of hepatic retinoid stores to meet tissue needs for retinoid, in vision and for retinoid-mediated transcription. Prolonged maintenance of an animal on a retinoid-deficient diet will eventually exhaust the HSC retinoid reserves, giving rise to a state of retinoid deficiency. It is clear from studies of Rbp$^{-/-}$ mice that RBP plays an essential role in mobilizing HSC retinoid stores [58,61,98]. In the absence of RBP, HSC retinoid stores cannot be mobilized. So it is clear that RBP is solely responsible for mobilization of these stores. How this is accomplished, though, remains to be established. As noted in "Hepatocytes" section, hepatocytes are the major cellular site of RBP synthesis in the body. There is no controversy regarding this point. There is, however, considerable controversy as to whether HSCs synthesize physiologically significant amounts of RBP or if HSCs synthesize very little or no RBP [35,48,50,79,87,90,99–108].

Blomhoff et al. have reported that primary rat HSC isolates possess significant levels of RBP protein and that HSCs express RBP mRNA [35,79,99–103]. These investigators have further reported that rat HSCs can take up and internalize RBP from the circulation [79,100,101,103]. These findings allowed Blomhoff et al. to propose an attractive model whereby RBP newly synthesized in HSCs directly mobilizes HSC retinoids into the circulation. These authors further proposed that circulating RBP taken up by HSCs might also be resecreted into the circulation after it had acquired retinol within the HSC [79,81–83,100,101,103]. In other words, these investigators propose that RBP synthesized in or internalized by HSCs can act directly to mobilize HSC retinoids into the circulation to meet the needs of peripheral tissues. Our laboratory, as well as others, however, has not been able to detect significant levels of RBP protein or RBP mRNA in primary HSC isolates obtained from rats or mice [50,89,90,104–107]. These findings cast doubt on whether RBP is truly synthesized in HSCs and on the idea that RBP synthesized in HSCs can account for retinol mobilization from these cells. Moreover, as mentioned earlier in the discussion of hepatocyte mobilization of retinoid, our studies of hRBP$^{-/-}$ mice (Rbp$^{-/-}$ mice that express hRBP in skeletal muscle) indicate that RBP expressed extrahepatically is not taken up by the liver and cannot draw on HSC retinoid stores [58,61,88]. Based collectively on these data, we have proposed that HSC retinoid stores can be mobilized from the liver only after retinol or retinyl ester has been transferred from HSC stores to hepatocytes, the cellular site of RBP synthesis. We propose that there is a reversal of the process that moves dietary retinoid from hepatocytes, where it is taken up in CRs, to HSCs for storage. This model is less attractive than the one proposed by Blomhoff et al. since it fails to explain how retinol or retinyl ester is transferred between the two cell types. However, our model best fits the findings from our laboratory and others. We have long felt that the simplest explanation for why RBP protein and mRNA were observed to be present in primary HSC isolates was that these isolates were contaminated with hepatocytes and/or hepatocyte-derived cell debris.

In summary, there is no consensus regarding how HSC retinoid stores are mobilized to meet tissue needs. It is clear that this process requires RBP, but whether this RBP is synthesized by HSCs or solely by hepatocytes remains to be resolved.

## WHAT IS NOT UNDERSTOOD

Perhaps the question in greatest need of resolution for fully understanding hepatic retinoid storage and metabolism concerns whether HSCs synthesize and secrete significant levels of RBP and whether this can provide an explanation for how HSC retinoids are mobilized. In the late 1980s, and through the mid-1990s, there were many publications focused on this issue [35,48,50,79,87,99–107], which concluded either that RBP is present and/or synthesized in HSCs or that RBP protein and mRNA cannot be found in HSCs. The data reported in each of the manuscripts supporting one position or the other are all individually convincing. This long-standing disagreement will not be resolved simply through new measures of HSC RBP protein or mRNA levels. New approaches for addressing this question will be required, possibly involving genetic manipulations of mice.

The molecular identity or identities of REH(s) present in HSCs and involved in mobilization of lipid droplet retinyl ester stores has/have not been determined. This, too, is an important and long-standing question, which is still not understood. Since REH activity is necessary to allow for HSC retinyl ester stores, this process and its regulation cannot be understood without the molecular identification of the enzyme(s) responsible for catalyzing hydrolysis. The present biochemical understanding of retinyl ester hydrolysis in HSCs is limited to the very basic understanding that there are enzymes present in liver homogenates that can catalyze the hydrolysis of retinyl ester in vitro and that a number of these enzymes are expressed in HSCs. Whether any or all of these enzymes play a physiologically significant role in catalyzing HSC retinyl ester hydrolysis remains to be established.

There is presently much research activity focused on the molecular composition of lipid droplets and the physiological processes that influence lipid droplet accumulation and dissolution. There is certainly a strong need to better understand the molecular properties and physiologic processes that regulate HSC lipid droplet formation and hydrolysis. Understanding of these properties and processes is fundamental for understanding how retinoids are stored in the liver and for understanding how retinoids are mobilized from the liver. What enzymes are important for HSC lipid droplet triglyceride and cholesteryl ester synthesis and hydrolysis? Are these processes coordinately regulated in conjunction with retinyl ester synthesis and hydrolysis? Without this information, we cannot truly understand retinoid storage and metabolism in the liver.

## OTHER HEPATIC NONPARENCHYMAL CELLS

As noted earlier in the section on hepatocyte retinoid storage "Role in Retinoid Storage" section, several groups that prepared and studied freshly isolated primary rat hepatic cells agreed that Kupffer and hepatic endothelial cells contain negligible amounts of retinoids and CRBPI, as well as REH activity, compared to hepatocytes or HSCs [49,50]. Hence, it was concluded that Kupffer and hepatic endothelial cells have insignificant or no role in hepatic retinoid storage and metabolism. More recent immunohistochemical studies undertaken by Nagatsuma et al., however, indicate that LRAT is expressed in both human and rat hepatic endothelial cells [108]. Based on this finding, Nagatsuma et al. raised the question as to whether LRAT has some as-yet-to-be-determined function in liver endothelial biology. It seems reasonable to suspect, as our understanding of liver cell interactions becomes more sophisticated and extensive, that hepatic Kupffer and endothelial cells will be found to have roles in hepatic retinoid storage and metabolism. These nonparenchymal cells may not act centrally in the bulk storage of hepatic retinoids, but they may act in some manner to facilitate or regulate storage and metabolism.

## ACKNOWLEDGMENT

The work discussed from the authors' laboratory was supported by grants from the National Institutes of Health [RC2 AA019413, R01 DK68437, and R01 DK079221].

## REFERENCES

1. Blaner, W. S. and Olson, J. A. (eds.) (1994) *Retinol and Retinoic Acid Metabolism*, Raven Press, Ltd., New York.
2. Friedman, S. L. (2008) *Physiol Rev* **88**, 125–172.
3. Geerts, A. (2001) *Semin Liver Dis* **21**, 311–335.
4. Blaner, W. S., O'Byrne, S. M., Wongsiriroj, N., Kluwe, J., D'Ambrosio, D. M., Jiang, H., Schwabe, R. F., Hillman, E. M., Piantedosi, R., and Libien, J. (2009) *Biochim Biophys Acta* **1791**, 467–473.
5. Shirakami, Y., Lee, S. A., Clugston, R. D., and Blaner, W. S. (2011) *Biochim Biophys Acta* **1821**, 124–136.
6. Goodman, D. W., Huang, H. S., and Shiratori, T. (1965) *J Lipid Res* **6**, 390–396.
7. Goodman, D. S. and Blaner, W. S. (1984) *Biosynthesis, Absorption, and Hepatic Metabolism of Retinol*, The Academic Press, New York.
8. Gregg, R. E., Zech, L. A., Schaefer, E. J., and Brewer, H. B., Jr. (1981) *Science* **211**, 584–586.
9. Brewer, H. B., Jr., Zech, L. A., Gregg, R. E., Schwartz, D., and Schaefer, E. J. (1983) *Ann Intern Med* **98**, 623–640.
10. van Bennekum, A. M., Kako, Y., Weinstock, P. H., Harrison, E. H., Deckelbaum, R. J., Goldberg, I. J., and Blaner, W. S. (1999) *J Lipid Res* **40**, 565–574.
11. Blaner, W. S., Obunike, J. C., Kurlandsky, S. B., al-Haideri, M., Piantedosi, R., Deckelbaum, R. J., and Goldberg, I. J. (1994) *J Biol Chem* **269**, 16559–16565.
12. Mjos, O. D., Faergeman, O., Hamilton, R. L., and Havel, R. J. (1975) *J Clin Invest* **56**, 603–615.
13. Redgrave, T. G. (1970) *J Clin Invest* **49**, 465–471.
14. Mahley, R. W. and Ji, Z. S. (1999) *J Lipid Res* **40**, 1–16.

15. Cooper, A. D. (1997) *J Lipid Res* **38**, 2173–2192.
16. Yu, K. C. and Cooper, A. D. (2001) *Front Biosci* **6**, D332–D354.
17. Scanu, A. (1966) *Science* **153**, 640–641.
18. Windler, E., Chao, Y., and Havel, R. J. (1980) *J Biol Chem* **255**, 5475–5480.
19. Shelburne, F., Hanks, J., Meyers, W., and Quarfordt, S. (1980) *J Clin Invest* **65**, 652–658.
20. Weisgraber, K. H., Mahley, R. W., Kowal, R. C., Herz, J., Goldstein, J. L., and Brown, M. S. (1990) *J Biol Chem* **265**, 22453–22459.
21. Hoeg, J. M., Demosky, S. J., Jr., Gregg, R. E., Schaefer, E. J., and Brewer, H. B., Jr. (1985) *Science* **227**, 759–761.
22. Cooper, A. D., Erickson, S. K., Nutik, R., and Shrewsbury, M. A. (1982) *J Lipid Res* **23**, 42–52.
23. Blaner, W. S., Prystowsky, J. H., Smith, J. E., and Goodman, D. S. (1984) *Biochim Biophys Acta* **794**, 419–427.
24. Harrison, E. H. and Gad, M. Z. (1989) *J Biol Chem* **264**, 17142–17147.
25. van Bennekum, A. M., Li, L., Piantedosi, R., Shamir, R., Vogel, S., Fisher, E. A., Blaner, W. S., and Harrison, E. H. (1999) *Biochemistry* **38**, 4150–4156.
26. Beisiegel, U., Weber, W., and Bengtsson-Olivecrona, G. (1991) *Proc Natl Acad Sci USA* **88**, 8342–8346.
27. Mulder, M., Lombardi, P., Jansen, H., van Berkel, T. J., Frants, R. R., and Havekes, L. M. (1993) *J Biol Chem* **268**, 9369–9375.
28. Ji, Z. S., Lauer, S. J., Fazio, S., Bensadoun, A., Taylor, J. M., and Mahley, R. W. (1994) *J Biol Chem* **269**, 13429–13436.
29. Shafi, S., Brady, S. E., Bensadoun, A., and Havel, R. J. (1994) *J Lipid Res* **35**, 709–720.
30. Medda, S. and Proia, R. L. (1992) *Eur J Biochem* **206**, 801–806.
31. Yan, B., Yang, D., Bullock, P., and Parkinson, A. (1995) *J Biol Chem* **270**, 19128–19134.
32. Gad, M. Z. and Harrison, E. H. (1991) *J Lipid Res* **32**, 685–693.
33. Mentlein, R. and Heymann, E. (1987) *Biochem J* **245**, 863–867.
34. Hagen, E., Myhre, A. M., Tjelle, T. E., Berg, T., and Norum, K. R. (1999) *J Lipid Res* **40**, 309–317.
35. Blomhoff, R., Eskild, W., Kindberg, G. M., Prydz, K., and Berg, T. (1985) *J Biol Chem* **260**, 13566–13570.
36. Harrison, E. H., Gad, M. Z., and Ross, A. C. (1995) *J Lipid Res* **36**, 1498–1506.
37. Harrison, E. H. (2000) *J Nutr* **130**, 340S–344S.
38. Sun, G., Alexson, S. E., and Harrison, E. H. (1997) *J Biol Chem* **272**, 24488–24493.
39. Schindler, R., Mentlein, R., and Feldheim, W. (1998) *Eur J Biochem* **251**, 863–873.
40. Sanghani, S. P., Davis, W. I., Dumaual, N. G., Mahrenholz, A., and Bosron, W. F. (2002) *Eur J Biochem* **269**, 4387–4398.
41. Linke, T., Dawson, H., and Harrison, E. H. (2005) *J Biol Chem* **280**, 23287–23294.
42. Holmes, R. S., Wright, M. W., Lauderkind, S. J., Cox, L. A., Hosokawa, M., Imai, T., Ishibashi, S. et al., (2010) *Mamm Genome* **21**, 427–441.
43. Schreiber, R., Taschler, U., Wolinski, H., Seper, A., Tamegger, S. N., Graf, M., Kohlwein, S. D., Haemmerle, G., Zimmermann, R., Zechner, R., and Lass, A. (2009) *J Lipid Res* **50**, 2514–2523.
44. Haemmerle, G., Lass, A., Zimmermann, R., Gorkiewicz, G., Meyer, C., Rozman, J., Heldmaier, G. et al. (2006) *Science* **312**, 734–737.
45. Kienesberger, P. C., Lee, D., Pulinilkunnil, T., Brenner, D. S., Cai, L., Magnes, C., Koefeler, H. C. et al. (2009) *J Biol Chem* **284**, 30218–30229.
46. Wei, S., Lai, K., Patel, S., Piantedosi, R., Shen, H., Colantuoni, V., Kraemer, F. B., and Blaner, W. S. (1997) *J Biol Chem* **272**, 14159–14165.

47. Strom, K., Gundersen, T. E., Hansson, O., Lucas, S., Fernandez, C., Blomhoff, R., and Holm, C. (2009) *FASEB J* **23**, 2307–2316.

48. Blomhoff, R., Helgerud, P., Rasmussen, M., Berg, T., and Norum, K. R. (1982) *Proc Natl Acad Sci USA* **79**, 7326–7330.

49. Blomhoff, R., Rasmussen, M., Nilsson, A., Norum, K. R., Berg, T., Blaner, W. S., Kato, M., Mertz, J. R., Goodman, D. S., Eriksson, U. et al. (1985) *J Biol Chem* **260**, 13560–13565.

50. Blaner, W. S., Hendriks, H. F., Brouwer, A., de Leeuw, A. M., Knook, D. L., and Goodman, D. S. (1985) *J Lipid Res* **26**, 1241–1251.

51. Batres, R. O. and Olson, J. A. (1987) *J Nutr* **117**, 874–879.

52. Kane, M. A., Chen, N., Sparks, S., and Napoli, J. L. (2005) *Biochem J* **388**, 363–369.

53. Kurlandsky, S. B., Gamble, M. V., Ramakrishnan, R., and Blaner, W. S. (1995) *J Biol Chem* **270**, 17850–17857.

54. Kanai, M., Raz, A., and Goodman, D. S. (1968) *J Clin Invest* **47**, 2025–2044.

55. Zovich, D. C., Orologa, A., Okuno, M., Kong, L. W., Talmage, D. A., Piantedosi, R., Goodman, D. S., and Blaner, W. S. (1992) *J Biol Chem* **267**, 13884–13889.

56. Tsutsumi, C., Okuno, M., Tannous, L., Piantedosi, R., Allan, M., Goodman, D. S., and Blaner, W. S. (1992) *J Biol Chem* **267**, 1805–1810.

57. Quadro, L., Hamberger, L., Colantuoni, V., Gottesman, M. E., and Blaner, W. S. (2003) *Mol Aspects Med* **24**, 421–430.

58. Quadro, L., Blaner, W. S., Hamberger, L., Novikoff, P. M., Vogel, S., Piantedosi, R., Gottesman, M. E., and Colantuoni, V. (2004) *J Lipid Res* **45**, 1975–1982.

59. Quadro, L., Hamberger, L., Gottesman, M. E., Colantuoni, V., Ramakrishnan, R., and Blaner, W. S. (2004) *Am J Physiol Endocrinol Metab* **286**, E844–E851.

60. Biesalski, H. K., Frank, J., Beck, S. C., Heinrich, F., Illek, B., Reifen, R., Gollnick, H., Seeliger, M. W., Wissinger, B., and Zrenner, E. (1999) *Am J Clin Nutr* **69**, 931–936.

61. Quadro, L., Blaner, W. S., Hamberger, L., Van Gelder, R. N., Vogel, S., Piantedosi, R., Gouras, P., Colantuoni, V., and Gottesman, M. E. (2002) *J Biol Chem* **277**, 30191–30197.

62. Yang, Q., Graham, T. E., Mody, N., Preitner, F., Peroni, O. D., Zabolotny, J. M., Kotani, K., Quadro, L., and Kahn, B. B. (2005) *Nature* **436**, 356–362.

63. Jaconi, S., Rose, K., Hughes, G. J., Saurat, J. H., and Siegenthaler, G. (1995) *J Lipid Res* **36**, 1247–1253.

64. Monaco, H. L., Rizzi, M., and Coda, A. (1995) *Science* **268**, 1039–1041.

65. Goodman, D. S. (1984) *N Engl J Med* **310**, 1023–1031.

66. van Bennekum, A. M., Wei, S., Gamble, M. V., Vogel, S., Piantedosi, R., Gottesman, M., Episkopou, V., and Blaner, W. S. (2001) *J Biol Chem* **276**, 1107–1113.

67. Noy, N., Slosberg, E., and Scarlata, S. (1992) *Biochemistry* **31**, 11118–11124.

68. Goodman, D. S. (1980) *Ann NY Acad Sci* **348**, 378–390.

69. Gaetani, S., Bellovino, D., Apreda, M., and Devirgiliis, C. (2002) *Clin Chem Lab Med* **40**, 1211–1220.

70. Suhara, A., Kato, M., and Kanai, M. (1990) *J Lipid Res* **31**, 1669–1681.

71. Bellovino, D., Morimoto, T., Tosetti, F., and Gaetani, S. (1996) *Exp Cell Res* **222**, 77–83.

72. Wei, S., Episkopou, V., Piantedosi, R., Maeda, S., Shimada, K., Gottesman, M. E., and Blaner, W. S. (1995) *J Biol Chem* **270**, 866–870.

73. Blaner, W. S. (2007) *Cell Metab* **5**, 164–166.

74. Kawaguchi, R., Yu, J., Honda, J., Hu, J., Whitelegge, J., Ping, P., Wiita, P., Bok, D., and Sun, H. (2007) *Science* **315**, 820–825.

75. Taneja, R., Bouillet, P., Boylan, J. F., Gaub, M. P., Roy, B., Gudas, L. J., and Chambon, P. (1995) *Proc Natl Acad Sci USA* **92**, 7854–7858.

76. Bouillet, P., Sapin, V., Chazaud, C., Messaddeq, N., Decimo, D., Dolle, P., and Chambon, P. (1997) *Mech Dev* **63**, 173–186.

77. Isken, A., Golczak, M., Oberhauser, V., Hunzelmann, S., Driever, W., Imanishi, Y., Palczewski, K., and von Lintig, J. (2008) *Cell Metab* **7**, 258–268.
78. Kim, Y. K., Wassef, L., Hamberger, L., Piantedosi, R., Palczewski, K., Blaner, W. S., and Quadro, L. (2008) *J Biol Chem* **283**, 5611–5621.
79. Senoo, H., Stang, E., Nilsson, A., Kindberg, G. M., Berg, T., Roos, N., Norum, K. R., and Blomhoff, R. (1990) *J Lipid Res* **31**, 1229–1239.
80. Tosetti, F., Campelli, F., and Levi, G. (1999) *Exp Cell Res* **250**, 423–433.
81. Blomhoff, R., Green, M. H., Green, J. B., Berg, T., and Norum, K. R. (1991) *Physiol Rev* **71**, 951–990.
82. Blomhoff, R., Green, M. H., and Norum, K. R. (1992) *Annu Rev Nutr* **12**, 37–57.
83. Senoo, H., Kojima, N., and Sato, M. (2007) *Vitam Horm* **75**, 131–159.
84. Ghyselinck, N. B., Bavik, C., Sapin, V., Mark, M., Bonnier, D., Hindelang, C., Dierich, A. et al. (1999) *EMBO J* **18**, 4903–4914.
85. Molotkov, A., Ghyselinck, N. B., Chambon, P., and Duester, G. (2004) *Biochem J* **383**, 295–302.
86. Blaner, W. S., Dixon, J. L., Moriwaki, H., Martino, R. A., Stein, O., Stein, Y., and Goodman, D. S. (1987) *Eur J Biochem* **164**, 301–307.
87. Blomhoff, R., Holte, K., Naess, L., and Berg, T. (1984) *Exp Cell Res* **150**, 186–193.
88. Okuno, M., Caraveo, V. E., Goodman, D. S., and Blaner, W. S. (1995) *J Lipid Res* **36**, 137–147.
89. D'Ambrosio, D. N., Walewski, J. L., Clugston, R. D., Berk, P. D., Rippe, R. A., and Blaner, W. S. (2011) *PLoS One* **6**, e24993.
90. Moriwaki, H., Blaner, W. S., Piantedosi, R., and Goodman, D. S. (1988) *J Lipid Res* **29**, 1523–1534.
91. Hendriks, H. F., Brekelmans, P. J., Buytenhek, R., Brouwer, A., de Leeuw, A. M., and Knook, D. L. (1987) *Lipids* **22**, 266–273.
92. Yumoto, S., Ueno, K., Mori, S., Takebayashi, N., and Handa, S. (1988) *Biomed Res* **2**, 147–160.
93. Wake, K. (1974) *J Cell Biol* **63**, 683–691.
94. Wake, K. (1980) *Int Rev Cytol* **66**, 303–353.
95. Straub, B. K., Stoeffel, P., Heid, H., Zimbelmann, R., and Schirmacher, P. (2008) *Hepatology* **47**, 1936–1946.
96. Londos, C., Brasaemle, D. L., Schultz, C. J., Segrest, J. P., and Kimmel, A. R. (1999) *Semin Cell Dev Biol* **10**, 51–58.
97. Girousse, A. and Langin, D. (2011) *Int J Obes (Lond)* [Epub ahead of print] PMID: 21673652.
98. Quadro, L., Blaner, W. S., Salchow, D. J., Vogel, S., Piantedosi, R., Gouras, P., Freeman, S., Cosma, M. P., Colantuoni, V., and Gottesman, M. E. (1999) *EMBO J* **18**, 4633–4644.
99. Andersen, K. B., Nilsson, A., Blomhoff, H. K., Oyen, T. B., Gabrielsen, O. S., Norum, K. R., and Blomhoff, R. (1992) *J Biol Chem* **267**, 1340–1344.
100. Lewis, K. C., Green, M. H., Green, J. B., and Zech, L. A. (1990) *J Lipid Res* **31**, 1535–1548.
101. Senoo, H., Smeland, S., Malaba, L., Bjerknes, T., Stang, E., Roos, N., Berg, T., Norum, K. R., and Blomhoff, R. (1993) *Proc Natl Acad Sci USA* **90**, 3616–3620.
102. Troen, G., Nilsson, A., Norum, K. R., and Blomhoff, R. (1994) *Biochem J* **300**(Pt 3), 793–798.
103. Malaba, L., Smeland, S., Senoo, H., Norum, K. R., Berg, T., Blomhoff, R., and Kindberg, G. M. (1995) *J Biol Chem* **270**, 15686–15692.
104. Yamada, M., Blaner, W. S., Soprano, D. R., Dixon, J. L., Kjeldbye, H. M., and Goodman, D. S. (1987) *Hepatology* **7**, 1224–1229.

105. Weiner, F. R., Blaner, W. S., Czaja, M. J., Shah, A., and Geerts, A. (1992) *Hepatology* **15**, 336–342.
106. Friedman, S. L., Wei, S., and Blaner, W. S. (1993) *Am J Physiol* **264**, G947–G952.
107. Sauvant, P., Sapin, V., Alexandre-Gouabau, M. C., Dodeman, I., Delpal, S., Quadro, L., Partier, A., Abergel, A., Colantuoni, V., Rock, E., and Azais-Braesco, V. (2001) *Int J Biochem Cell Biol* **33**, 1000–1012.
108. Nagatsuma, K., Hayashi, Y., Hano, H., Sagara, H., Murakami, K., Saito, M., Masaki, T. et al. (2009) *Liver Int* **29**, 47–54.

# 2 Metabolism of Vitamin A in White Adipose Tissue and Obesity

*Shanmugam M. Jeyakumar, Rumana Yasmeen, Barbara Reichert, and Ouliana Ziouzenkova*

## CONTENTS

## OBESITY

The World Health Organization (WHO) defined obesity as an excess of fat accumulation leading to adverse effects on health. Persons with body mass index (BMI) in the range of 25–30 kg/m$^2$ are considered overweight, and with a BMI ≥ 30 kg/m$^2$ as obese [1]. During the past two decades, over one billion adults worldwide have become overweight and 300 million are obese [2]. The prevalence of obesity has attained the proportions of an epidemic not only in developed but also in developing countries [3,4]. This trend was acknowledged by WHO term "globesity" [2]. Obesity is a chronic, highly prevalent, and refractory disease that poses increased risk for a variety of medical conditions including sleep apnea, diabetes, hypertension, dyslipidemia, cardiovascular disease, and some cancers [3,5]. Overall, obesity correlates with increased mortality [6]. However, a paradoxical decrease in mortality was also reported in obese patients with established heart failure or coronary heart disease, intracerebral hemorrhage, and in patients undergoing percutaneous coronary intervention [7,8]. This paradox is one of numerous phenomena that challenge researchers to understand how adipose tissue contributes to the regulation of systemic energy homeostasis and endocrine responses. Rigorous research continues to unravel the role of nutrition and physical exercise in obesity development and systemic control of energy homeostasis by multiple organs (Figure 2.1). Although the past decade marks the golden era for obesity research, the mechanisms underlying obesity remain incomplete, which prevents development of effective therapies beyond the control of nutrition and physical exercise.

Based on the developmental origin, genetics, physiology, and endocrine responses, researchers distinguish two types of adipose tissue: white and brown [9,10]. Brown

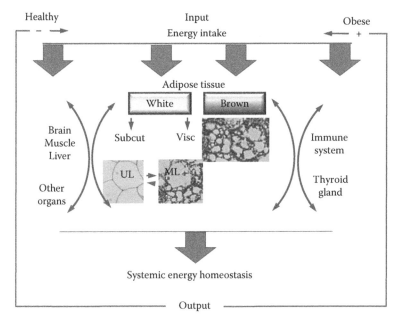

**FIGURE 2.1** Systemic control of energy homeostasis by multiple organs. Multiple organs, including white and brown adipose tissues, take up circulating nutrients according to their energy demands that are dictated by competitive gene programs for energy expenditure and storage. White adipose tissue (WAT) is comprised of unilocular (UL) and multilocular (ML) adipocytes. UL adipocytes express genes that promote energy uptake for storage. ML adipocytes express genes, such as *UCP1* that support energy expenditure for heat production (thermogenesis). Paraffin fixed WAT was stained with hematoxylin and eosin to show UL adipocytes and antibodies recognizing UCP1 protein to detect ML adipocytes. High proportions of ML adipocytes are associated with resistance to obesity in rodents and humans. All organs communicate their energy status and demands by cytokine release. Systemic energy demands are communicated to the central nervous system, which influences energy uptake. In obesity, UL adipocyte genetic programs are responsible for excessive energy storage that shifts energy balance and further rises demand for energy uptake.

adipose tissue (BAT) is an essential thermoregulatory organ in rodents [9]. In humans, BAT supports thermogenesis in infants. In adults, residual BAT is located in the neck and peri-vascular areas [11,12]. The energy dissipation through thermogenesis profoundly influences systemic energy homeostasis and is associated with resistance to obesity in rodents and humans [9,10]. White adipose tissue (WAT) is the primary tissue affected by obesity. The transplantation of WAT in animal models and liposuction surgeries in humans unequivocally demonstrated that WAT depots can have a marked impact on whole-body systemic responses [13]. Extensive work revealed several genes/gene products and signaling pathways in WAT that are key players in metabolic regulation [14]. The emerging data suggest that micronutrients, such as vitamin A, become a necessary functional component of WAT development upon their conversion into hormone-like molecules by specific enzymes [15,16]. The paradigm of WAT autocrine regulation through micronutrient–gene interactions offers new opportunities for understanding, preventing, and managing obesity. This chapter provides an overview of vitamin A metabolism in WAT and its impact on WAT biology and obesity.

## PROCESSES UNDERLYING WHITE ADIPOSE TISSUE REMODELING IN OBESITY

Obesity is a metabolic condition of imbalanced energy homeostasis associated with excessive accumulation of WAT [17]. In higher eukaryotes, WAT is a multi-depot organ supporting triglyceride (TG) storage for energy demands and signaling function by secreting inflammatory cytokines and adipose-specific cytokines known as adipokines. White adipose depots are classified based on their developmental origin and location into inguinal (subcutaneous), omental, and mesenchymal fat [18]. Visceral fat that surrounds specific organs, for instance, the heart, gonads, and kidney, is described as pericardial, perigonadal, and perirenal fat, respectively. Humans and rodents have different developmental origins of visceral fat, which is known as omental fat in humans. In the clinical reports, an "abdominal," "intra-abdominal," "central," or "trunk fat" description often refers to an omental fat depot, whereas "peripheral fat" describes inguinal fat. In humans and animals, males and females widely differ with respect to body weight gain, adipose tissue development, and adipose distribution [19,20]. Males accumulate more visceral fat in the abdominal region, while females (particularly premenopausal) deposit fat in subcutaneous regions [20]. Visceral and subcutaneous fat depots have disparate health effects [6,21,22]. Visceral fat accumulation is associated with increased risk for all-cause mortality in both lean and obese patient populations [6]. These pathological effects are associated with the pro-inflammatory properties of this fat depot and the release of a variety of pro-inflammatory cytokines and chemoattractants, which recruit inflammatory cells (reviewed in [21]). Some of them, including TNFα, IL6, and resistin, induce insulin resistance [23]. Subcutaneous fat generates adipokines, such as adiponectin, with anti-inflammatory and insulin-sensitizing properties [24]. The transcriptome analysis in multiple patients revealed different genetic signatures for inguinal and visceral fat that express distinct patterns of developmental genes

[18,25–28]. Mesenchymal fat shares similar characteristics with inguinal adipose tissues [18]. These data strengthen the concept that WAT is not a single distinct entity. White adipose depots have specific properties that are regulated by specific mechanisms. In their recent review, Drs. Kahn and Tran have characterized these depots as two separate tissues, visceral WAT and subcutaneous WAT [13].

WATs, both subcutaneous and visceral, also harbor multilocular adipocytes resembling thermogenic brown adipocytes (Figure 2.1). Multilocular adipocytes also contain numerous mitochondria involved in thermogenesis, and their lipids are organized as multilocular droplets accessible for lipolysis [9,29]. Despite their similar morphological features, multilocular adipocytes from white fat and brown adipocytes appear to derive from different lineages, whereas myocytes and brown adipocytes share a common precursor [30]. Uni- and multilocular adipocytes originate from mesenchymal progenitor cells and are capable of changing their morphology depending on their milieu [31]. Cold exposure and β-adrenergic stimulation markedly increase the proportion of multilocular adipocytes in visceral fat, whereas high-fat feeding enhances the number of unilocular adipocytes [31]. In white fat depots, the proportion of uni- and multilocular adipocytes, widely differs among species; moreover, sex, age, and other nutritional and hormonal factors also influence numbers of uni- versus multilocular cells [9]. An increase in multilocular adipocytes decreases adipocyte lipid load and improves metabolic responses, including insulin resistance and inflammation [9,10,13,30]. Whereas mechanisms initiating fat formation in specific depots are still unclear, major pathways leading to expansion of white fat have been discovered [32–34]. They include the following:

1. Increased number of differentiated adipocytes due to induced preadipocyte differentiation and/or reduced apoptosis
2. Increased adipocyte size due to increased lipogenesis and/or reduced lipolysis
3. Decreased numbers of thermogenic fat cells and/or reduction in their energy dissipation
4. Angiogenesis and recruitment of inflammatory cells and adipocyte precursor cells

In this chapter, we will discuss only white adipocyte–based processes, such as adipogenesis, lipolysis, and lipogenesis, and their intrinsic regulation through vitamin A metabolism in adipocytes. The ample and controversial topic of the regulation of adipose tissue by dietary vitamin A or administration of retinoids (retinol or retinoic acid [RA]) has been documented in a recent excellent review [35].

## ADIPOGENESIS

In humans, nearly 30 billion adipocytes are present during various stages of development, i.e., from infant to adolescent [36]. However, this number rises to 40–60 billion cells during obesity. Adipocytes are derived after differentiation of preadipocytes, which, in turn, originate from the multipotent mesenchymal stem cells.

Preadipocytes are committed to adipogenic lineage in a process known as determination [33]. Despite their overall stem cell morphology, preadipocytes cannot differentiate into chondrocytes, osteoblasts, or myocytes [14]. Preadipocytes comprise 15%–50% of all cells in white fat depots [37]. Preadipocytes express discrete genes, such as *Pref1* [38] and *Dact1*, which act through transcription factors Sox9 [39] and WNT/β-catenin signaling pathways [40]. These factors, however, are expressed in both mesenchymal stem cells and preadipocytes.

Nutrients and hormones can induce terminal differentiation of preadipocytes into mature adipocytes capable of lipid synthesis, storage, and production of adipokines [34]. This process, also known as adipogenesis, lasts from 5 to 7 days for mouse cells and 21–30 days for human cells in vitro. Adipogenesis is characterized by dynamic expression of multiple transcription factors [39–41]. The preadipocyte transcription factors, including *Sox9* and WNT/β-catenin, undergo suppression with the concomitant expression and activation of key regulators of transcription in adipogenesis, the CCAAT/enhancer-binding protein (C/EBP) family of bZIP transcription factors and PPARγ. *PPARγ* expression is a central event in adipogenesis. In the genetic absence of *PPARγ*, adipocytes cannot be formed in vitro or in vivo [42]. Ectopic *PPARγ* overexpression in preadipocytes leads to their differentiation into mature adipocytes with lipogenic properties and adipokine expression [43]. The importance of *C/EBPα*, *C/EBPβ*, and *C/EBPδ* was demonstrated in their knockout models, which developed markedly reduced WAT depots (reviewed in [34]). The remaining *C/EBPβ$^{-/-}$* and *C/EBPα$^{-/-}$* adipocytes, however, continued to express *PPARγ* and, thus, were dispensable for *PPARγ* regulation in vivo. Recent findings have identified alternative pathways responsible for adipogenesis induction and regulation of *PPARγ*. Transcription factors ZFP423, EBF1, and STAT5 have shown to be sufficient to induce *PPARγ* expression (reviewed in [16]). Other transcription factors, e.g., KLF5, are responsible for partial *PPARγ* expression or activation (SREBP1c and C/EBPα). Cell cycle-controlling mechanisms are at work during adipogenesis, for example deficient retinoblastoma protein (*pRb*) expression induces thermogenic gene expression in white preadipocytes [44]. Transcriptional regulation of adipogenesis has recently been thoroughly described [14,33,34].

Mechanistic studies on adipogenesis in mouse cell lines 3T3-L1 and 3T3-F442A are representative of subcutaneous fat. Visceral adipocytes express less *PPARγ* and a different combination of developmental *HOX* genes (reviewed in [16]). The different transcription factor network and the recruitment of specific transcriptional co-activators can change the outcome of differentiation mediated by *PPARγ* in brown, subcutaneous, and visceral adipose tissues [45]. In contrast to white adipocyte lineage, brown adipocyte lineage is characterized by expression *PRDM16* and *C/EBPβ* [10]. C/EBPβ and C/EBPδ activate the *UCP1* promoter, whereas PPARγ binds to the enhancer element for activation of *UCP1* expression. The genetic absence of *SRC1* co-activator, but recruitment of PGC1α to PPARγ, increases the number of multilocular thermogenic cells in adipose tissue [45,46]. In brown adipocytes, C/EBPβ is cAMP-sensitive and also participates in activation of PGC1α promoter [47,48]. Nonetheless, the studies in C/EBPβ-deficient mice reveal that C/EBPβ transcriptional events primarily control mobilization of fatty acids for thermogenesis [49]. A similar transcription cascade occurs in *pRb*-deficient adipocytes, in which

Foxc2 activation increases cAMP sensitivity and induces expression of thermogenic genes [44]. In contrast to the regulation of thermogenic genes, the transcription processes that lead to the diversification of properties between subcutaneous and visceral adipocytes have not been identified to date.

The number of adipocytes in white adipose depots depends on the number of pre-adipocytes and, mainly, on the endocrine and/or dietary stimuli altering differentiation or apoptosis. The vitamin A pathway is a remarkable example of a dietary factor that, in the presence of metabolizing enzymes, has the capacity to produce multiple signaling molecules that can govern differentiation, metabolism, and apoptosis in WAT in both an auto- and a paracrine fashion.

## LIPOLYSIS AND LIPOGENESIS IN WAT

Lipolysis and lipogenesis are key pathways determining mature adipocyte size and mass. Both processes function in conjunction with a family of lipid droplet proteins [50]. The diameter of a mature adipocyte ranges from 10 to 200 μm and contains 0.5–1 μg of fat, with a maximum of 4 μg. In a healthy human with normal weight, the mass of adipocytes accounts for approximately 20% of body weight [36]. WAT mass may vary from 2%–3% to 60%–70% of body weight, as with well-fit athletes or massively obese individuals, respectively [51]. Lipolysis and lipogenesis are inversely regulated by allosteric regulators, hormones, and nutrients activating signaling pathways and/or transcription factors.

WAT is a storage site of TGs, derived from circulating nonesterified fatty acids and/or through hydrolysis of TG-rich lipoproteins by lipoprotein lipase (LPL). LPL is considered to be the rate-limiting enzyme for TG hydrolysis from lipoproteins and their uptake into adipocytes. Adipocytes from visceral depots exhibit higher basal lipolytic activity than subcutaneous depots due to higher LPL activity. Further, in genetically obese rodent models and in human obesity, enhanced activity of LPL is reported [52–57]. Expression/activity of LPL during obesity, however, is still controversial. Various transporters, such as fatty acid transporter protein (FATP, e.g., CD36), the VLDL receptor of adipocytes, and fatty acid-binding proteins (FABPs), contribute to and further define lipid uptake and intracellular transport [58].

Intracellular lipolysis is an enzymatic process responsible for the hydrolysis of fatty acids from TG stored in cellular lipid droplets. Sequential TG lipolysis is mediated by different enzymes [59]. Adipose triglyceride lipase (ATGL) hydrolyzes the first fatty acid from triacylglycerol, followed by hydrolysis of the remaining diacylglycerol by HSL and monoacylglycerol by monoglyceride lipase (MGL) [60]. Genetic deficiency of *ATGL* prevents TG hydrolysis and increases adipose tissue in mice [59]. Moreover, it prevents wasting of WAT in experimental cachexia [61], suggesting the essentiality of this enzyme for the initiation of TG lipolysis. Mice studies of *HSL* deficiency revealed that these mice store diacylglycerol in adipose tissue and resist diet-induced obesity [62,63]. However, resistance to obesity has been recently attributed to another HSL function [64], namely, the hydrolysis of retinyl esters (discussed in vitamin A section). Evidence of the critical role of lipolysis in adipose tissue in healthy and obese subjects is widely documented in genetic animal models and human patients carrying genetic variants of the lipolytic genes [65,66]. Catecholamines are the most studied activators of WAT lipolysis in humans and rodents. They act through β- and α2-adrenergic receptors. In mice, this role is supported by β3-adrenergic receptors, whereas in humans, α2-adrenergic receptors mediate the effects of catecholamines [67]. Chimeric α2-adrenergic receptor-transgenic mice that lack β3-adrenergic receptors became susceptible to high-fat diet-induced obesity due to reduced fat mobilization capacity [67].

Adipocytes (including human adipocytes) express all necessary proteins for lipogenesis: transcription factors, such as *SREBP1*, *ChREBP*, and key enzymes *FAS*, *ACL*, and *SCD1*, although their contribution to de novo fatty acid synthesis has not been quantified [68–71]. SREBP1c appears to serve as a critical switch of lipid accumulation between white and brown depots, since transgenic nuclear expression of SREBP1c reduces white adipose depots, but increases lipid accumulation in BAT in a manner characteristic for WAT [72].

## VITAMIN A METABOLISM IN WHITE ADIPOSE TISSUE

### UPTAKE

WAT is the second largest storage depot for vitamin A in the body, after the hepatic tissue. Adipose tissue takes up vitamin A as retinol, retinyl esters, and, likely, RA by principally different pathways mediated by (1) LPL hydrolysis of TG-rich lipoproteins for delivery of retinol and retinyl esters, (2) lipocalin RBP4 for retinol delivery, and (3) circulating albumin and lipocalins, such as β-lactoglobulin (βLG) and lipocalin-type prostaglandin D synthase (L-PGDS) for delivery of retinol and RA (Figure 2.2).

The work from Goodman and Blaner laboratories demonstrated that postprandial retinol/retinyl esters are delivered to adipose tissue and other extrahepatic tissues from chylomicrons (about 25% of total retinyl esters are found in chylomicrons) [73,74]. In the vascular compartment of WAT, most of the TGs from the chylomicron are removed by LPL, resulting in the formation of chylomicron remnants [75,76]. LPL hydrolyzes retinol from retinyl esters of chylomicron remnants after completing hydrolysis of TGs [75]. In conjunction with LRP receptors, LPL also facilitates uptake of chylomicron remnants and retinyl esters in the chylomicron core in hepatocytes [77,78]; the contribution of the receptor-dependent process for retinyl ester delivery in adipose tissue, however, has not been reported. Under fasting conditions, the liver is the primary source (95%) of retinol that is secreted in the circulation bound to retinol-binding protein (RBP4) [79]. WAT tissue also expresses *RBP4* in a depot-specific manner, e.g., higher *RBP4* levels were found in visceral than in subcutaneous fat [80,81]. Brown fat in rodents expresses similar *RBP4* levels as subcutaneous adipose tissue [80]. The regulation of *RBP4* expression depends on glucose uptake by Glut4 in adipose tissue [82]. Increased *RBP4* expression in adipose tissue and RBP4 secretion into the circulation provoke insulin-resistant conditions in rodents and humans [81,82]. RBP4 secreted from adipose tissue is considered primarily as an adipokine that facilitates pathogenesis of type 2 diabetes and atherosclerosis [83], whereas its role in retinol metabolism in WAT

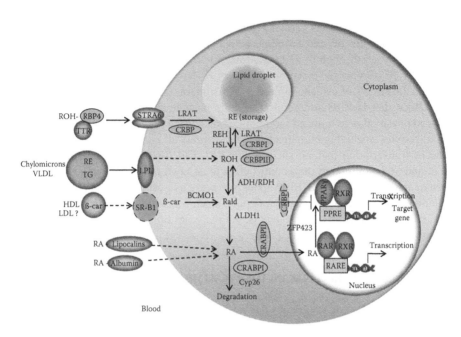

**FIGURE 2.2**  Vitamin A metabolism in WAT. Retinol carried by the retinol-binding protein (RBP4)-transthyretin (TTR) complex in blood enters cells by the receptor STRA6. CRBPI binds retinol, while LRAT converts retinol to retinyl esters for storage in lipid droplets. Lipoprotein lipase (LPL) mediates retinol delivery upon hydrolysis of retinyl esters from chylomicron remnants. In adipose tissue, hormone-sensitive lipase (HSL) is the major retinyl ester hydrolase. WAT is a RA-generating tissue, where retinol can be reversibly oxidized to retinaldehyde by either alcohol dehydrogenase (ADH) or retinol dehydrogenase/reductase (RDH) families of enzymes. Retinol can be transported intracellularly by CRBPI and/or CRBPIII. Retinaldehyde also binds to CRBPI in vitro; however, it is unknown if this protein can transport retinaldehyde to the nucleus. Retinaldehyde can repress PPARγ and RXRα, which inhibits the binding of PPARγ:RXRα heterodimers to PPAR response element (PPRE) and target gene transcription. Retinaldehyde is irreversibly oxidized to RA by the cytosolic aldehyde dehydrogenase 1 (ALDH1) family of enzymes. The retinoic acid (RA) produced can act in an autocrine fashion in adipocytes. Adipocytes express cellular RA-binding protein 2 (CRABP2), which may facilitate RA transport to the nucleus where RA binds to RA receptor (RAR). Activated RAR can heterodimerize with RXR and regulate target genes that contain RAR response element (RARE). RA can also act through transcription factor ZFP423 and activate expression of *PPARγ*. Cytochrome P450 (CYP26) catalyzes RA oxidation. Less explored pathways: circulating RA could be delivered into cells bound to lipocalins and albumin. β-Carotene, transported in HDL, could be taken up by the SR-B1 receptor. In cells, symmetric cleavage mediated by BCMO1 leads to production of two retinaldehyde molecules that can be oxidized to RA or reduced to retinol by ALDH1 or ADH/RDH families of enzymes, respectively.

remains incompletely understood. Kahn et al. have established association of *RBP4* expression in WAT with obesity [81,82]. More recent studies by Noy and associates strongly suggest that the link between RBP4 and fat formation is causal. *RBP4* expression is low in preadipocytes but is increased early in adipogenesis (day 2) and reaches a 16-fold induction on day 5 [84] (Figure 2.3). Suppression of *RBP4* mRNA

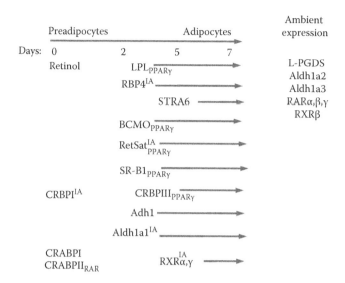

**FIGURE 2.3** Vitamin A metabolism during adipogenesis in 3T3-L1 preadipocytes. Differentiation of 3T3-L1, proceeding for 7 days (shown as arrow), is a classic model for adipogenesis. Stimulation of 3T3-L1 preadipocytes on day 0 leads to a sequential expression of transcription factors culminating in an induction of PPARγ, a master regulator of adipogenesis (day 2–3 after differentiation induction). Based on the published reports, genes participating in vitamin A metabolism were separated in three groups: (1) induced in preadipocytes, (2) induced during differentiation, and (3) induced throughout differentiation (ambient). Subscript "PPARγ" indicates vitamin A-metabolizing genes that are regulated by PPARγ. Superscript "IA" (influences adipogenesis) indicates vitamin A-metabolizing genes that significantly alter adipogenesis in genetic loss-of-function experiments.

expression decreased lipid accumulation in these cells [84]. RBP4 participation in fat formation in vivo was suggested based on genetic analysis in Japanese and Mongolian obese patients carrying a regulatory single-nucleotide polymorphism (SNP-803G>A) in RBP4 and showed a significantly higher BMI in Japanese men and women, and in Mongolian women [85].

The uptake of retinol-loaded RBP4 (holo-RBP4) is mediated by the STRA6 receptor [86], which mediates bidirectional retinol transport, e.g., inside and outside of the cell in a partnership with holo- or apo-RBP4, respectively [87]. Elegant studies by von Lintig and colleagues showed that STRA6-mediated transport of retinol from RBP-ROH in fibroblasts depends on LRAT, an enzyme catalyzing re-esterification of retinol [87]. This function is consistent with the topology of this enzyme as being associated with membranes [88]. The participation of LRAT in retinol metabolism appears somewhat different in adipose tissue, because in $LRAT^{-/-}$ mice, the retinol content is increased threefold in adipose tissue, whereas in liver, lung, and other tissues retinol content was decreased [89]. In adipose tissue, $STRA6$ expression is increased in the late phase (≥5 days) of adipogenesis [84]. The binding of RBP–ROH, but not RBP–retinaldehyde or RBP–RA, triggers phosphorylation of STRA6 by JAK2 and recruits and activates the transcription factor STAT5. This transcription factor activation upregulates expression of $SOCS3$ and $PPARγ$ in mature adipocytes [84]. In the future, adipose-specific mouse models deficient in RBP4 and/or STRA6 will establish functions of these proteins in WAT in vivo.

In addition to RBP4, few other members of lipocalin family of lipid-binding proteins can, potentially, participate in the retinoid transport to adipose tissue. β-lactoglobulin (β-LG) binds retinol (Kd = 44 nM) and RA (Kd = 39 nM) with high affinity; however, β-LG can also transport other lipophilic compounds and hormones [90,91]. L-PGDS is also a member of the lipocalin superfamily that binds RA with high affinity [92]. While RBP4 specifically binds retinol and β-LG preferentially transports RA, L-PGDS binds a range of lipophilic molecules [92]. L-PGDS also synthesizes 15-d-PGJ2, a natural ligand for PPARγ. In agreement with this potential role in PPARγ regulation, *L-PGDS* is expressed in human and rodent WAT and also in preadipocytes and adipocytes [93]. *L-PGDS* is expressed in a depot specific fashion and positively associated with BMI [94]. In contrast to the expected resistance to obesity, the genetic deficiency in L-PGDS results in severe metabolic syndrome in mice, which includes obesity, insulin resistance, and, notably, atherosclerosis [95,96]. This paradoxical finding warrants further examination of the potential contribution of L-PGDS in retinoid delivery and signaling in adipocytes. Unexpectedly, weak (Kd = ~3 μM) RA binding properties were discovered in apoprotein M, which was characterized as a lipocalin [97]. RA also binds to albumin [98]. Although under physiological conditions low concentrations of 13-*cis* and all-*trans* RA, are present in the circulation [99,100], their availability to adipocytes has not been investigated. Administration of RA has been extensively studied and found to have multiple effects on adipose tissue that were summarized by leaders in this field [35,101]. These findings will not be reviewed here, because the emphasis of this review is on the endogenously produced vitamin A metabolites. Endogenously produced RA and administered RA exert different metabolic effects in adipose tissue that were recently discussed in [16].

β-Carotene is another principal source of vitamin A in adipose tissue [102]. In humans, liver and adipose tissues are major stores of carotenoids. β-Carotene concentrations vary in the liver by individual from 2 to 150 μg/g liver and from 8 to 98 μg/g WAT [103]. β-Carotene is transported in lipoproteins [104], mainly in LDL (~75%), which determines the plasma concentration of this nutrient [105]. HDL contains less β-carotene than LDL. Harrison et al. demonstrated that scavenger receptor class B type I (SR-BI) facilitates β-carotene uptake by retinal pigment epithelial cells [106]. Adipocytes also express *SR-B1* and can regulate its translocation to the membrane and HDL uptake in response to insulin and starvation refeeding [107]. Moreover, SR-B1 is a PPARγ target gene [108]. Although participation of this pathway in the delivery of β-carotene to adipose tissue is currently unknown, SR-B1 could be one potential gate by which human adipose tissue is enriched with β-carotene. Mice and rats cleave β-carotene instantly to two retinaldehyde molecules by β-carotene 15,15′-carotene monooxygenase 1 (BCMO1) in the intestine [109] (Figure 2.2). Consequently, mouse and rat adipose tissue does not contain β-carotene and these animals belong to the "white fat animals" group. In contrast, humans are "yellow fat animals," because they accumulate β-carotene in adipocytes. Surprisingly, in spite of β-carotene absence in mice, murine adipocytes express β-carotene cleavage enzymes [102] that have potent effects on fat formation that will be discussed later.

## Production and Intracellular Transport of Vitamin A Metabolites in WAT

### Retinol

Intracellular retinol enters three main metabolic routes in WAT: (1) re-esterification and storage in lipid droplets. The retinol bound to a cellular retinol binding protein (CRBP) can (2) directly participate in the regulation of signaling pathways by binding to specific kinases, or (3) undergo enzymatic conversion to major metabolites retinaldehyde and RA (discussed in "Retinaldehyde" and Retinoic Acid" sections in this review), or other retinoids.

In regard to retinol metabolism, adipose tissue has distinct characteristics. Adipose tissue contains large concentrations of retinol that comprise 50%–75% of total retinol/retinyl ester stores, whereas in liver, lung, kidney, pancreas, and eye, retinyl ester is the predominant form [80,89]. In addition to WAT, brain and muscle are the only other two tissues where retinol exceeds retinyl ester content [89]. Studies by Liu and Gudas convincingly demonstrate that LRAT (EC2.3.1.135) is the key enzyme responsible for retinol esterification (~75%) in adipose tissue [89]. Mouse adipose stores are depleted of retinol to undetectable levels in $LRAT^{-/-}$ mice on a vitamin A-deficient diet, indicating that retinol uptake through the RBP4/STRA6 pathway depends almost exclusively on LRAT in adipose tissue [89], in agreement with mechanistic studies by von Lintig and associates [87]. At present, it is unclear what enzyme is responsible for 25% of retinyl esters in adipose tissue; however, WAT expresses low levels of ACAT, capable of retinol esterification [110]. This pathway could play an important role in retinyl ester storage, since in the context of a regular diet a regular diet, WAT showed a paradoxical effect unique to this tissue: an increase in retinyl esters in $LRAT^{-/-}$ mice [111]. This response is associated with a threefold increase in cytosolic retinol-binding protein type III (CRBPIII) [111], which is known to facilitate retinol esterification by LCAT and ACAT in the liver and mammary gland microsomes, respectively [112]. In agreement with CRBPIII/ACAT functions, $LRAT^{-/-}$ mice maintain concentrations of retinyl esters in the milk of lactating mice at levels seen in wild-type mice. Interestingly, vascular endothelial cells in WAT, but not in the liver also express more CRBPIII [113]. Further investigations are needed to dissect the CRBPIII/ACAT contribution to retinyl ester production in WAT. In spite of impaired retinol/retinyl ester levels, no adipose tissue development abnormalities were reported in these $LRAT^{-/-}$ mice [89,111,114]. Whether obesogenic diets can alter adipose tissue demands for retinoids and reveal different WAT formation in wild-type and $LRAT^{-/-}$ mice awaits further investigation.

The fundamental role of retinyl ester mobilization by retinyl hydrolases in the regulation of adipose depots has been established. Unlike other tissues (liver, kidney) storing retinyl esters in stellate cells, WAT accumulates retinyl esters primarily in the lipid droplets. Several proteins possessing retinyl hydrolase activity, such as HSL, TGH (Ces1d), TGH2 (ces1f), and ES22, were identified in adipose tissue (reviewed by [115]). Remarkably, a pivotal role in retinyl hydrolysis is played by HSL, which is a lipase known to hydrolyze diacylglycerides [64,116]. $HSL^{-/-}$ mice lack retinyl ester hydrolase activity in WAT and, consequently, have

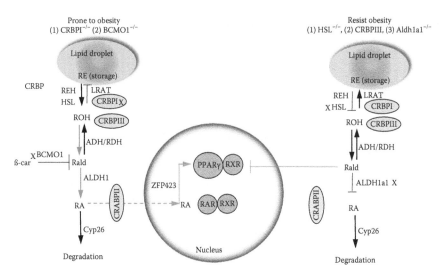

**FIGURE 2.4** Hypothetical description of altered vitamin A metabolism in adipocytes in genetic mouse models. This schematic describes only mouse models deficient in genes that participate in vitamin A metabolism in WAT and depend on retinoid production. *Rdh1*−/− and retinol saturase (RetSat) *RetSat*−/− mouse models are excluded from the discussion because *Rdh1* is not expressed in adipose tissue and effects of RetSat in adipocytes were not reproduced by dihydroretinol. *CRBPI*−/− and *BCMO1*−/− mice were prone to obesity on a high-fat diet. CRBPI promotes retinol esterification; genetic deficiency in this enzyme (X) may, hypothetically, increase obesity through production of endogenous RA (black arrows). RA can activate RAR- and/or ZFP423-mediated *PPARγ* expression. *BCMO1* deficiency can reduce production of retinaldehyde that can, potentially, inhibit PPARγ. *HSL*−/−, *CRBPIII*−/−, and *Aldh1a1*−/− mice were resistant to obesity on a high-fat diet. *HSL*−/− mice accumulate esterified retinol, suggesting diminished production of retinol, retinaldehyde, and RA. Deficiency in RA-dependent activation of RXR was suggested as a possible mechanism for reduced obesity in these mice. The impact of *CRBPIII* deficiency on vitamin A metabolism is not completely understood and, thus, is not discussed. *Aldh1a1* deficiency leads to increased retinaldehyde concentrations in adipose tissue and decreased production of RA. Retinaldehyde may suppress activity of PPARγ, whereas insufficient RA availability can account for the reduced *ZFP423* and *PPARγ* expression in these mice.

increased levels of retinyl esters and decreased levels of retinol, retinaldehyde, and all-*trans* RA (Figure 2.4). *HSL*−/− mice resist high-fat diet-induced obesity due to impaired adipogenesis and increased expression of thermogenic proteins in their WAT [117]. These changes are associated with impaired expression of *PPARγ* and its co-regulator *RIP140*, as well as cell cycle regulating *pRb*. These alterations in WAT appear to result from a deficiency in multiple vitamin A metabolites, since RA administration can only partially restore WAT formation in *HSL*−/− mice on a high-fat diet. RA administration was sufficient to restore *pRb*, *RXRβ*, and *RXRγ* levels to those seen in WT mice, whereas *PPARγ* and *RIP140* expressions remained suppressed [117]. Together, these studies by Holm and colleagues demonstrate that mobilization of retinol from retinyl esters by HSL and its further metabolic processing to RA is critical for the obesogenic responses of

WAT to a high-fat diet; the absence of retinyl ester-hydrolyzing activity of HSL leads to the preferential formation of BAT [64,117].

Hydrolyzed retinol (6–7 μg retinol/g adipose tissue) is found at greater levels in adipocytes than in preadipocytes (0.65–0.8 μg retinol/$10^6$ cells) [80]. Retinol is transported in adipocytes by CRBPs that belong to the intracellular lipid binding protein family (iLBP), which also includes FABP. WAT expresses *CRBPI*, *CRBPII*, and *CRBPIII* [111]. CRBPs transport both all-*trans* retinol and all-*trans* retinal; however, CRBPI has greater affinity (100 times more) for retinol than does CRBPII, while both show similar affinity for retinaldehyde [118–121]. Whereas CRBPII plays a critical role in retinol transport in the intestine [122], CRBPI and CRBPIII play important, albeit different roles in adipogenesis and fat tissue development in WAT [123,124]. *CRBPI* is expressed in preadipocytes [123]. *CRBPI* shRNA-mediated knockdown in 3T3-LI preadipocytes enhances adipocyte differentiation and TG accumulation. Ectopic expression of *CRBPI* does not change TG levels, in line with a CRBPI transporter protein function that is limited by the availability of hydrolyzed retinol. Moreover, studies in single and double *CRBPI* and *ADH1* knockout mice underscore CRBPI's function in retinol delivery for esterification [125]. Therefore, mice lacking *CRBPI* can mobilize more retinol for retinaldehyde and RA production. *CRBPI*$^{-/-}$ mice had increased WAT weight, particularly epididymal fat mass, compared to wild type [123]. Accordingly, expression of adipogenic genes, such as *PPARγ* and its downstream target genes *FABP4* (*aP2*), adiponectin, and LPL was elevated in WAT of *CRBPI*$^{-/-}$ compared to WT mice. These *CRBPI*$^{-/-}$ mice displayed improved glucose tolerance and insulin sensitivity compared to their counterpart expressing *CRBPI*.

In contrast, CRBPIII is a late-phase retinol transporter. Its expression is induced by master regulator of adipogenesis *PPARγ* [124]. CRBPIII seems to promote adipogenesis and reduce thermogenesis, since *CRBPIII*$^{-/-}$ mice on a high-fat diet display a moderate decrease in WAT and increase in thermogenesis. On the other hand, CREBPIII also assists fatty acid transport from adipose tissue. Deficiency in CRBPIII markedly reduces lipolysis and release of fatty acids into circulation [124]. Potentially, in *CRBPIII*$^{-/-}$ mice, CRBPI could compensate for retinol transport in adipocytes and mislead the identification of CRBPIII's function. This, and the other gaps in the understanding of the CRBPs function in WAT, namely, "What retinoid(s) is responsible for CRBPI or CRBPIII effects in WAT?" and "What transcription or signaling pathways are regulated by retinoids that are delivered by CRBPI or CRBPIII?," remains to be addressed. The studies by Vogel et al. [123,124] emphasize the versatile role of CRBPs in the control of adipogenesis and fat accumulation.

Retinol is a major substrate for ADH and RDH enzymes oxidizing this metabolite to retinaldehyde (retinal), which can be subsequently converted to RA by ALDH enzymes. Growing evidence suggests, however, that retinol also has a specific signaling role that is independent of the conversion of this metabolite to retinaldehyde or RA. Comprehensive work by Hammerling and colleagues has conclusively demonstrated the binding sites for retinol in a cysteine-rich regulatory domain on three different serine/threonine kinases: PKCα, PKCδ, and c-Raf [126–128]. The retinol-mediated signaling by these kinases regulates oxidative phosphorylation in the mitochondria of cancer cells, but without a doubt, these fundamental pathways are expected to play a role in WAT expressing these kinases [129–131]. PKCδ regulates mitochondrial uncoupling in rodent WAT [130]. Expression of *PKCδ* and *c-Raf* also accompanies differentiation of 3T3-L1 cells. Interestingly, in *CRBPI$^{-/-}$* mice, changes in WAT formation were not associated with the changes in the expression levels of RA-synthesizing pathway enzymes, ADH and ALDH [123]; furthermore, they showed elevated thermogenesis. These observations were in line with the proposed retinol function, although it is not clear how retinol is delivered to the mitochondria.

### (R)-All-*Trans*-13,14-Dihydroretinol/Retinol Saturase

Adipogenesis is associated with the expression of retinol saturase, which is regulated by *PPARγ* [132]. Retinol saturase was identified and characterized by Palczewski et al. as an enzyme converting retinol to (R)-all-*trans*-13,14-dihydroretinol (13,14-dihydroretinol) [133]. In adipogenesis in vitro, the genetic ablation of retinol saturase inhibits adipogenesis through a mechanism dependent on the induction of expression and activity. Unexpectedly, this effect was not dependent on 13,14-dihydroretinol production [132]. In tissue, 13,14-dihydroretinol can undergo further conversion to 13,14-dihydroretinal and 13,14-dihydroretinoic acid by short- and medium-chain alcohol dehydrogenases (SDR/RDH) and retinaldehyde dehydrogenases [134]; however, the role of these derivatives in the adipogenic effects of retinol saturase remains unstudied. The controversial finding that retinol saturase increases diet-induced obesity in vivo by induction of *PPARγ* expression [135] raises even more question about putative enzymatic products of this enzyme and their input on transcriptional regulation. The similar retinol and retinyl ester levels in retinol saturase-deficient and WT mice also suggest a minor role of this pathway in retinol utilization [135]. Undoubtedly, the decreased retinol saturase levels in obese patients [132] highlight the important role of this pathway in obesity.

### Retinaldehyde

Retinol can be reversibly oxidized to retinaldehyde by NAD+ or NADP+-dependent ADH and SDR/RDH enzyme superfamilies, whose members have wide substrate specificities [136]. ADH1 (isozymes A, B1, B2, C1), ADH2, ADH3, ADH4, and approximately 17 different SDR/RDH enzymes can oxidize retinol (reviewed in [136]). ADH and SDR/RDH enzymes catalyze oxidation of free unbound retinol under these in vitro conditions. CRBPI favors retinaldehyde reduction to retinol and esterification [136]. Retinaldehyde binds efficiently to CRBPI and RBP4 [121]. WAT expresses higher levels of *ADH1* enzyme in lean compared to Ob/Ob mice [121]. *ADH1* expression accompanied adipocyte differentiation [121]. Our lab showed retinaldehyde concentrations of nearly 0.1–1.0 μM in mouse WAT that were measured after derivatization of retinaldehyde with hydroxylamine to a respective oxime [121]. As expected, the highest concentrations of retinaldehyde in WAT (1.0 μM) were found in mice with impaired retinaldehyde dehydrogenase catabolism, namely, in *Aldh1a1$^{-/-}$* mice. The roles of retinaldehyde-generating ADH and SDR/RDH enzymes in WAT are largely unexplored. It is worth mentioning that *RDH* enzymes expressed outside of WAT can also influence WAT formation. *RDH1* is an example of an enzyme that is not expressed in WAT, preadipocytes, or adipocytes

(Ziouzenkova et al., unpublished results). Nonetheless, $RDH1^{-/-}$ mice develop larger mesenteric, femoral, and inguinal fat pads on a vitamin A–deficient diet than WT mice, while weights of other depots, such as retroperitoneal and epididymal, remain unaltered [136]. Although the underlying mechanism for $RDH1$ deficiency effects in WAT is unknown, these data illustrate systemic effects of enzymes in vitamin A metabolic pathways on adipose tissue formation/development.

Retinaldehyde could also be produced by central cleavage of β-carotene that is mediated by β-carotene 15,15′-monooxygenase 1 (BCMO1). $BCMO1$ expression is induced during adipogenesis in 3T3-L1 adipocytes starting on the third day of differentiation [102], probably through PPARγ-dependent activation of PPRE in the promoter region of BCMO1 [137]. $BCMO1$ is also expressed in WAT in mice [138]. In the genetic absence of $BCMO1$, mice develop obesity and liver steatosis on a high-fat diet [109]. When $BCMO1^{-/-}$ mice were fed a diet supplemented in β-carotene, their body fat was not altered, whereas age- and sex-matched WT mice have shown a significant adipose tissue reduction in inguinal, gonadal, and retroperitoneal WAT [138]. These metabolic changes were linked to suppressed expression levels of $PPAR\gamma$, $RXR$, and their target gene $LPL$ [138]. It is plausible that BCMO1-dependent β-carotene cleavage suppresses intake of postprandial fatty acids by LPL. Studies from $BCMO1$ knockout mice have clearly unraveled the close relation of vitamin A metabolism to the regulation of $PPAR\gamma$ expression, although the nature of the retinoid derivative mediating these responses is elusive. In fact, BCMO1 produces two molecules of retinaldehyde that could be further converted either in retinol or in RA, leaving a possibility for participation of all these metabolites in adipose tissue formation.

Retinaldehyde is the principal substrate for enzymatic RA production [139]. Similar to RA, retinaldehyde binds to an RAR receptor, although its affinity to this receptor is 500 times lower than that of RA [140]. In addition, retinaldehyde can target specific molecular pathways independent of its role in RA production. Our recent studies showed that retinaldehyde can bind to recombinant PPARγ in vitro [121]. Furthermore, in preadipocyte NIH-3T3 cultures, retinaldehyde, but not RA, inhibits PPARγ ligand-binding domain (LBD) activity in the presence of PPARγ ligands [121]. Remarkably, retinaldehyde also inhibits RXR activated by 9-*cis* RA in similar transactivation assays with RXR-LBD [121]. Corresponding to the inhibition of the PPARγ and RXRα heterodimer, retinaldehyde stimulation severely inhibits adipogenesis in 3T3-L1 adipocytes due to the downregulation of PPARγ target genes.

The possibility of a rapid conversion of retinaldehyde to RA or retinol led to a long-standing debate about the presence of this metabolite in tissue. Insight into retinaldehyde formation came from Duester et al., who created $Aldh1a1^{-/-}$ mice and showed increased levels of retinaldehyde in plasma and liver in $Aldh1a1^{-/-}$ compared to WT mice (Figure 2.4) [141,142]. Subsequently, several groups have detected retinaldehyde in WAT of WT, $Aldh1a1^{-/-}$, and $HSL^{-/-}$ mice [117,121,143]. Consistent with increased retinaldehyde levels and possible inhibition of PPARγ:RXR transcriptional activity by this metabolite, WAT mass in $Aldh1a1^{-/-}$ was reduced on a regular and a high-fat diet [15,121]. Given that Aldh1a1 oxidizes retinaldehyde to RA, $Aldh1a1^{-/-}$ adipocytes have reduced RA production [15]. Thus, retinaldehyde functions in $Aldh1a1^{-/-}$ mice should be studied in conjunction with deficient RA production that also influences WAT formation and contributes to dissimilar formation of subcutaneous and visceral depots [15].

## Retinoic Acid

RA is produced predominantly by a family of cytosolic, NADPH-dependent aldehyde dehydrogenase enzymes that comprises three members: ALDH1a1, ALDH1a2, and ALDH1a3. RA can also be generated by NADP(H)-dependent cytosolic aldo-keto reductases (AKR), such as AKR1B3 (aldose reductase), AKR1B7 (mouse has deferent protein), AKR1B8 (fibroblast-growth factor 1-regulated protein), and AKR1B9 (Chinese hamster ovary reductase) from all-*trans* retinaldehyde in vitro [144]. The AKR1B subfamily, especially AKRB3, is expressed in adipocytes and suppresses adipogenesis. This effect, however, is mediated by F2α, synthesized by this enzyme [145]. RA synthesis related to the activity of this enzyme in WAT has not been reported.

Recent work from our laboratory showed that all *Aldh1* enzymes are expressed in 3T3-L1 pre-adipocytes. *Aldh1a1* is expressed at higher levels than *Aldh1a2* and *Aldh1a3* in adipocytes; moreover, only *Aldh1a1* expression is increased in the course of adipogenesis [15] (Figure 2.3). Genetic deficiency in *Aldh1a1* decreases 70% of RA generation during differentiation and blocks differentiation [15]. In 3T3-L1 adipocytes, *Aldh1a1* expression is inversely regulated at high RA concentrations (≥100 nM), which suggests negative feedback regulation [15]. *Aldh1a1* is also the predominantly expressed enzyme from this family in both human and mouse WATs. All Aldh1 enzymes appear to act in concert for RA generation in adipocytes in vitro and in mouse fat in vivo. Even so, the expression of a specific subset of genes was different in *Aldh1a3$^{-/-}$* compared to *Aldh1a1$^{-/-}$* adipocytes and may indicate specific functions among members of Aldh1 family [15].

The concerted generation of RA by Aldh1 enzymes appears to regulate the formation of different fat depots. We found that WAT expresses both *Aldh1a1* and *Aldh1a3* enzymes in a depot-specific manner, e.g., the subcutaneous fat depot expresses more *Aldh1a1* and *Aldh1a3* than does visceral fat [15]. The significance of different RA generation in WAT depots was highlighted in studies in *Aldh1a1$^{-/-}$* female mice that developed less visceral than subcutaneous fat compared to WT mouse groups. In fact, RA generated by Aldh1 enzymes was necessary for the induction of key transcription factors in adipogenesis, *ZFP423* and *PPARγ* (discussed later). In humans, omental (visceral) depots express higher levels of *Aldh1a2* and *Aldh1a3* than does subcutaneous fat [15], probably due to the different developmental origin of mouse and human WAT.

In WAT, RA is transported by cellular RA-binding proteins I (CRABPI) and II (CRABPII) that are expressed in this tissue [146]. CRABPI transports RA to metabolizing enzymes in the cytoplasm, thereby modulating RA catabolism to 4-oxo RA [147]. CRABPI is activated by binding of thyroid hormones to holo-thyroid hormone receptors/retinoid receptors that, in turn, bind to a thyroid response element in the CRABPI promoter [148]. *CRABPI* expression is inhibited during 3T3-L1 differentiation due to the interaction of thyroid receptor with receptor-interacting protein 140 (RIP140) [149]. *CRABPII* expression is regulated by RA-mediated activation of RAR/RXR heterodimers that bind in humans to distant upstream retinoic acid response element (RARE): DR1 in mice and DR1 and DR2 in humans [150].

Dissection of CRABPII functions in mammary carcinoma cells revealed that CRABPII transports RA to the RARs [151]. *CRABPII* is inhibited during the 3T3-L1 adipocyte differentiation by C/EBP and GR transcription factors activated by IBMX and dexamethasone present in a hormonal cocktail used in inducing adipogenesis in vitro [152]. The dependence of adipogenesis on CRBPI and CRBPII has not been examined.

RA isomers exert a wide range of biological and pharmacological activities through direct and indirect regulation of multiple transcription factors in adipogenesis (reviewed in [16]). All RA isomers serve as ligands for the RAR family of nuclear receptors (RARα, RARβ, and RARγ) and, thereby, directly induce transcription [153,154]. 9-*cis* RA was early identified as a ligand for retinoid X receptors (RXRα, RXRβ, and RXRγ) [155]; however, it is still unknown whether this RA isomer is produced in WAT. RARs bind as RAR:RXR heterodimers to the specific DNA target sequences containing RAR response elements, or RAREs, in promoter/enhancer regions of RA target genes. RAREs are polymorphic and can contain direct repeat motifs separated by 5, 3, 2, and 1 nucleotides known as DR5, DR3, DR2, or DR1 in different genes [156]. RXRs heterodimerize with the majority of nuclear receptors, including the vitamin D receptor, thyroid hormone receptor, and, notably, PPARγ. Some RXR-containing heterodimers, such as TR:RXR, VDR:RXR, and RAR:RXR, are nonpermissive. They are activated by the partner's ligand, but not by the RXR ligand (9-*cis* RA) alone. Other RXR heterodimers, including PPAR:RXR, are permissive and can be activated by ligands of either partner [153,154]. In the presence of both ligands, permissive heterodimers are synergistically activated. All RAR and RXR isoforms are expressed in the course of 3T3-L1 differentiation, where all RAR and RXRβ are constitutively expressed, and RXRα and RXRγ are expressed in the late and very late phase, respectively [41]. It is unclear whether activation of these transcription factors is mediated by endogenous RA production in adipocytes or exogenous (para- or endocrine) RA delivery. RARE reporter activation during 3T3-L1 differentiation suggests that autocrine RA production could activate RAR-dependent pathways in a manner dependent on the expression of Aldh1 enzymes [15].

PPARδ was identified as a third nuclear receptor activated by direct binding of atRA at concentrations higher than 100 nM [157]. RA also binds to recombinant PPARα, PPARδ, and PPARγ in vitro with Kd 1, 0.77, and 0.2 μm, respectively [157]. In adipocytes, atRA activation of PPARδ acts through PDK1 [101]. Berry and Noy suggest that pharmacological treatment with RA prevents obesity through mechanisms involving both RAR and PPARγ pathways [101]. Given that genetic deficiency in PPARδ is not associated with an obesity phenotype [158], more research is needed to understand RA hierarchy in the activation of PPARδ, RARs, and RXRs, or combinations of these receptors in vivo.

Many transcription factors in adipogenesis are RAR target genes and are under indirect control of RA. The RARE sequence was identified in some HOX genes [159,160]. RA also induces crosstalk between RAR and WNT/β-catenin pathways by RARE-dependent and -independent means. In MCF-7 cells, β-catenin potentiates RARE activation by RA-bound RAR, although RA-activated RAR inhibits activation of β-catenin-LEF/TCF reporter activity [161]. Thus, RA-activated RAR influences large numbers of transcription factors that lack RARE. A pioneering work by Lazar and colleagues demonstrates that RA participation in the inhibition of the C/EBP family of transcription factors depends on RARα/RARγ but not RARE [162]. Subsequent work by Wiper-Bergeron laboratory shows that C/EBPβ inhibition by RA was dependent on Smad3 nuclear localization stimulated by RA. This RA effect on Smad3 was dependent on MAPK activation by RA [163]. Studies in *Smad3*$^{-/-}$ mice reveal the interaction of Smad3 with other transcription factors (PPARδ and PPARγ) that contributed to the suppression of diet-induced obesity in these mice [164]. Notably, other transcription factors involved in the induction of PPARγ, such as KLF5 and EBF1, are also regulated by RA, even though the influence of RA on these pathways was not studied in adipogenesis (reviewed in [16]).

Our studies show that endogenous RA is essential for the tandem expression of *ZFP423* and *PPARγ* [15]. Genetic deficiency of the *Aldh1* enzymes in adipocytes reduces 99% of *ZFP423* and *PPARγ* expression in adipocytes, although the molecular basis for this interaction is unknown. Ectopic *Aldh1* expression or stimulation with RA in *Aldh1*$^{-/-}$ adipocytes can in part recover both *ZFP423* and *PPARγ* expressions. This Aldh1-dependent mechanism for ZFP423 and PPARγ regulation plays an important role in the formation of different fat depots. Studies in *Aldh1a1*$^{-/-}$ mice reveal that 70% of *ZFP423* and *PPARγ* expression in visceral fat depends on the Aldh1a1, while only 40% of *PPARγ* expression in subcutaneous fat was affected by the genetic loss of *Aldh1a1*. These differences were linked to the higher expression of *Aldh1a3* in subcutaneous versus visceral fat that partially compensates for RA generation [15]. Future studies will examine whether different formation of fat depots in males and females as well as in different species or in ethnic populations in humans is also mediated by different autocrine RA production in these fat depots.

In addition to adipogenesis, RA is expected to influence many pathways in mature adipocytes, including lipolysis, lipogenesis, cytokine production, and thermogenesis. The majority of these effects were identified in studies with administered RA and were discussed in excellent reviews [35]. A pilot investigation of *Aldh1a1*$^{-/-}$ adipocytes showed the reduced expression of lipogenic *FAS* [15], whereas expression of lipases *LPL* or *ATGL* was not affected (Ziouzenkova and colleagues, unpublished results). Treatment of *Aldh1a1*$^{-/-}$ adipocytes with RA recovered *FAS* expression [15] suggesting that endogenous RA can participate in the regulation of lipogenic processes in adipocytes. Recent genetic studies using inducible adipose-specific *RXRα* knockout models demonstrate the marked role of this receptor in starvation-induced lipolysis in mature adipocytes [165]. The involvement of autocrine RA production in metabolic processes and transcription regulation in mature adipocytes awaits examination.

## Oxidized RA Metabolites and Apocarotenoids

RA is metabolized by cytochrome P450 enzymes (CYP26A1, CYP26B1). The oxidized metabolites and degradation products of RA, 13-*cis*-3-hydroxy-retinoic acid, all-*trans*-3-hydroxy-retinoic acid, and 4-oxo-retinoic acid and dihydro-retinoic acid, were detected in plasma and tissues [166]. These metabolites activate RAR [167,168]. The functions of CYP26A1 and B1 enzymes and oxidized RA derivatives in adipose tissue have not been reported.

Although the majority of β-carotene undergoes central cleavage mediated by BCMO1, WAT expresses another β-carotene-9′,10′-oxygenase (BCDO2), which converts β-carotene into β-10′-apocarotenal and β-ionone [138]. *BCMO1*$^{-/-}$ mice showed increased expression of *BCDO2* in adipocytes and β-10′-apocarotenal accumulated as the major β-carotene derivative. The consumption of a regular diet supplemented with β-carotene reduces adipose tissue mass in WT compared to *BCMO1*$^{-/-}$ mice. Although these data support a major role of central cleavage producing retinaldehyde in adipocytes, the expression of *BCDO2* in adipocytes also presumes a role for this enzyme in adipocytes.

The molecular effects of apocarotenals differ from those of retinoids. Apocarotenals are poor activators of RAR [169,170]. Two apocarotenals, β-apo-14′ carotenal [169] and β-apo-13′ carotenone [171] inhibit RXRα activation in vitro. In adipocytes, β-apo-14′ carotenal suppresses adipocyte differentiation through a mechanism dependent on the inhibition of both PPARγ and RXRα [121]. The paths for β-apo-14′ carotenal production in adipose tissue have not been identified yet, although homogenates from other tissues have been shown to possess enzymatic activity capable of yielding β-apo-14′ carotenal [172]. Regardless, investigation of apocarotenoids′ role in the inhibition of nuclear receptors may provide clues for the nutrient involvement in the deregulation of nuclear receptors in metabolic disorders and design of therapeutic drugs.

## SUMMARY

Animal models deficient in vitamin A–metabolizing enzymes provide unequivocal evidence of retinoid essentiality in the formation and function of WAT, especially in response to diet (Figure 2.4). The participation of specific retinoids in the regulation of signaling pathways in WAT requires more investigation.

Many attempts were made to utilize retinoids for therapeutic purposes by using oral, intravenous, and ectopic applications of retinoids. Currently, these treatments are limited to acne/psoriasis treatments by 13-*cis* RA and acute myeloid leukemia by all-*trans* RA, due to serious side effects known as "RA syndrome" [173]. RA syndrome manifests as dyslipidemia and obesity in some patients, supporting RA's role in adipose tissue. Identification of key proteins responsible for physiologic retinoid production in WAT opens a new avenue in the management of safe and specific retinoid production that may help to treat metabolic disorders associated with the pathologic formation of WAT depots.

## ABBREVIATIONS

| | |
|---|---|
| ACC | Acetyl-CoA carboxylase |
| Adh | Alcohol dehydrogenase |
| AMPK | AMP activated protein kinase |
| Aldh | Aldehyde dehydrogenase |
| ATGL | Adipose triglyceride lipase [annotated as patatin-like phospholipase domain-containing protein A2] |
| atRA | All-*trans* retinoic acid |
| ARAT | Acyl CoA:retinol acyl transferase |
| BAT | Brown adipose tissue |

| | |
|---|---|
| BCMO1 | β-Carotene 15,15′-monooxygenase 1 |
| βLG | Beta-Lactoglobulin |
| C/EBP | CCAAT enhancer-binding proteins |
| ChREBP | Carbohydrate response element-binding protein |
| CRABP-II | Cellular retinoic acid-binding protein II |
| CRBPIII | Cytosolic retinol-binding protein type III (CRBPIII) |
| Dact1 | Dapper1/Frodo1 |
| DNL | De novo lipogenesis |
| DR1 | Direct repeat motif 1, peroxisome proliferator-activated receptor response element (PPRE) |
| DR5 | Direct repeat motif 1, retinoic acid response element (RARE) |
| EBF1 | Early B-cell factor |
| ES22 | Esterase 22 (carboxylesterase 1E) |
| FABP | Fatty acid-binding protein |
| FAS | Fatty acid synthase |
| GR | Glucocorticoid receptor |
| Foxc2 | Forkhead transcription factor c2 |
| HOX | Homeobox proteins |
| HSL | Hormone-sensitive lipase |
| IBMX | 3-Isobuty-1-methylzanthine |
| KLF | Krüppel-like transcription factor |
| L-PGDS | Lipocalin-type prostaglandin (PG) D synthase |
| LPL | Lipoprotein lipase |
| LRP | Low density lipoprotein (LDL) receptor-related protein |
| MAPK | p38 Mitogen-activated protein kinase |
| MGL | Monoglyceride lipase |
| PDK1 | Phosphoinositide-dependent protein kinase 1 |
| PKC | Protein kinase C |
| Pref1 | Preadipocyte factor-1 |
| 15-d-PGJ2 | 15-Deoxy-delta (12,14)-Prostaglandin J2 |
| PGC1α | PPARγ co-activator-1α |
| PPARγ | Peroxisome proliferator-activated receptor γ |
| PPRE | Peroxisome proliferator-activated receptor response element (DR1) |
| pRb | Retinoblastoma protein |
| RAR | Retinoic acid receptor |
| RARE | Retinoic acid response element (DR5) |
| RA | Retinoic acid |
| Rald | Retinaldehyde/retinal |
| Raldh | Retinaldehyde dehydrogenase |
| RBP | Retinol-binding protein |
| RDH | Retinol dehydrogenase |
| SCD1 | Stearoyl-Coenzyme A desaturase |
| SDR | Short-chain dehydrogenase/reductase |
| SR-BI | Scavenger receptor class B type I |
| SOCS3 | Suppressor of cytokine signaling |

| | |
|---|---|
| Sox9 | Sex-determining region Y-box 9 |
| SRC1 | Steroid-receptor co-activator-1 |
| SREBP1c | Sterol regulatory element-binding protein-1c |
| TGs | Triglycerides/triacylglycerols |
| TGH | Triacylglycerol hydrolase |
| TR | Thyroid receptor |
| UCP | Uncoupling protein |
| VDR | Vitamin D receptor |
| VLDL | Very low density lipoprotein |
| WAT | White adipose tissue |
| WHO | World Health Organization |
| WNT | Wingless-type Wnt modulator MMTV integration |
| ZFP423 | Zinc finger protein 423, transcription factor |

## ACKNOWLEDGMENTS

The research was supported by ICMR-International Fellowship (S.M.J), College of Education and Human Ecology Dissertation fellowship (R.Y.), Alpha Omega Alpha Honor Medical Society 2011 Carolyn L. Kuckein Student Research Fellowship (B.R.), and College of Education and Human Ecology Seed Grant and Food Innovation Center Seed Grant 013000 (O.Z.).

## REFERENCES

1. Shamai L, Lurix E, Shen M et al. Association of body mass index and lipid profiles: Evaluation of a broad spectrum of body mass index patients including the morbidly obese. *Obes Surg* 2011; 21: 42–47.
2. WHO. Obesity: *Preventing and Managing the Global Epidemic*. Geneva, Switzerland: WHO, 2000; p. 256.
3. Klein S, Burke LE, Bray GA et al. Clinical implications of obesity with specific focus on cardiovascular disease: A statement for professionals from the American Heart Association Council on Nutrition, Physical Activity, and Metabolism: Endorsed by the American College of Cardiology Foundation. *Circulation* 2004; 110: 2952–2967.
4. Asia Pacific Cohort Studies Collaboration. The burden of overweight and obesity in the Asia-Pacific region. *Obes Rev* 2007; 8: 191–196.
5. Whitlock G, Lewington S, Sherliker P et al. Body-mass index and cause-specific mortality in 900 000 adults: Collaborative analyses of 57 prospective studies. *Lancet* 2009; 373: 1083–1096.
6. Zhang C, Rexrode KM, van Dam RM, Li TY, Hu FB. Abdominal obesity and the risk of all-cause, cardiovascular, and cancer mortality: Sixteen years of follow-up in US women. *Circulation* 2008; 117: 1658–1667.
7. Hastic CE, Padmanabhan S, Slack R et al. Obesity paradox in a cohort of 4880 consecutive patients undergoing percutaneous coronary intervention. *Eur Heart J* 2010; 31: 222–226.
8. Kim BJ, Lee SH, Ryu WS, Kim CK, Lee J, Yoon BW. Paradoxical longevity in obese patients with intracerebral hemorrhage. *Neurology* 2011; 76: 567–573.

9. Cannon B, Nedergaard J. Brown adipose tissue: Function and physiological significance. *Physiol Rev* 2004; 84: 277–359.
10. Kajimura S, Seale P, Kubota K et al. Initiation of myoblast to brown fat switch by a PRDM16-C/EBP-beta transcriptional complex. *Nature* 2009; 460: 1154–1158.
11. Zingaretti MC, Crosta F, Vitali A et al. The presence of UCP1 demonstrates that metabolically active adipose tissue in the neck of adult humans truly represents brown adipose tissue. *FASEB J* 2009; 23: 3113–3120.
12. Nedergaard J, Bengtsson T, Cannon B. Three years with adult human brown adipose tissue. *Ann NY Acad Sci* 2010; 1212: E20–E36.
13. Tran TT, Kahn CR. Transplantation of adipose tissue and stem cells: Role in metabolism and disease. *Nat Rev Endocrinol* 2010; 6: 195–213.
14. Rosen ED, MacDougald OA. Adipocyte differentiation from the inside out. *Nat Rev Mol Cell Biol* 2006; 7: 885–896.
15. Reichert B, Yasmeen R, Jeyakumar S et al. Concerted action of aldehyde dehydrogenases influences depot-specific fat formation. *Mol Endocrinol* 2011; 25: 799–809s.
16. Yasmeen R, Jeyakumar S, Reichert B, Yang F, Ziouzenkova O. The contribution of vitamin A to autocrine regulation of fat depots. *Biochim Biophys Acta* 2012; 1821: 190–197.
17. Rosen ED, Spiegelman BM. Adipocytes as regulators of energy balance and glucose homeostasis. *Nature* 2006; 444: 847–853.
18. Tchkonia T, Giorgadze N, Pirtskhalava T et al. Fat depot-specific characteristics are retained in strains derived from single human preadipocytes. *Diabetes* 2006; 55: 2571–2578.
19. Medrikova D, Jilkova ZM, Bardova K, Janovska P, Rossmeisl M, Kopecky J. Sex differences during the course of diet-induced obesity in mice: Adipose tissue expandability and glycemic control. *Int J Obes (Lond)* 2012; 36(2): 262–272.
20. Thomas EL, Parkinson JR, Frost GS et al. The missing risk: MRI and MRS phenotyping of abdominal adiposity and ectopic fat. *Obesity (Silver Spring)* 2012; 20: 76–87.
21. Lumeng CN, Saltiel AR. Inflammatory links between obesity and metabolic disease. *J Clin Invest* 2011; 121: 2111–2117.
22. Wajchenberg BL. Subcutaneous and visceral adipose tissue: Their relation to the metabolic syndrome. *Endocr Rev* 2000; 21: 697–738.
23. Ahima RS, Lazar MA. Adipokines and the peripheral and neural control of energy balance. *Mol Endocrinol* 2008; 22: 1023–1031.
24. Scherer PE. Adipose tissue: From lipid storage compartment to endocrine organ. *Diabetes* 2006; 55: 1537–1545.
25. Tchkonia T, Lenburg M, Thomou T et al. Identification of depot-specific human fat cell progenitors through distinct expression profiles and developmental gene patterns. *Am J Physiol Endocrinol Metab* 2007; 292: E298–E307.
26. Tchkonia T, Tchoukalova YD, Giorgadze N et al. Abundance of two human preadipocyte subtypes with distinct capacities for replication, adipogenesis, and apoptosis varies among fat depots. *Am J Physiol Endocrinol Metab* 2005; 288: E267–E277.
27. Tran TT, Yamamoto Y, Gesta S, Kahn CR. Beneficial effects of subcutaneous fat transplantation on metabolism. *Cell Metab* 2008; 7: 410–420.
28. Lefebvre AM, Laville M, Vega N et al. Depot-specific differences in adipose tissue gene expression in lean and obese subjects. *Diabetes* 1998; 47: 98–103.
29. Mrosovsky N. Body fat: What is regulated? *Physiol Behav* 1986; 38: 407–414.
30. Seale P, Kajimura S, Spiegelman BM. Transcriptional control of brown adipocyte development and physiological function of mice and men. *Genes Dev* 2009; 23: 788–797.
31. Himms-Hagen J, Melnyk A, Zingaretti MC, Ceresi E, Barbatelli G, Cinti S. Multilocular fat cells in WAT of CL-316243-treated rats derive directly from white adipocytes. *Am J Physiol Cell Physiol* 2000; 279: C670–C681.
32. Tontonoz P, Spiegelman BM. Fat and beyond: The diverse biology of PPARgamma. *Annu Rev Biochem* 2008; 77: 289–312.

33. Lefterova MI, Zhang Y, Steger DJ et al. PPARgamma and C/EBP factors orchestrate adipocyte biology via adjacent binding on a genome-wide scale. *Genes Dev* 2008; 22: 2941–2952.
34. Farmer SR. Transcriptional control of adipocyte formation. *Cell Metab* 2006; 4: 263–273.
35. Bonet ML, Ribot J, Palou A. Lipid metabolism in mammalian tissues and its control by retinoic acid. *Biochim Biophys Acta* 2012; 1821: 177–189.
36. Kawada T, Takahashi N, Fushiki T. Biochemical and physiological characteristics of fat cell. *J Nutr Sci Vitaminol (Tokyo)* 2001; 47: 1–12.
37. Kirkland JL, Hollenberg CH, Kindler S, Gillon WS. Effects of age and anatomic site on preadipocyte number in rat fat depots. *J Gerontol* 1994; 49: B31–B35.
38. Smas CM, Sul HS. Pref-1, a protein containing EGF-like repeats, inhibits adipocyte differentiation. *Cell* 1993; 73: 725–734.
39. Wang Y, Sul HS. Pref-1 regulates mesenchymal cell commitment and differentiation through Sox9. *Cell Metab* 2009; 9: 287–302.
40. Lagathu C, Christodoulides C, Virtue S et al. Dact1, a nutritionally regulated preadipocyte gene, controls adipogenesis by coordinating the Wnt/beta-catenin signaling network. *Diabetes* 2009; 58: 609–619.
41. Fu M, Sun T, Bookout AL et al. A nuclear receptor atlas: 3T3-L1 adipogenesis. *Mol Endocrinol* 2005; 19: 2437–2450.
42. Rosen ED, Sarraf P, Troy AE et al. PPAR gamma is required for the differentiation of adipose tissue in vivo and in vitro. *Mol Cell* 1999; 4: 611–617.
43. Tontonoz P, Hu E, Spiegelman BM. Stimulation of adipogenesis in fibroblasts by PPAR gamma 2, a lipid-activated transcription factor. *Cell* 1994; 79: 1147–1156.
44. Hansen JB, Jorgensen C, Petersen RK et al. Retinoblastoma protein functions as a molecular switch determining white versus brown adipocyte differentiation. *Proc Natl Acad Sci USA* 2004; 101: 4112–4117.
45. Wang Z, Qi C, Krones A et al. Critical roles of the p160 transcriptional coactivators p/CIP and SRC-1 in energy balance. *Cell Metab* 2006; 3: 111–122.
46. Puigserver P, Wu Z, Park CW, Graves R, Wright M, Spiegelman BM. A cold-inducible coactivator of nuclear receptors linked to adaptive thermogenesis. *Cell* 1998; 92: 829–839.
47. Karamanlidis G, Karamitri A, Docherty K, Hazlerigg DG, Lomax MA. C/EBPbeta reprograms white 3T3-L1 preadipocytes to a Brown adipocyte pattern of gene expression. *J Biol Chem* 2007; 282: 24660–24669.
48. Karamitri A, Shore AM, Docherty K, Speakman JR, Lomax MA. Combinatorial transcription factor regulation of the cyclic AMP-response element on the Pgc-1alpha promoter in white 3T3-L1 and brown HIB-1B preadipocytes. *J Biol Chem* 2009; 284: 20738–20752.
49. Carmona MC, Hondares E, Rodriguez de la Concepcion ML et al. Defective thermoregulation, impaired lipid metabolism, but preserved adrenergic induction of gene expression in brown fat of mice lacking C/EBPbeta. *Biochem J* 2005; 389: 47–56.
50. Ducharme NA, Bickel PE. Lipid droplets in lipogenesis and lipolysis. *Endocrinology* 2008; 149: 942–949.
51. Frayn KN, Karpe F, Fielding BA, Macdonald IA, Coppack SW. Integrative physiology of human adipose tissue. *Int J Obes Relat Metab Disord* 2003; 27: 875–888.
52. Berman DM, Nicklas BJ, Ryan AS, Rogus EM, Dennis KE, Goldberg AP. Regulation of lipolysis and lipoprotein lipase after weight loss in obese, postmenopausal women. *Obes Res* 2004; 12: 32–39.
53. Hikita M, Bujo H, Yamazaki K et al. Differential expression of lipoprotein lipase gene in tissues of the rat model with visceral obesity and postprandial hyperlipidemia. *Biochem Biophys Res Commun* 2000; 277: 423–429.

54. Llado I, Pons A, Palou A. Effects of fasting on lipoprotein lipase activity in different depots of white and brown adipose tissues in diet-induced overweight rats. *J Nutr Biochem* 1999; 10: 609–614.

55. McCarty MF. Modulation of adipocyte lipoprotein lipase expression as a strategy for preventing or treating visceral obesity. *Med Hypotheses* 2001; 57: 192–200.

56. Moustaid N, Jones BH, Taylor JW. Insulin increases lipogenic enzyme activity in human adipocytes in primary culture. *J Nutr* 1996; 126: 865–870.

57. Nicklas BJ, Rogus EM, Colman EG, Goldberg AP. Visceral adiposity, increased adipocyte lipolysis, and metabolic dysfunction in obese postmenopausal women. *Am J Physiol* 1996; 270: E72–E78.

58. Duncan RE, Ahmadian M, Jaworski K, Sarkadi-Nagy E, Sul HS. Regulation of lipolysis in adipocytes. *Annu Rev Nutr* 2007; 27: 79–101.

59. Zechner R, Kienesberger PC, Haemmerle G, Zimmermann R, Lass A. Adipose triglyceride lipase and the lipolytic catabolism of cellular fat stores. *J Lipid Res* 2009; 50: 3–21.

60. Zimmermann R, Strauss JG, Haemmerle G et al. Fat mobilization in adipose tissue is promoted by adipose triglyceride lipase. *Science* 2004; 306: 1383–1386.

61. Das SK, Eder S, Schauer S et al. Adipose triglyceride lipase contributes to cancer-associated cachexia. *Science* 2011; 333: 233–238.

62. Haemmerle G, Zimmermann R, Hayn M et al. Hormone-sensitive lipase deficiency in mice causes diglyceride accumulation in adipose tissue, muscle, and testis. *J Biol Chem* 2002; 277: 4806–4815.

63. Wang SP, Laurin N, Himms-Hagen J et al. The adipose tissue phenotype of hormone-sensitive lipase deficiency in mice. *Obes Res* 2001; 9: 119–128.

64. Strom K, Gundersen TE, Hansson O et al. Hormone-sensitive lipase (HSL) is also a retinyl ester hydrolase: Evidence from mice lacking HSL. *FASEB J* 2009; 23: 2307–2316.

65. Schweiger M, Lass A, Zimmermann R, Eichmann TO, Zechner R. Neutral lipid storage disease: Genetic disorders caused by mutations in adipose triglyceride lipase/PNPLA2 or CGI-58/ABHD5. *Am J Physiol Endocrinol Metab* 2009; 297: E289–E296.

66. Johansen CT, Gallinger ZR, Wang J et al. Rare ATGL haplotypes are associated with increased plasma triglyceride concentrations in the Greenland Inuit. *Int J Circumpolar Health* 2010; 69: 3–12.

67. Boucher J, Castan-Laurell I, Le Lay S et al. Human alpha 2A-adrenergic receptor gene expressed in transgenic mouse adipose tissue under the control of its regulatory elements. *J Mol Endocrinol* 2002; 29: 251–264.

68. Le Lay S, Lefrere I, Trautwein C, Dugail I, Krief S. Insulin and sterol-regulatory element-binding protein-1c (SREBP-1C) regulation of gene expression in 3T3-L1 adipocytes. Identification of CCAAT/enhancer-binding protein beta as an SREBP-1C target. *J Biol Chem* 2002; 277: 35625–35634.

69. Claycombe KJ, Jones BH, Standridge MK et al. Insulin increases fatty acid synthase gene transcription in human adipocytes. *Am J Physiol* 1998; 274:R1253–R1259.

70. Jeyakumar SM, Vajreswari A, Giridharan NV. Vitamin A regulates obesity in WNIN/Ob obese rat; independent of stearoyl-CoA desaturase-1. *Biochem Biophys Res Commun* 2008; 370: 243–247.

71. Letexier D, Pinteur C, Large V, Frering V, Beylot M. Comparison of the expression and activity of the lipogenic pathway in human and rat adipose tissue. *J Lipid Res* 2003; 44: 2127–2134.

72. Shimomura I, Hammer RE, Richardson JA et al. Insulin resistance and diabetes mellitus in transgenic mice expressing nuclear SREBP-1c in adipose tissue: Model for congenital generalized lipodystrophy. *Genes Dev* 1998; 12: 3182–3194.

73. Goodman DW, Huang HS, Shiratori T. Tissue distribution and metabolism of newly absorbed vitamin a in the rat. *J Lipid Res* 1965; 6: 390–396.

74. van Bennekum AM, Kako Y, Weinstock PH et al. Lipoprotein lipase expression level influences tissue clearance of chylomicron retinyl ester. *J Lipid Res* 1999; 40: 565–574.
75. Blaner WS, Obunike JC, Kurlandsky SB et al. Lipoprotein lipase hydrolysis of retinyl ester. Possible implications for retinoid uptake by cells. *J Biol Chem* 1994; 269: 16559–16565.
76. Goldberg IJ. Lipoprotein lipase and lipolysis: Central roles in lipoprotein metabolism and atherogenesis. *J Lipid Res* 1996; 37: 693–707.
77. Skottova N, Savonen R, Lookene A, Hultin M, Olivecrona G. Lipoprotein lipase enhances removal of chylomicrons and chylomicron remnants by the perfused rat liver. *J Lipid Res* 1995; 36: 1334–1344.
78. Chang S, Maeda N, Borensztajn J. The role of lipoprotein lipase and apoprotein E in the recognition of chylomicrons and chylomicron remnants by cultured isolated mouse hepatocytes. *Biochem J* 1996; 318(Pt 1): 29–34.
79. Quadro L, Blaner WS, Salchow DJ et al. Impaired retinal function and vitamin A availability in mice lacking retinol-binding protein. *EMBO J* 1999; 18: 4633–4644.
80. Tsutsumi C, Okuno M, Tannous L et al. Retinoids and retinoid-binding protein expression in rat adipocytes. *J Biol Chem* 1992; 267: 1805–1810.
81. Kloting N, Graham TE, Berndt J et al. Serum retinol-binding protein is more highly expressed in visceral than in subcutaneous adipose tissue and is a marker of intra-abdominal fat mass. *Cell Metab* 2007; 6: 79–87.
82. Yang Q, Graham TE, Mody N et al. Serum retinol binding protein 4 contributes to insulin resistance in obesity and type 2 diabetes. *Nature* 2005; 436: 356–362.
83. von Eynatten M, Humpert PM. Retinol-binding protein-4 in experimental and clinical metabolic disease. *Expert Rev Mol Diagn* 2008; 8: 289–299.
84. Berry DC, Jin H, Majumdar A, Noy N. Signaling by vitamin A and retinol-binding protein regulates gene expression to inhibit insulin responses. *Proc Natl Acad Sci USA* 2011; 108: 4340–4345.
85. Munkhtulga L, Nagashima S, Nakayama K et al. Regulatory SNP in the RBP4 gene modified the expression in adipocytes and associated with BMI. *Obesity (Silver Spring)* 2010; 18: 1006–1014.
86. Kawaguchi R, Yu J, Honda J et al. A membrane receptor for retinol binding protein mediates cellular uptake of vitamin A. *Science* 2007; 315: 820–825.
87. Isken A, Golczak M, Oberhauser V et al. RBP4 disrupts vitamin A uptake homeostasis in a STRA6-deficient animal model for Matthew-Wood syndrome. *Cell Metab* 2008; 7: 258–268.
88. Moise AR, Golczak M, Imanishi Y, Palczewski K. Topology and membrane association of lecithin: Retinol acyltransferase. *J Biol Chem* 2007; 282: 2081–2090.
89. Liu L, Gudas LJ. Disruption of the lecithin:retinol acyltransferase gene makes mice more susceptible to vitamin A deficiency. *J Biol Chem* 2005; 280: 40226–40234.
90. Futterman S, Heller J. The enhancement of fluorescence and the decreased susceptibility to enzymatic oxidation of retinol complexed with bovine serum albumin, -lactoglobulin, and the retinol-binding protein of human plasma. *J Biol Chem* 1972; 247: 5168–5172.
91. Dufour E, Haertle T. Binding of retinoids and beta-carotene to beta-lactoglobulin. Influence of protein modifications. *Biochim Biophys Acta* 1991; 1079: 316–320.
92. Inoue K, Yagi N, Urade Y, Inui T. Compact packing of lipocalin-type prostaglandin D synthase induced by binding of lipophilic ligands. *J Biochem* 2009; 145: 169–175.
93. Jowsey IR, Murdock PR, Moore GB, Murphy GJ, Smith SA, Hayes JD. Prostaglandin D2 synthase enzymes and PPARgamma are co-expressed in mouse 3T3-L1 adipocytes and human tissues. *Prostaglandins Other Lipid Mediat* 2003; 70: 267–284.
94. Quinkler M, Bujalska IJ, Tomlinson JW, Smith DM, Stewart PM. Depot-specific prostaglandin synthesis in human adipose tissue: A novel possible mechanism of adipogenesis. *Gene* 2006; 380: 137–143.

95. Ragolia L, Palaia T, Hall CE, Maesaka JK, Eguchi N, Urade Y. Accelerated glucose intolerance, nephropathy, and atherosclerosis in prostaglandin D2 synthase knock-out mice. *J Biol Chem* 2005; 280: 29946–29955.

96. Tanaka R, Miwa Y, Mou K et al. Knockout of the l-pgds gene aggravates obesity and atherosclerosis in mice. *Biochem Biophys Res Commun* 2009; 378: 851–856.

97. Ahnstrom J, Faber K, Axler O, Dahlback B. Hydrophobic ligand binding properties of the human lipocalin apolipoprotein M. *J Lipid Res* 2007; 48: 1754–1762.

98. Smith JE, Milch PO, Muto Y, Goodman DS. The plasma transport and metabolism of retinoic acid in the rat. *Biochem J* 1973; 132: 821–827.

99. Tang GW, Russell RM. 13-*cis*-retinoic acid is an endogenous compound in human serum. *J Lipid Res* 1990; 31: 175–182.

100. Folman Y, Russell RM, Tang GW, Wolf DG. Rabbits fed on beta-carotene have higher serum levels of all-trans retinoic acid than those receiving no beta-carotene. *Br J Nutr* 1989; 62: 195–201.

101. Berry DC, Noy N. All-*trans*-retinoic acid represses obesity and insulin resistance by activating both peroxisome proliferation-activated receptor beta/delta and retinoic acid receptor. *Mol Cell Biol* 2009; 29: 3286–3296.

102. Lobo GP, Amengual J, Li HN et al. Beta,beta-carotene decreases peroxisome proliferator receptor gamma activity and reduces lipid storage capacity of adipocytes in a beta,beta-carotene oxygenase 1-dependent manner. *J Biol Chem* 2010; 285: 27891–27899.

103. Parker RS. Carotenoids in human blood and tissues. *J Nutr* 1989; 119: 101–104.

104. Krinsky NI, Cronwell DG, Oncley JL. The transport of vitamin A and carotenoids in human plasma. *Arch Biochem Biophys* 1958; 73: 233–246.

105. Ziouzenkova O, Winklhofer-Roob BM, Puhl H, Roob JM, Esterbauer H. Lack of correlation between the alpha-tocopherol content of plasma and LDL, but high correlations for gamma-tocopherol and carotenoids. *J Lipid Res* 1996; 37: 1936–1946.

106. During A, Doraiswamy S, Harrison EH. Xanthophylls are preferentially taken up compared with beta-carotene by retinal cells via a SRBI-dependent mechanism. *J Lipid Res* 2008; 49: 1715–1724.

107. Yvan-Charvet L, Bobard A, Bossard P et al. In vivo evidence for a role of adipose tissue SR-BI in the nutritional and hormonal regulation of adiposity and cholesterol homeostasis. *Arterioscler Thromb Vasc Biol* 2007; 27: 1340–1345.

108. Ahmed RA, Murao K, Imachi H et al. Human scavenger receptor class B type 1 is regulated by activators of peroxisome proliferators-activated receptor-gamma in hepatocytes. *Endocrine* 2009; 35: 233–242.

109. Hessel S, Eichinger A, Isken A et al. CMO1 deficiency abolishes Vitamin A production from beta-carotene and alters lipid metabolism in mice. *J Biol Chem* 2007; 282: 33553–33561.

110. Uelmen PJ, Oka K, Sullivan M, Chang CC, Chang TY, Chan L. Tissue-specific expression and cholesterol regulation of acylcoenzyme A:cholesterol acyltransferase (ACAT) in mice. Molecular cloning of mouse ACAT cDNA, chromosomal localization, and regulation of ACAT in vivo and in vitro. *J Biol Chem* 1995; 270: 26192–26201.

111. O'Byrne SM, Wongsiriroj N, Libien J et al. Retinoid absorption and storage is impaired in mice lacking lecithin:retinol acyltransferase (LRAT). *J Biol Chem* 2005; 280: 35647–35657.

112. Randolph RK, Winkler KE, Ross AC. Fatty acyl CoA-dependent and -independent retinol esterification by rat liver and lactating mammary gland microsomes. *Arch Biochem Biophys* 1991; 288: 500–508.

113. Caprioli A, Zhu H, Sato TN. CRBP-III:lacZ expression pattern reveals a novel heterogeneity of vascular endothelial cells. *Genesis* 2004; 40: 139–145.

114. Batten ML, Imanishi Y, Maeda T et al. Lecithin-retinol acyltransferase is essential for accumulation of all-*trans*-retinyl esters in the eye and in the liver. *J Biol Chem* 2004; 279: 10422–10432.

115. Schreiber R, Taschler U, Preiss-Landl K, Wongsiriroj N, Zimmermann R, Lass A. Retinyl ester hydrolases and their roles in vitamin A homeostasis. *Biochim Biophys Acta* 2011; 1821: 113–123.

116. Wei S, Lai K, Patel S et al. Retinyl ester hydrolysis and retinol efflux from BFC-1beta adipocytes. *J Biol Chem* 1997; 272: 14159–14165.

117. Strom K, Hansson O, Lucas S et al. Attainment of brown adipocyte features in white adipocytes of hormone-sensitive lipase null mice. *PLoS ONE* 2008; 3: e1793.

118. Li E, Demmer LA, Sweetser DA, Ong DE, Gordon JI. Rat cellular retinol-binding protein II: Use of a cloned cDNA to define its primary structure, tissue-specific expression, and developmental regulation. *Proc Natl Acad Sci USA* 1986; 83: 5779–5783.

119. Li E, Norris AW. Structure/function of cytoplasmic vitamin A-binding proteins. *Annu Rev Nutr* 1996; 16: 205–234.

120. Ghyselinck NB, Bavik C, Sapin V et al. Cellular retinol-binding protein I is essential for vitamin A homeostasis. *EMBO J* 1999; 18: 4903–4914.

121. Ziouzenkova O, Orasanu G, Sharlach M et al. Retinaldehyde represses adipogenesis and diet-induced obesity. *Nat Med* 2007; 13: 695–702.

122. E X, Zhang L, Lu J et al. Increased neonatal mortality in mice lacking cellular retinol-binding protein II. *J Biol Chem* 2002; 277: 36617–36623.

123. Zizola CF, Frey SK, Jitngarmkusol S, Kadereit B, Yan N, Vogel S. Cellular retinol-binding protein type I (CRBP-I) regulates adipogenesis. *Mol Cell Biol* 2010; 30: 3412–3420.

124. Zizola CF, Schwartz GJ, Vogel S. Cellular retinol-binding protein type III is a PPARgamma target gene and plays a role in lipid metabolism. *Am J Physiol Endocrinol Metab* 2008; 295: E1358–E1368.

125. Molotkov A, Ghyselinck NB, Chambon P, Duester G. Opposing actions of cellular retinol-binding protein and alcohol dehydrogenase control the balance between retinol storage and degradation. *Biochem J* 2004; 383: 295–302.

126. Hoyos B, Imam A, Chua R et al. The cysteine-rich regions of the regulatory domains of Raf and protein kinase C as retinoid receptors. *J Exp Med* 2000; 192: 835–845.

127. Acin-Perez R, Hoyos B, Gong J et al. Regulation of intermediary metabolism by the PKCdelta signalosome in mitochondria. *FASEB J* 2010; 24: 5033–5042.

128. Hoyos B, Jiang S, Hammerling U. Location and functional significance of retinol-binding sites on the serine/threonine kinase, c-Raf. *J Biol Chem* 2005; 280: 6872–6878.

129. Bosch RR, Janssen SW, Span PN et al. Exploring levels of hexosamine biosynthesis pathway intermediates and protein kinase C isoforms in muscle and fat tissue of Zucker diabetic fatty rats. *Endocrine* 2003; 20: 247–252.

130. Kayali AG, Austin DA, Webster NJ. Rottlerin inhibits insulin-stimulated glucose transport in 3T3-L1 adipocytes by uncoupling mitochondrial oxidative phosphorylation. *Endocrinology* 2002; 143: 3884–3896.

131. Zmuidzinas A, Gould GW, Yager JD. Expression of c-raf-1 and A-raf-1 during differentiation of 3T3-L1 preadipocyte fibroblasts into adipocytes. *Biochem Biophys Res Commun* 1989; 162: 1180–1187.

132. Schupp M, Lefterova MI, Janke J et al. Retinol saturase promotes adipogenesis and is downregulated in obesity. *Proc Natl Acad Sci USA* 2009; 106: 1105–1110.

133. Moise AR, Kuksa V, Imanishi Y, Palczewski K. Identification of all-*trans*-retinol:all-*trans*-13,14-dihydroretinol saturase. *J Biol Chem* 2004; 279: 50230–50242.

134. Moise AR, Kuksa V, Blaner WS, Baehr W, Palczewski K. Metabolism and transactivation activity of 13,14-dihydroretinoic acid. *J Biol Chem* 2005; 280: 27815–27825.

135. Moise AR, Lobo GP, Erokwu B et al. Increased adiposity in the retinol saturase-knockout mouse. *FASEB J* 2010; 24: 1261–1270.

136. Pares X, Farres J, Kedishvili N, Duester G. Medium- and short-chain dehydrogenase/ reductase gene and protein families: Medium-chain and short-chain dehydrogenases/ reductases in retinoid metabolism. *Cell Mol Life Sci* 2008; 65: 3936–3949.

137. Boulanger A, McLemore P, Copeland NG et al. Identification of beta-carotene 15, 15′-monooxygenase as a peroxisome proliferator-activated receptor target gene. *FASEB J* 2003; 17: 1304–1306.

138. Amengual J, Gouranton E, van Helden YG et al. Beta-carotene reduces body adiposity of mice via BCMO1. *PLoS ONE* 2011; 6:e20644.

139. Duester G. Retinoic acid synthesis and signaling during early organogenesis. *Cell* 2008; 134: 921–931.

140. Repa JJ, Hanson KK, Clagett-Dame M. All-*trans*-retinol is a ligand for the retinoic acid receptors. *Proc Natl Acad Sci USA* 1993; 90: 7293–7297.

141. Fan X, Molotkov A, Manabe S et al. Targeted disruption of Aldh1a1 (Raldh1) provides evidence for a complex mechanism of retinoic acid synthesis in the developing retina. *Mol Cell Biol* 2003; 23: 4637–4648.

142. Molotkov A, Duester G. Genetic evidence that retinaldehyde dehydrogenase Raldh1 (Aldh1a1) functions downstream of alcohol dehydrogenase Adh1 in metabolism of retinol to retinoic acid. *J Biol Chem* 2003; 278: 36085–36090.

143. Kane MA, Folias AE, Napoli JL. HPLC/UV quantitation of retinal, retinol, and retinyl esters in serum and tissues. *Anal Biochem* 2008; 378: 71–79.

144. Ruiz FX, Moro A, Gallego O et al. Human and rodent aldo-keto reductases from the AKR1B subfamily and their specificity with retinaldehyde. *Chem Biol Interact* 2011; 191: 199–205.

145. Fujimori K, Ueno T, Nagata N et al. Suppression of adipocyte differentiation by aldo-keto reductase 1B3 acting as prostaglandin F2alpha synthase. *J Biol Chem* 2010; 285: 8880–8886.

146. Kato M, Blaner WS, Mertz JR, Das K, Kato K, Goodman DS. Influence of retinoid nutritional status on cellular retinol- and cellular retinoic acid-binding protein concentrations in various rat tissues. *J Biol Chem* 1985; 260: 4832–4838.

147. Boylan JF, Gudas LJ. The level of CRABP-I expression influences the amounts and types of all-*trans*-retinoic acid metabolites in F9 teratocarcinoma stem cells. *J Biol Chem* 1992; 267: 21486–21491.

148. Wei LN, Lee CH, Filipcik P, Chang L. Regulation of the mouse cellular retinoic acid-binding protein-I gene by thyroid hormone and retinoids in transgenic mouse embryos and P19 cells. *J Endocrinol* 1997; 155: 35–46.

149. Park SW, Huang WH, Persaud SD, Wei LN. RIP140 in thyroid hormone-repression and chromatin remodeling of Crabp1 gene during adipocyte differentiation. *Nucleic Acids Res* 2009; 37: 7085–7094.

150. Astrom A, Pettersson U, Chambon P, Voorhees JJ. Retinoic acid induction of human cellular retinoic acid-binding protein-II gene transcription is mediated by retinoic acid receptor-retinoid X receptor heterodimers bound to one far upstream retinoic acid-responsive element with 5-base pair spacing. *J Biol Chem* 1994; 269: 22334–22339.

151. Budhu AS, Noy N. Direct channeling of retinoic acid between cellular retinoic acid-binding protein II and retinoic acid receptor sensitizes mammary carcinoma cells to retinoic acid-induced growth arrest. *Mol Cell Biol* 2002; 22: 2632–2641.

152. Berry DC, Soltanian H, Noy N. Repression of cellular retinoic acid-binding protein II during adipocyte differentiation. *J Biol Chem* 2010; 285: 15324–15332.

153. Germain P, Chambon P, Eichele G et al. International Union of Pharmacology. LX. Retinoic acid receptors. *Pharmacol Rev* 2006; 58: 712–725.

154. Germain P, Chambon P, Eichele G et al. International Union of Pharmacology. LXIII. Retinoid X receptors. *Pharmacol Rev* 2006; 58: 760–772.

155. Heyman RA, Mangelsdorf DJ, Dyck JA et al. 9-*cis* retinoic acid is a high affinity ligand for the retinoid X receptor. *Cell* 1992; 68: 397–406.

156. Leid M, Kastner P, Chambon P. Multiplicity generates diversity in the retinoic acid signalling pathways. *Trends Biochem Sci* 1992; 17: 427–433.

157. Shaw N, Elholm M, Noy N. Retinoic acid is a high affinity selective ligand for the peroxisome proliferator-activated receptor beta/delta. *J Biol Chem* 2003; 278: 41589–41592.

158. Lee CH, Olson P, Hevener A et al. PPARdelta regulates glucose metabolism and insulin sensitivity. *Proc Natl Acad Sci USA* 2006; 103: 3444–3449.

159. Langston AW, Thompson JR, Gudas LJ. Retinoic acid-responsive enhancers located 3′ of the Hox A and Hox B homeobox gene clusters. Functional analysis. *J Biol Chem* 1997; 272: 2167–2175.

160. Serpente P, Tumpel S, Ghyselinck NB et al. Direct crossregulation between retinoic acid receptor {beta} and Hox genes during hindbrain segmentation. *Development* 2005; 132: 503–513.

161. Easwaran V, Pishvaian M, Salimuddin, Byers S. Cross-regulation of beta-catenin-LEF/TCF and retinoid signaling pathways. *Curr Biol* 1999; 9: 1415–1418.

162. Schwarz EJ, Reginato MJ, Shao D, Krakow SL, Lazar MA. Retinoic acid blocks adipogenesis by inhibiting C/EBPbeta-mediated transcription. *Mol Cell Biol* 1997; 17: 1552–1561.

163. Marchildon F, St-Louis C, Akter R, Roodman V, Wiper-Bergeron NL. Transcription factor Smad3 is required for the inhibition of adipogenesis by retinoic acid. *J Biol Chem* 2010; 285: 13274–13284.

164. Tan CK, Leuenberger N, Tan MJ et al. Smad3 deficiency in mice protects against insulin resistance and obesity induced by a high-fat diet. *Diabetes* 2011; 60: 464–476.

165. Imai T, Jiang M, Chambon P, Metzger D. Impaired adipogenesis and lipolysis in the mouse upon selective ablation of the retinoid X receptor alpha mediated by a tamoxifen-inducible chimeric Cre recombinase (Cre-ERT2) in adipocytes. *Proc Natl Acad Sci USA* 2001; 98: 224–228.

166. Wyss R, Bucheli F, Hartenbach R. Determination of 13-*cis*-3-hydroxyretinoic acid, all-*trans*-3-hydroxyretinoic acid and their 4-oxo metabolites in human and animal plasma by high-performance liquid chromatography with automated column switching and UV detection. *J Pharm Biomed Anal* 1998; 18: 761–776.

167. Schuchardt JP, Wahlstrom D, Ruegg J et al. The endogenous retinoid metabolite S-4-oxo-9-*cis*-13,14-dihydro-retinoic acid activates retinoic acid receptor signalling both in vitro and in vivo. *FEBS J* 2009; 276: 3043–3059.

168. Moise AR, Alvarez S, Dominguez M et al. Activation of retinoic acid receptors by dihydroretinoids. *Mol Pharmacol* 2009; 76: 1228–1237.

169. Ziouzenkova O, Orasanu G, Sukhova G et al. Asymmetric cleavage of beta-carotene yields a transcriptional repressor of retinoid X receptor and peroxisome proliferator-activated receptor responses. *Mol Endocrinol* 2007; 21: 77–88.

170. Marsh RS, Yan Y, Reed VM, Hruszkewycz D, Curley RW, Harrison EH. {beta}-Apocarotenoids do not significantly activate retinoic acid receptors {alpha} or {beta}. *Exp Biol Med (Maywood)* 2010; 235: 342–348.

171. Eroglu A, Hruszkewycz DP, Curley RW, Jr., Harrison EH. The eccentric cleavage product of beta-carotene, beta-apo-13-carotenone, functions as an antagonist of RXRalpha. *Arch Biochem Biophys* 2010; 504: 11–16.

172. Dmitrovskii AA, Gessler NN, Gomboeva SB, Ershov Yu V, Bykhovsky V. Enzymatic oxidation of beta-apo-8′-carotenol to beta-apo-14′-carotenal by an enzyme different from beta-carotene-15,15′-dioxygenase. *Biochemistry (Mosc)* 1997; 62: 787–792.

173. De Botton S, Dombret H, Sanz M et al. Incidence, clinical features, and outcome of all *trans*-retinoic acid syndrome in 413 cases of newly diagnosed acute promyelocytic leukemia. The European APL Group. *Blood* 1998; 92: 2712–2718.

# Part II

*Carotenoids and Vitamin A
in Skin, Eye, and Ear Diseases*

# 3 Retinoids in the Treatment of Psoriasis

*Neil Borkar and Alan Menter*

## CONTENTS

Psoriasis is a chronic inflammatory disorder of the skin affecting approximately 1%–2% of the world's population, although certain areas of the world have a lower prevalence rate. The highest reported incidences of psoriasis are in Denmark (2.9%) and the Faroe Islands (2.8%),[1] whereas in Japan[2] and Asia, the prevalence rate is closer to 0.4%. The disease tends to wax and wane in severity throughout an individual's lifetime, with a variable age of onset. The prevalence of psoriasis is equal between males and females.

The most common clinical presentation for this disease is that of plaque psoriasis, which presents as discrete, erythematous papules and plaques with an adherent silvery scale. Common locations for psoriasis are on the scalp, elbows, knees, lower posterior trunk, and nails. These lesions are commonly pruritic.

Psoriasis has a strong genetic component with a polygenic or multifactorial pattern.[3] Established triggers for psoriasis are trauma, infection, and certain medications, although in most cases, the specific trigger factor cannot be identified. Genetic predisposition to psoriasis has been associated with certain HLA subtypes of HLA-B13, HLA-B17, HLA-B37, HLA-Bw16, and HLA-Cw6,[4] with a total of 19 genes identified.

Psoriasis was initially seen as a pure abnormality in epidermal growth. Hence, research was initially targeted at only the biology and pathophysiology of the keratinocyte. Subsequently, due to the anecdotal discovery of the effect of cyclosporine,[5] which inhibits T-cells and IL-2, it was shown that there was a very complex, underlying immune-mediated cascade for this disease. An overexpression of type 1 helper T cell (Th1) cytokines such as IL-2, IL-6, IL-8, IL-12, IFN-$\gamma$, and tumor necrosis factor (TNF)-$\alpha$ has been demonstrated, with the overexpression of IL-8 leading to the accumulation of neutrophils.

In addition, it is now known that a distinctly separate T cell from the Th1 cell, called Th17, is stimulated by IL-23, to release IL-17 and IL-22, the latter of which results in the proliferation of epidermal keratinocytes and increased dermal inflammation.[6]

The term "psoriasis" encompasses a large variety of different clinical presentations, collectively under the heading psoriasis vulgaris. While the clinical phenotypes of psoriasis are distinct, the classification system of these variants is not a standardized, accepted system, with the inclusion of even archaic descriptive terms such as "elephantine" and "geographic."[7]

## PLAQUE

Plaque psoriasis is the most common form, affecting approximately 90% of patients.[1] This form of psoriasis presents as well-demarcated erythematous plaques. Histologically, this is demonstrated by psoriasiform epidermal hyperplasia, parakeratosis with intracorneal neutrophils, hypogranulosis, spongiform pustules, an infiltrate of neutrophils and lymphocytes in the epidermis and dermis, along with an expanded dermal papillary vasculature[8] (Figure 3.1). Patients may have involvement ranging from only a few plaques to numerous lesions covering almost the entire body surface. The plaques are irregular, round to oval in shape, and most often located on the scalp, trunk, buttocks, and limbs, with a predilection for extensor surfaces, such as the elbows and knees. Smaller plaques or papules may coalesce into larger lesions, especially on

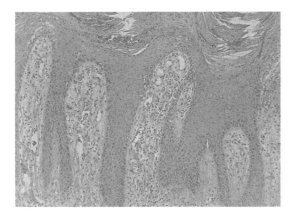

**FIGURE 3.1**    Histopathology of psoriasis.

the legs and trunk. Psoriatic plaques typically have a dry, silvery-white scale and tend to be symmetrically distributed over the body (Figure 3.2). Approximately 80% of those affected with psoriasis have mild-to-moderate disease, with 20% having moderate-to-severe psoriasis affecting more than 5% of the body surface area (BSA) or affecting crucial body areas such as the hands, feet, face, or genitals.

## INVERSE

Inverse psoriasis is characterized by lesions in the skin folds. The lesions tend to be erythematous with minimal induration or scale. Common locations include the axillary, genital, perineal, intergluteal, and inframammary areas. Flexural surfaces such as the antecubital fossae can exhibit similar lesions (Figure 3.3B), as well as secondary candidiasis, which is a common occurrence, especially in the groin, buttocks, or breast folds.

## ERYTHRODERMIC

Erythrodermic psoriasis can develop gradually from chronic plaque disease or acutely with little preceding psoriasis. Generalized erythema covering nearly the entire BSA with varying degrees of scaling is seen (Figure 3.2E). Altered thermoregulatory properties of erythrodermic skin may lead to chills and hypothermia, and fluid loss may lead to dehydration. Fluid retention can precipitate congestive heart failure. Fever and malaise are common.

## PUSTULAR

Pustular psoriasis may be generalized or localized. The acute generalized variety (termed the "von Zumbusch variant") is an uncommon severe form of psoriasis accompanied by fever and toxicity and consists of widespread pustules on an erythematous background (Figures 3.2D and 3.3C), frequently in association with erythroderma. Cutaneous lesions characteristic of psoriasis vulgaris may be present before, during, or after an acute pustular episode. There is also a localized pustular

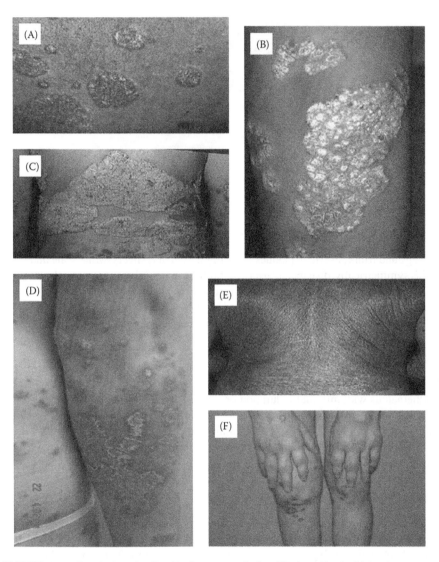

**FIGURE 3.2** Psoriasis. (A) Small plaque psoriasis. (B) Localized thick plaque-type psoriasis. (C) Large plaque psoriasis. (D) Inflammatory localized psoriasis. (E) Erythrodermic psoriasis. (F) Psoriasis and psoriatic arthritis. (From Menter, A. et al., *J. Am. Acad. Dermatol.*, 58(5), 826–850, 2008.)

variant of psoriasis involving the palms and soles, with or without evidence of classic plaque-type disease elsewhere on the body.

## GUTTATE

Guttate psoriasis is characterized by dew-drop like, 1- to 10-mm, pink papules, usually with a fine scale. This variant of psoriasis tends to be common in individuals younger than 30 years, and is found primarily on the trunk and the proximal

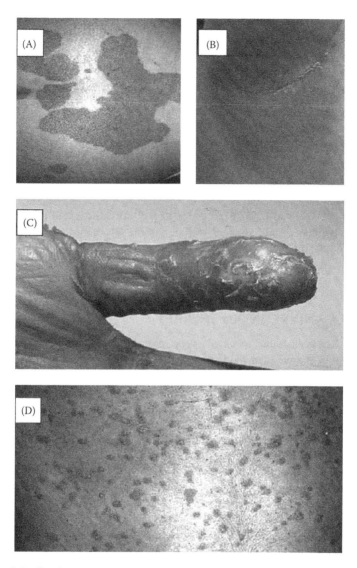

**FIGURE 3.3** Psoriasis. (A) Thin plaque-type psoriasis. (B) Inverse-type psoriasis. (C) Pustular-type psoriasis. (D) Guttate-type psoriasis. (From Menter, A. et al., *J. Am. Acad. Dermatol.*, 58(5), 826–850, 2008.)

extremities, and occurs in less than 2% of patients with psoriasis. A history of upper respiratory infection with group A beta-hemolytic streptococci often precedes guttate psoriasis, especially in younger patients, by 2–3 weeks. This sudden appearance of papular lesions may be either the first manifestation of psoriasis in a previously unaffected individual or an acute exacerbation of long-standing plaque psoriasis (Figure 3.3D).

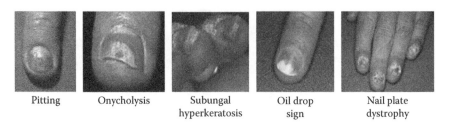

| Pitting | Onycholysis | Subungal hyperkeratosis | Oil drop sign | Nail plate dystrophy |

**FIGURE 3.4**    Nail psoriasis. (From Menter, A. et.al., *J. Am. Acad. Dermatol.*, 58(5), 826, 2008.)

## NAIL DISEASE (PSORIATIC ONYCHODYSTROPHY)

Fingernails are involved in approximately 50% of all patients and toenails in 35% of psoriatic patients. These changes include pitting, onycholysis, subungual hyperkeratosis, and the oil-drop sign (Figure 3.4). Patients with psoriatic arthritis are considered to have a higher incidence of nail changes. Psoriasis of the nails responds poorly, if not at all, to topical therapy.

## VITAMIN A AND RETINOID THERAPY

Vitamin A has been found to have a very important application with therapeutic effectiveness in the medical field of dermatology, especially in hyperproliferative and inflammatory skin disorders. In mammals, vitamin A exists in the forms of retinol (alcohol form), retinal (aldehyde form), and retinoic acid (acid form). In the blood, retinol is bound to retinol-binding protein.

In the 1920s, it was found that a deficiency of vitamin A resulted in epidermal hyperkeratosis (thickening of the stratum corneum) of the skin, squamous metaplasia of mucous membranes, and various keratinization disorders.[9] A condition called phrynoderma, a distinct form of follicular hyperkeratosis (characterized by excessive development of keratin in hair follicles), was found in African prisoners who also had night blindness and xerophthalmia.[10] When they were treated with cod-liver oil, which contains vitamin A, both the skin lesions and the night blindness improved.[11] These findings of vitamin A having therapeutic benefit in skin disorders involving abnormal keratinization spawned research to investigate its use in the treatment of psoriasis and other hyperproliferative and hyperkeratotic skin disorders.

Initially, vitamin A was administered orally to treat certain skin disorders. With the large dose required, the toxic adverse effect profile, which will be detailed later in this chapter, was a huge limitation in these patients, leading to the need to find agents that had a better therapeutic index and less likelihood of toxicity.

In the 1980s, the aromatic retinoid (second generation), etretinate, was approved for the treatment of psoriasis. In 1998, etretinate was replaced with acitretin, also a second-generation retinoid but with a shorter half-life.

Natural and synthetic compounds derived from retinol, retinal, or retinoic acid are called retinoids. Retinoids are pharmaceutical compounds that are structural and functional analogs of vitamin A. Chemically, retinoids are defined as a class of compounds that consist of four isoprenoid units joined in a head-to-tail manner.[12]

## TABLE 3.1
## Biologic Effects of Retinoids

Modulation of proliferation and cell differentiation

Anti-keratinization

Alteration of cellular cohesiveness

Immunologic and anti-inflammatory effects

Inhibition of tumor promotion and malignant cell growth

Induction of apoptosis

Effects on extracellular matrix components

*Source:* Monfrecola, G. and Baldo, A., *J. Rheumatol. Suppl.*, 83, 71, 2009.

Retinoids are commonly used in the treatment of psoriasis. Numerous studies have been able to document the role of retinoids in regulating growth, differentiation, proliferation, and apoptosis of epidermal cells (Table 3.1), specifically of epidermal keratinocytes. The mechanism of action of retinoids and physiologic functions of retinoic acid is mainly exhibited through nuclear retinoid receptors. There are three isomers of retinoid acid: all-*trans* retinoic acid (tretinoin), 9-*cis* retinoic acid (alitretinoin), and 13-*cis* retinoic acid (isotretinoin).

The retinoid receptor is composed of α, β, and γ subunits and can either be of the all-*trans* retinoic acid receptor (RAR) type or the 9-*cis* retinoic acid receptor (RXR) type. These subunits bind to retinoid response elements in the promoter region leading to transcriptional activation of genes. RARs bind both all-*trans* retinoic acid and 9-*cis* retinoic acid. On the other hand, 9-*cis* retinoid acid binds to RXR receptors. Previous to binding of the ligand to the RAR of the RAR/RXR heterodimer, transcriptional corepressor complexes result in chromatin condensation and transcriptional silencing (Figure 3.5).

Dietary retinyl esters are cleaved to retinol in the intestinal lumen after which they are absorbed, re-esterified, and transported to the liver via chylomicrons.[13] Retinol in the circulation is taken up by peripheral cells and converted to ATRA. ATRA then binds to cellular retinoic acid-binding proteins (CRABPs), which is then transported to the nucleus where it binds to RARs, which are ligand-dependent transcription factors. The RARs are heterodimered with retinoid X receptors (RXRs), which are bound to DNA sequences called retinoic acid response elements (RAREs) that are within the promoter region of target genes and gene transcription is initiated (Figure 3.6 and Tables 3.2 and 3.3).

## RETINOIC ACID PHYSIOLOGY

### ANATOMY OF THE SKIN

The epidermis itself is composed of several layers (Figure 3.7). The innermost basal layer is composed of mitotically active columnar-shaped keratinocytes and is in direct contact with the basement membrane. It is in this layer that keratinocytes commence the cell cycle and commit to terminal differentiation. Just above the basal

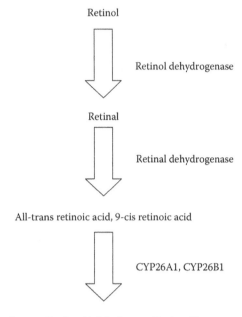

Retinol

⬇ Retinol dehydrogenase

Retinal

⬇ Retinal dehydrogenase

All-trans retinoic acid, 9-cis retinoic acid

⬇ CYP26A1, CYP26B1

4-oxo-retinoic acid, 4-hydroxy-retinoic acid

**FIGURE 3.5** Vitamin A Metabolism. (From Vahlquist, A. and Duvic, M., *Retinoids and Carotenoids in Dermatology*, Informa Healthcare, New York, 2007.)

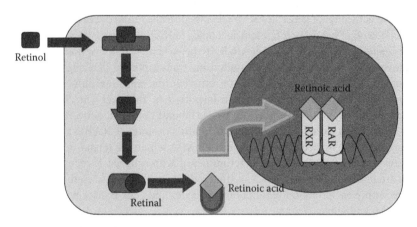

**FIGURE 3.6** Retinoic acid transport.

cell layer is the spiny cell layer, due to the high amount of keratin and desmosomes, thus giving the cell margins a spiny appearance, and the granular cell layer in which prominent keratohyalin granules are present bound to keratin filaments giving the cell a granular appearance. The outermost layer of the epidermis is the cornified layer (stratum corneum), where the keratinocyte's nuclei and organelles have under-

## TABLE 3.2
### Retinoid Characteristics

Tend to be hydrophobic compounds
Are tightly bound to plasma proteins
Chemically unstable in the presence of oxidizing compounds and ultraviolet (UV) radiation
Long elimination half-life when administered systemically

*Source:* Vahlquist, A. and Duvic, M., *Retinoids and Carotenoids in Dermatology*, Informa Healthcare, New York, 2007.

## TABLE 3.3
### Generations of Retinoids

| First-Generation Retinoids | Second-Generation Retinoids | Third-Generation Retinoids |
| --- | --- | --- |
| Tretinoin (all-*trans* retinoic acid) | Alitretinoin (9-*cis* retinoic acid) | Adapalene |
| Isotretinoin (13-*cis* retinoic acid) | Acitretin | Tazarotene |
| | Etretinate | Bexarotene |

*Source:* Bolognia, J. et al., *Dermatology*, Saunders, New York, 2008.

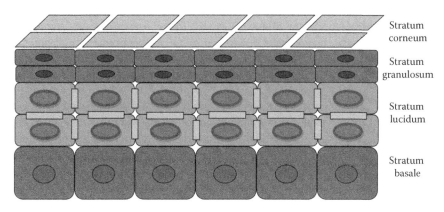

Stratum corneum
Stratum granulosum
Stratum lucidum
Stratum basale

**FIGURE 3.7** Layers of the epidermis.

gone apoptosis, leaving behind the keratin filaments and keratohyalin granules. It is the epidermis where the clinical characteristics of psoriasis can be seen.

Below the epidermis is the dermis, which is mainly composed of collagen (a dense, fibrous connective tissue), elastic fibers, and specialized glycoproteins called glycosaminoglycans.

Beneath the dermis is the subcutis, which is mainly composed of adipocytes, large blood vessels, and nerves.

## Tazarotene

Tazarotene was the first synthetic topical retinoid approved for the treatment of psoriasis by the U.S. Food and Drug Administration in 1997. It is hydrolyzed to the active form of the drug, tazarotenic acid, and has a high affinity for the $\gamma$ and $\beta$ subtypes of the nuclear retinoic acid receptors (RARs) (Figure 3.8).

Clinical trials have shown that when used as monotherapy, topical tazarotene 0.05% and 0.1% once or twice a day is as effective as fluocinonide 0.05% twice a day in the treatment of psoriasis.

Tazarotene is currently available in a topical gel and cream formulations at concentrations of 0.05% and 0.1%. Large-scale multicenter double-blind trials have demonstrated that tazarotene significantly improves psoriasis compared with placebo.[16] The typical dosage is the application of the gel daily for up to 12 weeks for mild-to-moderate plaque psoriasis. The systemic absorption of the drug is very minimal.

Keratinocyte proliferation and differentiation are reduced by tazarotene treatment through decreased expression of epidermal genes of type I keratinocyte transglutaminase, epidermal growth factor receptor, MRP-8, and scalp/elafin, which are usually overexpressed in keratinocytes.[17] Tazarotene also induces tazarotene-induced genes (TIG-1, TIG-2, and TIG-3), which are thought to play a role in proliferation.

In addition, tazarotene has been found to have an anti-inflammatory effect by reducing the expression of inflammatory markers including ICAM-1 and HLA-DR on cells in psoriasis lesions and increasing the reexpression of profilaggrin, a marker of terminal differentiation in keratinocytes.

In clinical trials, the combination of tazarotene with a mid- to high-potency corticosteroid has been found to be highly effective.[18] In one particular study comparing tazarotene + placebo, tazarotene + low-potency steroid, tazarotene + mid-potency steroid, and tazarotene + high-potency steroid, the combination seemed to be better in the reduction of erythema and scaling, as well as an improvement in the median time to initial success of treatment.[19]

The combination of tazarotene with phototherapy has also been well studied. In one study, bilateral target plaques were randomized to receive two of the following, one on each plaque once daily for 14 days: tazarotene 0.1% gel, vehicle gel, or no treatment. Thereafter, the same treatments were continued three times per week, plus UVB phototherapy three times per week, for an additional 67 days.[20] The tazarotene plus UVB phototherapy combination was found to be significant in the reduction of plaque

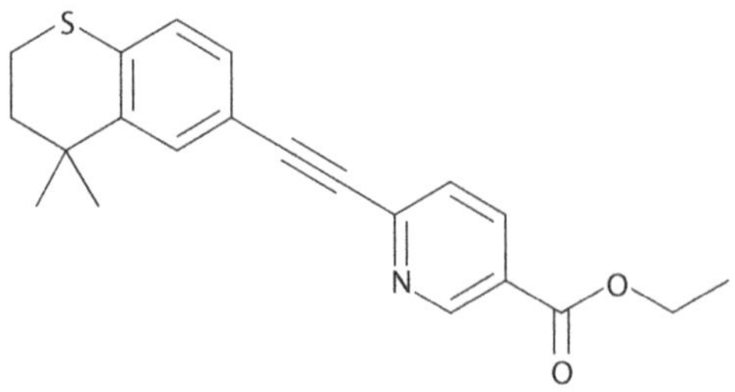

**FIGURE 3.8**   Tazarotene. (From Wolverton, S., *Comprehensive Dermatologic Drug Therapy*, Saunders, New York, 2007.)

elevation and scaling. Most importantly, the median cumulative UVB light exposure required for at least 50% improvement from baseline was much lower with the combination of tazarotene plus UVB versus the UVB with vehicle or UVB phototherapy alone.[20]

## ACITRETIN AND ETRETINATE

Etretinate was originally released in 1986 for the treatment of psoriasis, but was phased out and withdrawn from the market in 1998, due to its pharmacokinetic profile, and has largely been replaced by acitretin (Figures 3.9 and 3.10).

Acitretin, a vitamin A derivative, has been used to treat psoriasis since the early 1980s. Acitretin is the major metabolite and the pharmacologically active compound of etretinate.[21] Acitretin is currently the preferred systemic retinoid due to its pharmacokinetics, in comparison to etretinate. Both of these agents are second-generation aromatic systemic retinoids.

The mean half-life of acitretin is 2 days, whereas the mean half-life of etretinate is 120 days. Since etretinate is highly lipophilic, it tends to be stored in adipose tissue in its ester form, thus resulting in a longer elimination half-life along with being present long after therapy with etretinate has ended.[22] Even with acitretin, there is always the possibility that it can be converted back to etretinate. This is especially true in those with high alcohol intake.[23] Due to this and its effect on women with childbearing potential, it is recommended that women remain on oral contraceptive therapy for at least 2 years after the end of treatment with acitretin. Acitretin is eliminated both hepatically and renally.

As monotherapy, acitretin frequently shows rapid responses within 1–2 weeks in patients with erythrodermic and pustular psoriasis, and has also been shown to be effective in palmoplantar psoriasis. Higher doses of acitretin give better responses but are limited by toxicity factors, especially mucocutaneous adverse effects.

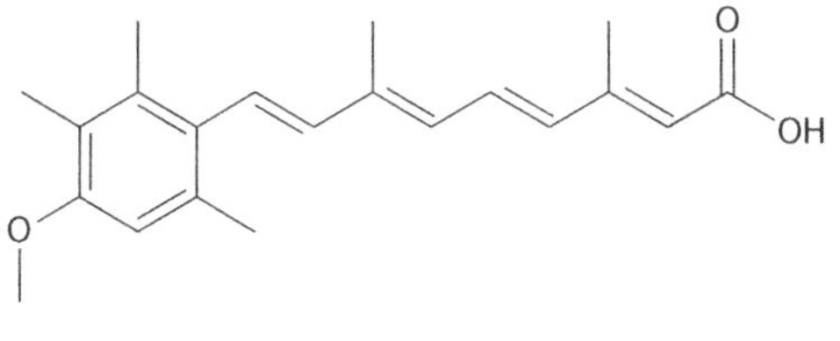

**FIGURE 3.9** Acitretin. (From Wolverton, S., *Comprehensive Dermatologic Drug Therapy*, Saunders, New York, 2007.)

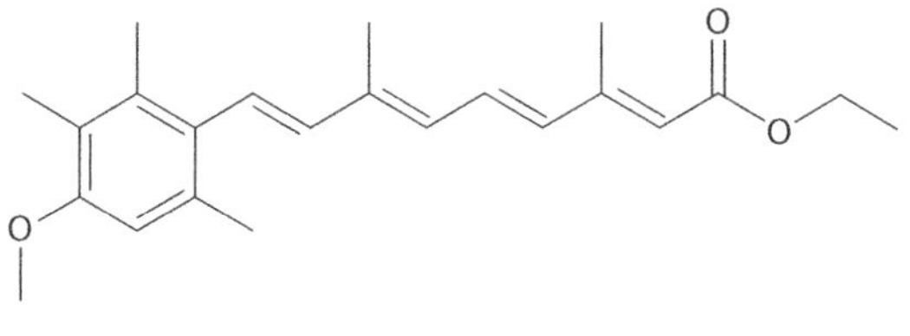

**FIGURE 3.10** Etretinate. (From Wolverton, S., *Comprehensive Dermatologic Drug Therapy*, Saunders, New York, 2007.)

In contrast, those with chronic plaque psoriasis may take 3–4 months to see any effectiveness, with just 50%–60% of patients achieving at least 50% improvement in PASI scores after 8 weeks of treatment, with the mean PASI improvement after 12 weeks being 70%–75%. Therefore, for this form of the disease, it is recommended that acitretin be given as a combination with other therapies. The use of acitretin in combination with other psoriasis treatments, particularly topical corticosteroids and calcipotriene, is a common practice. Acitretin is co-prescribed with the biologics, likely because of the relative lack of overlapping effects on immune function. The immune-sparing method of action of acitretin makes combination treatment with the systemic agents an attractive treatment option, especially in patients where further immunosuppression is unwarranted.

A multicenter, randomized, double-blind placebo-controlled study of calcipotriol, a topical vitamin D analog, ointment (50 mg/g) with acitretin therapy showed an additional improvement in PASI, redness, elevation, and scale in comparison to placebo. Clearance and improvement was achieved with a much lower dose of acitretin, in those receiving the combination.[24]

Numerous studies have also shown acitretin to be effective in combination with NB-UVB or photochemotherapy (PUVA), increasing response rates, decreasing the total number of treatments, and thus the cumulative dose of UV exposure. This combination also results in a lower acitretin dose to be needed. In a randomized double-blind study, acitretin with PUVA treatment was found to be significantly effective to placebo and PUVA with respect to a decrease in lesional scores, number of PUVA exposures, and total dose of UVA until remission.[25] In studies using broadband UVB or narrowband UVB in combination with acitretin, PASI scores improved greater than 70%. It is recommended that prior to the combination, the patient be on 2 weeks of treatment with low-dose acitretin.[26]

Vascular endothelial growth factor (VEGF) is a signaling protein that stimulates angiogenesis and is actually produced by epidermal keratinocytes with the VEGF gene being expressed on chromosome 6. Certain polymorphisms of VEGF are associated with an increased susceptibility to psoriasis, specifically at +405 and −460. These polymorphisms are associated with early-onset psoriasis and are situated close to the functional activator site through which retinoids block VEGF production. However, it has also been shown that the VEGF −460 polymorphism has been shown to play a role in predicting response or nonresponse of psoriasis to acitretin.

Another study evaluated polymorphisms of the apolipoprotein E gene (APOE) as predictors of response to acitretin 9. Although the frequency of the APOE e4 allele (+3937C/+4075C) was higher in patients with chronic plaque and guttate psoriasis than in controls, there was no significant difference in the frequency of alleles in acitretin responders and nonresponders.

## ALITRETINOIN

Alitretinoin (9-*cis*-retinoic acid) is a retinoid that binds the $\alpha$, $\beta$, and $\gamma$ subtypes of both RAR and RXR. It is currently available orally, and in a 0.1% gel, and is FDA approved for use in the treatment of Kaposi's sarcoma[27] as well as used for the treatment of chronic hand dermatitis. However, there has been a case report

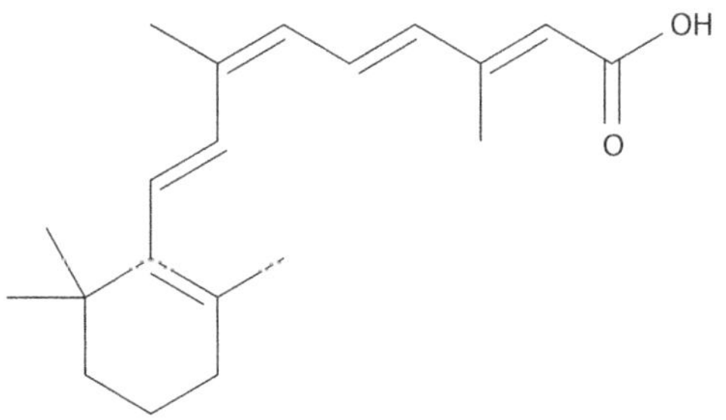

**FIGURE 3.11**   Alitretinoin. (From Wolverton, S., *Comprehensive Dermatologic Drug Therapy*, Saunders, New York, 2007.)

of the treatment of severe palmoplantar hyperkeratotic psoriasis with oral alitretinoin 30 mg/day and etanercept, a TNF-α inhibitor, 50 mg/week subcutaneously for 4 weeks,[28] after treatment failures of methotrexate, calcipotriol, and topical corticosteroids. Palmoplantar psoriasis tends to be recalcitrant to treatment and harder to manage due to constant use of the hands and feet by the patient, especially those in manual labor occupations. It is thought that the Koebner phenomenon, where new psoriasis lesions tend to develop along areas of trauma, is what makes the disease very difficult to treat with conventional agents (Figure 3.11).

## BEXAROTENE

Bexarotene (4-[1-(3,5,5,8,8-pentamethyltetralin-2-yl)ethenyl] benzoic acid) is a synthetic oral retinoid that binds selectively to the nuclear RXR ligand. Because of the specificity for the RXR ligand, this retinoid agent is sometimes referred to as a rexinoid. This RXR receptor is an effective target, as there are four times as many RXR receptors as RAR receptors in the skin.[29] It is currently available topically in a 1% gel and in oral form (Figure 3.12).

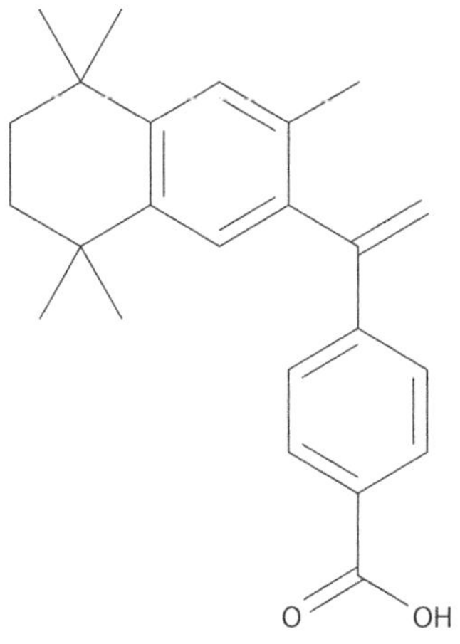

**FIGURE 3.12**   Bexarotene.

Currently, bexarotene has been approved in the treatment of refractory early-stage, as well as advanced-stage cutaneous T-cell lymphoma. Due to its ability to induce apoptosis as well as its antiproliferative effect on the epidermis, bexarotene has been studied in early clinical trials for psoriasis. Bexarotene is currently available as a 1% gel, as well as orally.

In a Phase II clinical trial, 24 patients with mild-to-moderate stable psoriasis, with 15% or less BSA, applied bexarotene 1% gel starting at once per day up to four times daily as tolerated and effective, up to 24 weeks.[30] Fifteen of the twenty-four patients achieved at least 50% improvement in PGA score at two or more consecutive visits, and six patients achieved clearing of their psoriasis of 90% or more. In another Phase II multicenter clinical trial, oral bexarotene was found to be effective in 29 patients with plaque psoriasis for 12 weeks at 1, 2, 0.5, and 3 mg/kg/day.[31]

Oral bexarotene is highly bound to plasma proteins (>99%) and metabolized through the cytochrome P4503 3A4 enzyme pathway, with a $T_{max}$ of 2 h and a half-life of 7 h.[29]

Common adverse events associated with topical bexarotene are rash, pruritus, contact dermatitis, and pain at the application site, with less common adverse effects being infection, headache, edema, and hyperlipidemia, specifically elevated triglycerides. Common adverse events associated with oral bexarotene are central hypothyroidism, leukopenia, agranulocytosis, diffuse skeletal hyperostosis, hypertriglyceridemia, pancreatitis, and elevated liver function tests.

## RETINOID ADVERSE EFFECTS

### TOPICAL RETINOID ADVERSE EFFECTS

Topical retinoids are found to easily penetrate the epidermis; however, following topical application, no significant increase in retinoids in plasma has been found to occur. For example, after the topical application of 0.1% tazarotene, the plasma concentrations were lower than or similar to endogenous tretinoin, thus implying a negligible fetal teratogenicity potential.[32] The most common adverse effect of topical retinoids is local skin irritation encompassing erythema, dryness, pruritus, scaling, and burning, as well as increased photosensitivity at the site of application. Thus, the use of topical tazarotene has to be very carefully considered with very careful application to plaques and avoiding peripheral spread beyond the plaque, thus limiting retinoid-induced irritation (Figure 3.13).

Adverse effects of tazarotene are erythema, scaling, pruritus, burning, stinging, dryness, and irritation, which are dose dependent. As retinoids are teratogenic, pregnancy is a contraindication.

### SYSTEMIC RETINOID ADVERSE EFFECTS

#### Teratogenicity

A major adverse effect of systemic retinoid use is their teratogenic potential in women of childbearing potential, referred to as retinoid embryopathy. This includes abnormalities of the central nervous system, cardiovascular system, as well as craniofacial

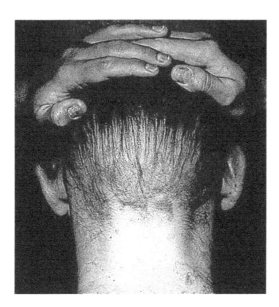

**FIGURE 3.13** Tazarotene-induced perilesional irritation. (From Wolverton, S., *Comprehensive Dermatologic Drug Therapy*, Saunders, New York, 2007.)

and limb abnormalities (Table 3.4). There is also the increased risk of potential spontaneous abortion and fetal death. Thus, women of childbearing potential must avoid systemic acitretin and be counseled on appropriate birth control measures.

## Mucous Membranes

Dry lips, or cheilitis, is a common adverse effect of systemic retinoid use, as well as dry mouth, and nasal dryness. Xerosis of the skin, with pruritus and peeling, especially of the palms and soles, is also common. Xerophthalmia and corneal ulceration can also occur. The effect on skin and mucous membranes is dose dependent.

## Hair and Nails

Telogen effluvium (diffuse or localized hair loss) tends to be more common with acitretin versus etretinate and is less common with the use of oral bexarotene. This adverse effect is reversible and dose related. Nail fragility is common with accompanying paronychia, onycholysis, and onychorrhexis (Figure 3.14).

## Musculoskeletal

Muscle pain and cramps have been observed in patients taking etretinate or acitretin. Increased muscle tone as well as axial muscle rigidity and myopathy were reported to be related to etretinate and acitretin therapy, respectively.[33]

Bone pain has been observed in patients treated with retinoids, without visible structural abnormalities or sequelae. There have been reports of systemic retinoids being implicated in the formation of diffuse hyperostoses of the spine as well as calcification of tendons and ligaments.[34]

Osteoporosis has been observed with hypervitaminosis A and after long-term therapy with etretinate.[35]

## TABLE 3.4
## Retinoid Embryopathy

*Craniofacial*
Atresia of the ear canal, microtia, anotia
Microphthalmia
Microcephaly
Asymmetry of the face
Cleft palate
Micrognathia
Anomalies of the thymus and parathyroid glands

*Cardiovascular*
Transposition of the great vessels
Tetralogy of Fallot
Truncus arteriosus communis
Supracristal ventricular septal defect
Interrupted aortic arch
Retroesophageal right subclavian artery
Hypoplastic aorta

*Central nervous system*
Hydrocephalus
Cortical agenesis
Cerebellar hypoplasia
Mental retardation
Difficulty with visual-motor integration

*Other*
Limb anomalies
Anal and vaginal atresia

*Source:*  Bolognia, J. et al., *Dermatology*, Saunders,
           New York, 2008.

**FIGURE 3.14**  Paronychia of the nails due to systemic retinoid use. (From Wolverton, S., *Comprehensive Dermatologic Drug Therapy*, Saunders, New York, 2007.)

## Central Nervous System

The most important change to be aware of with systemic retinoids is pseudotumor cerebri, which manifests with headache, nausea, vomiting, and visual changes, along with papilledema.

## Ophthalmologic

Although uncommon, systemic retinoid use has been associated with blepharoconjunctivitis, corneal opacities, decreased night vision, and bacterial conjunctivitis, as well as dry eyes.

## Thyroid

Central hypothyroidism has been reported to occur in association with oral bexarotene in about 40% of patients.[36] This adverse effect is reversible upon cessation of the agent.

## Lipids

Systemic retinoid therapy frequently elevates lipid levels, most commonly of triglycerides, as well as an increase in total cholesterol and LDL cholesterol. Discontinuation of the retinoid is recommended if fasting triglycerides reach 800 mg/dL. This adverse effect can be decreased by dose reduction, modification of dietary and lifestyle factors (alcohol and smoking), as well as starting lipid-lowering agents such as statins and/or fenofibrates to lower the risk for not only retinoid-induced hyperlipidemia but also pancreatitis.[37]

## Hepatic

Mild transient elevation of liver function tests is most commonly associated with the use of etretinate or acitretin, and less commonly associated with the use of oral bexarotene. In one prospective, open-label, 2 year multicenter study, 128 adults with chronic stable psoriasis were treated with oral acitretin (25–75 mg/day) for four 6 month intervals, with liver biopsies being performed before and after completion of the study. In histological comparison of the liver biopsies, 49 patients (59%) demonstrated no change, 20 patients (24%) showed improvement, and 14 patients (17%) showed worsening, with the latter changes being mild and not of clinical significance. No significant correlation was found between liver function test abnormalities, cumulative acitretin dose, and changes in liver biopsy status.[38]

Monitoring of liver function tests is recommended, with discontinuation of the agent, for greater than threefold elevations in transaminases.

## Hematologic

A high occurrence of leukopenia is associated with the use of oral bexarotene, which results mainly from neutropenia, and has been found to be dose related. This adverse effect resolved upon reduction of the dose or discontinuation of the drug.

## RETINOIC ACID METABOLISM-BLOCKING AGENTS

Relatively newer agents in the treatment of psoriasis are the retinoic acid metabolism-blocking agents (RAMBAs). Examples of these agents are liarozole, talarozole, and rambazole. These agents are not currently available for clinical use, although liarozole

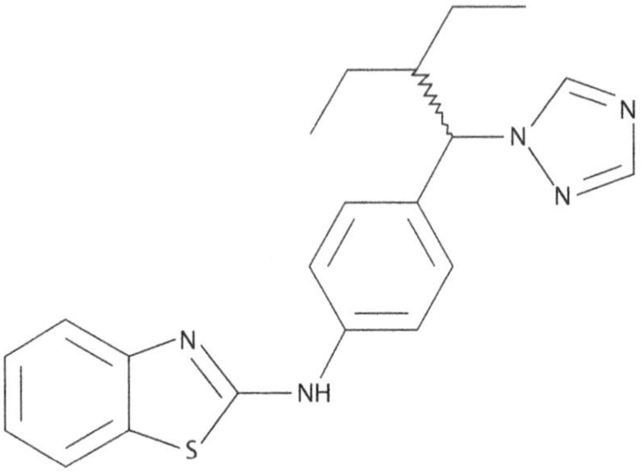

**FIGURE 3.15** Talarozole. (From Wolverton, S., *Comprehensive Dermatologic Drug Therapy*, Saunders, New York, 2007.)

has been granted orphan drug status for the treatment of congenital ichthyosis by the European Commission and the U.S. Food and Drug Administration (Figure 3.15).

RAMBAs are imidazole-containing compounds that increase endogenous levels of all-*trans*-retinoic acid levels in blood and skin by inhibiting its metabolism and breakdown in the cytochrome P450 system, specifically by blocking the CYP-dependent 4-hydroxylation of all-*trans*-RA. This inhibition increases the levels of retinoic acid in both plasma and skin. By increasing the presence of all-*trans*-RA, the hope is to increase the efficacy of retinoids locally, while avoiding excessive systemic retinoid exposure, thus minimizing dose-limiting side effects. RAMBAs, themselves, are not retinoids.

Besides serving as an adjuvant in retinoid therapy, RAMBAs have been shown to have an anti-keratinizing effect. For example, when dosed subchronically (5–20 mg/kg, once daily for 3 days) to ovariectomized rats, liarozole reversed the vaginal keratinization induced in these animals by estrogenic stimulation. Dose–response experiments indicated that the anti-keratinizing effect of liarozole was as potent as that of retinoic acid. However, in a large, double-blind, randomized clinical study, 20 patients with severe plaque psoriasis were treated with either liarozole or acitretin for 12 weeks and were found to have comparable results in decreased severity as well as in decreased inflammatory, keratinocyte proliferation and differentiation markers.[39]

In a double-blind, randomized, placebo-controlled trial, 15 patients with palmoplantar psoriasis were treated with 75 mg of oral liarozole twice daily for 12 weeks.[40] Using the Palmoplantar Psoriasis Severity Index, the study found a statistically significant improvement in the severity scores and the number of fresh pustules after treatment. Additionally, hematologic tests, liver function tests, and serum cholesterol and triglycerides were not significantly different between the liarozole and placebo groups.[41]

In an open-label study of 19 patients, in the Netherlands, 26% of the patients showed a 50% reduction in PASI (PASI 50) after 8 weeks of treatment with 1 mg daily of oral talarozole for moderate-to-severe psoriasis.[42]

In a Phase IIa open-label clinical study, six patients were treated with rambazole, 1 mg, once daily, for 8 weeks; plaque severity was reduced by 34% from week 0, as well as improvement in keratinocyte proliferation and differentiation markers.[43]

## CONCLUSION

Over the past 40 years, the field of dermatology has made enormous strides in the understanding of vitamin A and the role it plays in the cell and molecular biology of the skin. What started as an observation of the skin manifestations of vitamin A deficiency has now led us to several generations of synthetic retinoids, which have been developed in order to more effectively treat a variety of skin diseases, including psoriasis, and other disorders of keratinization.

Even with the advent of newer agents, retinoids, available both topically and systemically, have still been shown to be an effective treatment modality, either as a monotherapy, and particularly in combination with systemic biologic agents, phototherapy, or photochemotherapy (PUVA), in the treatment of psoriasis. However, its use remains limited due to adverse side effects and teratogenicity in women of childbearing potential, thus requiring frequent and often cumbersome laboratory and clinical monitoring.

With continuing developments in carotenoid and retinoid translational research, further developments in therapeutic retinoid agents for dermatological skin diseases such as psoriasis are likely.

## REFERENCES

1. Wolff K et al. *Fitzpatrick's Dermatology in General Medicine*. New York: McGraw-Hill, 2008.
2. Furue M, Yamazaki S. Prevalence of dermatological disorders in Japan: A nationwide, cross-sectional, seasonal, multicenter, hospital-based study. *J Dermatol* 2011; 38(4): 310–320.
3. Burns T, Breathnach S et al. *Rook's Textbook of Dermatology*. Oxford, U.K.: Blackwell Publishing, 2010.
4. Vahlquist A, Duvic M. *Retinoids and Carotenoids in Dermatology*. New York: Informa Healthcare, 2007.
5. Mueller W, Herrmann B. Cyclosporin A for psoriasis. *N Engl J Med* 1979; 301: 555.
6. Lee E, Trepicchio WL, Oestreicher JL, Pittman D, Wang F, Chamian F et al. Increased expression of interleukin 23 p19 and p40 in lesional skin of patients with psoriasis vulgaris. *J Exp Med* 2004; 199: 125–130.
7. Griffiths CE, Christophers E. A classification of psoriasis vulgaris according to phenotype. *Br J Dermatol* 2007; 156(2): 258–262.
8. Menter A, Gottlieb A, et al. Guidelines of care for the management of psoriasis and psoriatic arthritis: Section 1. Overview of psoriasis and guidelines of care for the treatment of psoriasis with biologics. *J Am Acad Dermatol* 2008; 58(5): 826–850.
9. Wolbach S, Howe P. Tissue changes following deprivation of fat soluble A vitamin. *J Exp Med* 1925; 42: 753–777.
10. Nicholls L. Phrynoderma: A condition due to vitamin deficiency. *Indian Med Gazette* 1933; 68: 681–687.
11. Maronn M, Allen DM, Esterly NB. Phrynoderma: A manifestation of vitamin A deficiency? The rest of the story. *Pediatr Dermatol* 2005; 22(1): 60–63.
12. Vahlquist A, Duvic M. *Retinoids and Carotenoids in Dermatology*. Informa Healthcare, New York, 2007.
13. O'Reilly K, Bailey SJ, Lane MA. Retinoid-mediated regulation of mood: Possible cellular mechanisms. *Exp Biol Med* 2008; 233(3): 251–258.

14. Monfrecola G, Baldo A. Retinoids and phototherapy for psoriasis. *J Rheumatol Suppl* 2009; 83: 71–72.
15. Vahlquist. *Carotenoids and Retinoids in Dermatology.* Informa Healthcare, New York, 2007, p. 58.
16. Marks R. The role of tazarotene in the treatment of psoriasis. Br J Dermatol. 1999 Apr: 140 Suppl 54: 24–8.
17. Duvic M, Nagpal S, Asano AT, Chandraratna RA. Molecular mechanisms of tazarotene action in psoriasis. *J Am Acad Dermatol* 1997; 37(2 Pt 3): S18–S24.
18. Green L, Sadoff W. A Clinical evaluation of tazarotene 0.1% gel, with and without a high- or mid-high-potency corticosteroid, in patients with stable plaque psoriasis. J Cutan Med Surg. 2002 Mar-Apr; 6(2): 95–102.
19. Lebwohl MG, Breneman DL, Goffe BS et al. Tazarotene 0.1% gel plus corticosteroid cream in the treatment of plaques psoriasis. *JAAD* 1998; 39(Pt 1): 590–596.
20. Koo JY, Lowe NJ et al. Tazarotene plus UVB phototherapy in the treatment of psoriasis. *JAAD* 2000; 43(5 Pt 1): 821–828.
21. Bolognia J et al. *Dermatology.* New York: Saunders, 2008.
22. Katz H, Waalen J, Leach E. Acitretin in psoriasis: An overview of adverse effects. J Am Acad Dermatol 1999; 41(3 Pt 2): S7–S12.
23. Larsen FG, Jakobsen P, Knudsen J, Weismann K, Kragballe K, Nielsen-Kudsk F. Conversion of acitretin to etretinate in psoriatic patients is influenced by ethanol. *J Invest Dermatol* 1993; 100(5): 623–627.
24. van de Kerkhof PC, Cambazard F et al. The effect of addition of calcipotriol ointment (50 micrograms/g) to acitretin therapy in psoriasis. *Br J Dermatol.* 1998; 138(1): 84–89.
25. Saurat JH, Geiger JM. Randomized double-blind multicenter study comparing acitretin-PUVA, etretinate-PUVA and placebo-PUVA in the treatment of severe psoriasis. *Dermatologica* 1988; 177(4): 218–224.
26. Young HS, Summers AM, Read IR, Fairhurst DA, Plant DJ, Campalani E, Smith CH, Barker JN, Detmar MJ, Brenchley PE, Griffiths CE. Interaction between genetic control of vascular endothelial growth factor production and retinoid responsiveness in psoriasis. *J Invest Dermatol* 2006; 126(2): 453–459.
27. Scheinfeld N. Schools of pharmacology: Retinoid update. *J Drugs Dermatol* 2006; 5(9): 921–922.
28. Meyer V, Goerge T, Luger TA, Beissert S. Successful treatment of palmoplantar hyperkeratotic psoriasis with a combination of etanercept and alitretinoin. *J Clin Aesthet Dermatol* 2011; 4(4): 45–46.
29. Vahlquist A, Duvic M. *Retinoids and Carotenoids in Dermatology.* New York: Informa Healthcare, 2007.
30. Breneman D, Sheth P, Berger V, Naini V, Stevens V. Phase II clinical trial of bexarotene gel 1% in psoriasis. *J Drugs Dermatol* 2007; 6(5): 501–506.
31. Smit JV, Franssen ME, de Jong EM, Lambert J, Roseeuw DI, De Weert J, Yocum RC, Stevens VJ, van De Kerkhof PC. A phase II multicenter clinical trial of systemic bexarotene in psoriasis. *J Am Acad Dermatol* 2004; 51(2): 249–256.
32. Menter A. Pharmacokinetics and safety of tazarotene. *J Am Acad Dermatol* 2000; 43: S31–S35.
33. Lister RK et al. Acitretin-induced myopathy. *Br J Dermatol* 1996; 134: 989.
34. Carey BM et al. Skeletal toxicity with isotretinoin therapy: A clinico-radiological evaluation. *Br J Dermatol* 1988; 119: 609.
35. DiGiovanna JJ, Sollitto RB, Abangan DL et al. Osteoporosis is a toxic effect of long-term etretinate therapy. *Arch Dermatol* 1995; 131: 1263–1267.
36. Sherman SI, Gopal J, Haugen BR et al. Central hypothyroidism associated with retinoid X receptor selective ligands. *N Engl J Med* 1999; 340: 1075–1079.

37. Talpur R, Ward S et.al. Optimizing bexarotene therapy for cutaneous T-cell lymphoma. *J Am Acad Dermatol* 2002; 47: 672–684.

38. Roenigk HH Jr, Callen JP et al. Effects of acitretin on the liver. *J Am Acad Dermatol* 1999; 41(4): 584–588.

39. Kuijpers AL, Van Pelt JP. The effects of oral liarozole on epidermal proliferation and differentiation in severe plaque psoriasis are comparable with those of acitretin. *Br J Dermatol* 1998; 139(3): 380–389.

40. Bhushan M, Burden AD, McElhone K, James R, Vanhoutte FP, Griffiths CE. Oral liarozole in the treatment of palmoplantar pustular psoriasis: A randomized, double-blind, placebo-controlled study. *Br J Dermatol* 2001; 145(4): 546–553.

41. Bhushan M, Burden AD. Oral liarozole in the treatment of palmoplantar pustular psoriasis: A randomized, double-blind, placebo-controlled study. *Br J Dermatol* 2001; 145(4): 546–553.

42. Verfaille CJ, Thissen CACB, Bovenschen HJ, Mertens J, Steijlen PM, van de Kerkhof PCM. Oral R115866 in the treatment of moderate to severe plaque type psoriasis. *J Eur Acad Dermatol Venereol* 2007; 21(8): 1038–1046.

43. Bovenschen HJ, Otero ME, Langewouters AM, van Vlijmen-Willems IM, van Rens DW, Seyger MM, van de Kerkhof PC. Oral retinoic acid metabolism blocking agent Rambazole for plaque psoriasis: An immunohistochemical study. *Br J Dermatol* 2007; 156(2): 263S–270S.

# 4 Age-Related Macular Degeneration and Carotenoids

*Nilesh Kalariya, Werner G. Siems,*
*Satish K. Srivastava, Kota V. Ramana,*
*and Frederik J.G.M. van Kuijk*

## CONTENTS

## AGE-RELATED MACULAR DEGENERATION

Age-related macular degeneration (AMD) is a degenerative condition of the human macula characterized by the dysfunction and disruption of photoreceptors secondary to an atrophic and/or a neovascular event. AMD is the leading cause of irreversible vision loss in people older than 50 years of age in the United States and European Union and currently affects more than 4 million Americans (Friedman et al. 2004; Klein et al. 2004). It is thought that this number will increase to 8 million people by the year 2020. Several epidemiological studies have shown an increase in the prevalence of AMD with increasing age (Hawkins et al. 1999; Smith et al. 2001). Due to the increasing longevity and demographic shift toward an elderly population, AMD prevalence is predicted to rise significantly with subsequent increase in socioeconomic problems too. The existing therapeutic interventions are not available to all of the AMD sufferers. Hence, the best approaches to minimize the impact of this vision disorder on quality of life are delay, prevention, modification, or arrest of degenerative progression.

There are numerous risk factors that include, but are not limited to, genetics, ethnicity, and family history (Hyman and Neborsky 2002; Klein and Francis 2003), iris color (Nicolas et al. 2003), drusen (Pauleikhoff et al. 1990), dietary lipids and cholesterol (Seddon et al. 2003), risk factors for atherosclerosis (Wachter et al. 2004),

C-reactive protein (Seddon et al. 2004), smoking (Christen et al. 1996), and sunlight (Cruickshanks 2001) for this debilitating ocular disease. Other studies have shown an increase in exudative AMD following cataract surgery suggesting that cumulative light exposure may play a role in the development of the disease (de Jong and Lubsen 2004). Other risk factors include macular pigment (Snodderly 1995; Hammond et al. 1997; Landrum et al. 1997; Beatty et al. 2000) and low dietary intake of antioxidants (Tsang et al. 1992; Mares-Perlman et al. 1995). It is thought that cigarette smoke exposes the individual to a variety of pro-oxidants, which cause a decrease in the amount of macular pigment in the retina (Hammond et al. 1996) and the levels of antioxidants in the blood, with an increase in lipid peroxidation products (Handelman et al. 1996). Therefore, smoking, macular pigment, and pro-oxidants and their dietary intake may be associated with a risk for AMD through a common mechanism of oxidative stress.

Despite such high prevalence of AMD and its enormous cost to society, the pathobiochemical mechanisms responsible for AMD remain poorly understood. Research so far has implicated oxidative stress in accelerating aging and degenerative diseases, including AMD. Retinal pigment epithelial (RPE) cells play a crucial role in the development of AMD, which is characterized by the accumulation of drusen between the RPE cells and Bruch's membrane (Mullins et al. 2000), and these changes have been well characterized in clinicopathological studies (Spraul and Grossniklaus 1997). Drusen contains saturated lipids, cholesterol, and various protein components (Mullins et al. 2000; Curcio et al. 2001; Haimovici et al. 2001; Crabb et al. 2002). One proposed cause of AMD is RPE dysfunction, resulting in the formation of abnormal lipid molecules, which are discharged into the inner side of Bruch's membrane as drusen and basal laminar deposits (Anderson et al. 2001). The cause of altered metabolism in aged RPE cells is unknown, but disruption of these processes has been shown to result in degeneration of the sensory retina in experimental animal models, and there is evidence that such defects in the RPE contribute to AMD in humans (Finnemann et al. 2002). Oxidative stress has been shown to play an important role in the development of RPE cellular dysfunction (Finnemann et al. 2002; Liang and Godley 2003). RPE cells are susceptible to oxidative damage due to the high partial oxygen pressure from the underlying choriocapillaris, intense light exposure, and high concentration of polyunsaturated fatty acids (PUFA) in the photoreceptor outer segments. Various reactive oxygen intermediates may be formed, which include hydrogen peroxide, hydroxyl radicals, and superoxide—all capable of contributing to RPE dysfunction (Schütt et al. 2000). For protection against this oxidative stress, RPE cells have high concentrations of antioxidant-enzymes, which decrease with increasing age in humans (De La Paz et al. 1996). When the oxidative stress overwhelms the antioxidant defense of the RPE, cell dysfunction may occur, which may kill the cells (Schütt et al. 2000). Anderson et al. (2004) showed that age-related products such as beta amyloids accumulate in the cytoplasm of RPE associated with neovascular AMD, thereby supporting the role of oxidative stress in AMD.

The levels of certain antioxidants and antioxidant enzymes are higher in the RPE than in other ocular tissues. The age-related decline of antioxidant systems in RPE cells may further increase the susceptibility of aged RPE to oxidative injury (De La Paz et al. 1996). The level of antioxidants also affects retinal degeneration in animal models (Boulton et al. 2001). Case–control studies have demonstrated a reduced risk for AMD in human subjects with high dietary intake and high serum levels of antioxidant nutrients

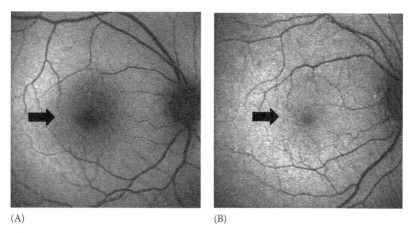

(A)                                                    (B)

**FIGURE 4.1**    Autofluorescence images of retinal macular region are shown in subjects with high (A) and low (B) macular pigment.

(Seddon et al. 1994). This further supports the importance of oxidative defense in AMD. Oxidative stress induced liquid peroxidation is a major cause of cytotoxicity in ocular tissues. Lipid peroxidation has been implicated in aging and degenerative diseases such as atherosclerosis, Alzheimer's, arthritis, cancer, cataract, and AMD (Knight 1995).

Several defense mechanisms exist in human retinal tissues to remove lipid hydroperoxides or prevent their formation. Very crucial molecules in human retina to fight against oxidative stress are carotenoids/xanthophyll particularly lutein and zeaxanthin (Figure 4.1). Carotenoids may quench singlet oxygen, which could neutralize it before it reacts with the double bonds of PUFA (Krinsky 1979; Kirschfeld 1982; Burton and Ingold 1984) and scavenges of lipid peroxyl free radicals to break the lipid peroxidation chain reaction. Furthermore, carotenoids quench oxidized vitamin E radicals, which restore the antioxidant capacity. Zeaxanthin stereoisomers are present at high levels in the macula and are the most effective carotenoid quenchers of vitamin E radicals (Schalch et al. 1999).

## MACULAR CAROTENOIDS

The first evidence of macular carotenoid existence and its identification as xanthophylls was provided by Wald (1945). A preliminary identification by Bone et al. (1985) showed that lutein and zeaxanthin predominantly constitute the human macular pigment. Subsequently, this result was confirmed by Bone et al. and Handelman et al. in 1988. Bone et al. (1997) also described the occurrence of stereoisomers, such as meso-zeaxanthin in the human retina.

Carotenoids found in the human body are characterized by $C_{40}H_{56}$-isoprenoid backbone and are subdivided into carotenes and their oxygenated derivatives, xanthophylls (Figure 4.2). Lutein and zeaxanthin are xanthophylls and are distinguished from other carotenoids by the presence of two hydroxyl groups, one on either side of the molecule. As shown in Figure 4.2, although lutein and zeaxanthin are constitutional isomers, four subtle differences exist in their biochemical structures.

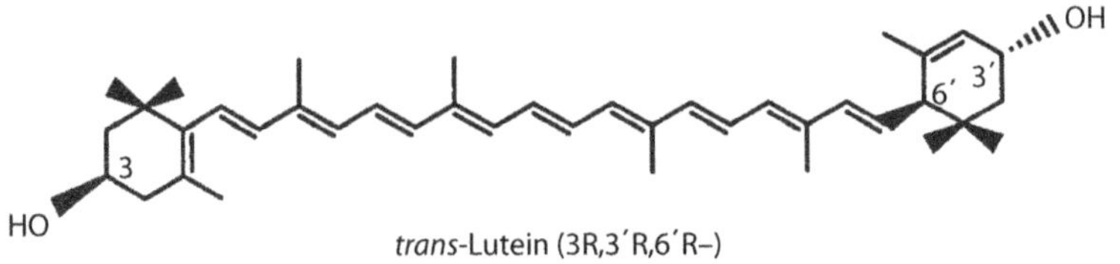


*trans*-Lutein (3R,3′R,6′R-)

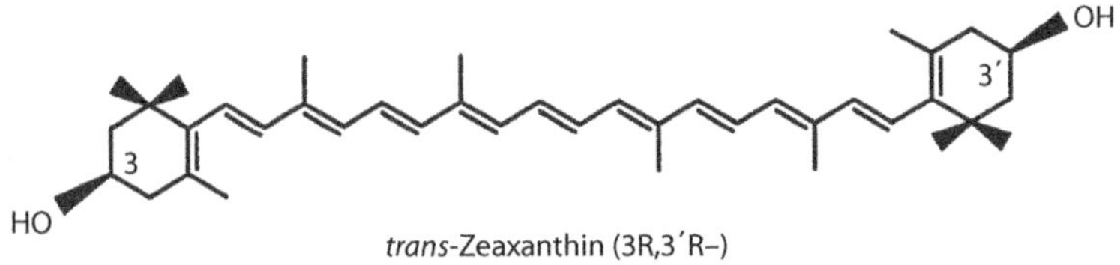


*trans*-Zeaxanthin (3R,3′R-)

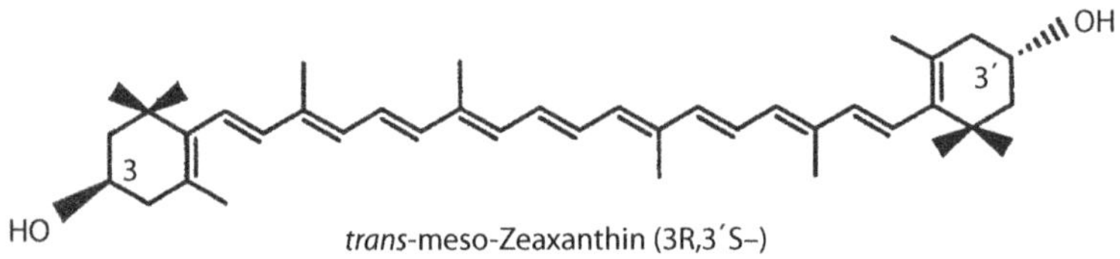


*trans*-meso-Zeaxanthin (3R,3′S-)

**FIGURE 4.2** Structures of lutein, zeaxanthin, and meso-zeaxanthin. (Adapted from Krinsky, N., *J. Nutr.*, 132, 540S, 2002. With permission.)

(1) There is a difference in the position of the double bond present within the six-carbon (ionone) ring located on the right side of the carbon chain. (2) Lutein has two different types of ionone rings, i.e., β-type and ε-type, whereas zeaxanthin has one type of ionone ring, i.e., β-type. (3) The 3′-hydroxyl group of the ε-ionone ring in lutein is folded back from the horizontal plane, while the corresponding group in zeaxanthin projects forward from the horizontal plane. (4) Lutein has the presence of three stereocenters while zeaxanthin has two (Bone et al. 1993). In human retinal macula, lutein exists as a single stereoisomer (3R, 3′R, 6′R-β, ε-carotene-3, 3′-diol), whereas zeaxanthin exists as two different isomers (3R, 3′R-β, β-carotene-3, 3′-diol: zeaxanthin and 3R, 3′S-β, β-carotene-3, 3′-diol: meso-zeaxanthin).

It is reported that estimated concentration of macular carotenoids rises to almost 1 mM within the central macula (Landrum and Bone 2001). Lutein, zeaxanthin, and meso-zeaxanthin represent about 36%, 18%, and 18% of total retinal carotenoids, respectively. These carotenoids are mainly localized in the plexiform layers of the retina (Snodderly et al. 1984) and in the outer segments of the photoreceptors (Sommerburg et al. 1999; Rapp et al. 2000). The concentration of these carotenoids is highest in the central fovea, but their absolute quantity is greater in the nonmacular retina.

Since human body cannot synthesize macular carotenoids, the sole source of these compounds is diet, particularly fruits and vegetables. Lutein and zeaxanthin are commonly found in food sources with yellow or green color, such as maize, egg yolk, orange, spinach, kale, etc. (Sommerburg et al. 1998; Semba and Dagnelie 2003). Macular carotenoids exist as protein complexes called carotenoproteins in natural food sources. Heating food denatures the carotenoproteins and facilitates the absorption of lutein and zeaxanthin, thus improving their bioavailability (Alves-Rodrigues and Shao 2004). Chylomicrons containing lutein and zeaxanthin that are formed in stomach facilitate absorption in the duodenum. Once absorbed, lutein and zeaxanthin reach the liver via the portal circulation to be packaged as plasma lipoproteins for subsequent release into the systemic circulation. It has been suggested that they are carried predominantly by the Apo E containing high-density lipoprotein in the systemic circulation. This fact is further strengthened by the existence of Muller cells in the retina, which are capable of synthesizing Apo E lipoproteins (Shanmugaratnam et al. 1997).

The capture and stabilization of lutein and zeaxanthin in the retinal macular region has remained elusive until recently. The abundance of cytosolic tubulin in the axonal layer of the retina may account for the heavy concentration of macular pigment within the Henle's layer. However, the uptake of the membrane fraction is thought to be actively mediated by highly specific xanthophyll-binding proteins (Yemelyanov et al. 2001). Over the past decade, Bernstein laboratory has identified and characterized several carotenoid-binding proteins from human retina including a pi isoform of glutathione S-transferase (GSTP1) as a zeaxanthin-binding protein, a member of the steroidogenic acute regulatory domain (StARD) (MLN64) family as a lutein-binding protein, and tubulin as a less specific, but higher-capacity site for carotenoid deposition (Bhosale et al. 2004, 2009; Li et al. 2011).

## FUNCTIONS OF MACULAR CAROTENOIDS

Blue light filtration and antioxidant capacity are two major functions of the macular carotenoids that provide protection to human retina against age-related degenerative processes. The first evidence of light-induced retinal damage was reported by Ham et al. (1979). The magnitude of light-induced retinal damage is a function of wavelength, and with decreasing wavelength, there is an exponential rise in the retinal injury. The generation of reactive oxygen species (ROS) in human retina is attributed to photosensitization by short-wavelength blue light (Ruffolo et al. 1984). Lutein and zeaxanthin provide protection against ROS production in photoreceptors by absorbing blue light. Thus, the blue light–filtering property of the macular carotenoids can be considered as an antioxidant function too. The blue light–filtering property of the macular carotenoids is based on three aspects. (1) The absorbance spectrum of macular carotenoids peaks at 460 nm (Figure 4.3), which corresponds to the wavelength of blue light. (2) Due to highest concentration of macular carotenoids in the prereceptorial axon layers of the retina, blue light is absorbed before it reaches the photoreceptors. (3) The macular carotenoids are distributed throughout the photoreceptor cell, and hence each receptor cell is able to filter blue light (Snodderly et al. 1984). Lutein is considered to have greater filtering efficacy than zeaxanthin due to the difference in the orientation of the respective molecules (Krinsky et al. 2003). However, zeaxanthin is more efficient in preventing retinal damage upon prolonged UV-exposure. Recently, lutein and zeaxanthin were shown to be more efficient blue light filters than other carotenoids (Junghans et al. 2001). Why lutein and zeaxanthin were selected from more than 600 carotenoids to be the macular pigment remains to be elucidated. Snodderly et al. (1991) and Siems et al. (1999) have suggested that it is related to their resistance to oxidative breakdown.

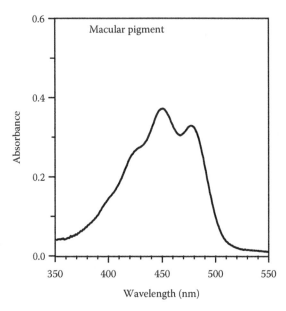

**FIGURE 4.3**   Absorption spectrum of macular pigment.

The first evidence of antioxidative property of macular carotenoids was reported by Khachik et al. (1997) who demonstrated the existence of oxidative products of macular carotenoids in the retina. Two underlying phenomena (physical and chemical) define the antioxidant property of the macular carotenoids. Under physical phenomenon, the carotenoid molecules quench singlet oxygen primarily by accepting the excitation energy from singlet oxygen (Semba and Dagnelie 2003). The quenching ability of a macular carotenoid is attributed to the number of conjugated double bonds (Semba and Dagnelie 2003). This accepted energy from singlet oxygen is dissipated by carotenoid through rotational and vibrational interactions and relaxes into its ground state (Semba and Dagnelie 2003). Under this physical mechanism, the structure of the carotenoid molecule does not alter. Under chemical phenomenon, the carotenoid molecule either donates an electron to the free radical or attracts free radicals to pair a single electron through covalent bond. This ability protects the cell molecules such as lipids, proteins, and its DNA from oxidative damage (Eperjesi and Beatty 2005). However, in doing so, the structure of carotenoid tends to alter.

## MACULAR CAROTENOIDS AND AMD

In order to understand the biologic role for dietary lutein and zeaxanthin in the eye, researchers have attempted to correlate dietary carotenoid intake with reduced risk for AMD. There are quite a few pioneering studies that helped to provide a basis for the hypothesis that dietary carotenoids may have a protective effect against AMD. The National Health and Nutritional Examination Survey (NHANES) showed an inverse relationship between diets high in fruits and vegetables and risk for AMD (Goldberg et al. 1988). In 1998, Sommerburg et al. suggested that fruits and vegetables are high in various types of carotenoids including lutein and zeaxanthin, which may have a protective role in AMD reduction. The results of the Eye Disease Case–Control Study (1993) revealed a significant inverse relationship between total serum carotenoids and AMD risk. Subsequent studies also found a similar significant relationship (Mares-Perlman et al. 2001; Snellen et al. 2002; Gale et al. 2003). The National Health and Nutritional Examination Survey III (NHANES III) found a significant association between the consumption of lutein and zeaxanthin and the risk of advanced AMD in the youngest age group at risk for this condition (Mares-Perlman et al. 2001). The Carotenoids in Age-Related Macular Degeneration Study revealed that lutein- and zeaxanthin-rich diets may protect against intermediate AMD in female patients (<75 years of age) (Moeller et al. 2006). The Blue Mountain Eye Study also reported reduced risk of AMD upon higher dietary lutein and zeaxanthin intake over 5 and 10 years (Tan et al. 2008). It is noteworthy to mention that a recently published meta-analysis study revealed that dietary lutein and zeaxanthin are not significantly associated with a reduced risk of early AMD, whereas an increase in the intake of these carotenoids may be protective against late AMD (Ma et al. 2012). Likewise, an association between early age-related maculopathy and low intake of lutein in the diet has also been demonstrated (Snellen et al. 2002).

Moreover, the relationship between lutein intake, serum lutein, and macular pigment optical density (MPOD) was also studied in a group of 278 healthy volunteers (Curran-Celentano et al. 2001). The result of this study showed that both serum lutein and dietary lutein significantly correlate with MPOD in a positive manner—the higher the level in serum or in the diet, the higher the MPOD. These results are consistent with another study by Bernstein et al. (2002), who correlated the use of lutein supplements with MPOD in AMD patients. They found a significantly higher MPOD in AMD patients consuming lutein supplement (≥4 mg/day) relative to those not supplementing. These studies provided an ample amount of evidence that dietary sources of lutein are associated with reduced AMD risk and increased MPOD. These observations were strengthened by evidence on relationship between retinal lutein concentration and AMD risk. In 2001, Bone et al. measured the actual concentration of lutein and zeaxanthin from control and AMD patients' donor eyes. The study analysis revealed 82% lower risk of having AMD with higher concentration of lutein relative to those with the lowest concentration (Bone et al. 2001). This study was the first of its kind to specifically examine the relationship between lutein and zeaxanthin concentrations in the macular region and AMD risk in humans. Thus, there are substantial number of studies that demonstrated beneficial relationship between xanthophylls and low risk of AMD.

On the other hand, there are quite a few reports that suggest that xanthophyll supplementation as monotherapy or in combination with some other nutritional supplements either has no effects or incurred an adverse effect, i.e., triggered progression of AMD or other systemic effects. There are also several epidemiological studies that did not find an association between dietary or serum concentrations of lutein/zeaxanthin and AMD risk (Vanden Langenberg et al. 1998; Dasch et al. 2005; van Leeuwen et al. 2005; Cho et al. 2008). Two cohort studies with 1968 and 2335 participants, respectively, could not demonstrate any significant relationship between dietary intake of lutein and zeaxanthin and the risk of AMD (Mares-Perlman et al. 1996; Flood et al. 2002). Besides, there are quite a few studies that did not find any relationship between serum levels of lutein and/or zeaxanthin and AMD (Saunders et al. 1993; Mares-Perlman et al. 1995; Simonelli et al. 2002; Cardinault et al. 2005). It was also reported that increased intake of lutein, zeaxanthin, and omega-3 fatty acids was associated with the progression of AMD (Vu et al. 2006; Robman et al. 2007). It has been argued that the lack of beneficial association observed in these studies could be due to various factors. For example, fewer subjects in the study have been argued to demonstrate lack of beneficial effects due to less statistical power to detect small differences between cases and controls (Saunders et al. 1993; Mares-Perlman et al. 1995; Simonelli et al. 2002; Cardinault et al. 2005). Moreover, many studies assessed the dietary intake of lutein and zeaxanthin with the help of a food frequency questionnaire that does not take into account the effect of cooking on nutrient content, food interactions affecting bioavailability, and the influence of memory and social desirability, which may affect the precise assessment of dietary intake of lutein and zeaxanthin. Recently, Cho et al. (2008) also reported that antioxidant effect of lutein/ zeaxanthin may be overwhelmed by smoking-related oxidative stress in the retina of AMD patients, and therefore, a preventive role could not be established.

Thus, due to the lack of conclusive results of several clinical studies, recommendation for routine carotenoids and other nutritional supplementation for primary prevention of AMD is questionable. This suspicion is based on the fact that therapeutic use of carotenoids in the past has met with discouraging results and consequences. The VITamins And Lifestyle (VITAL) cohort study examined association of supplemental beta-carotene, retinol, vitamin A, lutein, and lycopene with lung cancer risk among participants, aged 50–76 years in Washington State (Satia et al. 2009). This study revealed that longer duration of use of individual beta-carotene, retinol, and lutein supplements was associated with statistically significantly elevated risk of total lung cancer and histologic cell types. The study concluded that long-term use of individual beta-carotene, retinol, and lutein supplements should not be recommended for lung cancer prevention, particularly among smokers. The results of ATBC and CARET studies revealed increased mortality upon carotenoid supplementation among lung cancer patients who were smokers (Albanes et al. 1996; Omenn et al. 1996). It is noteworthy that these studies revealed a common phenomenon, which is the presence of smoking-associated systemic oxidative stress. This is important as these studies found detrimental effects of carotenoid supplementation on individuals who were smokers. These results imply that carotenoids might be detrimental for tissues or organs experiencing overwhelming oxidative stress. It is imperative to note that smoking has been considered as one of the major factors for AMD pathogenesis. Moreover, it has been proposed that the detrimental effects of carotenoids could be incurred by carotenoid oxidative products (Yeh and Wu 2006). Hence, it is essential to understand the possible toxic effects of carotenoids and/or their products while advocating for xanthophyll supplementation for reducing AMD risks.

## CAROTENOID CLEAVAGE PRODUCTS

Due to variable epidemiological results showing suppressive, neutral, and sometimes progressive effects of xanthophylls as supplementary carotenoids for AMD, it is necessary to study the structure and function of xanthophyll metabolites such as carotenoid cleavage products, which have been reported to incur toxicity. Carotenoids, besides acting as antioxidants, are also reported to act as pro-oxidants under high oxygen tension, high carotenoid concentration, and imbalanced intracellular redox status (Palozza et al. 2003). Under oxidative stress, while scavenging ROS, carotenoids undergo oxidation and generate a variety of oxidized products in vitro as well as in vivo (Handelman et al. 1991; Hurst et al. 2004; Prasain et al. 2005). The presence of carotenoid cleavage products (CCPs) has been demonstrated in human and monkey ocular tissues including the retina (Khachik et al. 1997; Bernstein et al. 2006; Bhosale and Bernstein 2005; Prasain et al. 2005). Recent studies have also shown that carotenoid supplementation in humans and monkeys leads to significant increase in the levels of their metabolites in serum and ocular tissues (Khachik et al. 2006a,b). Various carotenoid metabolites have been identified in human eye tissues, particularly in the retina (Table 4.1). Besides, there are many other cleavage products believed to exist in ocular tissues but could not be identified due to either lack of sensitive technique or undetectable concentration with the existing methods.

The physiological role of such CCP is poorly understood. However, several cellular as well as animal studies have demonstrated cytotoxic and/or genotoxic effects of carotenoid metabolites (Salgo et al. 1999; Nara et al. 2001; Siems et al. 2002; Liu et al. 2004; Hurst et al. 2005; Alija et al. 2006; Kalariya et al. 2008). Recently, van Helden et al. (2009) demonstrated that β-carotene metabolites can enhance inflammation-induced oxidative DNA damage in lung epithelial cells. Genotoxic effects of carotenoid cleavage products (CCPs) in human RPE cells have also been demonstrated (Kalariya et al. 2009). Autoxidized carotenoids were reported to oxidize DNA in fibroblasts (Yeh and Hu 2001) and promote apoptosis in murine tumor cells (Mo and Elson 1999)

**TABLE 4.1**
**Some Identified Xanthophyll Metabolites from Human Eye Tissues**

| No. | Xanthophyll Metabolites | References |
| --- | --- | --- |
| 1. | (3R,3′S,6′R)-Lutein (3′-epilutein) | Khachik et al. (1997); Bernstein et al. (2001) |
| 2. | 3-Hydroxy-β,ε-caroten-3′-one | Khachik et al. (1997); Bernstein et al. (2001) |
| 3. | (cis)-3-Hydroxy-β,ε-caroten-3′-one | Khachik et al. (1997) |
| 4. | 3-Hydroxy-β-ionone | Prasain et al. (2005) |
| 5. | 3-Hydroxy-14′-apocarotenal | Prasain et al. (2005) |
| 6. | ε-Carotene-3,3′-diol | Khachik et al. (1997) |
| 7. | ε,ε-Carotene-3,3′-dione | Khachik et al. (1997) |
| 8. | ε,ε-Carotene-3,3′-diol (lactucaxanthin) | Khachik et al. (1997) |
| 9. | 3′-Hydroxy-ε,ε-caroten-3-one | Khachik et al. (1997) |
| 10. | 2,6-Cyclolycopene-1,5-diol | Khachik et al. (1997) |
| 11. | 3-Methoxyzeaxanthin | Bhosale et al. (2007) |

and in human HL60 cells (Nara et al. 2001). It was also suggested that the proapoptotic activities of β-carotene at high concentration were mediated by oxidized species (Palozza et al. 2001). Furthermore, it was shown that β-ionone (a degradation product of β-carotene) more potently induces drug-metabolizing cytochrome P450 isoforms than its α-analogue (Jeong et al. 2002). Sommerburg et al. (2003) showed that hypochlorous acid produced in neutrophils can promote carotenoid oxidation, thus providing a pathophysiological pathway for this mechanism. Thus, CCP may potentiate the toxicity of xenobiotics. The CCPs have also been reported to inhibit an enzyme (Siems et al. 2000). For example, $Na^+$-$K^+$-ATPase regulates the intracellular levels of $Na^+$ and $K^+$ and is inactivated by aldehydes that react with its lysine and/or cysteine residues. The CCP derived from autoxidation of β-carotene inhibited $Na^+$-$K^+$-ATPase activity more potently than 4-hydroxy-nonenal (4-HNE) (Siems et al. 2000). Thus, inhibition of ATPase activities by CCP could be important in cellular pathology, such as those of the RPE cells.

The photoreceptor outer segments containing lutein and zeaxanthin are digested by the RPE cells (LaVail 1983). It is possible that intracellular free radicals could oxidize both lipids and carotenoids, which may lead to the formation of CCP, in addition to lipid-derived aldehydes (LDAs) in the RPE cells. The toxic effects of free radicals may therefore be amplified and sustained by the formation of these reactive aldehydes and other products produced from carotenoid oxidation. Lutein cleavage products have been shown to cause DNA damage and apoptosis in RPE cells under in vitro conditions (Kalariya et al. 2008, 2009). The results of ATBC and CARET studies revealed increased mortality upon carotenoid supplementation among lung cancer patients who were smokers, and carotenoid metabolites could have been responsible for mediating such fatal outcomes (Albanes et al. 1996; Omenn et al. 1996). Recently, it is also suggested that antioxidant effect of lutein/zeaxanthin may be overwhelmed by smoking-related oxidative stress in the retina of AMD patients (Cho et al. 2008). These results indicate that high-dose carotenoid supplementation, especially under a high degree of oxidative stress in the tissues, caused by other factors such as smoking and aging, could possibly aggravate the pathological disease condition by overwhelming the antioxidants properties of the carotenoids rather than preventing the progression of the disease process.

Recently, Sommerburg et al. (2009) demonstrated that CCPs are able to modify proteins, and the 20S proteasome can recognize and degrade CCP-modified proteins preferentially. Moreover, they also demonstrated that supra-physiological level of CCP could lead to the formation of protein–CCP adducts that are able to inhibit the proteasome, which results into protein aggregates and eventual cellular toxicity-induced pathogenesis. It is essential to note that the chemical reactivity of carotenoid products with biomolecules is based on the conjugated polyene chain system containing alternate double and single bonds. It is a highly reactive and electron-rich system. The most striking characteristic of this conjugated system is π-electrons that efficiently delocalize over the entire length of the conjugated polyene chain (Britton 1995). However, upon oxidation, the resultant carotenoid products have polyene chain with varied length and end group such as aldehydes, ketones, etc. Various lutein-derived products, with varied polyene chain and end groups, have been identified in the human retina

(Khachik et al. 1997; Bernstein et al. 2001; Bhosale and Bernstein 2005; Prasain et al. 2005; Bhosale et al. 2007). Such structural conformation renders carotenoid products as highly reactive particularly with protein and DNA.

It has also been proposed that due to aldehydic nature, carotenoid-derived aldehydes might act like LDA. Aldehydes such as 4-HNE and others derived from PUFA are known to conjugate with sulfhydryl, lysyl, and histidine residues and eventually lead to structural modification of proteins (Uchida and Stadtman 1992a,b; Szweda et al. 1993; Guéraud et al. 2010). Since carotenoids also form highly reactive aldehydes, it is likely that excessive use of carotenoids could result in increased formation of CCP that could cause posttranslational protein modification. Recently, lutein and β-carotene oxidation products have been shown to conjugate with proteins via sulfhydryl and lysine residue in an in vitro model study (Figure 4.4) (Kalariya et al. 2011). Interestingly, a previous study (Sparrow et al. 2003) also demonstrated that A2E (a component of human retinal lipofuscin with structural analogy to carotenoids) oxidized products can damage RPE cells. Moreover, Wang et al. (2006) demonstrated that oxidation of A2E can generate aldehydes, ketones, and epoxides that could be highly reactive. Aldehydes, particularly from lipids, are highly reactive and toxic and play an important role in various pathological conditions (Uchida 2003). Unlike CCP, extensive research studies on lipid aldehydes have shown mechanism of toxicity (Guéraud et al. 2010). Lipid aldehydes such as 4-HNE are reported to react

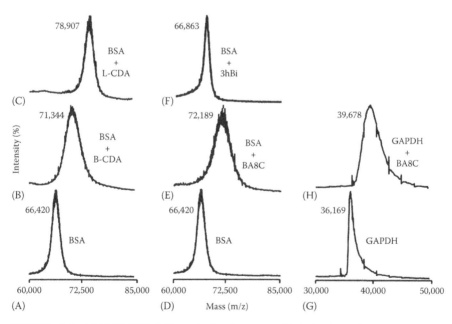

**FIGURE 4.4** MALDI-TOF MS analysis of carotenoid-derived aldehyde (CDA) conjugated protein. (A and D) Bovine serum albumin (BSA); (B) β-carotene-derived aldehyde (B-CDA) conjugated BSA; (C) lutein-derived aldehyde (L-CDA) conjugated BSA; (E) β-apo-8-carotenal (BA8C) conjugated BSA; (F) 3-hydroxy-β-ionone (3hBi) conjugated BSA; (G) glyceraldehyde-3-phosphate dehydrogenase (GAPDH); (H) BA8C-conjugated GAPDH. (Adapted from Kalariya, N.M. et al., *Biofactors*, 37, 104, 2011. With permission.)

with cysteine, lysine, and histidine moiety of protein (Uchida and Stadtman 1992a,b; Szweda et al. 1993; Guéraud et al. 2010). HNE and other $\alpha$, $\beta$-unsaturated aldehydes react rapidly with the sulfhydryl moiety in cysteine and glutathione (GSH). It was demonstrated that the intensity of this reactivity is based on the presence of a hydroxyl group in these aldehydes (Esterbauer et al. 1976). The electron-withdrawing 4-hydroxy group has been postulated to make C-3 of the 4-hydroxyalkenals more electropositive and as a result more susceptible to nucleophilic attack. In case of lutein as well as its oxidized products, a hydroxyl group is present on the cyclic ring. The hydroxyl group in lutein-derived product (e.g., 3-hydroxy-$\beta$-ionone, 3hBi) might withdraw an electron that could make its polyene chain more electropositive and susceptible to nucleophilic attack. Interestingly, 3hBi has been shown to conjugate with proteins efficiently (Figure 4.4) (Kalariya et al. 2011). Due to the lack of commercially available apocarotenals from lutein, Kalariya et al. (2011) used readily available $\beta$-apo-8-carotenal (BA8C, a $\beta$-carotene oxidation product, without hydroxyl group) to understand the interaction between carotenoid aldehyde and protein. In fact, BA8C could be an ideal candidate to study such interaction based on its three following characteristics: (1) the ability of $\pi$-electrons to delocalize across the polyene structure of BA8C, which might make one or more of the carbons across the polyene chain electropositive and subsequently susceptible to nucleophilic attack; (2) the presence of –CHO group in BA8C is crucial for the determination of radioactivity as $NaB(^3H)H_4$ could efficiently react with –CHO and form a radiolabeled alcohol; and (3) BA8C has been demonstrated to form $1,N^2$-etheno-2′-deoxyguanosine adducts (Marques et al. 2004), which established the fact that CCP such as BA8C could react with biomolecules. Using various techniques, Kalariya et al. (2011) have confirmed that CCP, such as BA8C, could form thioether linkage with proteins containing free sulfhydryl groups. Moreover, this study also revealed that CCP could also be conjugated with proteins other than thioether linkage, simultaneously. This observation is based on the fact that not all the radiolabeled BA8C molecules conjugated to glyceraldehyde-3-phosphate dehydrogenase (GAPDH) were retrieved after Raney nickel treatment. This observation was reinforced by the ESI-MS analysis of resultant products of BA8C–$N$-acetylcysteine (NAC) after treatment with $NaB(^3H)H_4$, which revealed various quasimolecular ions. These results also indicated that there could be multiple products rather than a single common product, resulting from thioether linkages. This fact also explains the plausible complex interaction between long polyene chain and cysteine moieties of the proteins. Moreover, due to apparent complexities in the interaction between CCP and protein molecules, the exact site on BA8C polyene chain where sulfhydryl moiety could form thioether linkage is yet to be determined. These results are in stark contrast with that of the 4-HNE–NAC or 4-HNE–GSH interactions in terms of retrieval of common products after Raney nickel treatment as well as conjugation sites on 4-HNE. Uchida and Stadtman (1992a) found a common radiolabeled product after Raney nickel treatment of 4-HNE–NAC and 4-HNE–GSH adducts. This result probably explains difference in the availability of conjugation sites on 4-HNE and CCP such as BA8C. The identified site on 4-HNE where sulfhydryl could form thioether linkage is C3 (Uchida and Stadtman 1992a). However, the precise identity of BA8C–NAC products formed and the number and exact conjugation sites on BA8C remain to be determined.

The authors also observed another phenomenon too. The results of MALDI-TOF and the incorporation of radioactive labels revealed that the number of BA8C molecules conjugated to GAPDH exceeds the number of available free sulfhydryl groups on GAPDH (GAPDH has four free sulfhydryl groups). It means that BA8C could also conjugate with GAPDH through a non-thioether linkage. Therefore, based on that fact, lipid aldehydes are known to conjugate with proteins through Schiff's base reaction besides thioether linkage. Szweda et al. (1993) proposed a mechanism that −CHO of 4-HNE could bind to ε-amino group of lysine. This interaction was confirmed by using 2,4-dinitrophenylhydrazine (DNP). Similarly, a Schiff's base reaction could also be possible between CCP and lysine. The advantage of using 2,4-DNP is that it could dissociate Schiff's base reaction between −CHO of BA8C and ε-amino group of lysine. The dissociated free BA8C hydrazone product could be separated from the protein. Moreover, if ε-amino group of lysine binds to double bond of BA8C, 2,4-DNP could bind to free carbonyl function of the protein-bound BA8C and form non-dissociable hydrazone derivative. Based on amino acid analysis and the carbonyl assay using 2,4-DNP, it was determined that approximately 4.6, 3.4, and 0.8 mol of lysine/mol of BSA, GAPDH, and α-crystalline respectively could react with double bond of BA8C (Kalariya et al. 2011). These results were further strengthened by ESI-MS analysis of BA8C–*N*-acetyllysine adducts that revealed the formation of quasimolecular ions (Kalariya et al. 2011). Consistent with this observation, recently, Sommerburg et al. (2009) also demonstrated protein modification by CCPs and protein aggregate formation in an in vitro study. Besides protein, adduct formation between DNA and BA8C, has also been demonstrated (Marques et al. 2004). Clearly, more investigation is required to understand the various possible mechanisms of adduct formation between CCP particularly from xanthophyll and cellular molecules such as proteins and/or DNA. As shown in Figure 4.5, Kalariya et al. (2011) proposed a possible mode of interaction between CCP and functional groups of protein.

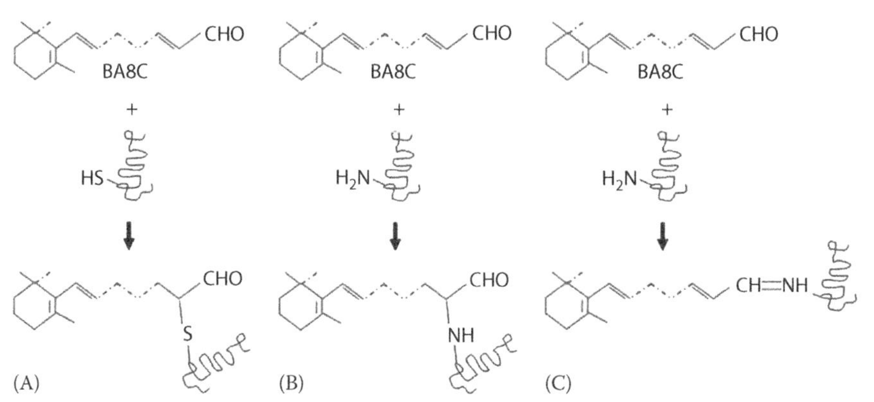

**FIGURE 4.5** Proposed mechanism of protein modification by carotenoid-derived aldehydes (CDAs). (A) Reaction of cysteine residue with double bond/alkyl group of β-apo-8-carotenal (BA8C). (B) Reaction of lysine residue with double bond/alkyl group of BA8C. (C) Reaction of lysine residue with −CHO group of BA8C. (Adapted from Kalariya, N.M. et al., *Biofactors*, 37, 104, 2011. With permission.)

Such interaction between CCP and protein and/or other biomolecules has great biologic significance as formation of protein–CCP adducts could inhibit the proteasomal system that might cause cellular toxicity and probably disease progression rather than suppression upon carotenoid supplementation. Moreover, the observation that lutein undergoes oxidation and forms a variety of oxidation products in human retina may be very important in terms of AMD. Since protein aggregation is a characteristic of drusen formation in AMD retina (Crabb et al. 2002), potential adduct formation between lutein products and protein in retina holds a profound significance in terms of therapeutic use of carotenoids for the treatment of AMD.

## CONCLUSION

Though various previous reports support the efficacy of xanthophylls for either prevention or suppression of AMD pathogenesis, not all studies have shown beneficial effects of xanthophylls. Based on the recent meta-analysis study, dietary lutein and zeaxanthin are not significantly associated with a reduced risk of early AMD, whereas an increase in the intake of these carotenoids may be protective against late AMD. Thus, a strong evidence to establish xanthophylls as a therapeutic agent for AMD intervention is lacking. Moreover, the therapeutic use of non-xanthophyll carotenoids, such as β-carotene, for other systemic diseases has resulted in undesirable outcomes thought to be incurred by carotenoid cleavage products. Therefore, while studying the controversial link between beneficial and harmful effects of xanthophyll on human retinal pathology, one should monitor the amount of xanthophyll uptake and generation of oxidized products levels in human retina but also explore the possible interaction of CCP with other biomolecules that might be helpful in establishing the true potential of xanthophyll as a therapeutic agent for AMD.

## ACKNOWLEDGMENT

This work was supported by Wilkins AMD Fund (to FJGMvK) and Research to Prevent Blindness.

## REFERENCES

Albanes D, Heinonen OP, Taylor PR et al. α-Tocopherol and β-carotene supplements and lung cancer incidence in the alpha-tocopherol, beta-carotene cancer prevention (ATBC) study: Effects of base-line characteristics and study compliance. *J Natl Cancer Inst* 1996; 88: 1560–1570.

Alija AJ, Bresgen N, Sommerburg O et al. β-Carotene breakdown products enhance genotoxic effects of oxidative stress in primary rat hepatocytes. *Carcinogenesis* 2006; 27: 1128–1133.

Alves-Rodrigues A, Shao A. The science behind lutein. *Toxicol Lett* 2004; 150: 57–83.

Anderson DH, Ozaki S, Nealon M et al. Local cellular sources of apolipoprotein E in the human retina and retinal pigmented epithelium: Implications for the process of drusen formation. *Am J Ophthalmol* 2001; 131: 767–781.

Anderson DH, Talaga KC, Rivest AJ et al. Characterization of beta amyloid assemblies in drusen: The deposits associated with aging and age-related macular degeneration. *Exp Eye Res* 2004; 78: 243–256.

Beatty S, Koh HH, Murray IJ, Henson DB, Boulton M. Macular pigment optical density and the risk for age-related macular degeneration. *Invest Ophthalmol Vis Sci* 2000; 41: S600.

Bernstein PS, Khachik F, Carvalho LS et al. Identification and quantitation of carotenoids and their metabolites in the tissues of the human eye. *Exp Eye Res* 2001; 72: 215–223.

Bernstein PS, Zhao DY, Wintch SW et al. Resonance Raman measurement of macular carotenoids in normal subjects and in age-related macular degeneration patients. *Ophthalmology* 2002; 109: 1780–1787.

Bhosale P, Bernstein PS. Quantitative measurement of 3′-oxolutein from human retina by normal-phase high-performance liquid chromatography coupled to atmospheric pressure chemical ionization mass spectrometry. *Anal Biochem* 2005; 345: 296–301.

Bhosale P, Larson AJ, Frederick JM et al. Identification and characterization of a Pi isoform of glutathione S-transferase (GSTP1) as a zeaxanthin-binding protein in the macula of the human eye. *J Biol Chem* 2004; 279: 49447–49454.

Bhosale P, Li B, Sharifzadeh M et al. Purification and partial characterization of a lutein-binding protein from human retina. *Biochemistry* 2009; 48: 4798–4807.

Bhosale P, Zhao da Y, Serban B, Bernstein PS. Identification of 3-methoxyzeaxanthin as a novel age-related carotenoid metabolite in the human macula. *Invest Ophthalmol Vis Sci* 2007; 48: 1435–1440.

Bone RA, Landrum JT, Fernandez L, Tarsis SL. Analysis of the macular pigment by HPLC: Retinal distribution and age study. *Invest Ophthalmol Vis Sci* 1988; 29: 843–849.

Bone RA, Landrum JT, Friedes LM et al. Distribution of lutein and zeaxanthin stereoisomers in the human retina. *Exp Eye Res* 1997; 64: 211–218.

Bone RA, Landrum JT, Hime GW, Cains A, Zamor J. Stereochemistry of the human macular carotenoids. *Invest Ophthalmol Vis Sci* 1993; 34: 2033–2040.

Bone RA, Landrum JT, Tarsis SL. Preliminary identification of the human macular pigment. *Vision Res* 1985; 25: 1531–1535.

Bone RA, Landrum JT, Mayne ST et al. Macular pigment in donor eyes with and without AMD: A Case–Control Study. *Invest Ophthalmol Vis Sci* 2001; 42: 235–240.

Boulton M, Rozanowska M, Rozanowski B. Retinal photodamage. *J Photochem Photobiol* 2001; 64: 144–161.

Britton G. Structure and properties of carotenoids in relation to function. *FASEB J* 1995; 9: 1551–1558.

Burton GW, Ingold KU. β-Carotene: An unusual type of lipid antioxidant. *Science* 1984; 224: 569–573.

Cardinault N, Abalain JH, Sairafi B et al. Lycopene but not lutein nor zeaxanthin decreases in serum and lipoproteins in age-related macular degeneration patients. *Clin Chim Acta* 2005; 357: 34–42.

Cho E, Hankinson SE, Rosner B, Willett WC, Colditz GA. Prospective study of lutein/zeaxanthin intake and risk of age-related macular degeneration. *Am J Clin Nutr* 2008; 87: 1837–1843.

Christen WG, Glynn RJ, Manson JE, Ajani UA, Buring JE. A prospective study of cigarette smoking and risk of age-related macular degeneration in men. *JAMA* 1996; 276: 1147–1151.

Crabb JW, Miyagi M, Gu X et al. Drusen proteome analysis: An approach to the etiology of age-related macular degeneration. *Proc Natl Acad Sci USA* 2002; 99: 14682–14687.

Cruickshanks KJ, Klein R, Klein BE, Nondahl DM. Sunlight and the 5-year incidence of early age-related maculopathy: The Beaver Dam Eye Study. *Arch Ophthalmol* 2001; 119: 246–250.

Curcio CA, Millican CL, Bailey T, Kruth HS. Accumulation of cholesterol with age in human Bruch's membrane. *Invest Ophthalmol Vis Sci* 2001; 42: 265–274.

Curran-Celentano J, Hammond BR Jr, Ciulla TA et al. Relation between dietary intake, serum concentrations, and retinal concentrations of lutein and zeaxanthin in adults in a Midwest population. *Am J Clin Nutr* 2001; 74: 79–80.

Dasch B, Fuhs A, Schmidt J et al. Serum levels of macular carotenoids in relation to age-related maculopathy: The muenster aging and retina study (MARS). *Graefes Arch Clin Exp Ophthalmol* 2005; 243: 1028–1035.

de Jong PT, Lubsen J. The standard gamble between cataract extraction and AMD. *Graefes Arch Clin Exp Ophthalmol* 2004; 242: 103–105.

De La Paz MA, Zhang J, Fridovich I. Antioxidant enzymes of the human retina: Effect of age on enzyme activity of macula and periphery. *Curr Eye Res* 1996; 15: 273–278.

Eperjesi F, Beatty S. *Nutrition and the Eye—A Practical Approach*. Toronto, Ontario, Canada: Elsevier; 2005, pp. 91–98.

Esterbauer H, Ertl A, Scholz, N. The reaction of cysteine with α,β-unsaturated aldehydes. *Tetrahedron* 1976; 32: 285–289.

Eye Disease Case–Control Study Group. Antioxidant status and neovascular age-related macular degeneration. *Arch Ophthalmol* 1993; 111: 104–109.

Finnemann SC, Leung LW, Rodriguez-Boulan E. The lipofuscin component A2E selectively inhibits phagolysosomal degradation of photoreceptor phospholipid by the retinal pigment epithelium. *Proc Natl Acad Sci USA* 2002; 99: 3842–3847.

Flood V, Smith W, Wang JJ, Manzi F, Webb K, Mitchell P. Dietary antioxidant intake and incidence of early age-related maculopathy: The Blue Mountains Eye Study. *Ophthalmology* 2002; 109: 2272–2278.

Friedman DS, O'Colmain BJ, Munoz B et al. Prevalence of age-related macular degeneration in the United States. *Arch Ophthalmol* 2004; 122: 564–572.

Gale CR, Hall NF, Phillips DI, Martyn CN. Lutein and zeaxanthin status and risk of age-related macular degeneration. *Invest Ophthalmol Vis Sci* 2003; 44: 2461–2465.

Goldberg J, Flowerdew G, Smith E, Brody JA, Tso MO. Factors associated with age-related macular degeneration. An analysis of data from the first National Health and Nutrition Examination Survey. *Am J Epidemiol* 1988; 128: 700–710.

Guéraud F, Atalay M, Bresgen N et al. Chemistry and biochemistry of lipid peroxidation products. *Free Radic Res* 2010; 44: 1098–1124.

Haimovici R, Gantz DL, Rumelt S, Freddo TF, Small DM. The lipid composition of drusen, Bruch's membrane, and sclera by hot stage polarizing light microscopy. *Invest Ophthalmol Vis Sci* 2001; 42: 1592–1599.

Ham WT Jr, Mueller HA, Ruffolo JJ Jr, Clarke AM. Sensitivity of the retina to radiation damage as a function of wavelength. *Photochem Photobiol* 1979; 29: 735–743.

Hammond BR, Johnson EJ, Russell RM et al. Dietary modification of human macular pigment density. *Invest Ophthalmol Vis Sci* 1997; 38: 1795–1801.

Hammond BR, Wooten BR, Snodderly DM. Cigarette smoking and retinal carotenoids: Implications for age-related macular degeneration. *Vision Res* 1996; 36: 3003–3009.

Handelman GJ, Dratz EA, Reay CC, van Kuijk FJGM. Carotenoids in the human macula and whole retina. *Invest Ophthalmol Vis Sci* 1988; 29: 850–855.

Handelman GJ, Packer L, Cross CE. Destruction of tocopherols, carotenoids, and retinol in human plasma by cigarette smoke. *Am J Clin Nutr* 1996; 63: 559–565.

Handelman GJ, van Kuijk FJGM, Chatterjee A, Krinsky NI. Characterization of products formed during the autoxidation of b-carotene. *Free Radic Biol Med* 1991; 10: 427–437.

Hawkins BS, Bird A, Klein R, West SK. Epidemiology of age-related macular degeneration. *Mol Vis* 1999; 5: 26.

van Helden YG, Keijer J, Knaapen AM et al. β-Carotene metabolites enhance inflammation-induced oxidative DNA damage in lung epithelial cells. *Free Radic Biol Med* 2009; 46: 299–304.

Hurst JS, Contreras JE, Siems WG, van Kuijk FJGM. Oxidation of carotenoids by heat and tobacco smoke. *Biofactors* 2004; 20: 23–35.

Hurst JS, Saini MK, Jin G, Awasthi YC, van Kuijk FGJM. Toxicity of oxidized β-carotene to cultured human cells. *Exp Eye Res* 2005; 81: 239–243.

Hyman L, Neborsky R. Risk factors for age-related macular degeneration: An update. *Curr Opin Ophthalmol* 2002; 13: 171–175.

Jeong HG, Chun YJ, Yun CH et al. Induction of cytochrome P450 1A and 2B by a- and b-ion-one in Sprague Dawley rats. *Arch Pharm Res* 2002; 25: 197–201.

Junghans A, Sies H, Stahl W. Macular pigments lutein and zeaxanthin as blue light filters studied in liposomes. *Arch Biochem Biophys* 2001; 391: 160–164.

Kalariya NM, Ramana KV, Srivastava SK, van Kuijk FJGM. Carotenoid derived aldehydes-induced oxidative stress causes apoptotic cell death in human retinal pigment epithelial cells. *Exp Eye Res* 2008; 86: 70–80.

Kalariya NM, Ramana KV, Srivastava SK, van Kuijk FJ. Genotoxic effects of carotenoid breakdown products in human retinal pigment epithelial cells. *Curr Eye Res* 2009; 34: 737–747.

Kalariya NM, Ramana KV, Srivastava SK, van Kuijk FJGM. Post-translational protein modification by carotenoid cleavage products. *Biofactors* 2011; 37: 104–116.

Khachik F, Bernstein PS, Garland DL. Identification of lutein and zeaxanthin oxidation products in human and monkey retinas. *Invest Ophthalmol Vis Sci* 1997; 38: 1802–1811.

Khachik F, de Moura FF, Chew EY et al. The effect of lutein and zeaxanthin supplementation on metabolites of these carotenoids in the serum of persons aged 60 or older. *Invest Ophthalmol Vis Sci.* 2006a; 47: 5234–5242.

Khachik F, London E, de Moura FF et al. Chronic ingestion of (3R, 30R, 60R)-lutein and (3R, 30R)-zeaxanthin in the female rhesus macaque. *Invest Ophthalmol Vis Sci.* 2006b; 47: 5476–5486.

Kirschfeld K. Carotenoid pigments: Their possible role in protecting against photooxidation in eyes and photoreceptor cells. *Proc R Soc Lond B Biol Sci* 1982; 216: 71–85.

Klein ML, Francis PJ. Genetics of age-related macular degeneration. *Ophthalmol Clin North Am* 2003; 16: 567–574.

Klein R, Peto T, Bird A, Vannewkirk MR. The epidemiology of age-related macular degeneration. *Am J Ophthalmol* 2004; 137: 486–495.

Knight JA. Diseases related to oxygen-derived free radicals. *Ann Clin Lab Sci* 1995; 25: 111–121.

Krinsky NI. Carotenoid protection against oxidation. *Pure Appl Chem* 1979; 51: 649–660.

Krinsky NI, Landrum JT, Bone RA. Biologic mechanisms of the protective role of lutein and zeaxanthin in the eye. *Annu Rev Nutr* 2003; 23: 171–201.

Landrum JT, Bone RA. Lutein, zeaxanthin, and the macular pigment. *Arch Biochem Biophys* 2001; 385: 28–40.

Landrum JT, Bone RA, Kilburn MD. The macular pigment: A possible role in protection from age-related macular degeneration. *Adv Pharmacol* 1997; 38: 537–556.

LaVail MM. Outer segment disc shedding and phagocytosis in the outer retina. *Trans Ophthalmol Soc UK* 1983; 103: 397–404.

van Leeuwen R, Boekhoorn S, Vingerling JR et al. Dietary intake of antioxidants and risk of age-related macular degeneration. *JAMA* 2005; 294: 3101–3107.

Li B, Vachali P, Frederick JM, Bernstein PS. Identification of StARD3 as a lutein-binding protein in the macula of the primate retina. *Biochemistry* 2011; 50: 2541–2549.

Liang F-Q, Godley BF. Oxidative stress-induced mitochondrial DNA damage in human retinal pigment epithelial cells: A possible mechanism for RPE aging and age-related macular degeneration. *Exp Eye Res* 2003; 76: 397–403.

Liu C, Russell RM, Wang XD. $\alpha$-Tocopherol and ascorbic acid decrease the production of $\beta$-apo-carotenals and increases the formation of retinoids from b-carotene in the lung tissues of cigarette smoke-exposed ferrets *in vitro*. *J Nutr* 2004; 134: 426–430.

Ma L, Dou H, Wu Y, Huang Y et al. Lutein and zeaxanthin intake and the risk of age-related macular degeneration: A systematic review and meta-analysis. *Br J Nutr* 2012; 107: 350–359.

Mares-Perlman JA, Brady WE, Klein R et al. Serum antioxidants and age-related macular degeneration in a population-based case-control study. *Arch Ophthalmol* 1995; 113: 1518–1523.

Mares-Perlman JA, Fisher AI, Klein R et al. Lutein and zeaxanthin in the diet and serum and their relation to age related maculopathy in the third national health and nutrition examination survey. *Am J Epidemiol* 2001; 153: 424–432.

Mares-Perlman JA, Klein R, Klein BE et al. Association of zinc and antioxidant nutrients with age-related maculopathy. *Arch Ophthalmol* 1996; 114: 991–997.

Marques SA, Loureiro AP, Gomes OF et al. Induction of 1,N(2)-etheno-20-deoxyguanosine in DNA exposed to beta-carotene oxidation products. *FEBS Lett* 2004; 560: 125–130.

Mo H, Elson CE. Apoptosis and cell-cycle arrest in human and murine tumor cells are initiated by isoprenoids. *J Nutr* 1999; 129: 804–813.

Moeller SM, Parekh N, Tinker L et al. CAREDS Research Study Group. Associations between intermediate age-related macular degeneration and lutein and zeaxanthin in the Carotenoids in Age-related Eye Disease Study (CAREDS): Ancillary study of the Women's Health Initiative. *Arch Ophthalmol* 2006; 124: 1151–1162.

Mullins RF, Russell SR, Anderson DH, Hageman GS. Drusen associated with aging and age-related macular degeneration contain proteins common to extracellular deposits associated with atherosclerosis, elastosis, amyloidosis, and dense deposit disease. *FASEB J* 2000; 14: 835–846.

Nara E, Hayashi H, Kotake M, Miyashita K, Nagao A. Acyclic carotenoids and their oxidation mixtures inhibit the growth of HL-60 human promyelocytic leukemia cells. *Nutr Cancer* 2001; 39: 273–283.

Nicolas CM, Robman LD, Tikellis G et al. Iris colour, ethnic origin and progression of age-related macular degeneration. *Clin Exp Ophthalmol* 2003; 31: 465–469.

Omenn GS, Goodman GE, Thornquist MD et al. Risk factors for lung cancer and for intervention effects in CARET, the beta-carotene and retinol efficacy trial. *J Natl Cancer Inst* 1996; 88: 1550–1559.

Palozza P, Calviello G, Serini S et al. b-Carotene at high concentrations induces apoptosis by enhancing oxy-radical production in human adenocarcinoma cells. *Free Radic Biol Med* 2001; 30: 1000–1007.

Palozza P, Serini S, Di Nicuolo F, Piccioni E, Calviello G. Prooxidant effects of b-carotene in cultured cells. *Mol Aspects Med* 2003; 24: 353–362.

Pauleikhoff D, Barondes MJ, Minassian D, Chisholm IH, Bird AC. Drusen as risk factors in age-related macular disease. *Am J Ophthalmol* 1990; 109: 38–43.

Prasain JK, Moore R, Hurst JS, Barnes S, van Kuijk FJGM. Electrospray tandem mass spectrometric analysis of zeaxanthin and its oxidation products. *J Mass Spectrom* 2005; 40: 916–923.

Rapp LM, Maple SS, Choi JH. Lutein and zeaxanthin concentrations in rod outer segment membranes from perifoveal and peripheral human retina. *Invest Ophthalmol Vis Sci* 2000; 41: 1200–1209.

Robman L, Vu H, Hodge A et al. Dietary Lutein, Zeaxanthin, and fats and the progression of age-related macular degeneration. *Can J Ophthalmol* 2007; 42: 720–726.

Ruffolo JJ, Ham WT Jr, Mueller HA, Millen JE. Photochemical lesions in the primate retina under conditions of elevated blood oxygen. *Invest Ophthalmol Vis Sci* 1984; 25: 893–898.

Salgo M, Cueto R, Winston G, Pryor W. β-Carotene and its oxidation products have different effects on microsome mediated binding of benzo[a]pyrene to DNA. *Free Radic Biol Med* 1999; 26: 162–173.

Satia JA, Littman A, Slatore CG, Galanko JA, White E. Long-term use of {beta}-carotene, retinol, lycopene, and lutein supplements and lung cancer risk: Results from the VITamins and lifestyle (VITAL) study. *Am J Epidem* 2009; 169: 815–828.

Saunders TA, Haines AP, Wormald R, Wright LA, Obeid O. Essential fatty acids, plasma cholesterol, and fat-soluble vitamins in subjects with age-related maculopathy and matched control subjects. *Am J Clin Nutr* 1993; 57: 428–433.

Schalch W, Dayhaw-Barker P, Barker FM. The carotenoids of the human retina. In: Taylor A, ed. *Nutritional and Environmental Influences on the Eye.* Boca Raton, FL: CRC Press; 1999, pp. 215–250.

Schütt F, Davies S, Kopitz J, Holz FG, Boulton ME. Photodamage to human RPE cells by A2-E, a retinoid component of lipofuscin. *Invest Ophthalmol Vis Sci* 2000; 41: 2303–2308.

Seddon JM, Ajani UA, Sperduto RD et al. Dietary carotenoids, vitamins A, C, and E, and advanced age-related macular degeneration. *JAMA* 1994; 272: 1413–1420.

Seddon JM, Cote J, Rosner B. Progression of age-related macular degeneration: Association with dietary fat, transunsaturated fat, nuts, and fish intake. *Arch Ophthalmol* 2003; 121: 1728–1737.

Seddon JM, Gensler G, Milton RC, Klein ML, Rifai N. Association between C-reactive protein and age-related macular degeneration. *JAMA* 2004; 291: 704–710.

Semba RD, Dagnelie G. Are lutein and zeaxanthin conditionally essential nutrients for eye health? *Med Hypotheses* 2003; 61: 465–472.

Shanmugaratnam J, Berg E, Kimerer L et al. Retinal Muller glia secrete apolipoproteins E and J which are efficiently assembled into lipoprotein particles. *Brain Res Mol Brain Res* 1997; 50: 113–120.

Siems WG, Sommerburg O, Hurst JS, van Kuijk FJGM. Carotenoid oxidative degradation products inhibit $Na^+$-$K^+$-ATPase. *Free Radic Res* 2000; 33: 427–435.

Siems WG, Sommerburg O, van Kuijk FJGM. Lycopene and β-carotene decompose more rapidly than lutein and zeaxanthin upon exposure to various pro-oxidants *in vitro*. *Biofactors* 1999; 10: 105–113.

Siems W, Sommerburg O, Schild L et al. β-Carotene cleavage products induce oxidative stress in vitro by impairing mitochondrial respiration. *FASEB J* 2002; 16: 1289–1291.

Simonelli F, Zarrilli F, Mazzeo S et al. Serum oxidative and antioxidant parameters in a group of Italian patients with age-related maculopathy. *Clin Chim Acta* 2002; 320: 111–115.

Smith W, Assink J, Klein R et al. Risk factors for age-related macular degeneration: Pooled findings from three continents. *Ophthalmology* 2001; 108: 697–704.

Snellen EL, Verbeek AL, van den Hoogen GW, Cruysberg JR, Hoyng CB. Neovascular age-related macular degeneration and its relationship to antioxidant intake. *Acta Ophthalmol Scand* 2002; 80: 368–371.

Snodderly DM. Evidence for protection against age-related macular degeneration by carotenoids and antioxidant vitamins. *Am J Clin Nutr* 1995; 62: 1448S–1461S.

Snodderly DM, Brown PK, Delori FC, Auran JD. The macular pigment. Absorbance spectra, localization, and discrimination from other yellow pigments in primate retinas. *Invest Ophthalmol Vis Sci* 1984; 25: 660–673.

Snodderly DM, Handelman GJ, Adler AJ. Distribution of individual macular pigment carotenoids in central retina of macaque and squirrel monkeys. *Invest Ophthalmol Vis Sci* 1991; 32: 268–279.

Sommerburg O, Karius N, Siems W et al. Proteasomal degradation of beta-carotene metabolite-modified proteins. *Biofactors* 2009; 35: 449–459.

Sommerburg O, Keunen JE, Bird AC, van Kuijk FJ. Fruits and vegetables that are sources for lutein and zeaxanthin: The macular pigment in human eyes. *Br J Ophthalmol* 1998; 82: 907–910.

Sommerburg O, Langhans CD, Arnhold J et al. β-carotene cleavage products after oxidation mediated by hypochlorous acid: A model for neutrophil-derived degradation. *Free Radic Biol Med* 2003; 35: 1480–1490.

Sommerburg O, Siems WG, Hurst JS et al. Lutein and zeaxanthin are associated with photoreceptors in the human retina. *Curr Eye Res* 1999; 19: 491–495.

Sparrow JR, Vollmer-Snarr HR, Zhou J et al. A2E-epoxides damage DNA in retinal pigment epithelial cells. Vitamin E and other antioxidants inhibit A2E epoxide formation. *J Biol Chem* 2003; 278: 18207–18213.

Spraul CW, Grossniklaus HE. Characteristics of drusen and Bruch's membrane in postmortem eyes with age-related macular degeneration. *Arch Ophthalmol* 1997; 115: 267–273.

Szweda LI, Uchida K, Tsai L, Stadtman ER. Inactivation of glucose-6-phosphate dehydrogenase by 4-hydroxy-2-nonenal. Selective modification of an active-site lysine. *J Biol Chem* 1993; 268: 3342–3347.

Tan JSL, Wang JJ, Flood V et al. Dietary antioxidants and the long term incidence of age-related macular degeneration: The Blue Mountain Eye Study. *Ophthalmology* 2008; 115: 334–341.

Tsang NC, Penfold PL, Snitch PJ, Billson F. Serum levels of antioxidants and age-related macular degeneration. *Doc Ophthalmol* 1992; 81: 387–400.

Uchida K. 4-Hydroxy-2-nonenal: A product and mediator of oxidative stress. *Prog Lipid Res* 2003; 42: 318–343.

Uchida K, Stadtman ER. Selective cleavage of thioether linkage in proteins modified with 4-hydroxynonenal. *Proc Natl Acad Sci USA* 1992a; 89: 5611–5615.

Uchida K, Stadtman ER. Modification of histidine residues in proteins by reaction with 4-hydroxynonenal. *Proc Natl Acad Sci USA* 1992b; 89: 4544–4548.

Vanden Langenberg GM, Mares-Perman JA, Klein R et al. Associations between antioxidant and zinc intake and the 5-year incidence of early age-related maculopathy in the Beaver Dam Eye Study. *Am J Epidemiol* 1998; 148: 204–214.

Vu HT, Robman L, McCarty CA, Taylor HR, Hodge A. Does dietary lutein and zeaxanthin increase the risk of age related macular degeneration? The Melbourne Visual Impairment Project. *Br J Ophthalmol* 2006; 90: 389–393.

Wachter A, Sun Y, Dasch B et al. Munster age- and retina study (MARS). Association between risk factors for arteriosclerosis and age-related macular degeneration. *Ophthalmologe* 2004; 101: 50–53.

Wald G. Human vision and the spectrum. *Science* 1945; 101: 653.

Wang Z, Keller LM, Dillon J, Gaillard ER. Oxidation of A2E results in the formation of highly reactive aldehydes and ketones. *Photochem Photobiol* 2006; 82: 1251–1257.

Yeh SL, Hu ML. Induction of oxidative DNA damage in human foreskin fibroblast Hs68 cells by oxidized β-carotene and lycopene. *Free Radic Res* 2001; 35: 203–213.

Yeh SL, Wu SH. Effects of quercetin on beta-apo-80-carotenal-induced DNA damage and cytochrome P1A2 expression in A549 cells. *Chem Biol Interact* 2006; 163: 199–206.

Yemelyanov AY, Katz NB, Bernstein PS. Ligand-binding characterization of xanthophyll carotenoids to solubilized membrane proteins derived from human retinas. *Exp Eye Res* 2001; 72: 381–392.

# 5 Lutein, Zeaxanthin, and Vision across the Life Span

*Billy R. Hammond, Jr.*

## CONTENTS

## INTRODUCTION

There is increasing recognition among carotenoid researchers that the 20 or so individual carotenoids that enter the blood stream tend to target specific systems within the body. Lycopene, e.g., tends to preferentially accumulate in the prostate gland of men and beta-carotene within the corpus luteum of women. Perhaps one of the more noteworthy examples of exclusivity, however, is the primate eye (including the retina, retinal pigment epithelium, ciliary tissue, crystalline lens, etc; Bernstein et al., 2001), which accumulates only lutein (L), zeaxanthin (Z), and the intermediary isomer, meso-Z (Bone et al., 1988; Bone et al., 1997). This accumulation is quite variable across individuals (e.g., Hammond et al., 1997). The primary driver appears to be the dietary intake of carotenoid-rich foods (even in infancy, see Figure 5.1), which can range from average levels that are quite low in the United States (about 1–2 mg/day of L and Z) to levels that, in some cultures, would be considered megadosing (mean intake of about 20 mg/day of L and Z in Fiji Islanders; Le Marchand et al., 1993). Once ingested, the probability that the carotenoids will actually reach their target tissue is moderated by numerous factors. These variables, roughly segregated, reflect a balance of deleterious (largely pro-oxidant) and protective factors: deleterious variables tend to lower pigment levels (smoking, obesity, diabetic status, etc.), whereas more salubrious factors (such as a healthy diet) appear to allow more direct uptake (i.e., no systemic reduction) within the target tissue (reviewed by Hammond and Johnson, 2002). Whether high or low, however, the eye accumulates just these two isomers with distinct specificity: e.g., in adult retina, Z concentrates

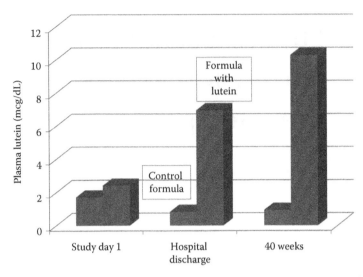

**FIGURE 5.1**    Plasma lutein levels in preterm infants given formula that was supplemented with carotenoids compared to a control formula lacking carotenoids. (Based on data derived from Rubin, L.P. et al., *J Perinatol* 7, 2011.)

in the very center whereas L is spread more evenly throughout the retinal tissue (Bone et al., 1997, 1988).

The selectivity in accumulation of retinal L and Z appears to be motivated by matching the multifunctionality of the pigments to the diverse needs of the tissue. For example, lutein and zeaxanthin are strong antioxidants (Z is about twice as potent a lipid-based antioxidant as L, which may explain the differential distribution within the retina; see Table 5.1). The retina is the most metabolically active tissue

**TABLE 5.1**
**Lutein and Zeaxanthin Have Different Antioxidant Properties Depending on the Chemical Structure and Type of Free Radical**

| Carotenoid | Tocopheryl Radical Cation[a] | Singlet Oxygen[b] |
|---|---|---|
| 3R,3′R-Zeaxanthin | 26.4 | 12.6 |
| Lycopene | 13.5 | 16.8 |
| β-Carotene | 10.2 | 13.5 |
| Canthaxanthin | 8.8 | 13.2 |
| Lutein | 5.3 | 6.6 |

*Source:*   From Bohm et al., *J. Am. Chem. Soc.* 119, 621, 1997. Conn et al., *Free Radic. Res. Comm.*, 16, 401, 1992.

Second-order rate constants $M^{-1}\ s^{-1}$.

[a]   In hexane.

[b]   In benzene.

in the body and, hence, highly susceptible to damage due to reactive oxygen species (e.g., Delmelle, 1977). Both carotenoids are also lipophilic, and the retina (as well as brain) contains very high concentrations of lipids like docosahexaenoic acid (DHA, $C22:6$ n-3), especially within the membranes of the vulnerable outersegments (where levels can reach as high as about 50%; e.g., Anderson, Benolken et al., 1974). DHA is the most oxidizable lipid in the body, and some evidence suggests synergistic uptake of LZ and DHA within the retina (e.g., Johnson et al., 2008b). The yellow LZ pigments are most highly concentrated in the inner layers of the retina in and around the foveal depression (e.g., Snodderly et al., 1984). Hence, they form a filter that screens shortwave (blue) visible light before it impinges on the foveal cones. Cones are concentrated in the central retina and are most responsible for fine detailed acuity, color perception, and generally good vision during the day (photopic sensitivity). Rods, not screened by L and Z, are found in the more peripheral portions of the retina and mediate the high sensitivity necessary for vision at night (scotopic sensitivity). Although past data have linked higher levels of LZ to preservation of both photopic and scotopic sensitivity (e.g., Hammond et al., 1998a), it is likely that this effect was based on a protective mechanism (e.g., preventing rod loss due to damage or age would promote maintenance of scotopic sensitivity). As a filter though, L and Z primarily screen the foveal cones (although the relatively small number of foveal rods that are screened may lead to improved mesopic acuity; Kvansakul et al., 2006) and therefore would be expected to influence mostly vision that is mediated by those cones. This filtering action can be quite significant in that individuals can have retinal L and Z levels that are so high as to block the majority of shortwave light from reaching the central photoreceptors (e.g., Hammond et al., 1997a). It is likely that this purely optical effect can have very significant effects on visual function throughout life (reviewed by Hammond, 2008).

The other significant presence of L and Z in the visual system is within the brain itself (Craft et al., 2004; Vishwanathan et al., 2011; see Figure 5.2). Of course, visual perception occurs within the brain and not the eyes per se. The eye is simply a detector that turns light waves into neural signals that are later interpreted by higher neural structures. L and Z have been identified in all of the brain tissues involved in this processing including visual, parietal, and frontal cortices (Craft et al., 2004; Vishwanathan et al., 2011). Although L and Z are not the sole carotenoids within the brain, they are found in much greater concentrations than the other carotenoids (like beta-carotene or lycopene) despite equal dietary intake (e.g., Vishwanathan et al., 2011). Analogous to their specific effects on the retina, this selectivity suggests some function for the pigments in higher-order neural processing. Data that have recently emerged are consistent with this possibility (e.g., Hammond and Wooten, 2005; Johnson et al., 2008a; Renzi and Hammond, 2010a). Hence, L and Z may be involved in visual function at every stage of the process: as early as the crystalline lens (promoting clarity; Hammond et al., 1997b) and as late as association cortices (Johnson et al., 2008a; Renzi and Hammond, 2010; Vishwanathan et al., 2011).

When considering the action of carotenoids within the visual system, the confluence of available empirical evidence suggests that the effects are pleiotropic: to wit, the carotenoids likely serve many roles ranging from protecting vulnerable lipid-rich cells from light-initiated damage due to reactive oxygen species to more

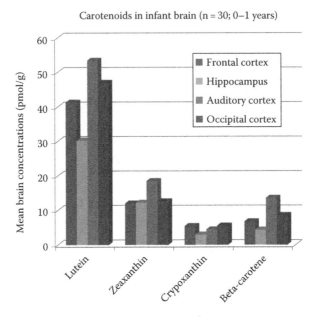

Carotenoids in infant brain (n = 30; 0–1 years)

**FIGURE 5.2**   Carotenoid concentrations in different brain sections of 30 newborn infants. (Based on data derived from Vishwanathan, R. et al., *FASEB J.*, 25, 344.1, 2011.)

immediate overt effects on visual function that are based on filtering mechanisms (e.g., Stringham and Hammond, 2007, 2008) and cellular processes (e.g., Stahl et al., 1997). Such a diversity of effects may be more meaningful in some stages of life compared to others.

## EFFECTS OF L AND Z DURING VISUAL DEVELOPMENT: INFANCY AND CHILDHOOD

Probably the two most well-known functions of the xanthophylls are their ability to serve as antioxidants and blue-light filters (e.g., Sondderly, 1995).* Both functions have special relevance very early in life (e.g., Zimmer and Hammond, 2007; Hammond et al., 2008). Infant retinas, e.g., are especially vulnerable to oxidative stress. This is likely due to two main factors: increased light stress due to very clear anterior lenses (e.g., Dillon et al., 2004) and the high oxygen utilization of the developing visual system (e.g., Hardy et al., 2000). The metabolic activity of the retina, for instance, is increased due to the rapid maturation of visual areas during the first year of life. This is especially true of the macula. At infancy, the macula often does not display the characteristic depression seen in adulthood, and cone photoreceptors

---

* It is common to refer to light from the shortwave end of the visible spectrum as "blue" and the cones that process this light as "blue cones." This description, although a useful mnemonic, leads to the misperception that was stated earlier: it suggests that visual perception is within the eye. As originally noted by Sir Isacc Newton, neither the rays are "colored" nor are the cones. They simply respond to shortwave light. The brain creates the perception of hue.

are both much wider in girth and dramatically shorter than comparable cones in adulthood (e.g., Yuodelis and Hendrickson, 1986). Indeed, macroscopically, it is hard to discriminate the fovea and cones from the peripheral retina and rods. The rapid change in morphology seen over the first year requires an equally robust blood supply to furnish sufficient oxygen to fuel this nearly frenetic growth. The blood supply, particularly, from the choroid, however, is also not fully mature and regu lated (e.g., Hardy et al., 2000). Taken together, the oxygen stress on the infant retina is high.

This is one reason why increased oxygen stress can be so harmful in situations where babies must be administered supplemental oxygen (e.g., Lucey and Dangman, 1984). The lungs are one of the last things to develop in utero. Hence, premature babies are often born in respiratory distress. This requires these babies to be placed in oxygenated incubators. The increased oxygen, however, must be carefully titrated because, in combination with the lack of tissue maturity, it puts the babies at high risk of developing an eye disease called retinopathy of prematurity (i.e., retrolental fibroplasia; Lucey and Dangman, 1984). Reducing the oxygen stress, and antioxidant therapy, is a common approach to this condition (clinical trials assessing the efficacy of L and Z are just now beginning; see Rubin et al., 2011).

In general, carotenoids are likely important antioxidants that reduce the probability of oxidative damage when babies, both full term and preterm, are at this most vulnerable stage. This may be one reason that breast milk accumulates carotenoids (even concentrating it in the early months) in a form that is highly bioavailable to infants (Bettler et al., 2010). This recognition has also spawned the addition of L and Z to infant formula. This addition is very recent. Many children have been raised on infant formula containing no lutein or zeaxanthin (or, for that matter, none of the hundreds of other nutrients that breast milk tends to contain).

Another possible function of importance to babies is the filtering of shortwave light. Such light is known to be highly damaging to retinal tissue (e.g., Barker et al., 2011). The energy of light is inversely proportional to wavelength. Ultraviolet light would be quite harmful (as it was before ultraviolet chromophores were added to intraocular implants; Werner et al., 1989), but it does not reach the retina because it is absorbed by the cornea and lens. Light past about 500 nm reaches the retina but is not energetic enough to damage retinal tissue except through thermal mechanisms. Light in the absorbance band of macular pigment (MP), however, 400–520 nm, reaches the retina, and there is sufficient energy to initiate photochemical effects. Infants have an exceptional clear crystalline lens that transmits even higher amounts of this damaging waveband of light (e.g., Dillon et al., 2004).

The axiom that "you are what you eat" is perhaps most true when the body is developing. Lipid membranes, e.g., are composed of a mix of both rigid saturated fats and more fluid omega fatty acids. Unless, of course, the diet is absent in these compounds. In the absence of sufficient exogenous intake of n-3 fatty acids (DHA), for instance, they are likely replaced with the less desirable n-6 fatty acids (docosapentaenoic acid; DPA n-6, C22:5 n-6) (e.g., Moriguchi et al., 2001). It is likely possible that the structure of the macula could be affected by the absence of L and Z, which are, when present, major components of the tissue. Leung et al. (2004) raised monkeys on diets lacking n-3 fatty acids and carotenoids and compared these monkeys with others raised on a normal diet where these ingredients were included in the chow. Leung et al. (2004) found that the monkeys on the deficient chow displayed distinct changes in the morphology of their macula. Recently, Rubin et al. (2011) conducted an analogous study on human preterm infants. Rubin et al. (2011) compared babies on regular breast milk, a control formula with no L and Z, or the same formula with L and Z added. These authors found that the preterm infants supplemented with L and Z were more like the breast-fed infants: they had lower levels of systemic inflammatory stress and improvements in retinal development (as assessed by electroretinogram) when compared to matched infants on a control formula with no L and Z (see Figure 5.3).

Although L and Z are important for the protection and, possibly, maturation of the visual system, it seems conspicuous that their absence does not cause more overt issues. After all, many generations of babies have been raised on diets that contain either no or trace amounts of L and Z (until recently, no infant formulas contained a significant amount of L and Z). Indeed, many adults have been measured that contain very small amounts of L and Z within their retina (e.g., Hammond and Caruso-Avery, 2000). One likely reason is simply based on the mechanisms of evolution by natural selection. Traits are selected that are most important to producing offspring. Hence, we have likely evolved mechanisms that are designed to keep us functioning optimally long enough

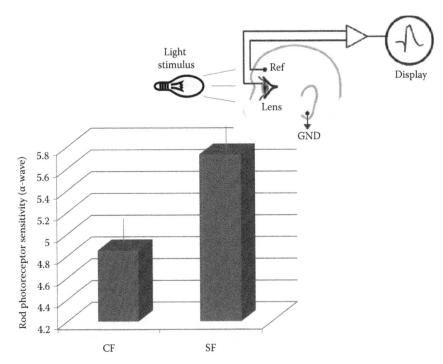

**FIGURE 5.3** Electroretinograms from preterm infants fed formulas with supplemented formula (SF) or without control formula (CF) carotenoid supplementation. Infants lacking carotenoids had, on average, reduced alpha waves in their ERGs.

to reproduce. Some of these mechanisms are as simple as cellular redundancy. Early damage likely does not manifest until relatively late in life. By analogy, teens can spend long amounts of time in the sun without suffering visible skin damage. They are, in fact, damaging their skin, but the effects are not obvious until later in life. Rods are lost in great number throughout life, but it is only in the fifth to sixth decade that significant loss in scotopic sensitivity is observed (e.g., Hammond et al., 1998). The damage that characterizes most degenerative disease, in fact, begins as soon as life itself begins, but the system compensates to keep the organism functional as long as possible. The key to health in later life is to ameliorate damage as much as possible throughout life in order to preserve function later.

## EFFECTS OF L AND Z ON THE VISUAL FUNCTION OF HEALTHY ADULTS

When considering how diet influences any function, visual or otherwise, it is important to consider both who we are and who we were. Many visual problems that influence adults in modern life were likely rare during most of human evolution. For example, nearsightedness (myopia) is largely a modern visual problem (e.g., Guggenheim et al., 2000) often caused by having an ocular orbit of excessive axial length. Myopia in the modern world has become pandemic. In the United States, it has nearly doubled in prevalence in the last 20–30 years (going from 25% in 1975 to about 43% in the early 2000s; Vitale et al., 2009). In some countries, like Singapore, the majority of the population is now myopic (e.g., Seet et al., 2001). For most of our history, however, humans did not dwell in cities but, rather, lived as hunters and gatherers. Such groups did not perform the kind of near-work (like reading) that compromises much of the visual activity of modern life. Activities like near-work and lack of natural visual signals combined with exercise has been posited as the primary etiological factors in most cases of refractive error (e.g., He et al., 2009). It is not therefore surprising that the mechanisms to accumulate retinal L and Z did not evolve to improve problems that are largely associated with modern life, like refractive error (e.g., Neelam et al., 2006). Most of the activity of hunter and gatherers, like agrarian groups, involved seeing objects at a distance, mostly outside, and primarily during the daytime under natural sunlight. Concomitantly, the pigments tend to influence visual function in a way that appears linked to the conditions under which they evolved: e.g., intense glaring light from sources that are similar to the sun (as originally suggested by Walls and Judd, 1933).

For example, in a series of studies (Stringham et al., 2003, 2011, Stringham and Hammond, 2007, 2008), Stringham and others have found that MP reduces photostress recovery time, glare disability, and glare discomfort. The mechanism appears quite simple. MP absorbs from about 400–520 nm (a full third of the visible spectrum). When a subject is exposed to a bright blinding flash of white light, vision is lost for a period until the visual system recovers (photostress recovery that probably occurs due to the subsequent recombination of photopigment and adaptation effects).

Having a dense yellow filter in front of your cones prevents this blinding light from impinging on the cones, so less recovery is needed.

Filtering blue light may have special ecological significance since it is largely blue light that tends to degrade vision through the atmosphere. Visual range (how far one can see) is largely limited, for instance, by the scattering of light within the atmosphere. This scattering is greatest for the most energetic portions of the visible spectrum according to Rayleigh's simple equation ($\lambda^{-4}$; scattering is directly proportional to frequency). One hypothesis for MP function is that it could increase visual range outdoors by filtering this highly scattered blue light (the visibility hypothesis; Wooten and Hammond, 2002; Hammond et al., 2012). Yellow carotenoids are a common choice as intraocular filters throughout nature (e.g., Walls and Judd, 1933). It is perhaps not a coincidence that the dominant wavelength of sky light (460 nm) exactly corresponds to the peak absorption of L and Z. Empirical data have shown that MP improves contrast relations when mid-wave targets are detected on blue backgrounds (e.g., Renzi and Hammond, 2010).

Effects based on optical improvements are likely not the only means by which L and Z influence visual function. Model studies have shown that the pigments can directly influence cellular interactions, e.g., by influencing gap junction communications (e.g., Stahl et al., 1997). Although the exact mechanisms by which L and Z might influence neural activity are not clear, there is some evidence to suggest that the pigments may have this function (Johnson et al., 2008a; Renzi and Hammond, 2010). First, the pigments are found within central nervous tissues such as visual cortex, cerebellum, hippocampus, auditory cortex, pons, frontal cortex, etc. (e.g. Craft et al., 2004). This accumulation seems to favor lutein over the other carotenoids (Vishwanathan et al., 2011). The presence of L and Z throughout the visual system suggests that these pigments serve some function that is post-receptoral. Preliminary data show that levels within the retina are highly correlated with levels within the brain (Vishwanathan et al., 2011). Using retinal levels as a proxy estimate of brain levels, laboratory studies have found relations to visual processing speed (Hammond and Wooten, 2005; Renzi and Hammond, 2010) and numerous cognitive measures (Johnson et al., 2008a). Vision is, ultimately, in the brain. The eyes serve as detectors for processing and transduction of light signals that are then used by higher-level areas to create our visual perceptions. It is possible that L and Z influence every stage of this process.

## EFFECTS OF L AND Z ON THE AGING AND DISEASED VISUAL SYSTEM

It is possibly the case that the most significant effects of L and Z on the visual system of the elderly are simply prophylactic (i.e., it prevents damage over decades). As noted, such effects start as early as infancy and extend through life. Although difficult to assess retrospectively (past diet is notoriously difficult to quantify), or prospectively (such studies usually follow subjects a relatively short time late in life), there appears to be a growing consensus that the pigments serve this function (e.g., Bernstein et al., 2009). There is, in fact, overwhelming evidence that the pigments are protective (e.g., Snodderly, 1995). The real issue is whether they are protective enough to matter when considering all of the other factors, both positive and negative, involved in the aging process and degenerative disease (see recent data from Berrow et al., 2011). Despite this lack of certainty, L and Z supplements are widely used as supplements designed to protect eyes from degenerative change. This practice seems especially common in the elderly who fear, perhaps, the rising risk of degenerative disease.

Is this practice advisable? One could certainly argue that neurological loss cannot be reversed. Once neurons are lost, they are not replaced. There is evidence, however, that neural loss is not linear throughout the life span. The elderly, like infants, may be more vulnerable than middle-aged adults to damage due to light and oxygen. One reason for this is that the elderly have a higher concentration of photosensitizers than younger adults (e.g., Wu et al., 2006). Damage due to light and oxygen can take place only in the presence of a photosensitizer. The elderly have much higher levels of lipofuscin, a potent photosensitizer in the presence of blue light. L and Z have been shown to deactivate A2E (e.g., Kim et al., 2006), the most toxic portion of lipofuscin. Also, like infants, the elderly tend to be under higher levels of inflammatory stress, and L and Z are known to be potent anti-inflammatory agents (Izumt-Nagai et al., 2007; Ohgami et al., 2003). All such effects are exacerbated in diseased tissue and tend to cause a cascade of effects that eventually lead to total loss. One example is the neovascularization that represents the worst and most blinding outcome of macular degeneration or diabetic retinopathy. Some data suggest that L and Z are most strongly associated with the neovascular form of age related macular degeneration (AMD) (Eye Disease Case–Control Study Group, 1992, 1993) and are lowest in patients with the most severe grades of diabetic retinopathy (e.g., Davies and Moreland, 2002). Animal models show that the pigments are anti-angiogenic. If the pigments are also palliative (improving visual function in patients with AMD; Richer et al., 2004, 2011) and safe (both are GRAS approved and no toxicological effects have been found at even very high doses; Alves-Rodriguez and Shao, 2004), then supplementation (via either carotenoid-rich foods, Hammond et al., 1997, or purified supplements) for vulnerable populations seems reasonable.

## EFFECT OF OTHER CAROTENOIDS ON THE VISUAL SYSTEM

L and Z have been available in supplement form only for about a decade. Prior to that time, most of the epidemiology focused on the possible protective effects of beta-carotene. Beta-carotene was better characterized in food-frequency questionnaires and tended to be more sponsored by industry that had products that contained beta-carotene. For example, the first large nutritional intervention conducted by the National Eye Institute in the United States used beta-carotene as the carotenoid, which was targeted in a prospective trial designed to determine if late-stage supplementation could reduce the probability of developing AMD.

AREDS enrolled 3557 subjects (55–80 years) across 11 sites within the United States (Age-Related Eye Disease Study Research Group, 2001). Subjects were divided into disease categories based on their visual health at baseline (classified ophthalmologically). Patients across categories were randomly assigned to four different treatments: placebo, antioxidants, zinc, or antioxidants + zinc. The supplements contained (amount/day): 500 mg vitamin C, 400 IU dl-α-tocopherol acetate, 15 mg β-carotene, 80 mg zinc with 2 mg copper as cupric oxide.* At years 1 and 5, this supplementation had yielded the expected increases in serum vitamin C, vitamin E, and β-carotene, and increases in serum zinc in the zinc groups (~18% increase) with ≤3% change in serum copper, as well as decreases in serum L+Z (decreased about 22% in the antioxidants groups and about 7% in the placebo group). The two primary outcomes were progression to an AMD event (treatment for choroidal neovascularization such as photocoagulation, or photographic documentation of geographic atrophy involving center of macula, non-drusenoid retinal pigment epithelial detachment, serous or hemorrhagic retinal detachment, hemorrhage under the retina or retinal pigment epithelium, and/or subretinal fibrosis) or visual acuity loss. The largest result was that the combination of the antioxidant cocktail and zinc reduced the risk of disease progression by 25% for patients with intermediate drusen or advanced AMD in one eye (from 16% risk over 5 years to 12% risk). Risk of vision loss was also reduced (around 19%). Zinc alone was associated with a 21% reduction of disease progression and a 11% reduction in vision loss. The antioxidant cocktail alone was associated with a 17% reduction of disease progression and a 10% reduction in vision loss. Very little effect of supplementation was seen for subjects with less advanced AMD (Categories 1 and 2). Although the experimental design of this study was optimal (a double-blind, placebo-controlled clinical trial), the intervention and the outcomes measured were not optimal. For example, AREDS was begun before an appreciation of the role of L and Z and DHA on the development of AMD. AREDS II (sometimes called the Lutein/Zeaxanthin, Omega-3 Supplementation trial) was designed to correct these deficiencies (the study should conclude around 20,114; see www.areds2.org) by using the same rigorous experimental design but by adding L, Z, DHA, and EPA to the formulation (and now excluding BC).

Two other carotenoids have been studied for possible influences on the visual system, both negative and positive. With respect to the latter, canthaxanthin (a carotenoid once used as an oral supplement for the purpose of coloring the skin in a way analogous to tanning) when taken in large doses was found to create (reversible) crystalline deposits within the retina (canthaxanthin retinopathy; Köpcke et al., 1995). With respect to the former, astaxanthin has been investigated (e.g., Nagaki et al., 2006) for a possible role in reducing eye fatigue (asthenopia). Like the other xanthophylls (L and Z), astaxanthin is a potent anti-inflammatory (Ogami et al., 2002) and antioxidant (Palozza and Krinsky, 1992). A direct mechanism (other than actions such as reducing systemic blood pressure, etc.), however, for improving eye fatigue is not clear (although such effects as improving ciliary blood flow and accommodative ability have been suggested; Nagaki et al., 2006).

## CONCLUSION

It is likely that carotenoids, especially L and Z, have effects upon the visual system that are pleiotropic. This is schematized in Figure 5.4. At the level of the eye, L and Z, for instance, likely improve lenticular and retinal health. This is an effect that

---

* Copper was added to prevent copper deficiency anemia, which can be induced by zinc supplementation. Cupric oxide, however, is chemically inert and unavailable for absorption in nonruminant animals (e.g., humans; Johnson et al., 1998). Hence, its addition probably had no effect on either limiting zinc toxicity or contributing to the overall effect of the zinc supplement.

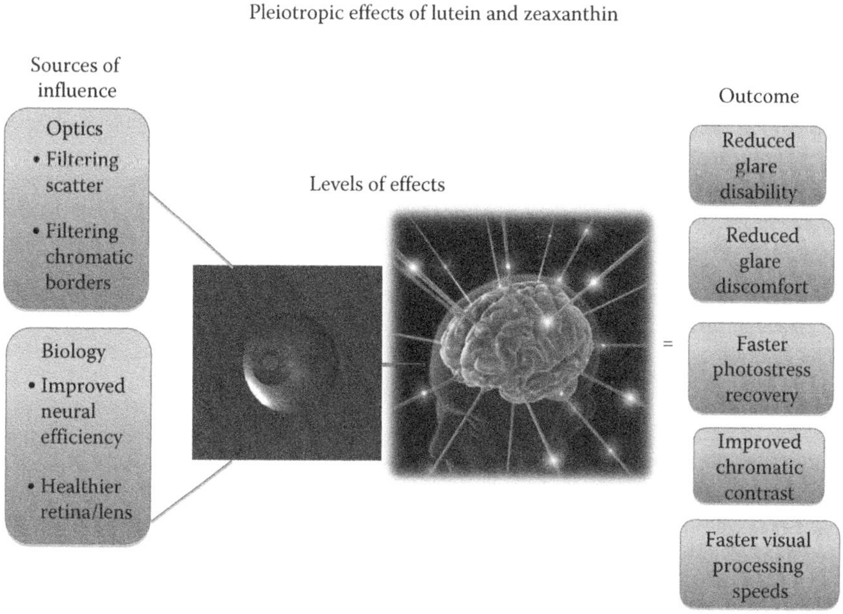

Pleiotropic effects of lutein and zeaxanthin

**FIGURE 5.4** The multifunctionality of lutein and zeaxanthin: these carotenoids likely affect processing and health throughout the multiple stages of the visual system.

is clearly important across the life span but may have special significance during two stages of life: infancy/childhood when the visual system may be particularly vulnerable to damage, and later in life, e.g., early in the (often covert) visual disease process. It is also likely that, within the eye, L and Z could improve visual function as an optical filter, reducing intraocular scattered light (Kvansakul et al., 2006), improving glare disability and discomfort, shortening photostress recovery, and contrast relations. This is likely an important palliative for patients who are suffering from visual problems (e.g., Richer et al., 2004, 2011) and for normal adults facing visual challenges (such as driving under glaring conditions). Taken together, L and Z may have differing and important roles throughout the stages of life. This is apropos since they also likely have differing roles throughout the multiple stages of the visual system itself. L and Z are present in the brain, and they may influence post-receptoral processing.

## REFERENCES

Age-Related Eye Disease Study Research Group. (2001). A randomized, placebo-controlled, clinical trial of high-dose supplementation with vitamins C and E, beta carotene, and zinc for age-related macular degeneration and vision loss. AREDS Report No. 8. *Arch Ophthalmol* 119:1417–1436.

Alves-Rodrigues A and Shao A. (2004). The science behind lutein. *Toxicol Lett* 150:57–83.

Anderson RE, Benolken RM et al. (1974). Proceedings: Polyunsaturated fatty acids of photoreceptor membranes. *Exp Eye Res* 18(3):205–213.

Barker II, FM, Snodderly DM, Johnson EJ, Schalch W, Koepcke W, Gerss J, & Neuringer M. (2011). Nutritional manipulation of primate retinas, V: Effects of lutein, zeaxanthin, and n–3 fatty acids on retinal sensitivity to blue-light–induced damage. *Invest Ophthalmol Vis Sci* 52(7):3934–3942.

Bernstein PS, Delori FC, Richer S, van Kuijk FJ, & Wenzel AJ. (2010). The value of measurement of macular carotenoid pigment optical densities and distributions in age-related macular degeneration and other retinal disorders. *Vis Res* 50(7):716–728.

Bernstein PS, Khachik F, Carvalho LS, Muir GJ, Zhao DY, & Katz NB. (2001). Identification and quantitation of carotenoids and their metabolites in the tissues of the human eye. *Exp Eye Res* 72(3):215–223.

Berrow E, Bartlett Eperjesi H, Eperjesi F, & M Gibson J. (2011). Risk factors for age-related macular disease. *Eur Ophthal Rev* 5(2):143–153.

Bettler J, Zimmer JP, Neuringer M, and DeRusso PA. (2010). Serumlute in concentrations in healthy term infants fed human milk or infant formula with lutein. *Eur J Nutr* 49:45–51.

Bone RA, Landrum JT, Fernandez L, and Tarsis SL. (1988). Analysis of the macular pigment by HPLC: Retinal distribution and age study. *Invest Ophthalmol Vis Sci* 29:843–849.

Bone RA, Landrum JT, Friedes LM, Gomez CM, Kilburn MD, Menendez E, Vidal I, and Wang W. (1997). Distribution of lutein and zeaxanthin stereoisomers in the human retina. *Exp Eye Res* 64:211–218.

Craft NE, Haitema HB, Garnett KM, Fitch KA, and Dorey CK. (2004). Carotenoid, tocopherol and retinol concentrations in the elderly human brain. *J Nutr Health Aging* 8(3):156–162.

Davies NP and Moreland AB. (2002). Color matching in diabetes: Optical density of the crystalline lens and macular pigments. *Invest Ophthalmol Vis Sci* 43:281–284.

Delmelle M. (1977). Retinal damage by light: Possible implication of singlet oxygen. *Biophys Struct Mech* 3:195.

Dillon J, Zhenga L, Merriama JC, and Gaillard ER. (2004). Transmission of light to the aging human retina: Possible implications for age related macular degeneration. *Exp Eye Res* 79:753–759.

Eye Disease Case–Control Study Group. (1992). Risk factors for neovascular age-related macular degeneration. *Arch Ophthalmol* 110(12):1701–1708.

Eye Disease Case–Control Study Group. (1993). Antioxidant status and neovascular age-related macular degeneration. *Arch Ophthalmol* 111(1):104–109.

Guggenheim JA, Kirov G, and Hodson SA. (2000). The heritability of high myopia: A reanalysis of Goldschmidt's data. *J Med Genet* 37:227–231.

Hammond BR. (2008). Possible role for dietary lutein and zeaxanthin in visual development. *Nutr Rev* 66(12):695–702.

Hammond BR and Caruso-Avery M. (2000). Macular pigment optical density in a Southwestern sample. *Invest Ophthalmol Visual Sci* 41(6):1492–1497.

Hammond BR and Johnson MA. (2002). Dietary prevention and treatment of age-related macular degeneration. *Recent Res Dev Nutr* 5:43–68.

Hammond BR, Johnson EJ, Russell RM, Krinsky NI, Yeum KJ, Edwards RB, and Snodderly DM. (1997). Dietary modification of human macular pigment density. *Invest Ophthalmol Vis Sci* 38:1795–1801.

Hammond BR, Wenzel AJ, Luther MS, Rivera RO, King SJ, and Choate ML. (1998a). Scotopic sensitivity: Relation to age, dietary patterns, and smoking status. *Opt Vis Sci* 75:867–872.

Hammond BR and Wooten BR. (2005). CFF thresholds: Relation to macular pigment optical density. *Ophthalmic Physiol Opt* 25:315–319.

Hammond BR, Wooten BR, Engles M, & Wong JC. (2012). The influence of filtering by the macular carotenoids on contrast sensitivity measured under simulated blue haze conditions. *Vision Res* 63:58–62.

Hammond BR, Wooten BR, and Snodderly DM. (1997). Density of the human crystalline lens is related to the macular pigment carotenoids, lutein and zeaxanthin. *Opt Vis Sci* 74(7):499–504.

Hammond BR, Wooten BR, and Snodderly DM. (1998b). Preservation of visual sensitivity of older subjects: Association with macular pigment density. *Invest Ophthalmol Vis Sci* 39:397–406.

Hardy P, Dumont I et al. (2000). Oxidants, nitric oxide and prostanoids in the developing ocular vasculature: A basis for ischemic retinopathy. *Cardiovasc Res* 47(3):489–509.

He M, Zheng Y, and Xiang F. (2009). Prevalence of myopia in urban and rural children in mainland China. *Optom Vis Sci* 86:40–44.

Izumi-Nagai K, Nagai N, Ohgami K, Satofuka S, Ozawa Y, Tsubota K, Umezawa K, Ohno S, Oike Y, and Ishida S. (2007). Macular pigment lutein is antiinflammatory in preventing choroidal neovascularization. *Arterioscler Thromb Vasc Biol* 27:2555–2562.

Johnson EJ, Chung H-Y, Caldarella SM, and Snodderly DM. (2008b). The influence of supplemental lutein and docosahexaenoic acid on serum, lipoproteins, and macular pigmentation. *Am J Clin Nutr* 87:1521–1529.

Johnson EJ, McDonald K, Caldarella SM, Chung HY, Troen AM, and Snodderly DM. (2008a). Cognitive findings of an exploratory trial of docosahexaenoic acid and lutein supplementation in older women. *Nutr Neurosci* 11:75–83.

Johnson MA, Smith MM, and Edmonds JT. (1998). Copper, iron, zinc, and manganese in dietary supplements, infant formulas, and ready-to-eat breakfast cereals. *Am J Clin Nutr* 67:1035S–1040S.

Kim SR, Nakanishi K, Itagaki Y, & Sparrow JR. (2006). Photooxidation of A2-PE, a photoreceptor outer segment fluorophore, and protection by lutein and zeaxanthin. *Exp Eye Res* 82(5):828–839.

Köpcke W, Barker FM, and Schalch W. (1995). Canthaxanthin deposition in the retina: A biostatistical evaluation of 411 patients. *J Toxicol Cut Ocular Toxicol* 14:89–104.

Kvansakul J, Rodriguez-Carmona M, Edgar DF et al. (2006). Supplementation with the carotenoids lutein or zeaxanthin improves human visual performance. *Ophthalmic Physiol Opt* 26:362–371.

Le Marchand L, Hankin JH, Kolonel LN, Beecher GR, Wilkens LR, and Zhao LP. (1993). Intake of specific carotenoids and lung cancer risk. *Cancer Epidemiol Biomarkers Prev* 2:183–187.

Leung IY, Sandstrom MM, Zucker CL, Neuringer M, and Snodderly DM. (2004). Nutritional manipulation of primate retinas, II: Effects of age, n-3 fatty acids, lutein, and zeaxanthin on retinal pigment epithelium. *Invest Ophthalmol Vis Sci* 45(9):3244–3256.

Lucey JL and Dangman B. (1984). A reexamination of the role of oxygen in retrolental fibroplasia. *Pediatrics* 73:82–96.

Moriguchi T, Loewke J, Garrison M, Catalan JN, and Salem N. (2001). Reversal of docosahexaenoic acid deficiency in the rat brain, retina, liver, and serum. *J Lipid Res* 42(3):419–427.

Nagaki Y et al. (2006). The supplementation effect of astaxanthin on accommodation and asthenopia. *J Clin Therap Med* 22(1):41–54.

Neelam K, Nolan J, Loane E, Stack J, O'Donovan O et al. (2006). Macular pigment and ocular biometry. *Vision Res* 46:2149–2156.

Ohgami K, Shiratori K, Kotake S, Nishida T, Mizuki N, Yazawa K, & Ohno S. (2003). Effects of astaxanthin on lipopolysaccharide-induced inflammation in vitro and in vivo. *Invest Ophthalmol Vis Sci* 44(6):2694–2701.

Palozza P and Krinsky NI. (1992). Astaxanthin and canthaxanthin are potent antioxidants in a membrane model. *Arch Biochem Biophys* 297:291–295.

Renzi L and Hammond BR. (2010a). The relation between the macular carotenoids, lutein and zeaxanthin, and temporal vision. *Ophthalmic Physiol Opt* 30(4):351–357.

Renzi L and Hammond BR. (2010b). The effect of macular pigment on heterochromatic luminance contrast. *Exp Eye Res* 91(6):896–900.

Richer S, Stiles W, Graham-Hoffman K, Levin M et al. (2011). Randomized, double-blind, placebo-controlled study of zeaxanthin and visual function in patients with atrophic age-related macular degeneration. *Optometry* 82:667–680.

Richer S, Stiles W, Statkute L, Pulido J, Frankowski J, Rudy D et al. (2004). Double-masked, placebo-controlled, randomized trial of lutein and antioxidant supplementation in the intervention of atrophic age-related macular degeneration: The Veterans LAST Study (Lutein Antioxidant Supplementation Trial). *Optometry* 75(4):216–230.

Rubin LP, Chan GM, Barrett-Reis BM, Fulton AB, Hansen RM, Ashmeade TL, ... & Adamkin DH. (2011). Effect of carotenoid supplementation on plasma carotenoids, inflammation and visual development in preterm infants. *J Perinatology*, 32(6), 418–424.

Seet B, Wong TY, Tan DTH et al. (2001). Myopia in Singapore: Taking a public health approach. *Br J Ophthalmol* 85:521–526.

Snodderly DM. (1995). Evidence for protection against age-related macular degeneration by carotenoids and antioxidant vitamins. *Am J Clin Nutr* 62:1448S–1461S.

Snodderly DM, Auran JD, and Delori FC. (1984). The macular pigment. II. Spatial distribution in primate retinas. *Invest Ophthalmol Vis Sci* 25:674–685.

Stahl W, Nicolai S, Briviba K et al. (1997). Biological activities of natural and synthetic carotenoids: Induction of gap junctional communication and singlet oxygen quenching. *Carcinogenesis* 18:89–92.

Stringham JM, Fuld K, and Wenzel AJ. (2003). Action spectrum for photophobia. *J Opt Soc Am A* 20:1852–1858.

Stringham JM, Garcia PV, Smith PA, McLin LN, & Foutch BK. (2011). Macular pigment and visual performance in glare: benefits for photostress recovery, disability glare, and visual discomfort. *Invest Ophthalmol Vis Sci* 52(10):7406–7415.

Stringham J and Hammond BR. (2007). The glare hypothesis of macular pigment function. *Optom Vis Sci* 84:859–864.

Stringham J and Hammond BR. (2008). Macular pigment and visual performance under glare conditions. *Optom Vis Sci* 85:82–88.

Vishwanathan R, Neuringer M, Schalch W, and Johnson E. (2011). Lutein (L) and zeaxanthin (Z) levels in retina are related to levels in the brain. *FASEB J* 25:344.1.

Vitale S, Sperduto RD, and Ferris FL. (2009). Increased prevalence of myopia in the United States between 1971–1972 and 1999–2004. *Arch Ophthalmol* 127:1632–1639.

Walls GL and Judd HD. (1933). Intra-ocular color filters of vertebrates. *Br J Ophthalmol* 17:641–725.

Werner JS, Steele VG, & Pfoff DS. (1989). Loss of human photoreceptor sensitivity associated with chronic exposure to ultraviolet radiation. *Ophthalmol* 96(10):1552–1558.

Wooten BR and Hammond BR. (2002). Macular pigment: Influences on visual acuity and visibility. *Prog Retinal Eye Res* 21:225–240.

Wu J, Seregard S, and Algvere PV. (2006). Photochemical damage of the retina. *Surv Ophthalmol* 51(5):461–481.

Yuodelis C and Hendrickson AE. (1986). A qualitative and quantitative analysis of the human fovea during development. *Vision Res* 26:847–855.

Zimmer P and Hammond BR. (2007). Lutein and zeaxanthin and the developing retina. *Clin Ophthalmol* 1:181–189.

# 6 Assuring Vitamin A Adequacy to Prevent Eye and Ear Disorders

*Theodor Sauer, Susan Emmett, and Keith P. West, Jr.*

## CONTENTS

## INTRODUCTION

Clinical disorders of the eye (xerophthalmia) and ear represent responses of two different organ systems to systemic depletion of essential, pleiotropic vitamin A. Disruptions in both sensory systems were first attributed to vitamin A deficiency (VAD) in the early twentieth century[1–3] but have coursed strikingly different historical paths in their clinical recognition, investigative pursuit, and public health attribution as VA deficiency disorders. The contrast is evident in today's knowledge of xerophthalmia as the world's leading cause of preventable childhood blindness,[4] with estimates of magnitude updated and monitored regularly by the World Health Organization (WHO),[5] and hearing loss due to VAD remaining in experimental science and largely conjectured in human populations based on far fewer clinical and epidemiological reports[6] and trials.[7,8] Fortunately, both disorders are prevented, if to varied degrees, with adequate preschool VA prophylaxis. Not knowing the latter effect, however, may lead to an understated public health benefit of preventing this deficiency, especially in low-income countries.

As an essential nutrient, VA is required from embryonic development throughout postnatal life influencing cellular differentiation, maturation, energy production, growth, and apoptosis via its roles as a regulatory ligand in gene transcription, mitochondrial metabolism, and photoreceptor activation.[9] Multiple consequences of

VAD and health responses to its repletion can be expected, though require careful investigation to discern. We review here the importance of VA in preventing xerophthalmia and, likely, the severity of ear infection. Both effects are mostly, if not entirely, relevant to undernourished societies where VA interventions have been shown to reduce child,[4] infant,[10] and, in some settings, maternal[11] mortality, providing a nexus for better understanding the urgency when xerophthalmia occurs at any age, beyond its potentially blinding consequences, and mechanisms that may explain roles by which VA protects the ear and hearing.

## EYE: XEROPHTHALMIA

Adequacy of VA protects the eye from at least two major clinically distinct disorders: (1) night blindness, via its role in maintaining the "visual cycle" that enables light detection and perception under low-lit conditions, and (2) keratinizing metaplasia on the conjunctiva and cornea that leads to "Bitot's spots" and corneal lesions, respectively, through its effects on cell differentiation. Both sets of conditions, while regulated by different mechanisms, are jointly classified by the WHO as xerophthalmia[12] (Table 6.1).

## CLINICAL PRESENTATION AND TREATMENT

Xerophthalmia literally means dry eye (xeros—dry; ophthalmia—inflamed eye); however, the term has come to represent the full gamut of ocular signs and symptoms of VAD. Classically, these stages of xerophthalmia follow a well-described pattern

### TABLE 6.1
### WHO/IVACG Classifications and Minimum Prevalence Criteria for Xerophthalmia and Vitamin A Deficiency as a Public Health Problem

| Ocular Symptom/Sign | Minimum Prevalence | Period of High Risk |
|---|---|---|
| Children 1–5 years of age | | |
| Night blindness (XN) | >1.0% | 2–6 years |
| Conjunctival xerosis (X1A) | — | — |
| Bitot's spots (X1B) | >0.5% | 2–6 years |
| Cornea xerosis (X2)/corneal ulceration (X3A)/keratomalacia (X3B) | >0.01% | 1–3 years |
| Xerophthalmic corneal scar (XS) | >0.05% | >1 years |
| Deficient serum retinol (<0.70 μmol/L) | >15.0% | < 5 years |
| Pregnant or lactating women | | |
| Night blindness (XN) during most recent pregnancy | >5.0% | Third trimester |
| Low serum retinol (<1.05 μmol/L) | >20.0% | Third trimester |

*Source:* Adapted from Sommer, A. and Davidson, F.R., *J. Nutr.*, 132, 2845S, 2002; and West, K.P., Jr., *J. Nutr.*, 132, 2857S, 2002.

Characters in parentheses denote the WHO classification scheme for xerophthalmia.

that parallels a decline in VA status.[4] The earliest and most common symptom of VAD is *night blindness (XN)*. The VA derivative, retinal, mediates vision transduction as the ligand for rhodopsin (visual purple), a G-protein-coupled receptor located on the outer segments of rod photoreceptors in the retina. On detecting light, retinal undergoes isomerization, a process that induces a conformational change in rhodopsin that initiates transmission of a cascade of neurochemical signals along the optic nerve, forming a visual image in the brain. Adequate VA nurture is maintained in the retina by the high affinity of retinal pigment epithelium (RPE) for circulating VA and by the recycling of retinoid molecules to and from photoreceptor outer segments within the RPE, known as the visual or retinoid cycle.[13] During dark adaptation, the eye increases levels of rhodopsins to improve scotopic (low illumination) vision. While initial dark adaptation is mediated by cones (the photoreceptors responsible for color and high-resolution vision) after approximately 7 min, rod sensitivity surpasses that of cone (rod-cone break point) and continues to increase several fold over the following ~30 min.[13] Early VAD slows rod-mediated dark adaptation and delays the rod-cone break point. With increasingly severe VA deprivation, threshold sensitivity is reduced first in the rods and eventually in the cones.[4] The ensuing condition of night blindness is recognized in areas of endemic VAD, defined by local terms that typically translate to "twilight blindness" or "chicken eyes" (chickens genetically lack rods), referring to being sighted in the daytime but unable to see after sunset to sunrise.[4] Night blindness typically resolves within 24–48 h of large-dose VA treatment (Table 6.2).

As VA status declines, the conjunctiva of the eye undergoes epithelial metaplasia and keratinization leading to xerosis (*X1A*), likely reflecting defective VA-dependent processes of gene transcription and cell differentiation.[14] Conjunctival xerosis is

---

**TABLE 6.2**

**Xerophthalmia Treatment and Prevention Schedules**

| Age | Treatment at Diagnosis[a] | Prevention Dosage | Prevention Frequency |
|-----|---------------------------|-------------------|----------------------|
| <6 months | 50,000 IU | 50,000 IU | Once within 3 days after birth[b] |
| 6–11 months | 100,000 IU | 100,000 IU | Every 4–6 months |
| 12–59 months | 200,000 IU | 200,000 IU | Every 4–6 months |
| Women | By severity of eye signs[c] | 200,000 IU | Two doses 24 h apart and 6 weeks after delivery |

*Source:* Adapted from Ross, D., *J. Nutr.*, 132, 2902S, 2002.

[a] Treat cases of xerophthalmia and measles on days 1 and 2; give another dose on day 14. For severe malnutrition, give 1 dose on day 1 for prophylaxis.

[b] Based on three trials in South Asia, to date, showing an ~20% reduction in infant mortality following oral supplementation shortly after birth.

[c] For women, give 200,000 IU only for corneal xerophthalmia on days 1, 2, and 14; for night blindness or Bitot's spots, give 10,000 IU per day or 25,000 IU per week for 12 weeks or longer.

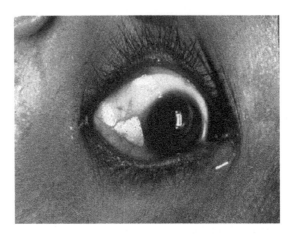

**FIGURE 6.1**    Bitot's spot. (Courtesy of Dr. Randolf Whitfield.)

an irritating keratinizing metaplasia of the conjunctival epithelium that typically precedes corneal involvement.[15] As a mild change, X1A tends to be unreliably recognized, although it can be confirmed by conjunctival impression cytology, a technique involving removal of the superficial epithelial layers by application of a Millipore filter to the conjunctiva, examined by light microscopy. Histologically, abnormal tissue is characterized by large, fragmented, and keratinized epithelial cells with absent goblet cells and occasionally a granular cell layer.[16,17]

Chronic keratinization can lead to the formation of readily detectable whitish, foamy, or bubbly patches of friable, desquamated keratin and bacteria on the conjunctival surface, known as a *Bitot's spots* (*X1B*; Figure 6.1).[12] The lesions are nearly always temporal and nasal when more severe. Often young children with X1B report being night blind,[18] representing a further degree of severity than either condition alone.[4] Bitot's spots typically begin to recede within days and disappear within weeks following treatment with large oral dose VA as recommended by WHO (Table 6.2), although the lesions may also persist "in situ" in older aged children, possibly reflecting a different etiology.[16,18]

Corneal xerophthalmia usually begins as a superficial *punctate keratopathy (PK)* in the inferior nasal quadrant, visible only under fluorescein dye. PK is found in 60% of cases of night blindness and responds rapidly to VA treatment.[18,19] In more severe deficiency, the cornea undergoes xerosis (*X2*), presenting as a dry "non-wettable" surface on hand light exam, typically extending from the inferior quadrant and involving both eyes. Stromal edema may lead the cornea to appear hazy. Severe xerosis gains a "treebark" appearance as the epithelium cornifies.[4,15] Although a stage that can deteriorate quickly, corneal xerosis responds readily to VA treatment (Table 6.2).

*Corneal Ulceration and Keratomalacia* (softening or necrosis of the cornea) *(X3)* are potentially blinding lesions, for which immediate treatment with VA may save sight and improve a child's chance of survival.[12] Corneal ulcers due to VAD are typically "punched out," cylindrical defects up to 2 mm wide, involving ¼–½ of the thickness of the cornea and found in the nasal or inferior quadrants, peripheral to the visual axis. VA treatment will heal the cornea, leaving an opaque

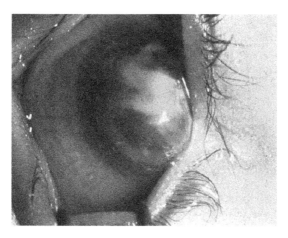

**FIGURE 6.2**   Keratomalacia. (Courtesy of Dr. Alfred Sommer.)

stromal scar (leukoma). An ulcer that perforates the cornea may, on healing, leave an adherent leukoma, associated with a prolapsed iris. Necrosis involves full-thickness dissolution of the cornea, presenting as a gray-to-yellow opaque and edematous lesion that may involve part or the entire cornea. Ulcers and necrosis that involve less than one-third of the corneal surface are classified *X3A*. With VA treatment, these lesions typically contract on healing to spare vision. *X3B* (Figure 6.2) involves more than a third of the cornea, often with stromal sloughing and thinning, in which perforation can cause a loss of intraocular content and a shrunken globe.[4,12] VA treatment will not reverse ocular damage, but will heal the eye and improve a child's chance of survival. Subsequent scarring (*corneal scar or XS*) can affect vision, depending on the size, shape, and location of the scar in relation to the visual axis. Severe, blinding scars can include protuberous staphylomas (uveal tissue) or a phthisical (shrunken) eye.

A rarely investigated ocular sign of VAD is a *xerophthalmic fundus (XF)* or colloquially known as "fundus specks." These discrete yellow-white dots in the fundus, detected on ophthalmoscopic exam, can lead to a mottling of the retina.[20] While XF has no apparent effect on vision, the lesions are speculated to result from disrupted outer rod segments with potential cone and RPE involvement. XF is reversed or diminished after receipt of VA.[20]

## EAR: POTENTIAL HEARING LOSS

While less completely established and understood than for eye disease, compelling evidence suggests that VA plays a significant role in guiding the development and protecting the inner and middle ear from severe infection, either of which may protect hearing. Global estimates of hearing loss remain imprecise, due to insufficient population data; however, recent attempts to model prevalence suggest that there are nearly 540 million hearing-impaired persons of all ages worldwide, most of whom live in low-to-middle-income countries,[21] where access to hearing care is limited. In children, hearing loss often results in permanent speech and language deficits, which can make progression in school particularly challenging.[22,23] Adults with hearing loss are often economically isolated, with incomes 40%–45% lower than the hearing population.[24] Hearing loss can be socially isolating at all ages, in all cultures, resulting in ostracization from the community. Laboratory and epidemiological evidence suggests that preventable nutritional deficiencies may be predisposing factors to hearing loss via diverse mechanisms.[26] Given the substantial impact of hearing impairment on quality of life, that correcting VA deficiency, could protect ear health and prevent hearing loss, raises this possibility to a public health imperative that warrants intensified study.

## GESTATIONAL VITAMIN A DEFICIENCY

Extensive investigation in animal models has outlined the mechanisms by which VAD may affect otic development in utero. Hindbrain development is critical to the formation of a normal otic capsule and inner ear structures. In both rats and quail, it has been noted that the hindbrain incompletely develops in VAD, resulting in immature otic-like vesicles or dysmorphic orthotopic vesicles.[25,26] The loss of hindbrain segmentation and incidence of otic vesicle abnormalities occur in a dose-dependent manner, worsening with more severe levels of deficiency. These fetal otic dysmorphisms can be prevented, however, by titrating a certain level of retinoic acid (RA) back into the diets of VA-deficient mothers.[25] It is hypothesized that RA's critical effects on hindbrain development are related to the homeobox (Hox) genes, a gene family integrally involved in hindbrain patterning that contains RA response elements (RAREs), upstream sequences to which RA receptors bind and activate transcription.[26]

Not only is the amount of VA in the diet critical to fetal otic development, but the timing of its presence is also fundamental. In the mouse, there is a well-defined developmental window in which RA-controlled gene expression is altered and fetal otic abnormalities emerge from exposure to a VA-deficient state.[27] We suspect that there is a similar developmental window of susceptibility to VAD in human pregnancy.

There are also intriguing correlations between RA-controlled genes and hearing loss in humans. RA has a direct effect on the expression of FGF 3 and 10, two fibroblast growth factors required for normal patterning and development of the inner ear.[28] In humans, homozygous mutations of the FGF3 gene have been associated with congenital sensorineural deafness and microtia, and mutations in FGF10 are associated with LADD (lacrimo-auriculo-dento-digital) syndrome, which frequently includes sensorineural hearing loss and external ear anomolies.[29,30] Two indirect RA gene targets, Dlx5 and Dlx6, are further downstream in the otic gene expression pathway and are controlled by FGF 3 and 10 expression.[31–33] Deletions of Dlx5 on chromosome 7q21 have been associated with split-hand/split-foot malformation, a syndrome often accompanied by sensorineural hearing loss.[27,32]

Given the mounting evidence of the critical role of RA in otic development in the animal literature, and genetic studies linking deletions in RA-controlled genes to hearing loss in humans, there is reason to believe that VAD in pregnancy could result in abnormalities in inner ear development and consequently produce hearing loss in affected offspring. This hypothesis has yet to be tested. And while studies on the effects of developmental timing and severity of VAD remain the domain of animal research, it is possible to approach the causality of nutrient restoration, from a public health perspective. This could be done via studies that follow and examine children born to mothers who have participated in placebo-controlled, randomized, antenatal VA supplementation trials in VA-deficient settings, such as those that have been carried out in Nepal,[34] Ghana,[35] and Bangladesh[36] over the past 15 years.

## POSTNATAL VITAMIN A DEFICIENCY

Aside from its role in ear development in utero, VAD has also been shown to contribute to hearing loss postnatally. Chronic otitis media (OM) is one of the most common causes of hearing loss in children and young adults.[26] The mucociliary clearance mechanisms of the middle ear lining are known to be critical in preventing and recovering from OM.[37] However, it has been well established in animal studies that this mucosal lining is severely disrupted with depletion in VA nutriture, resulting in increased incidence of middle ear infection.[38,39]

Unlike developmental perturbances, the relationships between VAD, chronic OM, and hearing loss are increasingly being evaluated in population studies. VA-deficient children exhibit higher risk of having acute OM,[40,41] and risk of progression to chronic disease, attributed to compromised immune and epithelial defenses to infection.[41] Measles provides an example. VA supplementation can markedly lower fatality of children with severe measles, reduce complications, and accelerate recovery.[4] Acute suppurative OM is a frequent complication of measles infection, attributed to eustachian tube obstruction and bacterial infection associated with inflammation.[42] In Kenya, a randomized trial of children with measles demonstrated that patients who received VA had a significantly lower risk of developing OM than those who did not receive a supplement.[7] While these studies did not evaluate hearing loss, given the strong predisposing role of childhood OM to hearing loss, and the likely involvement of VA nutriture in modulating ear infection, causal pathways are plausible. This hypothesis was recently strengthened by findings from a follow-up study of 2,378 young persons, aged 15–23 years, in rural Nepal who had been randomized as preschoolers to receive 4 monthly VA supplementation versus placebo for a 16 month period. In this setting, where ~6% of subjects had hearing loss (based on a pure tone average threshold of $\geq$30 db in the worse ear),[43] and 20% had known episodes of early childhood ear discharge, itself associated with a sixfold higher risk of hearing loss, VA receipt reduced the risk of hearing loss by 42% (95% confidence interval: 8%–63%) compared to controls.[8] The impact was sufficient to reduce by 1% (17% relative reduction) the prevalence of hearing loss from all causes in this young adult age group. This is the first trial to demonstrate a protective effect of VA supplementation against hearing loss associated with middle ear infection. While additional studies are needed to further define the protective relationships between VA, middle ear infection, and hearing loss, this study suggests that there is likely to be additional public health benefit, beyond known reductions in xerophthalmia, nutritional blindness, and mortality,[4] gained by programs that control VAD in undernourished, underserved societies.

## EXTENT OF VITAMIN A DEFICIENCY

The burden of preventable xerophthalmia, and fraction of hearing loss attributable to VAD, in a population is determined largely by the extent and severity of chronic, dietary deficit, compounded by infectious diseases. Global estimates of the prevalence of xerophthalmia and VAD (i.e., serum retinol <0.70 µmol/L), periodically updated by WHO, indicate that 33%, or 190 million, of preschool children in nutritionally vulnerable, low-income countries are VA deficient, of whom 5.17 m (0.9% of all) have night blindness.[5] Approximately one-third of all deficient preschoolers live in India.[44] Few current estimates exist on Bitot's spots or corneal disease, because of a paucity of childhood ophthalmologic surveys conducted globally over the past decade, possibly motivated by a high, sustained coverage of preschool children in undernourished settings with semiannual, large-dose (200,000 IU) VA supplementation, which is ~90% efficacious in preventing xerophthalmia.[45] Yet, while semiannual VA prophylaxis has prevented nutritional blindness and reduced child mortality,[4] hyporetinolemia persists at the same global level, affecting some 190 million children, as estimated two decades ago.[46] This reveals a general failure, to date, of food-based strategies to reach, improve, and sustain normal VA intakes and status in high-risk populations.[47] That diets of the poor are generally low in VA is also suggested by current WHO estimates that ~19 million pregnant women in low-income countries are VA deficient (i.e., serum retinol <0.70 µmol/L), of whom 9.75 m have gestational night blindness,[5] figures that have risen over the past decade,[48] likely due to improved data. A high prevalence and number of VA deficient and night blind gravida may, in chronically undernourished settings, have health and survival consequences for mothers.[11] There may also be health consequences for offspring gestationally deprived of VA, reflected by a lower lung capacity[49] and levels of circulating natural antibodies for the school-aged years.[50] Whether hearing loss is a latent effect of developmental VA insufficiency remains a pressing but unanswered public health research question.

### PREVENTION

Approaches to prevent VAD have been developed and have continued to evolve over the past several decades. These have included (1) periodic, large-dose VA supplementation to preschool-aged children, currently the most widespread, direct, and effective means of controlling VAD disorders; (2) fortification of food items with VA, widely practiced in industrialized countries for over 60 years, but still lagging in most developing countries for lack of processed, fortifiable food vehicles that penetrate markets and diets of the poor; and (3) dietary approaches that increase usual intakes of food sources of preformed VA (animal source foods) and provitamin A carotenoids (dark green leafy and orange-yellow vegetables and fruits), typically through homestead food production (gardening, small animal husbandry, etc.) and nutrition education.[45] Each has advantages with respect to coverage, impact, resources needed, and sustainability. Periodic large-dose VA supplementation can reduce xerophthalmia, child mortality,[51] and possibly hearing loss[8] but appears unable to reliably normalize VA status (i.e., serum retinol level).[47] This strategy, however, has "bought time" for countries to develop longer-term approaches. Yet, with few exceptions, longer-term dietary achievements have yet to be accomplished. Sugar fortification with VA in Guatemala[4] and Central America has been a wide success, providing a template and motivation for fortifying other staple foods such as wheat flour,[52] but most successes to date have been pilot in nature. Food production approaches are gaining welcomed resources and attention as potential local solutions, as has the development and testing of beta-carotene-enhanced staple crops, such as Golden Rice,[53] and other biofortified crops that offer promise of easing stresses of key deficiencies in the future.[54]

## ACKNOWLEDGMENT

Preparation of this chapter was gratefully supported by the Bill and Melinda Gates Foundation (Grant GH614) and the Sight and Life Research Institute, Baltimore, MD.

## REFERENCES

1. Bloch CE. Blindness and other diseases in children arising from deficient nutrition (lack of fat-soluble A factor). *Am J Dis Child* 1924; 27: 139–148.
2. Wolbach SB, Howe PR. Tissue changes following deprivation of fat-soluble a vitamin. *J Exp Med* 1925; 42(6): 753–777.
3. Clausen SW. The effects of moderate deficiency of vitamins. *Bull NY Acad Sci* 1934; 10: 471–482.
4. Sommer A, West KP Jr. *Vitamin A Deficiency: Health, Survival, and Vision.* New York: Oxford University Press, 1996.
5. World Health Organization. Global prevalence of vitamin A deficiency in populations at risk 1995–2005. WHO Global Database on Vitamin A Deficiency. Geneva, Switzerland; 2009. Available at: http://www.who.int/nutrition/publications/micronutrients/vitamin_a_deficiency/9789241598019/en/index.html. Accessed February 2, 2012.
6. Elemraid MA, MacKenzie IJ, Fraser WD, Brabin BJ. Nutritional factors in the pathogenesis of ear disease in children: A systematic review. *Ann Trop Paediatr* 2009; 29: 85–99.
7. Ogaro FO, Orinda VA, Onyango FE, Black RE. Effect of vitamin A on diarrhoeal and respiratory complications of measles. *Trop Geogr Med* 1993; 45: 283–286.
8. Schmitz J, West KP, Khatry SK, Wu L, LeClerq SC, Karna SL, Katz J, Sommer A, Pillion J. Vitamin A supplementation in preschool children and risk of hearing loss as adolescents and young adults in rural Nepal: Randomized trial cohort follow-up study. *BMJ* 2012; 344: d7962
9. Blomhoff R, Blomhoff HK. Overview of retinoid metabolism and function. *J Neurobiol* 2006; 66: 606–630.
10. Klemm RDW, Labrique AB, Christian P, Rashid MR, Shamim AA, Katz J, Sommer A, West KP Jr. Newborn vitamin A supplementation reduced infant mortality in rural Bangladesh. *Pediatrics* 2008; 122: e242–e250.
11. Christian P, West KP Jr., Khatry SK, Kimbrough-Pradhan E, LeClerq SC, Katz J, Shrestha SR, Dali SM, Sommer A. Night blindness during pregnancy and subsequent mortality among women in Nepal: Effects of vitamin A and β-carotene supplementation. *Am J Epidemiol* 2000; 152: 542–547.
12. Sommer A. *Vitamin A Deficiency and Its Consequences: A Field Guide to Detection and Control*, 3rd edn. Geneva, Switzerland: World Health Organization, 1995.
13. Lamb TD, Pugh EN. Phototransduction, dark adaptation, and rhodopsin regeneration. The Proctor Lecture. *Invest Ophthalmol Vis Sci* 2006; 47: 5138–5152.

14. Pfahl M, Chytil F. Regulation of metabolism by retinoic acid and its nuclear receptors. *Annu Rev Nutr* 1996; 16: 257–283.
15. Sommer A. Effects of vitamin A deficiency on the ocular surface. *Ophthalmology* 1983; 90: 592–600.
16. Sommer A, Green WR, Kenyon KR. Bitot's spots responsive and nonresponsive to vitamin A: Clinicopathologic correlations. *Arch Ophthalmol* 1981; 99: 2014–2027.
17. Amedee-Manesme O, Luzeau R, Wittepenn J, Hanck A, Sommer A. Impression cytology detects subclinical vitamin A deficiency. *Am J Clin Nutr* 1988; 47: 875–878.
18. Emran N, Tjakrasudjatma S. Clinical characteristics of vitamin A responsive and nonresponsive Bitot's spots. *Am J Ophthalmol* 1980; 90: 160–171.
19. Sommer A, Emran N, Tamba T. Vitamin A-responsive punctate keratopathy in xerophthalmia. *Am J Ophthalmol* 1979; 87: 330–333.
20. Sommer A, Tjakrasudjatma S, Djunaedi E, Green WR. Vitamin A-responsive panocular xerophthalmia in a healthy adult. *Arch Ophthalmol* 1978; 96: 1630–1634.
21. Stevens G, Flaxman S, Brunskill E, Mascarenhas M, Mathers CD, Finucane M. On behalf of the Global Burden of Disease Hearing Loss Expert Group. Global and regional hearing impairment prevalence: An analysis of 42 studies in 29 countries. *Eur J Publ Health* 2011; 1–2; doi:10.1093/eurpub/ckr17.
22. Moeller PM. Early intervention and language development in children who are deaf and hard of hearing. *Pediatrics* 2000; 106: e43.
23. Kennedy CR, McCann DC, Campbell MJ, Law CM, Mullee M, Petrou S, Watkin P, Worsfold S, Ming Yuen H, Stevenson J. Language ability after early detection of permanent childhood hearing impairment. *N Engl J Med* 2006; 354: 2131–2141.
24. Olusanya BO, Ruben RJ, Parving A. Reducing the burden of communication disorders in the developing world: An opportunity for the Millennium Development Project. *JAMA* 2006; 296: 441–444.
25. White JC, Highland M, Kaiser M, Clagett-Dame, M. Vitamin A deficiency results in the dose-dependent acquisition of anterior character and shortening of the caudal hindbrain of the rat embryo. *Dev Biol* 2000; 220: 263–284.
26. Maden M, Gale E, Kostetskii I, Zile M. Vitamin A-deficient quail embryos have half a hindbrain and other neural defects. *Curr Biol* 1996; 6: 417–426.
27. Frenz DA, Liu W, Cvekl A, Xie Q, Wassef L, Quadro L, Niederreither K, Maconochie M, Shanske A. Retinoid signaling in inner ear development: A "Goldilocks" phenomenon. *Am J Med Genet Part A* 2010; 152A: 2947–2961.
28. Alvarez Y, Alonso MT, Vendrell V, Zelarayan LC, Chamero P, Theil T, Bosl MR, Kato S, Maconochie M, Riethmacher D, Schimmang T. Requirements for Fgf3 and Fgf10 during inner ear formation. *Development* 2003; 130: 6329–6338.
29. Tekin M, Hismi BU, Fitoz S, Ozdag H, Cengiz FB, Sirmaci A, Aslan I, Inceoglu B, Yuksel-Konuk EB, Yilmaz ST et al. Homozygous mutations in fibroblast growth factor 3 are associated with a new form of syndromic deafness characterized by inner ear agenesis, microtia, and microdontia. *Am J Hum Genet* 2007; 80: 338–344.
30. Rohmann E, Brunner HG, Kayserili H, Uyguner O, Nurnberg G, Lew ED, Dobbie A, Eswawakumar VP, Uzumcu A, Ulubil-Emeroglu M et al. Mutations in different components of FGF signaling in LADD syndrome. *Nat Genet* 2006; 38: 414–417.
31. Liu W, Levi G, Shanske A, Frenz DA. Retinoic acid-induced inner ear teratogenesis caused by defective Fgf3/Fgf10-dependent Dlx5 signaling. *Birth Defects Res B Dev Reprod Toxicol* 2008; 83: 134–144.
32. Robledo RF, Lufkin T. *Dlx5* and *Dlx6* Homeobox genes are required for specification of the mammalian vestibular apparatus. *Genesis* 2006; 44: 425–437.
33. Acampora D, Merlo GR, Palaeri L, Zerega B, Postiglione MP, Mantero S, Bober E, Barbieri O, Simeone A, Levi G. Craniofacial, vestibular and bone defects in mice lacking *Distal-less*-related gene *Dlx5*. *Development* 1999; 126: 3795–3809.

34. West KP Jr., Katz J, Khatry SK, LeClerq SC, Pradhan EK, Shrestha SR, Connor PB, Dali SM, Christian P, Pokhrel RP, Sommer A, and the NNIPS-2 Study Group. Double blind, cluster randomized trial of low dose supplementation with vitamin A or β-carotene on mortality related to pregnancy in Nepal. *BMJ* 1999; 318: 570–575.
35. Kirkwood BR, Hurt L, Amenga-Etego S, Tawiah C, Zandoh C, Danso S, Hurt C, Edmond K, Hill Z, Ten Asbroek G, Fenty J, Owusu-Agyei S, Campbell O, Arthur P, ObaapaVitA Trial Team. Effect of vitamin A supplementation in women of reproductive age on maternal survival in Ghana (ObaapaVitA): A cluster-randomised, placebo-controlled trial. *Lancet* 2010; 375: 1640–1649.
36. West KP Jr., Christian P, Labrique AB, Rashid M, Shamim AA, Klemm RDW, Massie AE, Mehra S, Schulze KJ, Ali H, Ullah B, Wu LSF, Katz J, Banu H, Akhter H, Sommer A. Effects of vitamin A or beta-carotene supplementation on pregnancy-related mortality and infant mortality in rural Bangladesh: A cluster-randomized trial. *JAMA* 2011; 305: 1986–1995.
37. Sade J, Ar A. Middle ear and auditory tube: Middle ear clearance, gas exchange, and pressure regulation. *Otolaryngol Head Neck Surg* 1997; 116: 499–524.
38. Chole RA. Squamous metaplasia of the middle ear mucosa during vitamin A deprivation. *Otolaryngol Head Neck Surg* 1979; 87: 837–844.
39. Manning SC, Wright CG. Incidence of otitis media in vitamin A-deficient guinea pigs. *Otolaryngol Head Neck Surg* 1992; 107: 701–706.
40. Lloyd-Puryear M, Humphrey J, West KP, Aniol K, Mahoney F, Mahoney J et al. Vitamin A deficiency and anemia among Micronesian children. *Nutr Res* 1989; 9: 1007–1016.
41. Lasisi AO. The role of retinol in the etiology and outcome of suppurative otitis media. *Eur Arch Otorhinolaryngol* 2009; 266: 647–652; doi 10.1007/s00405-008-0794-6.
42. Perry RT, Halsey NA. The clinical significance of measles: A review. *J Infect Dis* 2004; 189(Suppl 1): S4–S16.
43. Schmitz J, Pillion J, LeClerq SC, Khatry SK, Wu LSF, Prasad R et al. Prevalence of hearing loss and ear morbidity among adolescents and young adults in rural southern Nepal. *Int J Audiol* 2010; 49: 388–394.
44. National Institute of Nutrition, Indian Council of Medical Research. Prevalence of micronutrient deficiencies. National Nutrition Monitoring Bureau (NNMB) Technical Report No. 22, Hyderabad, India: National Institute of Nutrition, 2003.
45. West KP Jr., Darnton-Hill I. Vitamin A deficiency. In *Nutrition and Health in Developing Countries*, 2nd edn. (eds. R. D. Semba and M. W. Bloem). Totowa, NJ: Humana Press, 2008.
46. Underwood BA. Vitamin A in human nutrition. Public health considerations. In *The Retinoids: Biology, Chemistry, and Medicine*, 2nd edn. (eds. M. B. Sporn, A. B. Roberts, and D. S. Goodman). New York: Raven Press Ltd, 1994; Vol. 4, pp. 211–227.
47. Palmer AC, West KP, Dalmiya N, Schultink W. The use and interpretation of serum retinol distributions in evaluating the public health impact of vitamin A programmes. *Public Health Nutr* 2012; 9: 1–15.
48. West KP Jr. Extent of vitamin A deficiency among preschool children and women of reproductive age. *J Nutr* 2002; 132: 2857S–2866S.
49. Checkley W, West KP Jr., Wise RA, Baldwin MR, Wu L, LeClerq SC, Christian P, Katz J, Tielsch JM, Khatry SK, Sommer A. Maternal Vitamin A Supplementation and Lung Function in Offspring. *N Engl J Med* 2012; 362: 1784–1794.
50. Palmer AC. Early life nutritional exposures and long-term programming of the immune system. Doctoral dissertation. Baltimore, MD: Johns Hopkins University, 2010.
51. Mayo-Wilson E, Imdad A, Herzer K, Yakoob MY, Bhutta ZA. Vitamin A supplements for preventing mortality, illness, and blindness in children aged under 5: Systematic review and meta-analysis. *BMJ* 2011; 343: d5094; doi: 10.1136/bmj.d5094.
52. Klemm RD, West KP Jr., Palmer AC, Johnson Q, Randall P, Ranum P, Northrop-Clewes C. Vitamin A fortification of wheat flour: Considerations and current recommendations. *Food Nutr Bull* 2010; 31(1 Suppl): S47–S61.

53. Tang G, Hu Y, Yin S, Wang Y, Dallal GE, Grusak MA, Russell RM. β-carotene in golden rice is as good as β-carotene in oil in providing vitamin A to children. *Am J Clin Nutr* 2012; 96: 658–664; doi:10.3945/ajcn.111.030775.

54. Nestel P, Bouis HE, Meenakshi JV, Pfeiffer W. Biofortification of staple food crops. *J Nutr* 2006; 136: 1064–1067.

55. Sommer A, Davidson FR. Assessment and control of vitamin A deficiency: The Annecy Accords. *J Nutr* 2002; 132: 2845S–2850S.

56. Ross D. Recommendations for vitamin A supplementation. *J Nutr* 2002; 132: 2902S–2906S.

# 7 Disturbed Accumulation and Abnormal Distribution of Macular Pigment in Retinal Disorders

*Tos T.J.M. Berendschot, Charbel Issa Peter, and Thomas Theelen*

## CONTENT

In primates, the fovea appears as a yellow spot, as noted already in 1762 by Buzzi [1]. A first thought was that this represented a retinal hole [2] or was due to postmortem changes [3,4]. However, in particular, after Wald's paper on the photochemistry of vision [5], it was clear that it is caused by a pigment, which is concentrated in the central area of the retina, the so-called macular pigment [6]. Its specific yellow color is caused by carotenoids [7], more in particular the stereo-isomers lutein and zeaxanthin [8,9] and the intermediate meso-zeaxanthin [10]. Humans are unable to synthesize these carotenoids and are therefore dependent upon dietary intake. Serum levels associated with the normal diet are far below the maximal levels achieved by supplementation. The normal Western diet contains 1.3–3 mg/day of lutein and zeaxanthin combined [11]. The highest concentration of lutein is found in food sources with yellow color, such as corn and egg yolk. Dark, leafy green vegetables, such as spinach and kale, are also good sources of lutein. The same is true for orange peppers, followed by egg yolk, corn, and orange juice for zeaxanthin [12–15]. In general, fruit and vegetables contain 7–10 times more lutein than zeaxanthin. Meso-zeaxanthin is virtually nonexistent in food sources originating from plants. Lutein serum levels and the amount of macular pigment correlate, in particular in men [16–20]. It has further been shown that the amount of macular pigment can be increased by a dietary modification [11,21–23] or by supplements [17,24–38] in healthy subjects as well as in subjects with a diseased macula [39–42]. Note that meso-zeaxanthin is primarily formed in the retina following conversion from lutein [10,43].

Some argue that the presence of the macular pigment in the fovea is merely the remains of an earlier form of color vision [44]. However, also many potential functions are attributed to the macular pigment, like reducing the consequences of chromatic aberration [45–47], minimizing stray light [48], preserving visual sensitivity in older subjects [49], and improving glare disability and photostress recovery [50–52]. In 1994, Seddon et al. observed an inverse association between a diet with a high content of the carotenoids lutein and zeaxanthin, and the prevalence of age-related macular degeneration (AMD) [53]. This pointed to even another potential role of macular pigment, which is almost exclusively composed of these two carotenoids [9,54], namely, in preventing AMD. AMD is a degenerative disease of the retina and, in the developed world, the leading cause of visual impairment in people 50 years and older [55–57]. In its end stage, it affects the macula that enables the central high-resolution visual acuity. Dysfunction of the macula results in the inability to see fine detail, read, and recognize faces. As a consequence, the disease has a great impact on the quality of life and large negative financial and economic consequences [58–63]. Recently, new treatments have emerged for AMD. Unfortunately, they are suitable only for the small proportion of people with the neovascular form. No treatments are available for geographic atrophy AMD, which makes it important to identify preventative factors for the development of AMD [64]. Apart from quitting smoking, macular pigment, which can be modified by diet, could well be an easy target to decrease susceptibility for AMD [65,66]. There are plausible arguments to assume that macular pigment exerts a protective effect in the retina [54]. First of all, it acts as a blue light filter, absorbing between 390 and 540 nm [67–70], thereby decreasing chances for photochemical light damage [71]. In addition, macular pigment is capable of scavenging free radicals [72]. Finally, lutein is capable to suppress inflammation [73]. Many papers have been published that study the relation between lutein, zeaxanthin, and macular pigment on the one hand and AMD on the other hand [33,74–93]. They are discussed in the chapter "Lutein and zeaxanthin and eye health: evidence from epidemiological studies."

In healthy eyes, the absorption of macular pigment peaks at the center of the fovea and decreases with eccentricity. An easy subjective measurement to observe macular pigment oneself is Maxwell's spot [94]: Staring at a bright yellow surface for a while and then at a blue surface, a dark spot can be seen, which shows the presence of macular pigment. The darker it is, the more macular pigment is present. Objectively, the macular pigment distribution can be quantified using fundus reflectance or autofluorescence [24,95–100]. Since the lens and the macular pigment are the only absorbers in this wavelength region, digital subtraction of the log of the resulting gray-level maps at the two wavelengths provides density maps of the sum of both absorbers. The macular pigment optical density is assumed to be negligible at a peripheral site. If then this site is used to provide an estimate for the lens density, the mean macular pigment optical density at the fovea can be calculated [24,99], and its spatial distribution can be imaged easily and with a high resolution. At first sight, the macular pigment optical density spatial distribution seems to decrease monotonously to very low values at an eccentricity of about 10°. This has for a long time been the common belief. However, doubts were already expressed in 1954 by Miles [101]. When studying the Maxwell's Spot [94], he found an additional ring in 14 of the 19 subjects that were able to see this phenomenon. Berendschot et al. [97] and Delori et al. [102] analyzed the spatial distribution in detail using reflectance and autofluorescence maps. They found a distinct ring pattern at a distance of about 0.7° of the fovea. In some subjects, the ring has an even larger optical density than the central peak. The distribution could be modeled with an exponentially decaying density as a function of eccentricity, in combination with a Gaussian distributed ring pattern that was visible in about 50% of all subjects [97]. Using the same technique, similar results were found by others [103–105]. Several authors studied the spatial distribution using the technique of heterochromatic flicker photometry. Although this technique has a low spatial resolution, they nevertheless also noted deviations from an exponential decreasing function with eccentricity between 0.5° and 1.0° showing up as valleys or flanking peaks [106–108]. The maximum value of the macular pigment optical density as well as the prominence of the ring seems to be independent of age [97,109,110]. Some authors found an association between the spatial profile of the macular pigment optical density and an individual's foveal architecture [111,112]. However, others did not observe such a relation [108,113], which may in part be due to a different measurement procedure [112]. Other biometric parameters do not correlate with macular pigment optical density [114].

The underlying cause of the bimodal distribution is still in debate. A possible explanation is the distribution of the macular pigments in different layers in retina. Snodderly et al. and Trieschmann et al. analyzed retinal sections [115,116]. One of their observations was that there appeared to be large interindividual differences. On average, the macular pigment was most dense in the outer plexiform layer (also known as the receptor axons or Henle fiber layer) [117]. However, they also found a high density in the inner plexiform layer [118]. Another cause might be the fact that lutein and zeaxanthin are not distributed evenly over the retina. In the central 0–2.3 mm, zeaxanthin predominates over lutein, whereas for eccentricities beyond 2.3 mm, lutein is the major carotenoid [9,119,120]. Figure 7.1 shows mean lutein and zeaxanthin profiles of 19 healthy individuals (13 women, 6 men, aged $26 \pm 8$ years) determined by spectral fundus reflectance [119]. Snodderly et al. used micro-densitometry to study possible associations between the spatial distribution of lutein and zeaxanthin and the spatial distribution of the macular pigment optical density [121]. They reported a macular pigment optical density distribution with a central peak with shoulders around 0.8° eccentricity. The main peak was associated with macular pigment along the receptor axons, whereas the shoulders followed the inner plexiform layer. However, both lutein and zeaxanthin reached their highest concentrations at the center of the fovea and declined monotonically with eccentricity. The discrepancy between these two findings was explained by variations in the orientation of the dichroic lutein and zeaxanthin as a function of eccentricity [122,123]. As mentioned earlier, the macular pigment optical density can be modified only by diet or supplementation. Most of the studies hereon measured only the peak macular pigment optical density. A supplementation study in rhesus monkeys however stated that the increase in macular pigment optical density was substantially higher in the periphery than in the central fovea [36]. On the other hand, if supplemented preferentially with zeaxanthin, the center optical density seems to show a larger increase than the periphery [41,119]. Note that high doses dietary supplementation of a single carotenoid may alter the assimilation of other carotenoids [124–126].

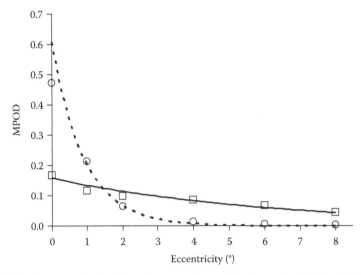

**FIGURE 7.1**   Mean lutein (circles) and zeaxanthin profiles (squares) of 19 healthy individuals, 13 women, 6 men, aged 26 ± 8 years, determined by spectral fundus reflectance. Lines are guides to the eye, based on an exponential decreasing function with eccentricity. (From van de Kraats, J. et al., *Invest. Ophthalmol. Vis. Sci.*, 49, 5568, 2008.)

More than 600 carotenoids exist in nature, of which 30–50 carotenoids are present in the human diet, and about 15 are detected in the serum [127]. However, in the human retina, only lutein, zeaxanthin, and meso-zeaxanthin are found [128]. Moreover, the spatial distribution of these carotenoids peaks in the fovea with up to millimolar concentrations, although different for lutein and zeaxanthin (see Figure 7.1) [8–10]. This hints at the presence of high-affinity binding proteins. Bernstein and his group took up the challenge to search these specific carotenoid-binding proteins. They first identified a pi isoform of glutathione S-transferase (GSTP1) as the zeaxanthin-binding protein in the macula of the human retina [129]. Only recently they identified StARD3 as a lutein-binding protein in the macula of the primate retina [130]. The distribution of the GSTP1 seems to mirror the distribution of zeaxanthin in the retina, with high concentrations in the inner and outer plexiform layers of the macula near the fovea [131]. On the other hand, StARD3 localizes to all neurons of monkey macular retina and especially axons and cone inner segments [130].

Although the specific deposition of lutein and zeaxanthin within the retina has been identified to be mediated by GSTP1 and StARD3, our understanding of the transport processes of these carotenoids from serum to the particular layers within the retina is still incomplete. It may well be that the uptake of carotenoids shares the transport pathway with cholesterol [132–136], and that HDL and the receptors of HDL such as SR-BI may be involved in this process for delivery to the RPE [137–139]. Li et al. speculated that carotenoids are delivered from the RPE to the retina by a pathway analogous to the one used for retinoid transport that employs inter-photoreceptor retinoid-binding protein to facilitate transport of hydrophobic ligands across the inter-photoreceptor space [130].

Studying diseased retinas can be very useful to improve our understanding of macular pigment physiology. Although in most retinal degenerations the spatial profile of the macular pigment optical density resembles that of healthy subjects, recently two diseases have been suggested to show an abnormal distribution of macular pigment or even a complete absence that are type 2 idiopathic macular telangiectasia and Sjögren–Larsson syndrome (SLS).

Type 2 idiopathic macular telangiectasia (MacTel) usually becomes symptomatic between the fifth and seventh decade. It is characterized by slow, but progressive loss of visual acuity, reading difficulties, and/or metamorphopsia [140–143]. Well-defined parafoveal scotomas develop with disease progression [144]. Morphologic and functional alterations, such as telangiectatic vessels with corresponding leakage on fluorescence angiography, foveal pseudocysts, and retinal pigment epithelial clumping, are most pronounced on the temporal macula [140,141,144–149]. Charbel Issa et al. noticed a parafoveal increase in confocal blue reflectance in 98% of eyes with type 2 idiopathic macular telangiectasia [147]. The area of this increased reflectance was found to be slightly larger than the area of hyperfluorescence in late-phase fluorescein angiography. Using autofluorescence and reflectance maps, they concluded that a reduction of the macular pigment optical density within the central retina with a surrounding ring-like structure of preserved macular pigment at about 6° eccentricity was evident in most eyes with macular telangiectasia type 2 [98]. Figure 7.2 shows maps of the spatial distribution of macular pigment optical density of 14 MacTel patients obtained by two-wavelength fundus autofluorescence, as explained earlier. They all show absence of macular pigment at the fovea, however, some macular pigment at about 6° eccentricity. It can be argued that the results showing reduced macular pigment optical density may have been influenced by unknown absorption or reflection of fluorescent substances, e.g., a supposed lack of macular pigment could as well be due to unknown absorbers in the retina. Moreover, it was also unclear, if the macular pigment was just preserved or even increased eccentric to its central reduction. In order to solve this question, Charbel Issa et al. measured 28 eyes of 14 MacTel patients using spectral fundus reflectometry and compared the outcome with two-wavelength fundus autofluorescence data [150]. Spectral fundus reflectometry is capable of distinguishing between the different spectral characteristics of the various absorbing and reflecting substances within the eye [95,151–153]. It allows quantification of the macular pigment optical density and distinguishes between its constituents lutein and zeaxanthin and as such is a suitable method to investigate the reduction or accumulation of macular pigment in retinal regions of interest [154]. In accordance with data obtained from autofluorescence maps, they found a reduced macular pigment optical density within the central 4° eccentricity with a climax of decrease temporal to the foveola. At 6°, the macular pigment optical density was not reduced. Further, MacTel patients showed a high zeaxanthin reduction, which again suggests that in MacTel type 2, there might be an inability to accumulate macular pigment in particular in the central retina.

SLS is an autosomal recessively inherited disorder, characterized by mental retardation, spastic diplegia, and congenital ichthyosis [155]. The disease is caused by an enzymatic defect in lipid metabolism, microsomal fatty aldehyde dehydrogenase (FALDH) deficiency. FALDH catalyzes the oxidation of many different medium- and long-chain fatty aldehydes into fatty acids. Its deficiency results in the

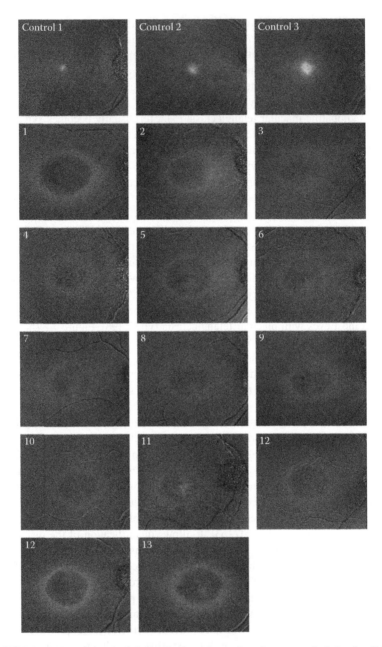

**FIGURE 7.2** Maps of the spatial distribution of macular pigment optical density (MPOD) obtained by two-wavelength fundus autofluorescence. The upper row shows maps of three controls subjects. All other maps are from patients with type 2 idiopathic macular telangiectasia (MacTel). The presented controls demonstrate the MPOD distribution that shows a foveal peak. In contrast, recordings in MacTel patients exhibit a reduced central MPOD signal and a surrounding ring-like structure. (Reprinted from Charbel Issa, P. et al., *Exp. Eye. Res.*, 89, 25, 2009. With permission.)

accumulation of fatty aldehydes and fatty alcohols in body tissues, which is considered the principal causative mechanism leading to the clinical symptoms. Patients with SLS also exhibit a typical crystalline juvenile macular dystrophy with onset in childhood, which is characterized by reduced visual acuity and photophobia [156]. Van der Veen et al. noticed that maculae of SLS patients lack the normal dark-yellowish appearance [157]. To decide whether an altered distribution or concentration of macular pigment caused this observation, they set out to measure 32 eyes of 16 patients with spectral fundus reflectometry and fundus autofluorescence. Due to limited cognitive capacities, photophobia, and inability for stable fixation due to spasticity, fundus reflectometry could be performed only in 19 eyes of 10 patients and autofluorescence in 9 eyes of 5 patients. The autofluorescence images lacked the typical attenuation in the macular region that is seen in healthy eyes with a normal macular pigment distribution (see Figure 7.3). In accordance with this observation,

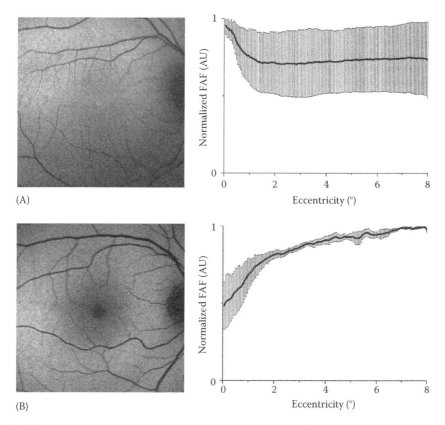

**FIGURE 7.3** Fundus autofluorescence characteristics in SLS. Fundus autofluorescence sample images and profiles of mean (error bars indicate standard deviation of mean) in (A) SLS patients and (B) healthy age-matched controls. The SLS patients show an autofluorescence maximum within 1° eccentricity, which is roughly 25% more than the peripheral levels. Instead, healthy individuals show central attenuation of autofluorescence of approximately 50% and highest values in the periphery. AU, arbitrary units. (Reprinted from van der Veen, R.L.P. et al., *Ophthalmology*, 117, 966, 2010. With permission.)

mean foveal macular pigment optical density determined by fundus reflectometry showed significantly lower values in SLS patients than in healthy individuals [157].

Apart from this apparent inability to accumulate macular pigment, there are other features that link MacTel and SLS, namely, the presence of yellow-white refractile crystals arranged along the Henle fibers and at the inner plexiform layer and central pseudocystic spaces in the macula [158–160]. Recently, Sallo et al. studied in detail the intraretinal crystalline deposits in a large cohort with type 2 MacTel using multiple imaging modalities [161] and found an association between the amount of crystals and both the fluorescein leakage and macular pigment loss.

Müller cell loss may explain the observed lack of macular pigment in the fovea. Recently, Powner et al. found a lack of Müller cell immunostaining in an eye of a MacTel patient within an area of the same extent as the loss of macular pigment [6]. As suggested by Gass, the Müller cell cone acts as a reservoir for the retinal xanthophylls that make up the macular pigment (see Figure 7.4) [162]. It is therefore possible that the central population of Müller cells forms its own cell entity, which is affected especially in MacTel and SLS. The fact that predominantly zeaxanthin is lost in both diseases points to the Müller cell cone to be the host for this carotenoid. The idea of Müller cell loss is consistent with the fact that Müller cell degeneration seemed to be involved in the development of the juvenile maculopathy in SLS [163]. Further, it is also corroborated by a reduced numbers of cell nuclei counted in the inner nuclear layer [6] and retinal thinning found in MacTel patients [164].

Another link between Müller cell loss and a low macular pigment optical density might be caused by the following. The amount of macular pigment depends on the intake of lutein and zeaxanthin. They are absorbed by intestinal mucosal cells and incorporated into chylomicrons that are secreted into the lymphatic system and subsequently enter the circulation [165]. Hepatic cells incorporate the carotenoids from the chylomicrons into lipoproteins that facilitate further transport of the carotenoids to the various body tissues [166]. Retinal Müller cells produce apolipoprotein-E (Apo-E), which is to known to play a part in lipid transport and binding of lipoproteins to target sites within the central nervous system and in the targeted uptake of

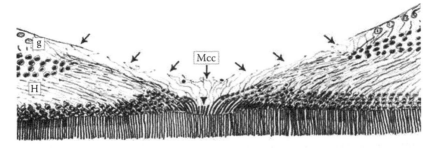

**FIGURE 7.4** Drawing of the anatomy of the fovea centralis showing Müller cell cone (Mcc) whose base (arrows) corresponds with the internal limiting membrane, and whose apex corresponds with the outer limiting membrane centrally (arrowhead). Henle nerve fiber layer (H) and foveal edge of the ganglion cell layer (g) are shown. (The drawing represents the interpretation by Gass of a photomicrograph from Yamada, E., *Arch. Ophthalmol.*, 82, 151, 1969; Reprinted from Gass, J.D., *Arch. Ophthalmol.*, 117, 821, 1999. With permission.)

the lipoproteins carrying lutein and zeaxanthin within the retina [167,168]. This is in accordance with the fact the reflective properties of the intraretinal crystals, which are located in the Henle layer and the inner plexiform layer, are compatible with those of lipids [161]. The crystals also show an association with a minimal level of retinal thickening in MacTel patients, consistent with Müller cell loss [161].

In MacTel, late-phase fluorescein angiography reveals a diffuse hyperfluorescence in the parafoveal area [156]. As yet, an unanswered question is what causes this late hyperfluorescence [169,170]. Although the leakage pattern on fluorescein angiography resembles that of cystoid macula edema [171], it lacks the typical cystoid petaloid appearance, suggesting a different pathophysiology [172]. Furthermore, in MacTel subjects, retinal thickness is found to be normal or even below [146,173]. Also, treatment with anti-VEGF seems less effective in MacTel than in cystoid macula edema. Müller cells extend branches that interdigitate with retinal neurons and glial cells, and their processes ensheathe the retinal capillaries [174]. By this, they form an important part of the inner blood–retina barrier [175,176]. Thus, it may well be that the late-phase hyperfluorescence in MacTel is caused by fluid that accumulates in intraretinal space devoid of Müller cells.

To conclude, macular pigment, which constitutes of the carotenoids lutein, zeaxanthin, and meso-zeaxanthin, peaks at the foveal center. Its distribution can be modeled with an exponentially decaying density as a function of eccentricity, in combination with a Gaussian distributed ring pattern at about $0.7°$ of the fovea. Most retinal degenerations show a macular pigment spatial profile that resembles that of healthy subjects. Hitherto, two diseases with macular pigment depletion are known: type 2 idiopathic macular telangiectasia and the SLS. The deposition of lutein and zeaxanthin within the retina has been identified to be mediated by GSTP1 and StARD3. Müller cells seem to play an important role in the transport of these carotenoids.

## REFERENCES

1. Buzzi F. Nuovo esperienze fatte sull' occhio umano. *Opuscoli scelti di Milano* 1782; 5: 87.
2. Soemmering S. De Foramine Centralis limbo luteo cincto retinae humane. *Soc Reg Sic Goetting* 1799; 13: 3.
3. Gullstrand F. Die farbe der Macula centralis retinae. *Graefe's Arch Clin Exp Ophthalmol* 1906; 62: 1–72.
4. Nordenson JW. Neue Argumente zur der Frage der Makulafarbe. *Graefe's Arch Clin Exp Ophthalmol* 1958; 160: 43–46.
5. Wald G. The photochemistry of vision. *Doc Ophthalmol* 1949; 3: 94–137.
6. Powner MB, Gillies MC, Tretiach M et al. Perifoveal müller cell depletion in a case of macular telangiectasia type 2. *Ophthalmol* 2010; 117: 2407–2416.
7. Wald G. Human vision and the spectrum. *Science* 1945; 101: 653–658.
8. Bone RA, Landrum JT, Tarsis SL. Preliminary identification of the human macular pigment. *Vision Res* 1985; 25: 1531–1535.
9. Bone RA, Landrum JT, Fernandez L, Tarsis SL. Analysis of the macular pigment by HPLC: Retinal distribution and age study. *Invest Ophthalmol Vis Sci* 1988; 29: 843–849.
10. Bone RA, Landrum JT, Hime GW, Cains A, Zamor J. Stereochemistry of the human macular carotenoids. *Invest Ophthalmol Vis Sci* 1993; 34: 2033–2040.

11. Hammond BR, Johnson EJ, Russell RM et al. Dietary modification of human macular pigment density. *Invest Ophthalmol Vis Sci* 1997; 38: 1795–1801.
12. Sommerburg O, Keunen JEE, Bird AC, van Kuijk FJGM. Fruits and vegetables that are sources for lutein and zeaxanthin: The macular pigment in human eyes. *Br J Ophthalmol* 1998; 82: 907–910.
13. Granado F, Olmedilla B, Herrero C, Perez-Sacristan B, Blanco I, Blazquez S. Bioavailability of carotenoids and tocopherols from *Broccoli*: In vivo and in vitro assessment. *Exp Biol Med* 2006; 231: 1733–1738.
14. Maiani G, Periago Caston MJ, Catasta G et al. Carotenoids: Actual knowledge on food sources, intakes, stability and bioavailability and their protective role in humans. *Mol Nutr Food Res* 2008; 53: S198–S203.
15. Thurnham DI. Macular zeaxanthins and lutein? A review of dietary sources and bioavailability and some relationships with macular pigment optical density and age-related macular disease. *Nutr Res Rev* 2007; 20: 163–179.
16. Broekmans WMR, Berendschot TTJM, Klöpping WA et al. Macular pigment density in relation to serum and adipose tissue concentrations of lutein and serum concentrations of zeaxanthin. *Am J Clin Nutr* 2002; 76: 595–603.
17. Johnson EJ, Neuringer M, Russell RM, Schalch W, Snodderly DM. Nutritional manipulation of primate retinas, III: Effects of lutein or zeaxanthin supplementation on adipose tissue and retina of xanthophyll-free monkeys. *Invest Ophthalmol Vis Sci* 2005; 46: 692–702.
18. Johnson EJ, Hammond BR, Yeum KJ et al. Relation among serum and tissue concentrations of lutein and zeaxanthin and macular pigment density. *Am J Clin Nutr* 2000; 71: 1555–1562.
19. Landrum JT, Bone RA, Chen Y, Herrero C, Llerena CM, Twarowska E. Carotenoids in the human retina. *Pure Appl Chem* 1999; 71: 2237–2244.
20. Loane E, Beatty S, Nolan JM. The relationship between lutein, zeaxanthin, serum lipoproteins and macular pigment optical density. *Invest Ophthalmol Vis Sci* 2009; 50: 1710.
21. Wenzel AJ, Gerweck C, Barbato D, Nicolosi RJ, Handelman GJ, Curran-Celentano J. A 12-Wk egg intervention increases serum zeaxanthin and macular pigment optical density in women. *J Nutr* 2006; 136: 2568–2573.
22. Burke JD, Curran-Celentano J, Wenzel AJ. Diet and serum carotenoid concentrations affect macular pigment optical density in adults 45 years and older. *J Nutr* 2005; 135: 1208–1214.
23. Chung HY, Rasmussen HM, Johnson EJ. Lutein bioavailability is higher from lutein-enriched eggs than from supplements and spinach in men. *J Nutr* 2004; 134: 1887–1893.
24. Berendschot TTJM, Goldbohm RA, Klöpping WA, van de Kraats J, van Norel J, van Norren D. Influence of lutein supplementation on macular pigment, assessed with two objective techniques. *Invest Ophthalmol Vis Sci* 2000; 41: 3322–3326.
25. Landrum JT, Bone RA, Joa H, Kilburn MD, Moore LL, Sprague KE. A one year study of the macular pigment: The effect of 140 days of a lutein supplement. *Exp Eye Res* 1997; 65: 57–62.
26. Bone RA, Landrum JT, Guerra LH, Ruiz CA. Lutein and zeaxanthin dietary supplements raise macular pigment density and serum concentrations of these carotenoids in humans. *J Nutr* 2003; 133: 992–998.
27. Snodderly DM, Chung HC, Caldarella SM, Johnson EJ. The influence of supplemental lutein and docosahexaenoic acid on their serum levels and on macular pigment. *Invest Ophthalmol Vis Sci* 2005; 46: 1766.
28. Tanito M, Obana A, Okazaki S, Ohira A, Gellermann W. Change of macular pigment density quantified with resonance raman spectrophotometry and autofluorescence imaging in normal subjects supplemented with oral lutein or zeaxanthin. *Invest Ophthalmol Vis Sci* 2009; 50: 1716.

29. Bhosale P, Zhao DY, Bernstein PS. HPLC measurement of ocular carotenoid levels in human donor eyes in the lutein supplementation era. *Invest Ophthalmol Vis Sci* 2007; 48: 543–549.

30. Bone RA, Landrum JT, Cao Y, Howard AN, varez-Calderon F. Macular pigment response to a supplement containing meso-zeaxanthin, lutein and zeaxanthin. *Nutr Metab (Lond)* 2007; 4: 12.

31. Zeimer M, Hense HW, Heimes B, Austermann U, Fobker M, Pauleikhoff D. The macular pigment: Short- and intermediate-term changes of macular pigment optical density following supplementation with lutein and zeaxanthin and co-antioxidants. The LUNA Study. *Ophthalmologe* 2009; 106: 29–36.

32. Köpcke W, Schalch W, LUXEA-study group. Changes in macular pigment optical density following repeated dosing with lutein, zeaxanthin, or their combination in healthy volunteers—Results of the LUXEA-study. *Invest Ophthalmol Vis Sci* 2005; 46: 1768.

33. Richer S, Devenport J, Lang JC. LAST II: Differential temporal responses of macular pigment optical density in patients with atrophic age-related macular degeneration to dietary supplementation with xanthophylls. *Optometry* 2007; 78: 213–219.

34. Rougier MB, Delyfer MN, Korobelnik JF. Measuring macular pigment in vivo. *J Fr Ophtalmol* 2008; 31: 445–453.

35. Johnson EJ, Chung HY, Caldarella SM, Snodderly DM. The influence of supplemental lutein and docosahexaenoic acid on serum, lipoproteins, and macular pigmentation. *Am J Clin Nutr* 2008; 87: 1521–1529.

36. Neuringer M, Sandstrom MM, Johnson EJ, Snodderly DM. Nutritional manipulation of primate Retinas, I: Effects of lutein or zeaxanthin supplements on serum and macular pigment in xanthophyll-free rhesus monkeys. *Invest Ophthalmol Vis Sci* 2004; 45: 3234–3243.

37. Schalch W, Cohn W, Barker FM et al. Xanthophyll accumulation in the human retina during supplementation with lutein or zeaxanthin—The LUXEA (lutein xanthophyll eye accumulation) study. *Arch Biochem Biophys* 2007; 458: 128–135.

38. Connolly EE, Beatty S, Loughman J, Howard AN, Louw MS, Nolan JM. Supplementation with all three macular carotenoids: Response, stability, and safety. *Invest Ophthalmol Vis Sci* 2011; 52: 9207–9217.

39. Koh HH, Murray IJ, Nolan D, Carden D, Feather J, Beatty S. Plasma and macular responses to lutein supplement in subjects with and without age-related maculopathy: A pilot study. *Exp Eye Res* 2004; 79: 21–27.

40. Trieschmann M, Beatty S, Nolan JM et al. Changes in macular pigment optical density and serum concentrations of its constituent carotenoids following supplemental lutein and zeaxanthin: The LUNA study. *Exp Eye Res* 2007; 84: 718–728.

41. Richer SP, Stiles W, Graham-Hoffman K et al. Randomized, double-blind, placebo-controlled study of zeaxanthin and visual function in patients with atrophic age-related macular degeneration: The zeaxanthin and visual function study (ZVF) FDA IND #78, 973. *Optometry* 2011; 82: 667–680.

42. Weigert G, Kaya S, Pemp B et al. Effects of lutein supplementation on macular pigment optical density and visual acuity in patients with age-related macular degeneration. *Invest Ophthalmol Vis Sci* 2011; 52: 8174–8178.

43. Bone RA, Landrum JT, Friedes LM et al. Distribution of lutein and zeaxanthin stereoisomers in the human retina. *Exp Eye Res* 1997; 64: 211–218.

44. Weale RA. Guest editorial: Notes on the macular pigment. *Ophthalmic Physiol Opt* 2007; 27: 1–10.

45. Schultze M. Zur Anatomie und Physiologie der Retina. *Arch Mickrosk Anat* 1866; 2: 165–286.

46. Engles M, Wooten BR, Hammond B. Macular pigment: A test of the acuity hypothesis. *Invest Ophthalmol Vis Sci* 2007; 48: 2922–2931.

47. McLellan JS, Marcos S, Prieto PM, Burns SA. Imperfect optics may be the eye's defence against chromatic blur. *Nature* 2002; 417: 174–176.

48. Wooten BR, Hammond BR. Macular pigment: Influences on visual acuity and visibility. *Prog Retin Eye Res* 2002; 21: 225–240.

49. Hammond BR, Wooten BR, Snodderly DM. Preservation of visual sensitivity of older subjects: Association with macular pigment density. *Invest Ophthalmol Vis Sci* 1998; 39: 397–406.

50. Stringham JM, Hammond BR, Jr. The glare hypothesis of macular pigment function. *Optom Vis Sci* 2007; 84: 859–864.

51. Stringham JM, Hammond BR. Macular pigment and visual performance under glare conditions. *Optom Vis Sci* 2008; 85: 82–88.

52. Stringham JM, Garcia PV, Smith PA, McLin LN, Foutch BK. Macular pigment and visual performance in glare: Benefits for photostress recovery, disability glare, and visual discomfort. *Invest Ophthalmol Vis Sci* 2011; 52: 7406–7415.

53. Seddon JM, Ajani UA, Sperduto RD et al. Dietary carotenoids, vitamins A, C, and E, and advanced age-related macular degeneration. Eye disease case-control study group. *JAMA* 1994; 272: 1413–1420.

54. Davies NP, Morland AB. Macular pigments: Their characteristics and putative role. *Prog Retin Eye Res* 2004; 23: 533–559.

55. Klaver CCW, Wolfs RC, Vingerling JR, Hofman A, de Jong PTVM. Age-specific prevalence and causes of blindness and visual impairment in an older population: The Rotterdam study. *Arch Ophthalmol* 1998; 116: 653–658.

56. Attebo K, Mitchell P, Smith W. Visual acuity and the causes of visual loss in Australia. The blue mountains eye study. *Ophthalmology* 1996; 103: 357–364.

57. Tielsch JM, Javitt JC, Coleman A, Katz J, Sommer A. The prevalence of blindness and visual impairment among nursing home residents in Baltimore. *N Engl J Med* 1995; 332: 1205–1209.

58. Soubrane G, Cruess A, Lotery A et al. Burden and health care resource utilization in neovascular age-related macular degeneration: Findings of a multicountry study. *Arch Ophthalmol* 2007; 125: 1249–1254.

59. Bandello F, Lafuma A, Berdeaux G. Public health impact of neovascular age-related macular degeneration treatments extrapolated from visual acuity. *Invest Ophthalmol Vis Sci* 2007; 48: 96–103.

60. Cruess A, Zlateva G, Xu X, Rochon S. Burden of illness of neovascular age-related macular degeneration in Canada. *Can J Ophthalmol* 2007; 42: 836–843.

61. Marback RF, Maia OD Jr., Morais FB, Takahashi WY. Quality of life in patients with age-related macular degeneration with monocular and binocular legal blindness. *Clinics* 2007; 62: 573–578.

62. Gupta OP, Brown GC, Brown MM. Age-related macular degeneration: The costs to society and the patient. [Miscellaneous]. *Curr Opin Ophthalmol* 2007; 18: 201–205.

63. Covert D, Berdeaux G, Mitchell J, Bradley C, Barnes R. Quality of life and health economic assessments of age-related macular degeneration. *Surv Ophthalmol* 2007; 52: S20–S25.

64. Evans JR. Primary prevention of age related macular degeneration. *BMJ* 2007; 335: 729.

65. Ho L, van Leeuwen R, Witteman JCM et al. Reducing the genetic risk of age-related macular degeneration with dietary antioxidants, zinc, and {omega}-3 fatty acids: The Rotterdam study. *Arch Ophthalmol* 2011; 129: 758–766.

66. Charbel Issa P, Victor Chong N, Scholl H. The significance of the complement system for the pathogenesis of age-related macular degeneration—Current evidence and translation into clinical application. *Graefe's Arch Clin Exp Ophthalmol* 2011; 249: 163–174.

67. Vos JJ. Literature review of human macular absorption in the visible and its consequence for the cone receptor primaries. Report. Institute for Perception TNO. 1972;17.

68. Bone RA, Landrum JT, Cains A. Optical density spectra of the macular pigment in vivo and in vitro. *Vision Res* 1992; 32: 105–110.
69. DeMarco P, Pokorny J, Smith VC. Full-spectrum cone sensitivity functions for X-chromosome-linked anomalous trichromats. *J Opt Soc Am A* 1992; 9: 1465–1476.
70. Sharpe LT, Stockman A, Knau H, Jägle H. Macular pigment densities derived from central and peripheral spectral sensitivity differences. *Vision Res* 1998; 38: 3233–3239.
71. Landrum JT, Bone RA, Kilburn MD. The macular pigment: A possible role in protection from age-related macular degeneration. *Adv Pharmacol* 1997; 38: 537–556.
72. Khachik F, Bernstein PS, Garland DL. Identification of lutein and zeaxanthin oxidation products in human and monkey retinas. *Invest Ophthalmol Vis Sci* 1997; 38: 1802–1811.
73. Izumi-Nagai K, Nagai N, Ohgami K et al. Macular pigment lutein is antiinflammatory in preventing choroidal neovascularization. *Arterioscler Thromb Vasc Biol* 2007; 27: 2555–2562.
74. Snellen EL, Verbeek AL, van den Hoogen GW, Cruysberg JR, Hoyng CB. Neovascular age-related macular degeneration and its relationship to antioxidant intake. *Acta Ophthalmol Scand* 2002; 80: 368–371.
75. Richer S, Stiles W, Statkute L et al. Double-masked, placebo-controlled, randomized trial of lutein and antioxidant supplementation in the intervention of atrophic age-related macular degeneration: The Veterans Last Study (Lutein Antioxidant Supplementation Trial). *Optometry* 2004; 75: 216–230.
76. Cho E, Seddon JM, Rosner B, Willett WC, Hankinson SE. Prospective study of intake of fruits, vegetables, vitamins, and carotenoids and risk of age-related maculopathy. *Arch Ophthalmol* 2004; 122: 883–892.
77. Gale CR, Hall NF, Phillips DI, Martyn CN. Lutein and zeaxanthin status and risk of age-related macular degeneration. *Invest Ophthalmol Vis Sci* 2003; 44: 2461–2465.
78. Mozaffarieh M, Sacu S, Wedrich A. The role of the carotenoids, lutein and zeaxanthin, in protecting against age-related macular degeneration: A review based on controversial evidence. *Nutr J* 2003; 2: 20.
79. Mares-Perlman JA, Millen AE, Ficek TL, Hankinson SE. The body of evidence to support a protective role for lutein and zeaxanthin in delaying chronic disease. Overview. *J Nutr* 2002; 132: 518S–524S.
80. Beatty S, Murray IJ, Henson DB, Carden D, Koh H, Boulton ME. Macular pigment and risk for age-related macular degeneration in subjects from a Northern European population. *Invest Ophthalmol Vis Sci* 2001; 42: 439–446.
81. Berendschot TTJM, Willemse-Assink JJM, Bastiaanse M, de Jong PTVM, van Norren D. Macular pigment and melanin in age-related maculopathy in a general population. *Invest Ophthalmol Vis Sci* 2002; 43: 1928–1932.
82. Bone RA, Landrum JT, Mayne ST, Gomez CM, Tibor SE, Twaroska EE. Macular pigment in donor eyes with and without AMD: A case-control study. *Invest Ophthalmol Vis Sci* 2001; 42: 235–240.
83. Beatty S, Stevenson M, Nolan JM, Woodside J, The CARMA Study Group, Chakravarthy U. Longitudinal relationships between macular pigment and serum lutein in patients enrolled in the CARMA clinical trial (carotenoids and co-antioxidants in age-related maculopathy). *Invest Ophthalmol Vis Sci* 2009; 50: 1719.
84. Nolan JM, Stack J, O' Donovan O, Loane E, Beatty S. Risk factors for age-related maculopathy are associated with a relative lack of macular pigment. *Exp Eye Res* 2007; 84: 61–74.
85. Loane E, Kelliher C, Beatty S, Nolan JM. The rationale and evidence base for a protective role of macular pigment in age-related maculopathy. *Br J Ophthalmol* 2008; 92: 1163–1168.
86. Chong EW, Wong TY, Kreis AJ, Simpson JA, Guymer RH. Dietary antioxidants and primary prevention of age related macular degeneration: Systematic review and meta-analysis. *BMJ* 2007; 335: 755.

87. Mares-Perlman JA. Too soon for lutein supplements. *Am J Clin Nutr* 1999; 70: 431–432.
88. O'Connell ED, Nolan JM, Stack J et al. Diet and risk factors for age-related maculopathy. *Am J Clin Nutr* 2008; 87: 712–722.
89. Carpentier S, Knaus M, Suh M. Associations between lutein, zeaxanthin, and age-related macular degeneration: An overview. *Crit Rev Food Sci Nutr* 2009; 49: 313–326.
90. Tsika CI, Kontadakis G, Makridaki M et al. Assessment of macular pigment optical density in patients with unilateral wet amd. *Invest Ophthalmol Vis Sci* 2009; 50: 1720.
91. Delcourt C, Carriere I, Delage M, Barberger-Gateau P, Schalch W. The POLA Study Group. Plasma lutein and zeaxanthin and other carotenoids as modifiable risk factors for age-related maculopathy and cataract: The POLA study. *Invest Ophthalmol Vis Sci* 2006; 47: 2329–2335.
92. Trumbo PR, Ellwood KC. Lutein and zeaxanthin intakes and risk of age-related macular degeneration and cataracts: An evaluation using the food and drug administration's evidence-based review system for health claims. *Am J Clin Nutr* 2006; 84: 971–974.
93. Nussbaum JJ, Pruett RC, Delori FC. Historic perspectives. Macular yellow pigment. The first 200 years. *Retina* 1981; 1: 296–310.
94. Maxwell JC. On the unequal sensibility of the foramen centrale to light of different colours. *Rep Br Assoc Adv Sci* 1856; 2: 12.
95. Berendschot TTJM, van Norren D. Objective determination of the macular pigment optical density using fundus reflectance spectroscopy. *Arch Biochem Biophys* 2004; 430: 149–155.
96. Ossewaarde-Van Norel J, van den Biesen PR, van de Kraats J, Berendschot TTJM, van Norren D. Comparison of fluorescence of sodium fluorescein in retinal angiography with measurements in vitro. *J Biomed Opt* 2002; 7: 190–198.
97. Berendschot TTJM, van Norren D. Macular pigment shows ringlike structures. *Invest Ophthalmol Vis Sci* 2006; 47: 709–714.
98. Helb HM, Charbel Issa P, van der Veen RLP, Berendschot TTJM, Scholl HP, Holz FG. Abnormal macular pigment distribution in type 2 idiopathic macular telangiectasia. *Retina* 2008; 28: 808–816.
99. Elsner AE, Burns SA, Delori FC, Webb RH. Quantitative reflectometry with the SLO. In: Naseman JE, Burk ROW (eds.), *Laser Scanning Ophthalmoscopy and Tomography*. Munich, Germany: Quintessenz-Verlag, 1990: pp. 109–121.
100. Wüstemeyer H, Mößner A, Jahn C, Wolf S. Macular pigment density in healthy subjects quantified with a modified confocal scanning laser ophthalmoscope. *Graefe's Arch Clin Exp Ophthalmol* 2003; 241: 647–651.
101. Miles WR. Comparison of functional and structural areas in human fovea. I. Method of entoptic plotting. *J Neurophysiol* 1954; 17: 22–38.
102. Delori FC, Goger DG, Keilhauer C, Salvetti P, Staurenghi G. Bimodal spatial distribution of macular pigment: Evidence of a gender relationship. *J Opt Soc Am A Opt Image Sci Vis* 2006; 23: 521–538.
103. Wolf-Schnurrbusch UEK, Wolf S, Volker D, Brinkmann C, Rothenbuehler SP, Delori FC. Spatial distribution of macular pigment and its relation to the fovea size. *Invest Ophthalmol Vis Sci* 2009; 50: 1724.
104. Dietzel M, Zeimer M, Heimes B, Pauleikhoff D, Hense HW. The ring-like structure of macular pigment in age-related maculopathy—Results from the muenster aging and retina study (MARS). *Invest Ophthalmol Vis Sci* 2011; 52: 8016–8024.
105. Wolf-Schnurrbusch UEK, Roosli N, Weyermann E, Heldner MR, Hohne K, Wolf S. Ethnic differences in macular pigment density and distribution. *Invest Ophthalmol Vis Sci* 2007; 48: 3783–3787.

106. Hammond BR, Wooten BR, Snodderly DM. Individual variations in the spatial profile of human macular pigment. *J Opt Soc Am A* 1997; 14: 1187–1196.
107. Kirby ML, Galea M, Loane E, Stack J, Beatty S, Nolan JM. Foveal anatomic associations with the secondary peak and the slope of the macular pigment spatial profile. *Invest Ophthalmol Vis Sci* 2009; 50: 1383–1391.
108. Nolan JM, Stringham JM, Beatty S, Snodderly DM. Spatial profile of macular pigment and its relationship to foveal architecture. *Invest Ophthalmol Vis Sci* 2008; 49: 2134–2142.
109. Berendschot TTJM, van Norren D. On the age dependency of the macular pigment optical density. *Exp Eye Res* 2005; 81: 602–609.
110. Ciulla TA, Hammond JR. Macular pigment density and aging, assessed in the normal elderly and those with cataracts and age-related macular degeneration. *Am J Ophthalmol* 2004; 138: 582–587.
111. Liew SH, Gilbert CE, Spector TD et al. Central retinal thickness is positively correlated with macular pigment optical density. *Exp Eye Res* 2006; 82: 915–920.
112. van der Veen RLP, Ostendorf S, Hendrikse F, Berendschot TTJM. Macular pigment optical density relates to foveal thickness. *Eur J Ophthalmol* 2009; 19: 836–841.
113. Kanis MJ, Berendschot TTJM, van Norren D. Interocular agreement in melanin and macular pigment optical density. *Exp Eye Res* 2007; 84: 934–938.
114. Neelam K, Nolan J, Loane E et al. Macular pigment and ocular biometry. *Vision Res* 2006; 46: 2149–2156.
115. Snodderly DM, Brown PK, Delori FC, Auran JD. The macular pigment. I. Absorbance spectra, localization, and discrimination from other yellow pigments in primate retinas. *Invest Ophthalmol Vis Sci* 1984; 25: 660–673.
116. Trieschmann M, van Kuijk FJ, Alexander R et al. Macular pigment in the human retina: Histological evaluation of localization and distribution. *Eye* 2008; 22: 132–137.
117. Segal J. Localisation du pigment maculaire de la retine. *C R Seances Soc Biol* 1950; 144: 1630–1631.
118. Snodderly DM, Auran JD, Delori FC. The macular pigment. II. Spatial distribution in primate retinas. *Invest Ophthalmol Vis Sci* 1984; 25: 674–685.
119. van de Kraats J, Kanis MJ, Genders SW, van Norren D. Lutein and zeaxanthin measured separately in the living human retina with fundus reflectometry. *Invest Ophthalmol Vis Sci* 2008; 49: 5568–5573.
120. van der Veen RLP, Berendschot TTJM, Makridaki M, Hendrikse F, Carden D, Murray IJ. Correspondence between retinal reflectometry and a flicker-based technique in the measurement of macular pigment spatial profiles. *J Biomed Opt* 2009; 14: 064046.
121. Snodderly DM, Handelman GJ, Adler AJ. Distribution of individual macular pigment carotenoids in central retina of macaque and squirrel monkeys. *Invest Ophthalmol Vis Sci* 1991; 32: 268–279.
122. Handelman GJ, Snodderly DM, Krinsky NI, Russett MD, Adler AJ. Biological control of primate macular pigment. Biochemical and densitometric studies. *Invest Ophthalmol Vis Sci* 1991; 32: 257–267.
123. Bone RA, Landrum JT. Dichroism of lutein: A possible basis for Haidinger's brushes. *Appl Opt* 1983; 22: 775–776.
124. Wang YM, Illingworth DR, Connor SL, Duell PB, Connor WE. Competitive inhibition of carotenoid transport and tissue concentrations by high dose supplements of lutein, zeaxanthin and beta-carotene. *Eur J Nutr* 2010; 49: 327–336.
125. van den Berg H. Carotenoid interactions. *Nutr Rev* 1999; 57: 1–10.
126. Tyssandier V, Cardinault N, Caris-Veyrat C et al. Vegetable-borne lutein, lycopene, and β-carotene compete for incorporation into chylomicrons, with no adverse effect on the medium-term (3-wk) plasma status of carotenoids in humans. *Am J Clin Nutr* 2002; 75: 526–534.

127. Khachik F, Spangler CJ, Smith JC, Jr, Canfield LM, Steck A, Pfander H. Identification, quantification, and relative concentrations of carotenoids and their metabolites in human milk and serum. *Anal Chem* 1997; 69: 1873–1881.
128. Khachik F, de Moura FF, Zhao DY, Aebischer CP, Bernstein PS. Transformations of selected carotenoids in plasma, liver, and ocular tissues of humans and in nonprimate animal models. *Invest Ophthalmol Vis Sci* 2002; 43: 3383–3392.
129. Bhosale P, Larson AJ, Frederick JM, Southwick K, Thulin CD, Bernstein PS. Identification and characterization of a Pi isoform of glutathione S-transferase (GSTP1) as a zeaxanthin-binding protein in the macula of the human eye. *J Biol Chem* 2004; 279: 49447–49454.
130. Li B, Vachali P, Frederick JM, Bernstein PS. Identification of StARD3 as a lutein-binding protein in the macula of the primate retina. *Biochemistry* 2011; 50: 2541–2549.
131. Li B, Vachali P, Bernstein PS. Human ocular carotenoid-binding proteins. *Photochem Photobiol Sci* 2010; 9: 1418–1425.
132. Loane E, Nolan JM, Beatty S. The respective relationships between lipoprotein profile, macular pigment optical density, and serum concentrations of lutein and zeaxanthin. *Invest Ophthalmol Vis Sci* 2010; 51: 5897–5905.
133. Waters D, Clark RM, Greene CM, Contois JH, Fernandez ML. Change in plasma lutein after egg consumption is positively associated with plasma cholesterol and lipoprotein size but negatively correlated with body size in postmenopausal women. *J Nutr* 2007; 137: 959–963.
134. Kirby ML, Harrison M, Beatty S, Greene I, Nolan JM. Changes in macular pigment optical density and serum concentrations of lutein and zeaxanthin in response to weight loss. *Invest Ophthalmol Vis Sci* 2009; 50: 1709.
135. Greene C, Waters D, Clark R, Contois J, Fernandez M. Plasma LDL and HDL characteristics and carotenoid content are positively influenced by egg consumption in an elderly population1. *Nutr Metab* 2006; 3: 6.
136. Vishwanathan R, Goodrow-Kotyla EF, Wooten BR, Wilson TA, Nicolosi RJ. Consumption of 2 and 4 egg yolks/d for 5 wk increases macular pigment concentrations in older adults with low macular pigment taking cholesterol-lowering statins. *Am J Clin Nutr* 2009; 90: 1272–1279.
137. Borel P. Genetic variations involved in interindividual variability in carotenoid status. *Mol Nutr Food Res* 2012; 56: 228–240.
138. Reboul E, Abou L, Mikail C et al. Lutein transport by Caco-2 TC-7 cells occurs partly by a facilitated process involving the scavenger receptor class B type I (SR-BI). *Biochem J* 2005; 387: 455–461.
139. Zerbib J, Seddon JM, Richard F et al. rs5888 variant of SCARB1 gene is a possible susceptibility factor for age-related macular degeneration. *PLoS One* 2009; 4: e7341.
140. Charbel Issa P, Scholl HPN, Helb HM, Holz FG. Idiopathic macular telangiectasia. In: Holz FG, Spaide RF (eds.), *Medical Retina*. Berlin, Germany: Springer, 2007: pp. 183–197.
141. Gass JD, Blodi BA. Idiopathic juxtafoveolar retinal telangiectasis. Update of classification and follow-up study. *Ophthalmology* 1993; 100: 1536–1546.
142. Finger RP, Charbel Issa P, Fimmers R, Holz FG, Rubin GS, Scholl HP. Reading performance is reduced by parafoveal scotomas in patients with macular telangiectasia type 2. *Invest Ophthalmol Vis Sci* 2009; 50: 1366–1370.
143. Charbel Issa P, Holz FG, Scholl HP. Metamorphopsia in patients with macular telangiectasia type 2. *Doc Ophthalmol* 2009; 119: 133–140.
144. Charbel Issa P, Helb HM, Rohrschneider K, Holz FG, Scholl HPN. Microperimetric assessment of patients with type 2 idiopathic macular telangiectasia. *Invest Ophthalmol Vis Sci* 2007; 48: 3788–3795.
145. Barthelmes D, Gillies MC, Sutter FK. Quantitative OCT analysis of idiopathic perifoveal telangiectasia. *Invest Ophthalmol Vis Sci* 2008; 49: 2156–2162.

146. Gaudric A, Ducos de LG, Cohen SY, Massin P, Haouchine B. Optical coherence tomography in group 2A idiopathic juxtafoveolar retinal telangiectasis. *Arch Ophthalmol* 2006; 124: 1410–1419.

147. Charbel Issa P, Berendschot TTJM, Staurenghi G, Holz FG, Scholl HP. Confocal blue reflectance imaging in type 2 idiopathic macular telangiectasia. *Invest Ophthalmol Vis Sci* 2008; 49: 1172–1177.

148. Yannuzzi LA, Bardal AM, Freund KB, Chen KJ, Eandi CM, Blodi B. Idiopathic macular telangiectasia. *Arch Ophthalmol* 2006; 124: 450–460.

149. Schmitz-Valckenberg S, Fan K, Nugent A et al. Correlation of functional impairment and morphological alterations in patients with group 2A idiopathic juxtafoveal retinal telangiectasia. *Arch Ophthalmol* 2008; 126: 330–335.

150. Charbel Issa P, van der Veen RLP, Stijfs A, Holz FG, Scholl HPN, Berendschot TTJM. Quantification of reduced macular pigment optical density in the central retina in macular telangiectasia type 2. *Exp Eye Res* 2009; 89: 25–31.

151. van de Kraats J, Berendschot TTJM, Valen S, van Norren D. Fast assessment of the central macular pigment density with natural pupil using the macular pigment reflectometer. *J Biomed Opt* 2006; 11: 064031–064037.

152. van de Kraats J, Berendschot TTJM, van Norren D. The pathways of light measured in fundus reflectometry. *Vision Res* 1996; 36: 2229–2247.

153. Berendschot TTJM, de Lint PJ, Norren D. Fundus reflectance-historical and present ideas. *Prog Retin Eye Res* 2003; 22: 171–200.

154. van de Kraats J, Kanis MJ, Genders SW, van Norren D. Lutein and zeaxanthin measured separately in the living human retina by reflectometry. *Invest Ophthalmol Vis Sci* 2008; 48: 5568–5573.

155. Sjögren T, Larsson T. Oligophrenia in combination with congenital ichthyosis and spastic disorders; a clinical and genetic study. *Acta Psychiatr Neurol Scand Suppl* 1957; 113: 1–112.

156. Willemsen MA, Cruysberg JR, Rotteveel JJ, Aandekerk AL, Van Domburg PH, Deutman AF. Juvenile macular dystrophy associated with deficient activity of fatty aldehyde dehydrogenase in Sjögren-Larsson syndrome. *Am J Ophthalmol* 2000; 130: 782–789.

157. van der Veen RLP, Fuijkschot J, Willemsen MA, Cruysberg JR, Berendschot TTJM, Theelen T. Patients with Sjögren-Larsson syndrome lack macular pigment. *Ophthalmology* 2010; 117: 966–971.

158. Gass JD, Oyakawa RT. Idiopathic juxtafoveolar retinal telangiectasis. *Arch Ophthalmol* 1982; 100: 769–780.

159. Moisseiev J, Lewis H, Bartov E, Fine SL, Murphy RP. Superficial retinal refractile deposits in juxtafoveal telangiectasis. *Am J Ophthalmol* 1990; 109: 604–605.

160. Abujamra S, Bonanomi MT, Cresta FB, Machado CG, Pimentel SL, Caramelli CB. Idiopathic juxtafoveolar retinal telangiectasis: Clinical pattern in 19 cases. *Ophthalmologica* 2000; 214: 406–411.

161. Sallo FB, Leung I, Chung M et al. Retinal crystals in type 2 idiopathic macular telangiectasia. *Ophthalmology* 2011; 118: 2461–2467.

162. Gass JD. Muller cell cone, an overlooked part of the anatomy of the fovea centralis: Hypotheses concerning its role in the pathogenesis of macular hole and foveomacular retinoschisis. *Arch Ophthalmol* 1999; 117: 821–823.

163. Fuijkschot J, Cruysberg JR, Willemsen MA, Keunen JEE, Theelen T. Subclinical changes in the juvenile crystalline macular dystrophy in Sjogren-Larsson syndrome detected by optical coherence tomography. *Ophthalmology* 2008; 115: 870–875.

164. Charbel Issa P, Helb HM, Holz FG, Scholl HPN. Correlation of macular function with retinal thickness in nonproliferative type 2 idiopathic macular telangiectasia. *Am J Ophthalmol* 2008; 145: 169–175.

165. Iqbal J, Hussain MM. Intestinal lipid absorption. *Am J Physiol—Endocrinol Metab* 2009; 296: E1183–E1194.

166. de Moura FF, Ho CC, Getachew G, Hickenbottom S, Clifford AJ. Kinetics of 14C distribution after tracer dose of 14C-lutein in an adult woman. *Lipids* 2005; 40: 1069–1073.
167. Amaratunga A, Abraham CR, Edwards RB, Sandell JH, Schreiber BM, Fine RE. Apolipoprotein E Is Synthesized in the retina by müller glial cells, secreted into the vitreous, and rapidly transported into the optic nerve by retinal ganglion cells. *J Biol Chem* 1996; 271: 5628–5632.
168. Shanmugaratnam J, Berg E, Kimerer L et al. Retinal Müller glia secrete apolipoproteins E and J which are efficiently assembled into lipoprotein particles. *Brain Res Mol Brain Res* 1997; 50: 113–120.
169. Gass JDM. *Retinal Capillary Diseases. Stereoscopic Atlas of Macular Disease.* St. Louis, MO: Mosby, 1997: 504–512.
170. Charbel Issa P, Holz FG, Scholl HPN. Findings in fluorescein angiography and optical coherence tomography after intravitreal bevacizumab in type 2 idiopathic macular telangiectasia. *Ophthalmology* 2007; 114: 1736–1742.
171. Chopdar A. Retinal telangiectasis in adults: Fluorescein angiographic findings and treatment by argon laser. *Br J Ophthalmol* 1978; 62: 243–250.
172. Albini TA, Benz MS, Coffee RE et al. Optical coherence tomography of idiopathic juxtafoveolar telangiectasia. *Ophthalmic Surg Lasers Imaging* 2006; 37: 120–128.
173. Paunescu LA, Ko TH, Duker JS et al. Idiopathic juxtafoveal retinal telangiectasis: New findings by ultrahigh-resolution optical coherence tomography. *Ophthalmology* 2006; 113: 48–57.
174. Sarthy V, Ripps H. Structural organization of retinal glia. In: Blakemore C (ed.), *The Retinal Müller Cell: Structure and Function.* New York: Kluwer Academic/Plenum Publishers, 2001: pp. 5–27.
175. Tout S, Chan-Ling T, Holländer H, Stone J. The role of Müller cells in the formation of the blood-retinal barrier. *Neuroscience* 1993; 55: 291–301.
176. Bringmann A, Pannicke T, Grosche J et al. Muller cells in the healthy and diseased retina. *Prog Retin Eye Res* 2006; 25: 397–424.
177. Yamada E. Some structural features of the fovea centralis in the human retina. *Arch Ophthalmol* 1969; 82: 151–159.

# 8 Retinol and Otitis Media

*Akeem Olawale Lasisi*

## CONTENTS

## INTRODUCTION

Retinol represents the functional molecule in the retinoid compounds referred to as vitamin A; they play essential role in a large number of physiological functions that encompass vision, growth, reproduction, hematopoiesis, and immunity [1]. Despite major advances in the knowledge of vitamin A biology, its deficiency is still a serious public health problem affecting an estimated 127 million preschool children and 7.2 million pregnant women worldwide [2]. The effects, which have been linked to the immunological functions, include increased risks of mortality and morbidity from measles, respiratory tract infections, diarrheal diseases [3], blindness, and anemia, among others [4–6]. In addition, vitamin A supplementation has been reported to reduce morbidity rates associated with pneumococcal disease by delaying the rate of colonization and the age of occurrence [7]. Among common infections caused by *Pneumococci* spp., otitis media is highly prevalent in developing countries. The annual incidence rate of acute otitis media (AOM) has been reported between 1.4 per 100,000 per year and 64.5 cases per 1,000 children [8,9]. It accounts for about one-third of the patients in the otorhinolaryngologic clinic and has been linked to delay in language development and poor school performance [9,10]. Malnutrition has featured prominently among the reports of the clinical risk factors, and this has prompted researches into the role of nutritional factors in the development and reversal of otitis media [10,11].

## VITAMIN A AND PATHOGENESIS OF OTITIS MEDIA

Otitis media is inflammation of the mucosa lining of the middle ear and its integuments; this suggests that epithelial integrity is a factor in the course of otitis media. Vitamin A and their derivatives are required for the maintenance of the normal epithelial mucociliary phenotype and secretion of mucus. Vitamin A compounds exert their effects via specific nuclear receptors, retinoic acid receptors, and retinoid X receptors, all of which are members of the steroid/thyroid receptor superfamily [10–14]. Physicochemical injuries alter the respiratory epithelium resulting in a multistep process to squamous metaplasia [15–17]. In vivo, vitamin A deficiency has been found to induce replacement of the normal pseudostratified mucociliary epithelium by a metaplastic stratified squamous epithelium [18,19], a process that can be reversed by a dietary vitamin A supplement [20,21]. Vitamin A inhibits the expression of squamous-related genes, such as keratin 13 in rabbit tracheobronchial cells [22], cholesterol sulfate in human epithelial cells [23], and transglutaminase and cornifin in human epidermal keratinocytes [24]. RA also induces a mucosecretory phenotype by activating both transcription of specific mucin genes, particularly MUC2, MUC5B, and MUC5AC, and mucin secretion [25,26]. All of these result in alteration of mucin secretion, which could be a factor in persistence of otorrhea and recurrent suppurative otitis media.

## CLINICAL EVIDENCE LINKING VITAMIN A AND OTITIS MEDIA

### VITAMIN A AND PREDISPOSITION TO NASOPHARYNGEAL COLONIZATION

A significant risk factor in the pathogenesis of AOM, which has also been found to be adversely affected by low retinol, is nasopharyngeal colonization [7,27]. Coles et al. [7] reported that the risk of nasopharyngeal colonization among infants aged 4 months who were not colonized by age 2 months was significantly reduced in the vitamin A group compared with the placebo group (odds ratio 0.51 [0.28, 0.92], P=0.02). The odds of colonization were 27% lower in the vitamin A–treated group than in the placebo group (odds ratio 0.73 [0.48, 1.1], P=0.13). The risk of colonization with penicillin-resistant isolates was 74% lower in the vitamin A–treated group than in the placebo group at 2 months of age. Hence, they concluded that neonatal vitamin A may play a role in lowering morbidity rates associated with pneumococcal disease by delaying the age at which colonization occurs. Studies in several countries with endemic vitamin A deficiency and high rates of pneumococcal disease have reported rapid and abundant pneumococcal colonization of the nasopharynx in early infancy [28,29]. The body of evidence suggests that reversal of vitamin A deficiency may reduce the rate of colonization and decrease associated morbidity rates. In addition to lowering the risk for pneumococcal infection in the individuals, decreasing carriage could lower the pool of infection in the community. Alternatively, supplementation may delay colonization, in which case it would lower morbidity rates in the age-group at highest risk. Under these conditions, vitamin A supplementation may present a cost-effective approach in the event that immunization is not effective or as an adjunct to vaccination for decreasing the risk of pneumococcal disease in infants in developing countries. Lasisi et al. [30] also studied the serum level of retinol among AOM patients. The study found a significantly low serum retinol in patients with Acute Suppurative Otitis Media (ASOM) who progressed to chronicity compared to those who showed evidence of healing. This suggests a possible role of hyporetinolemia in the potentiation of chronicity of suppurative otitis media. Similarly, Yilmaz et al. [31] compared blood levels of vitamin A among other antioxidants (β-carotene, α-tocopherol, laycopene, ascorbic acid, superoxide dismutase, glutathione peroxidase) and oxidation products (malondialdehyde) between group of children with otitis media with effusion who were to undergo bilateral ventilation tube insertion and adenoidectomy, and a control group made of otherwise healthy children. They found that in the study group, the blood antioxidant levels increased and oxidant levels decreased significantly after the operation. Hence, they concluded that oxidants and antioxidants played a significant role in the pathogenesis of otitis media with effusion in children.

In contrast, the work of Durand et al. [32] did not find direct correlation between otitis media and serum retinol status. They conducted an observational study of 200 children, ages 3–5 years, noting their baseline measurements of serum retinol concentrations and incidence of otitis media in the following year. They found that the serum retinol concentrations of the children ranged from 13 to 58 µg/dL, and the episodes of otitis media occurred in 22% of children during the follow-up period. However, children with low serum retinol concentrations did not have an excess of episodes of otitis media. Controlling for potential confounding variables such as duration of breast feeding, smoking in the household, illness, or live virus vaccination in the 2 weeks before serum collection, and day care attendance did not substantially alter this finding. Hence, they concluded that in the range of serum vitamin concentrations found in this population, the status of vitamin A and related compounds in children appeared to have no effect on the incidence of otitis media.

## Vitamin A and Clinical Course of Otitis Media

Animal study by Aladag et al. [33] revealed that epithelial integrity was significantly improved following the addition of parenteral single dose of 100,000 IU vitamin A to the treatment of otitis media in rats. In addition, pretreatment with vitamin A increases antioxidant enzyme activities and reduces formation of nitric oxide and malonyldialdehyde; hence, they recommended that vitamin A may be considered as an additional medicament for the medical treatment of AOM. Similarly, clinical studies among humans have corroborated the additive role of vitamin A deficiency in the development and persistence of suppurative otitis media. In a study of 23 patients, Cemek et al. [34] reported a significantly decreased serum level of retinol in the patients with AOM (P < 0.05). Lasisi et al. [30] found significant hyporetinolemia in patients with acute suppurative otitis media compared to controls (P = 0.000), suggesting a significant association between the deficiency of retinol and the development of ASOM. In addition, they found that there was greater tendency to resolution of AOM among the subjects with normal to high serum retinol compared to those with low levels. Furthermore, low serum vitamin A in the fetal life was found to be a significant correlation factor for the development of otitis media in the first year of life [30,35]. In their study on the fetal retinol on the development of early onset suppurative otitis media, the fetal serum retinol among subjects who developed otitis media in the first year of life was 0.95 µg/L compared to 1.08 µg/L in those who did not have otitis media.

The deficiency of the immunomodulatory functions associated with retinol has been reported as the critical factor in the predisposition to infections. Vitamin A deficiency induces an up-regulation of the T helper subset 1 cell (Th1)-mediated response [24,36]. Experiments in vitro and animal studies suggested that retinoids were important regulators of monocytic differentiation and functions. When added to monocytic, myelomonocytic, or dendritic cell line cultures, retinoic acid promotes cellular differentiation [37,38]. In addition, it influences the secretion of key cytokines produced by macrophages, including tumor necrosis factor, IL-1β, IL-6, and IL-12. All-trans-retinoic acid skewed the differentiation of human peripheral blood monocytes to IL-12-secreting dendritic cells [37–39]. In another study, it was reported that vitamin A inhibited lipopolysaccharide-induced IL-12 production by mouse macrophages and decreased the secretion of tumor necrosis factor in murine peripheral blood mononuclear cells, myelomonocytes, and macrophages [40–44]. On the other hand, retinoids appear to enhance the secretion of IL-1β and IL-6 by macrophages and monocytes [45–47]. The development of neutrophils in the bone marrow is controlled by retinoic acid receptor-modulated genes [48,49], and retinoic acid in cultures accelerates neutrophil maturation [49]. In addition, treatment with retinoic acid has been shown to restore the number of neutrophils and the superoxide-generating capacity in rats and calves [50–52]. Rahman et al. [53] examined the effect of vitamin A supplementation on cell-mediated immunity among infants younger than 6 months in Bangladesh. Their results showed that cell-mediated immunity responses were improved among infants with adequate serum retinol concentrations after supplementation, but there was no improvement among children with low serum retinol levels. The impact of vitamin A on circulating effectors of innate immunity, including acute-phase response proteins and the complement system, was studied in trials from Ghana, Indonesia, and South Africa. In the Ghana study of preschool children, large doses of vitamin A every 4 months for 1 year resulted in significantly increased serum amyloid A and C-reactive protein among children [53]. However, no effect was found on the C-reactive protein concentrations in the Indonesia study [54,55]. Plasma C3 complement was not affected by four doses of vitamin A administered within a 42 day period to South African children [56].

The inferences from various reports appear to suggest that supplementation of vitamin A in children may help in control of otitis media and its chronicity. Furthermore, this has been corroborated by findings from vitamin A supplementation studies [57–59]. Studies from England, South Africa, and Tanzania [57–59] have reported decrease in the risk of nutritional blindness and morbidity of infectious origin from measles, respiratory infections, severe diarrhea, HIV, and intestinal helminthiases following periodic vitamin A supplementation to children. The effects of vitamin A supplementation on child morbidity include a reduction in the severity of measles that could be correlated with the enhanced T-cell-dependent antibody production that was observed [60,61]. A decrease in the severity of measles morbidity could also explain an overall average reduction in measles-specific mortality of about 60% [62]. This could have a positive effect in the control of suppurative otitis media. The benefits of vitamin A on measles-related outcomes may go beyond the correction of underlying deficiencies and could actually represent adjuvant therapeutic effects [63]. In addition, the degree of depression of retinol was associated with severity of these illnesses. The beneficial effects of vitamin A supplementation among children could be mediated by a short-term increase in antibody production, possibly as a result of increased lymphocyte proliferation. However, more data are needed from human trials about the role of vitamin A supplementation in modulating the outcome and the course of otitis media.

## REFERENCES

1. Long KZ, Montoya Y, Hertzmark E, Santos IJ, Rosado JL. A double-blind, randomized, clinical trial of the effect of vitamin A and zinc supplementation on diarrheal disease and respiratory tract infections in children in Mexico City, Mexico. *Am J Clin Nutr* 2006; 83(3): 693–700.
2. West KP Jr. Extent of vitamin A deficiency among preschool children and women of reproductive age. *J Nutr* 2002; 132: 2857S–2866S.
3. Craft NE, Haitema T, Brindle LK, Yamini S, Humphrey JH, West KP. Retinol analysis in dried blood spots by HPLC. *J Nutr* 2000; 130: 882–885.
4. Kirby LT, Applegarth DA, Davidson AGF, Wong LTK, Hardwick DF. Use of a dried blood spot in immunoreactive-trypsin assay for detection of cystic fibrosis in infants. *Clin Chem* 1981; 27: 678–680.
5. Semba RD, Bloem MW. The anemia of vitamin A deficiency: Epidemiology and pathogenesis. *Eur J Clin Nutr* 2002; 56: 271–281.
6. Christian P, West KP Jr, Khatry SK, Kimbrough-Pradhan E, Le-Clerq SC, Katz J, Shrestha SR, Dali SM, Sommer A. Night blindness during pregnancy and subsequent mortality among women in Nepal: Effects of vitamin A and beta-carotene supplementation. *Am J Epidemiol* 2002; 152: 542–547.
7. Coles CL, Rahmathullah L, Kanungo R, Thulasiraj RD, Katz J, Santhosham M, Tielsch JM. Vitamin A supplementation at birth delays pneumococcal colonization in south Indian infants. *J Nutr* 2001; 131: 255–261.
8. Wang PC, Chang YH, Chuang LJ, Su HF, Li CY. Incidence and recurrence of acute otitis media in Taiwan's pediatric population. *Clinics (Sao Paulo)* 2011; 66(3): 395–399.
9. Homøe P, Jensen RG, Brofeldt S. Acute mastoiditis in Greenland between 1994–2007. *Rural Remote Health* 2010; 10(2): 1335.
10. Lasisi OA, Olaniyan FA, Muibi SA, Azeez AI, Idowu KG, Lasisi JT, Imam ZK, Yekinni TO, Olayemi O. Clinical and demographic risk factors associated with chronic suppurative otitis media. *Int J Paediatr Otorhinolaryngol* 2007; 71(10): 1549–1554.
11. Lasisi OA, Sulaiman OA, Afolabi OA. Socio-economic status and hearing loss in chronic suppurative otitis media in Nigeria. *Ann Trop Paediatr* 2007; 27: 291–296.
12. Aukrust P, Muller F, Ueland T, Svardal AM, Berge RK, Froland SS. Decreased vitamin A levels in common variable immunodeficiency: Vitamin A supplementation in vivo enhances immunoglobulin production and down regulates inflammatory responses. *Eur J Clin Invest* 2000; 30: 252–259.
13. Baeten JM, McClelland RS, Corey L, Overbaugh J, Lavreys L, Richardson BA, Wald A, Mandaliya K, Bwayo JJ, Kreiss JK. Vitamin A supplementation and genital shedding of herpes simplex virus among HIV-1-infected women: A randomized clinical trial. *J Infect Dis* 2004; 189: 1466–1471.
14. Baeten JM, McClelland RS, Overbaugh J, Richardson BA, Emery S, Lavreys L, Mandaliya K, Bankson DD, Ndinya-Achola JO, Bwayo JJ, Kreiss JK. Vitamin A supplementation and human immunodeficiency virus type 1 shedding in women: Results of a randomized clinical trial. *J Infect Dis* 2002; 185: 1187–1191.
15. Bahl RN, Kant BS, Molbak K, Ostergaard E, Bhan MK. Effect of vitamin A administered at Expanded Program on Immunization contacts on antibody response to oral polio vaccine. *Eur J Clin Nutr* 2002; 56: 321–325.

16. Beaton GH, Martorell R, Aronson KJ. Effectiveness of vitamin A supplementation in the control of young child morbidity and mortality in developing countries. ACC/SCN State-of-the-Art Series policy discussion paper no. 13. World Health Organization, Geneva, Switzerland, 1993.

17. Benn CS, Aaby P, Bale C, Olsen J, Michaelsen KF, George E, Whittle H. Randomised trial of effect of vitamin A supplementation on antibody response to measles vaccine in Guinea-Bissau, West Africa. *Lancet* 1997; 350: 101–105.

18. Benn CS, Balde A, George E, Kidd M, Whittle H, Lisse IM, Aaby P. Effect of vitamin A supplementation on measles-specific antibody levels in Guinea-Bissau. *Lancet* 2002; 359: 1313–1314.

19. Benn CS, Bale C, Sommerfelt H, Friis H, Aaby P. Hypothesis: Vitamin A supplementation and childhood mortality: Amplification of the non-specific effects of vaccines? *Int J Epidemiol* 2003; 32: 822–828.

20. Benn CS, Lisse IM, Bale C, Michaelsen KF, Olsen J, Hedegaard K, Aaby P. No strong long-term effect of vitamin A supplementation in infancy on CD4 and CD8 T-cell subsets. *Eur Arch Otorhinolaryngol* 2000; 123(20): 259–264.

21. Benn CS, Whittle H, Aaby P, Bale C, Michaelsen KF, Olsen J. Vitamin A and measles vaccination. *Lancet* 1995; 346: 503–504.

22. Bhaskaram P, Rao K. Enhancement in seroconversion to measles vaccine with simultaneous administration of vitamin A in 9-months-old Indian infants. *Indian J Pediatr* 1997; 64: 503–509.

23. Binka FN, Ross DA, Morris SS, Kirkwood BR, Arthur P, Dollimore N, Gyapong JO, Smith PG. Vitamin A supplementation and childhood malaria in northern Ghana. *Am J Clin Nutr* 1995; 61: 853–859.

24. Breitman TR, Selonick SE, Collins SJ. Induction of differentiation of the human promyelocytic leukemia cell line (HL-60) by retinoic acid. *Proc Natl Acad Sci* 1980; 77: 2936–2940.

25. Moon SK, Yoo JH, Kim HN, Lim DJ, Chung MH. Effects of retinoic acid, triiodothyronine and hydrocortisone on mucin and lysozyme expression in cultured human middle ear epithelial cells. *Acta Otolaryngol* 2000; 120: 944–949.

26. Hwang PH, Chan JM. Retinoic acid improves ciliogenesis after surgery of the maxillary sinus in rabbits. *Laryngoscope* 2006; 116(7): 1080–1085.

27. Berman S. Epidemiology of acute respiratory infections in children of developing countries. *Rev Infect Dis* 1991; 13(Suppl 6): S454–S462.

28. Lloyd-Evans N, O'Dempsey TJ, Baldeh I, Secka O, Demba E, Todd JE, Mcardle TF, Banya WS, Greenwood BM. Nasopharyngeal carriage of *Pneumococci* in Gambian children and in their families. *Pediatr Infect Dis J* 1996; 15(10): 866–871.

29. Nahed A, Walid A, Asmar B, Basim A, Ronald T, Shermine D, Ricardo G. Nasopharyngeal colonization with streptococcus pneumoniae in children receiving trimethoprim-sulfamethoxazole prophylaxis. *Pediatr Infect Dis J* 1999; 18(7): 647–649.

30. Lasisi OA, Olayemi O, Tongo O, Arinola OG, Bakare RA, Omilabu SA. Cord blood immunobiology and the development of early suppurative otitis media. *J Neonat Perinat Med* 2009; 2: 187–192.

31. Yilmaz T, Koçan EG, Besler HT, Yilmaz G, Gürsel B. The role of oxidants and antioxidants in otitis media with effusion in children. *Otolaryngol Head Neck Surg* 2004; 131(6): 797–803.

32. Durand AM, Sabino H Jr, Masga R, Sabino M, Olopai F, Abraham I. Childhood vitamin A status and the risk of otitis media. *Pediatr Infect Dis J* 1997; 16(10): 952–954.

33. Aladag I, Guven M, Eyibilen A, Sahin S, Köseoglu D. Efficacy of vitamin A in experimentally induced acute otitis media. *Int J Pediatr Otorhinolaryngol* 2007; 71(4): 623–628.

34. Cemek M, Dede S, Bayiroglu F, Caksen H, Cemek F, Yuca K. Oxidant and antioxidant levels in children with acute otitis media and tonsillitis: A comparative study. *Int J Pediatr Otorhinolaryngol* 2005; 69(6): 823–827.

35. Lasisi OA. The role of retinol in the aetiology and outcome of suppurative otitis media. *Eur Arch Oto-Rhino-Laryngol* 2009; 266: 647–652.

36. Geissmann F, Revy FP, Brousse N, Lepelletier Y, Folli C, Durandy A, Chambon P, Dy M. Retinoids regulate survival and antigen presentation by immature dendritic cells. *J Exp Med* 2003; 198: 623–634.

37. Jiang YJ, Xu TR, Lu B, Mymin D, Kroeger EA, Dembinski T, Yang X, Hatch GM, Choy PC. Cyclooxygenase expression is elevated in retinoic acid—Differentiated U937 cells. *Biochim Biophys Acta* 2003; 1633: 51–60.

38. Mohty M, Morbelli S, Isnardon D, Sainty D, Arnoulet C, Gaugler B, Olive D. All-trans retinoic acid skews monocyte differentiation into interleukin-12-secreting dendritic-like cells. *Br J Haematol* 2003; 122: 829–836.

39. Lasisi OA, Arinola OG, Olayemi O. Role of elevated immunoglobulin E levels in suppurative otitis media. *Ann Trop Paediatr* 2008; 28: 123–127.

40. Na SY, Kang BY, Chung SW, Han SJ, Ma X, Trinchieri G, Im SY, Lee JW, Kim TS. Retinoids inhibit interleukin-12 production in macrophages through physical associations of retinoid X receptor and NFkappaB. *J Biol Chem* 1999; 274: 7674–7680.

41. Kim BH, Kang KS, Lee YS. Effect of retinoids on LPS-induced COX-2 expression and COX-2 associated PGE(2) release from mouse peritoneal macrophages and TNF-alpha release from rat peripheral blood mononuclear cells. *Toxicol Lett* 2004; 150: 191–201.

42. Mou L, Lankford-Turner P, Leander MV, Bissonnette RP, Donahoe RM, Royal W. RXR-induced TNF-alpha suppression is reversed by morphine in activated U937 cells. *J Neuroimmunol* 2004; 147: 99–105.

43. Mathew JS, Sharma RP. Effect of all-trans-retinoic acid on cytokine production in a murine macrophage cell line. *Int J Immunopharmacol* 2000; 22: 693–706.

44. Motomura K, Ohata M, Satre M, Tsukamoto H. Destabilization of TNF-alpha mRNA by retinoic acid in hepatic macrophages: Implications for alcoholic liver disease. *Am J Physiol Endocrinol Metab* 2001; 281: E420–E429.

45. Hashimoto S, Hayashi S, Yoshida S, Kujime K, Maruoka S, Matsumoto K, Gon Y, Koura T, Horie T. Retinoic acid differentially regulates interleukin-1beta and interleukin-1 receptor antagonist production by human alveolar macrophages. *Leuk Res* 2008; 22: 1057–1061.

46. Matikainen S, Serkkola E, Hurme M. Retinoic acid enhances IL-1 beta expression in myeloid leukemia cells and in human monocytes. *J Immunol* 1991; 147: 162–167.

47. Arena A, Capozza AB, Delfino D, Iannello D. Production of TNF alpha and interleukin 6 by differentiated U937 cells infected with Leishmania major. *New Microbiol* 1997; 20: 233–240.

48. Abbas AK, Lichtman AH. *Cellular and Molecular Immunology*. Saunders, Philadelphia, PA, 2003.

49. Maun NA, Gaines P, Khanna-Gupta A, Zibello T, Enriquez L Goldberg L, Berliner N. G-CSF signaling can differentiate promyelocytes expressing a defective retinoic acid receptor: Evidence for divergent pathways regulating neutrophil differentiation. *Blood* 2004; 103: 1693–1701.

50. Ribeiro OG, Maria DA, Adriouch S, Pechberty S, Cabrera WH, Morisset J, Ibanez OM, Seman M. Convergent alteration of granulopoiesis, chemotactic activity, and neutrophil apoptosis during mouse selection for high acute inflammatory response. *J Leukoc Biol* 2003; 74: 497–506.

51. Cherian T, Varkki S, Raghupathy P, Ratnam S, Chandra RK. Effect of vitamin A supplementation on the immune response to measles vaccination. *Vaccine* 2003; 21: 2418–2420.

52. Higuchi H, Nagahata H. Effects of vitamins A and E on superoxide production and intracellular signalling of neutrophils in Holstein calves. *Can J Vet Res* 2000; 64: 69–75.

53. Rahman MM, Mahalanabis D, Alvarez JO, Wahed MA, Islam MA, Habte D, Khaled MA. Acute respiratory infections prevent improvement of vitamin A status in young infants supplemented with vitamin A. *J Nutr* 1996; 126: 628–633.

54. Filteau SM, Morris SS, Raynes JG, Arthur P, Ross DA, Kirkwood BR, Tomkins AM, Gyapong JO. Vitamin A supplementation, morbidity, and serum acute-phase proteins in young Ghanaian children. *Am J Clin Nutr* 1995; 62: 434–438.

55. Semba R, West MK, Natadisastra G, Eisinger W, Lan Y, Sommer A. Hyporetinolemia and acute phase proteins in children with and without xerophthalmia. *Am J Clin Nutr* 2000; 72: 146–153.

56. Zhao Z, Ross AC. Retinoic acid repletion restores the number of leukocytes and their subsets and stimulates natural cytotoxicity in vitamin A-deficient rats. *J Nutr* 1995; 125: 2064–2073.

57. Coutsoudis A, Broughton M, Coovadia HM. Vitamin A supplementation reduces measles morbidity in young African children: A randomized, placebo-controlled, double-blind trial. *Am J Clin Nutr* 1991; 54: 890–895.

58. Hussey GD, Klein M. A randomized, controlled trial of vitamin A in children with severe measles. *N Engl J Med* 1990; 323: 160–164.

59. Brown N, Roberts C. Vitamin A for acute respiratory infection in developing countries: A meta-analysis. *Acta Paediatr* 2004; 93: 1437–1442.

60. Fawzi W, Chalmers T, Herrera M, Mosteller F. Vitamin A supplementation and child mortality: A meta-analysis. *JAMA* 1993; 269: 898–903.

61. Fawzi W, Mbise R, Spiegelman D, Fataki M, Hertzmark E, Ndossi G. Vitamin A supplements and diarrheal and respiratory infections among children in Dar es Salaam, Tanzania. *J Pediatr* 2000; 137: 660–667.

62. Grotto I, Mimouni M, Gdalevich M, Mimouni D. Vitamin A supplementation and childhood morbidity from diarrhea and respiratory infections: A meta-analysis. *J Pediatr* 2003; 142: 297–304.

63. Butler JC, Havens PL, Day S, Chusid MJ, Sowell AL, Huf DL, Peterson DE, Bennin RA, Circo R, Davis JP. Measles severity and serum retinol (vitamin A) concentration among children in the United States. *Pediatrics* 1993; 91(6): 1176–1181.

# Part III

Carotenoids and Vitamin A
in Cancer

# 9 Carotenoids and Mutagenesis

*Peter M. Eckl, Avdulla Alija, Nikolaus Bresgen,
Ekramije Bojaxhi, Cornelia Vogl, Giuseppe Martano,
Hanno Stutz, Siegfried Knasmüller, Franziska Ferk,
Werner G. Siems, Claus-Dieter Langhans,
and Olaf Sommerburg*

## CONTENTS

## INTRODUCTION

Carotenoids are fat-soluble natural pigments, found mainly in plants, where they serve as an accessory light-gathering pigment and protect these organisms from toxic effects of oxygen [1]. Depending on the number of double bonds, several *cis/trans* configurations are possible and enable a large number of different carotenoids. At present, approximately 800 have been identified. Carotenoids can be subdivided into carotenes, consisting only of carbon and hydrogen, and xanthophylls, which are oxygen-containing derivatives of carotenes.

Human sources of β-carotene are fruits such as apricots and peaches, vegetables such as carrots, squash, and broccoli, fish such as salmon, and crustaceans such as lobster and shrimps.

After uptake, β-carotene is enzymatically cleaved in the intestinal mucosa by β-carotene 15,15' monooxygenase (BCDO1; [2,3]) at the central double bond. This symmetrical cleavage yields two molecules of vitamin A (retinol), which are further converted enzymatically into the vision pigment retinal. Therefore, β-carotene is also called provitamin A. β-Carotene can additionally be cleaved asymmetrically by β-carotene 9',10'dioxygenase (BCDO2; [4]) to apo-10'-carotenal and β-ionone.

Apart from its action as a provitamin, β-carotene has been demonstrated to affect signaling pathways, that is, it induces cell cycle arrest and apoptosis via down-regulation of cyclin A and Bcl-2 family proteins [5] and has also been shown to stimulate cell communication via gap junctions, an effect correlated with the ability to inhibit chemically induced neoplastic transformation [6,7].

β-Carotene has also been found to modulate immune function, that is, it stimulates blood neutrophil killing activity via increased myeloperoxidase and phagocytic activity, an enhanced antibody response, an increased mitogen-induced lymphocyte proliferation, and an increased respiratory burst [8].

## ANTIOXIDANT AND ANTIMUTAGENIC EFFECTS

Apart from the effects described earlier, β-carotene has been demonstrated to possess antioxidant activity in vitro, that is, it scavenges peroxyl radicals, in particular lipid peroxyl radicals, nitrogen dioxide ($NO_2^•$)-, thiyl (RS$^•$)-, and sulfonyl ($RSO_2^•$)-radicals, quenches singlet oxygen, and inhibits lipid peroxidation [9–15].

The antioxidant action of carotenoids is also reflected by their antimutagenic/antigenotoxic activity, in particular when the mutagen is acting via the formation of reactive oxygen species (ROS) such as ionizing radiation. For example, it has been demonstrated that β-carotene administered before whole-body γ-irradiation of rats significantly reduces formation of micronuclei in polychromatic and normochromatic erythrocytes [16] and that it significantly decreased micronucleus formation in mice splenocytes and reticulocytes after whole-body x-irradiation [17]. Similarly, β-carotene supplementation immediately after in vitro γ-irradiation of human lymphocytes significantly reduced the formation of micronuclei [18]. Several chemical mutagens can also cause the formation of ROS, that is, the metabolism of benzo(a)pyrene leads to a metabolic profile that induces benzo(a)pyrene quinones produced biologically by cytochrome P450 isozymes and peroxidases [19], and can undergo one-electron redox cycling with their semiquinone radicals resulting in the formation of ROS [20,21]. Due to its antioxidant activity, β-carotene can therefore be expected to inhibit the mutagenic effect of B(a)P. In fact, Raj and Katz [22] already demonstrated in 1985 that dietary supplementation of β-carotene can inhibit B(a)P-induced formation of micronuclei in bone marrow cells by 41%–61%.

Some antimutagenic effects cannot be solely explained by the antioxidant action of β-carotene. For example, β-carotene has been demonstrated to inhibit mitomycin C-, cyclophosphamide-, and dioxidine-induced chromosomal breaks in bone marrow cells of mice [22–24], to inhibit mitomycin C- and cyclophosphamide-induced clastogenicity in human hepatoma cells [25], and to reduce N-ethyl-N-nitrosourea (ENU) mutagenicity in rat spleen lymphocytes [26]. Furthermore, the application of the food dyes E160e (β-apo-8′-carotenal in an oil suspension) and E160a (β-carotene in an oil suspension) before or together with intraperitoneal administration of cyclophosphamide or dioxidine was shown to inhibit the clastogenicity of these mutagens in mouse bone marrow cells [27], and last, even solvent extracts of carotenoid-rich fruits applied in the Salmonella/microsome test together with aflatoxin B1, benzo(a)pyrene, 2-amino-3-methylimidazo[4,5-f]quinoline, and cyclophosphamide resulted in antimutagenic activities of many plant extracts, in particular the n-hexane extracts [28]. Antimutagenicity was mainly associated with the carotenoid-rich fractions (α-, β-carotene, lycopene), xanthophylls (β-cryptoxanthin, lutein), and also carotenol esters.

Apart from in vitro studies employing *Salmonella typhimurium* and cell lines, and in vivo animal experiments with rats and mice, there is also evidence for an antimutagenic potential of carotenoids from human populations such as the reduction with vitamin A, β-carotene, and canthaxanthin administration of the proportion of micronucleated buccal mucosal cells in Asian betel nut and tobacco chewers [29,30]. The antioxidant canthaxanthin had no protective activity in this study. Therefore, the authors concluded that vitamin A and β-carotene exert their inhibitory effect on formation of micronuclei by a mechanism not involving scavenging of free radicals [30].

One of the major mechanisms by which carotenoids can inhibit mutagenesis was first demonstrated by Astorg et al. [31], who investigated the effects of β-carotene and canthaxanthin on liver xenobiotic-metabolizing enzymes in the rat. While canthaxanthin increased the liver content of cytochrome P-450, the activity of NADH-cytochrome c reductase, some P-450-dependent enzymes (ethoxy-, methoxy-, pentoxy,- and benzoxyresorufin O-dealkylases), and the phase II enzymes UGT1, UGT2, and GST, neither β-carotene nor an excess of vitamin A induced significant changes of the enzyme activities measured. Further studies revealed that other carotenoids such as astaxanthin and apo-8′-carotenal are also potent inducers of liver cytochromes P4501A1 and 1A2 [32,33], that bixin, canthaxanthin, and astaxanthin are also inducers of rat lung and kidney xenobiotic metabolizing enzymes [34] and that the induction of xenobiotic-metabolizing enzymes depends on the activation of the Ah receptor [35], which acts as a transcription factor.

In order to test whether carotenoids can inhibit DNA damage and liver carcinogenesis via alterations of xenobiotic-metabolizing enzymes, Gradelet et al. [36] treated rats with β-carotene, apo-8′-carotenal, astaxanthin, canthaxanthin or lycopene, and the aflatoxin $AFB_1$. With the exception of lycopene, all carotenoids tested reduced the number and size of preneoplastic foci. As expected from previous studies with apo-8′-carotenal, astaxanthin, and canthaxanthin, this effect depends on the induction of xenobiotic-metabolizing enzymes shifting metabolism toward detoxification, which is additionally reflected by a reduction of DNA damage. β-Carotene had only marginal effects on $AFB_1$ metabolism and did not protect against DNA damage. Therefore, it was concluded that the anticarcinogenic effect is mediated by other mechanisms. A possible candidate for this mechanism could be the pregnane-X-receptor (PXR), a member of the nuclear hormone-receptor family, which is activated even by physiological concentrations of β-carotene [37,38] and transcriptionally regulates PXR dependent genes via ligand-activated heterodimers with retinoid-X receptor (RXR) [39], and influences predominantly the transcription of phase I xenobiotic-metabolizing genes such as the cytochrome P450 class CYP3A [40], phase II enzymes such as glutathione-S-transferases [41], and genes involved in the regulation of the excretion of xenobiotic substances such as the ABC transporters multidrug resistance protein (MDR) and multidrug resistance-associated protein (MRP) [42,43]. Another candidate is the peroxisome proliferator-activated receptor γ (PPARγ), which increases both at the mRNA and protein level significantly upon β-carotene treatment in vitro [44]. PPARγ activates p21, which is also up-regulated by β-carotene, and p21 inhibits cell cycle progression leading to $G_1$ arrest.

## PRO-OXIDANT AND MUTAGENIC EFFECTS

The antioxidant and antigenotoxic effects described earlier provide an explanation for the observation that the increased uptake of carotenoids or fruits and vegetables as primary source of carotenoids reduces the risk to develop certain types of degenerative disease. This effect was associated with cardiovascular diseases [45–47], age-dependent macula degeneration [48], cataract formation [49,50], and different types of cancer [51] such as mouth, pharynx, larynx, esophagus, stomach, colon, rectum, bladder, and cervix [52–55].

Apart from these observations, a beneficial role of β-carotene could not be supported by several cancer chemoprevention trials. In two major trials (the Alpha-Tocopherol Beta-Carotene Cancer Prevention Study and the Beta-Carotene and RETinol Efficacy Trial), the incidence of cancer and death from coronary artery disease was increased after β-carotene supplementation in both cigarette smokers and asbestos workers [56–58]. Leo and Lieber [59] further reported that β-carotene supplementation in smokers who additionally consume alcohol promotes pulmonary cancer, hepatotoxicity, and possibly also cardiovascular diseases. These results are in part supported by earlier findings by van Poppel et al. [60], who demonstrated a lack of protective effect of β-carotene on smoking-induced DNA damage in lymphocytes of heavy smokers. An explanation of this adverse effect was given by Wang and Russell [61], who reported that β-carotene decreases the level of retinoic acid in the lungs, which in turn reduces the inhibitory effect on activator protein-1. Consequently, lung cell proliferation and, potentially, tumor formation are enhanced. Wang and Russell [61] also suggested that β-carotene metabolites are responsible for the carcinogenic response in the lungs of smokers. Sommerburg et al. [62] further postulated that in certain tissues oxidative, nonenzymatic cleavage of carotenoids is carried out primarily by oxidants liberated by polymorphonuclear leukocytes (PMLs) and demonstrated that β-carotene is degraded in culture medium of activated PML and that hypochlorous acid treatment of β-carotene as a model for neutrophil-derived degradation caused the formation of volatile short-chain metabolites such as β-cyclocitral, β-ionone, 5,6-epoxy-β-ionone, dihydroactinidiolide, and 4-oxo-β-ionone as well as long-chain metabolites such as apo-8′-carotenal and apo-12′-carotenal. The cleavage products (CPs) obtained by hypochlorous acid degradation of β-carotene have been demonstrated to modify respiratory burst and induce apoptosis of human neutrophils [63] and to induce oxidative stress in vitro by impairing mitochondrial respiration [64,65].

Based on these observations, CPs and one of the major carotenals formed by hypochlorous acid degradation of β-carotene—apo-8′-carotenal—were tested in the highly sensitive primary rat hepatocyte assay by analysis of the endpoints mitotic index, necrosis, apoptosis, chromosomal aberrations, micronuclei, and sister chromatid exchanges [66]. The results indicate a significant genotoxic potential of both, CPs and apo-8′-carotenal at concentrations as low as 100 nM, in other words at pathophysiologically relevant levels of β-carotene and β-carotene CPs. β-Carotene itself did not cause significant changes relative to the control levels up to a concentration of 10 μM. These results were further

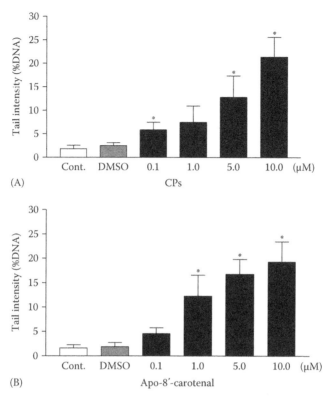

**FIGURE 9.1**    Effects of a 3 h CP (A) and apo-8′-carotenal (B) exposure on DNA-damage in cultures of primary rat hepatocytes. Per experimental point, three cultures were made in parallel, and from each culture, 50 cells were evaluated. Vitality of the cells was determined by the trypan blue exclusion method. Only cultures with a vitality of ≥80% were analyzed. Bars indicate means ± SD of three independent experiments. *P < 0.05 compared to the DMSO control.

confirmed by the results of the COMET assay [67], which detects DNA single- and double-strand breaks and apurinic sites, after exposure to CPs (Figure 9.1A) and apo-8′-carotenal (Figure 9.1B).

Since smokers who take supplements of β-carotene can be expected to be exposed to the genotoxic action of both CPs due to oxidative degradation of β-carotene [68] and oxidative stress from smoking [69–71], primary hepatocytes were also treated in parallel with CPs and oxidative stress by either hypoxia/reoxygenation or the redox cycling quinone 2,3-dimethoxy-1,4-naphthoquinone. The results of this investigation clearly indicate an enhancement of the genotoxic response compared to the treatment with CPs alone [72]. Interestingly, at concentrations of 1 and 10 µM CPs, highly damaged metaphases appeared at a frequency of 1/30 of the analyzed metaphases showing 10 or more aberrations. This effect appears to depend on a lack of cytotoxicity, since neither CPs nor CPs in combination with oxidative stress caused significant changes in the level of necrotic and apoptotic cells [66,72,73].

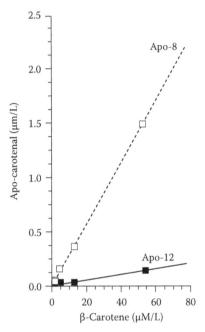

**FIGURE 9.2** Formation of long-chain β-carotene CPs in primary cultures of rat hepatocytes by oxidative stress after a 3 h supplementation with β-carotene (dissolved in DMSO) in the reoxygenation phase after 24 h of hypoxia. The applied water-dispersible β-carotene in soybean suspension (2% w/v) was a gift of Cognis Australia Pty Ltd. (Australia).

In a third approach, primary hepatocytes were treated with β-carotene for 3 h in the reoxygenation phase after 24 h of hypoxia (95% $N_2$, 5% $CO_2$). Thereafter, the incidence of chromosomal aberrations and micronucleated cells was determined. In addition, the concentrations of both β-carotene and the major carotenals formed were measured by HPLC [62]. As shown in Figure 9.2, hypoxia/reoxygenation caused a β-carotene-concentration-dependent increase in apo-8′-carotenal and apo-12′-carotenal, which was most prominent for apo-8′-carotenal with a ratio between apo-8′-carotenal and apo-12′-carotenal of about 10:1. Regression analysis of the data gave highly significant (P < 0.01) linear regressions for both apo-8′-carotenal (y = 0.028x + 0.002) and apo-12′-carotenal (y = 0.003x + 0.004) formation, whereas both, the induction of micronuclei and chromosomal aberrations (Figure 9.3) showed a saturation-type response, which also gave highly significant (P < 0.01) regression curves—micronucleated cells: y = 0.033ln(x) + 0.26; chromosomal aberrations: y = 1.081ln(x) + 14.46. These results confirm that oxidative stress leads to the formation of long-chain CPs and their formation correlates with genotoxic effects [73].

Primary hepatocytes that had been treated with β-carotene for 3 h in the reoxygenation phase after 24 h of hypoxia were also analyzed for induction of DNA damage in the COMET assay. The results are shown in Figure 9.4 and indicate a significant β-carotene dose-dependent increase in DNA damage. Interestingly, β-carotene alone at a concentration of 50 μM also induced a significant increase

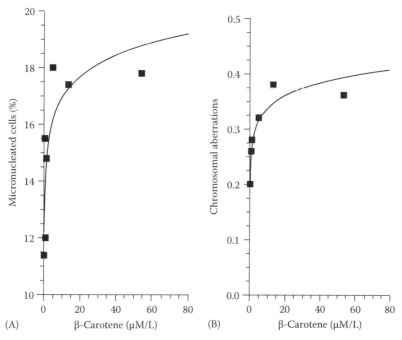

**FIGURE 9.3** Induction of micronucleated cells (A) and chromosomal aberrations (B) in primary cultures of rat hepatocytes by oxidative stress after a 3 h supplementation of β-carotene in the reoxygenation phase after 24 h of hypoxia.

in DNA damage compared to the soybean oil matrix plus DMSO, indicating that high concentrations of β-carotene are genotoxic even in the absence of an external oxidative stress. This observation can be explained by the findings of Paolini et al. [74], who demonstrated that β-carotene supplementation in rats causes up to 15-fold increases of the activities of cytochrome P450 enzymes, which are paralleled by increased generation of oxygen radicals, which was most prominent in the liver. Compared to the controls, a 30- to 40-fold increase was observed. Furthermore, Marques et al. [75] have shown that the treatment of calf thymus DNA with β-carotene or β-carotene oxidation products significantly increased the levels of 1,N2-etheno-2′-deoxyguanosine and 8-oxo-7,8-dihydro-2′-deoxyguanosine (a brief literature summary of the effects of carotenoids on DNA and chromosomes is given in Table 9.1).

Altogether, these findings are clear evidence that β-carotene is mutagenic under oxidative stress due to oxidative degradation to a range of CPs and that the mutagenic effect can most likely be attributed to the carotenals formed, in particular apo-8′-carotenal. Apart from the mutagenic effects observed, also cytotoxic effects were reported; that is, Hurst et al. [80] found that carotenoid-derived aldehydes prepared via hypochlorite bleaching of β-carotene exhibited a similar cytotoxic effect to K562 cells as they observed with 4-hydroxynonenal [81], a known cyto- and genotoxic lipid per-oxidation product [82]. In contrast, Alija et al. [66,72,73] did not observe any signifi-cant changes to both necrotic and apoptotic cells in primary hepatocyte cultures after

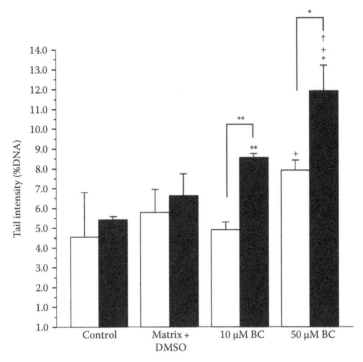

**FIGURE 9.4** Induction of DNA damage in primary cultures of rat hepatocytes by oxidative stress after a 3 h supplementation of β-carotene in the reoxygenation phase after 24 h of hypoxia. Bars represent the means ± SD of the tail intensities (%DNA) in normoxic control cultures (open bars) and cultures treated with hypoxia/reoxygenation (black bars); N ≥ 3; the differences observed after hypoxia/reoxygenation are significant at P < 0.005 (one-way ANOVA and Welsh Test); *P < 0.05, **P < 0.005 compared to the corresponding control; +P < 0.05 compared to treatment with the soybean matrix plus DMSO (Student's double-sided t-test for independent samples); †P < 0.05 compared to treatment with the soybean matrix plus DMSO (ANOVA post-hoc testing, Tamhane T2).

treatment with CPs. The authors hypothesize that the epidermal growth factor applied for proliferative stimulation of the cells antagonizes the cytotoxic effect and repeated their experiments in the absence of the growth factor [83]. Under these conditions, significant increases of apoptoses and necroses were found at a CP concentration of 1 μM highlighting the complex interactions most likely occurring in vivo. Based on the observations of the authors, the following scenario could be envisaged: CPs produced under oxidative stress induce cell death. As a consequence, there is compensatory cell proliferation driven by growth factors released by neighboring cells. These growth factors may antagonize the cytotoxic effect allowing even damaged cells to proceed through the cell cycle as evidenced by the appearance of heavily damaged metaphases in the study by Alija et al. [72] resulting eventually in mutated/initiated cells.

Summarizing, supplementation with β-carotene can result in antimutagenic as well as mutagenic effects depending on the level of exposure to oxidative stress (Figure 9.5) and the amount of β-carotene taken up.

**TABLE 9.1**

**Genotoxic Effects of β-Carotene and Cleavage Products**

| Carotenoids | Concentration | Genotoxic Effect | References |
|---|---|---|---|
| β-Carotene and β carotene oxidation products | | DNA adducts: 1,N2-etheno-2′-deoxyguanosine | [75] |
| β-Carotene and β-apo-8′-carotenal | 20 μM | Enhancement of DNA strand breaks induced by B(a)P | [76] |
| β-Carotene | 10 μM | Enhancement of DNA strand breaks induced by hydrogen peroxide | [77] |
| β-Carotene | 0.5–4 μM | Enhancement of tobacco smoke condensate–induced 8-hydroxyl-2′-deoxyguanosine | [78] |
| β-Carotene | 0.1–1 μM | Increase in hydrogen peroxide– and bleomycin-induced chromosomal aberrations | [79] |
| Retinol, retinal | 1–5 μM | Induction of 8-oxo-7,8-dihydro-2′-deoxyguanosine due to superoxide generation | [113] |
| β-Apo-8′-carotenal and a mixture of cleavage products prepared by hypochlorite bleaching of β-carotene | 0.1–10 μM | Chromosomal aberrations, micronuclei, sister chromatid exchanges; enhanced effects under oxidative stress | [66,72] |

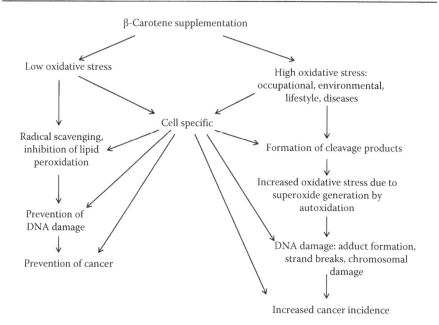

**FIGURE 9.5** Mechanisms of action of β-carotene.

## CONSEQUENCES FOR MEDICAL APPLICATIONS

The medical uses of β-carotene apart from those discussed earlier (prevention of cancer and cardiovascular disease) concern mainly light-sensitive skin diseases such as protoporphyria [84,85], polymorphic light eruptions [86], ultraviolet sensitivity [87], lupus erythematosus [88], or photoallergic drug reactions [89–91]. Furthermore, β-carotene has been demonstrated to prevent exercise-induced asthma [92], age-dependent macula degeneration [48,93], and cataract formation [49,50], and long-term supplementation was claimed to provide cognitive benefits [94]. In addition, indications for oral use of β-carotene include acne [90] and rheumatoid arthritis [95], the latter based on the inadequate intake of β-carotene and vitamin A in rheumatoid arthritis patients [96]. A further potential application relates to the prevention of psoriasis, which has been demonstrated to be inversely related to the intake of carrots, tomatoes, and the index of β-carotene intake [97]. Furthermore, animal studies indicate that β-carotene may be even anti-epileptogenic [98]. It has to be mentioned, however, that similar to cancer chemoprevention, not all studies gave unequivocal proof for beneficial effects, that is, studies on the prevention of cataract formation provided conflicting results [99–102].

Summarizing, a wide range of potential medical applications of β-carotene exists, and in addition, numerous β-carotene-containing nutritional supplements are available on the market without prescription.

Based on the wide range of applications, it is therefore necessary to take the adverse effects found in the cancer-chemoprevention trials into consideration. Mayne [103] concluded in 1996 that the pharmacological use of supplemental β-carotene for the prevention of cardiovascular disease and lung cancer, particularly in smokers, can no longer be recommended. The in vitro findings on the mutagenicity of CPs and β-carotene under oxidative stress further indicate that oxidative stress in general may pose a risk, and precautions should be taken when individuals environmentally or occupationally exposed to oxidative stress receive β-carotene supplementation. Exposures involving oxidative stress are not only related to asbestos, for which an increased risk for developing lung cancer has been demonstrated in the Beta-Carotene and Retinol Efficacy Trial [57,58]. Oxidative stress is also associated with air pollution in general [104], in particular particles such as environmental particles [105] ultrafine particulate pollutants [106], carbon nanotubes [107], and nanomaterials such as fullerenes [108] as well as metals [109]. Welders are exposed to oxidative stress due to the presence of transition metals such as iron and copper in the welding fume [110,111].

Exposure to oxidative stress is also associated with alcohol consumption [112]. Thus, precautions have to be taken when β-carotene supplementation is coupled with heavy alcohol consumption. According to Leo and Lieber [113], alcoholic liver injury results in vitamin A depletion by an increased mobilization from the liver [114,115] and by enhancing the activity of liver enzymes metabolizing vitamin A [59]. When ethanol is administered to rats together with β-carotene in an amount comparable to that applied in the CARET trial, the hepatotoxic effect of ethanol is potentiated resulting in characteristic liver lesions such as increased activity of liver enzymes in the plasma, autophagic vacuoles, alterations of the endoplasmic reticulum, and the mitochondria [116]. Interestingly, this investigation also revealed that ethanol-induced oxidative stress assessed by an increase in hepatic 4-hydroxynonenal and $F_2$-isoprostanes does not improve by administration of antioxidants although hepatic antioxidants (β-carotene and vitamin E) increased. This observation appears to contradict the finding that β-carotene is an inhibitor of lipid peroxidation [9–15]. It may, however, be explained by the findings of Paolini et al. [74] that β carotene supplementation of rats causes up to 15-fold increases of cytochrome P450 enzymes accompanied by an over-generation of oxygen radicals, and this increase is most prominent in the liver. Taking into consideration that ethanol-induced oxidative stress in the liver will also cause the formation of β-carotene CPs, these products, in particular apo-8′-carotenal [32], will in turn influence the expression of cytochromes P450 and thus the production of oxygen radicals. In addition, the superoxide radical has been demonstrated to be formed by the autoxidation of retinoids [117]. Under these circumstances, the pro-oxidant effect most likely overcompensates the antioxidant effects of β-carotene.

In this context, it is worth to note that cigarette smoke–induced oxidative eccentric cleavage of β-carotene is significantly decreased when α-tocopherol (50 µmol/L) and ascorbic acid (50 µmol/L) are incubated together with β-carotene (10 µmol/L) and lung postnuclear fractions of cigarette smoke–exposed ferrets [118]. This finding is confirmed by the observation that the cytotoxicity of CPs to primary rat hepatocytes can be significantly reduced by the application of Trolox, ascorbic acid, or N-acetyl-cysteine [83]. The mechanisms involved include inhibition of oxidative stress [119–121] and detoxification of aldehydes [122,123], in particular 4-hydroxynonenal [124]. These findings indicate that at least part of the toxic effect of β-carotene under oxidative stress could be avoided by simultaneous supplementation of additional antioxidants; however, this hypothesis requires further investigation and confirmation.

One of the most interesting questions relating to the use of β-carotene supplementation is exercise-induced oxidative stress. Physical exercise causes oxidative stress due to increased oxygen consumption, which is directly related to an increased formation of oxygen free radicals primarily from the mitochondrial electron transport chain [125,126]. As a consequence, there is lipid peroxidation and the oxidation of proteins, RNA, and DNA [124,127,128]. Increased oxidative DNA damage is observed, for example, following single bouts of exercise in untrained rats [129], after massive aerobic exercise in humans [130], and endurance exercise, that is, an ultramarathon run [131]. Antioxidant supplementation, that is, vitamin E plus vitamin C and a fruit-and-vegetable juice powder concentrate, was found to attenuate the rise in protein carbonyls but had no effect on the lipid peroxidation product malondialdehyde and 8-hydroxydeoxy-guanosine as a marker for oxidative DNA damage [132]. On the other hand, supplementation of Alaskan sled dogs with α-tocopherol, β-carotene, and lutein decreased oxidative DNA damage as measured by the plasma concentration of 7,8 dihydro-8-oxo-2′deoxyguanosine [133]. In a further study, kayakers received antioxidant capsules containing 136 mg of α-tocopherol, 200 mg of vitamin C, 15 mg of β-carotene, 1 mg of lutein, 200 µg of selenium, 15 mg of zinc, and 300 mg of magnesium (two capsules daily for 28 days), and markers of oxidative stress and inflammation were determined before and after exercise compared to a placebo group [134]. Based on the outcome, the authors concluded that antioxidant supplementation does not offer protection against exercise-induced lipid peroxidation and inflammation. Taking these findings together, there is no clear-cut evidence that antioxidants, in particular β-carotene, can attenuate oxidative damage during exercise.

In contrast, there is unequivocal evidence that exercise reduces oxidative DNA damage. For example, DNA migration (COMET assay) in peripheral white blood cells of subjects completing an incremental treadmill test until exhaustion was significantly lower in trained compared to untrained persons [135]. Similar results were obtained for oxidative DNA damage (8-hydroxydeoxy-guanosine) due to exercise in rats [127,136]. These beneficial effects are due to adaptation involving heat shock and oxidative stress responses [137], repair enzymes [138], and the expression of many other genes as reviewed by Radak et al. [127]. The authors conclude that regular moderate physical exercise/activity provides systemic beneficial effects, including improved physiological function, decreased incidence of disease, and a higher quality of life.

The question therefore arises, whether physical exercise–induced oxidative stress can cause the formation of β-carotene CPs and an associated increase in DNA damage in β-carotene-supplemented individuals and whether regular exercise leads to an altered rate of CP formation and DNA damage. Investigations addressing this question would help clarify whether different qualities of oxidative stress and/or adaptation to oxidative stress have a different impact on the pro-oxidant action of β-carotene.

Summarizing, medical applications and nutritional supplementation with β-carotene have to take into consideration lifestyle, occupational, environmental, and also disease-related exposure to oxidative stress. When β-carotene application is indicated, exposure to high levels of oxidative stress should be avoided. In case this is not possible, the application of additional antioxidants such as vitamin C or vitamin E may be useful to reduce oxidative stress and to detoxify aldehydic β-carotene CPs. Additionally, an improvement of the endogenous enzymatic antioxidant defense system by regular physical exercise can be recommended.

## ACKNOWLEDGMENT

Experimental work on β-carotene and cleavage products in primary rat hepatocytes was supported by grant P20096-B05 from the Austrian Science Foundation FWF.

## REFERENCES

1. Olsen JA. Carotenoids and human health. *Arch Latinoam Nutr* 1999; 49: 7S–11S.
2. von Lintig J, Vogt K. Filling the gap in vitamin A research: Molecular identification of an enzyme cleaving β-carotene to retinal. *J Biol Chem* 2000; 275: 11915–11920.
3. von Lintig J, Wyss A. Molecular analysis of vitamin A formation: Cloning and characterization of β-carotene 15,15ʹ-dioxygenases. *Arch Biochem Biophys* 2001; 385: 47–52.
4. Wyss A. Carotene oxygenases: A new family of double bond cleavage enzymes. *J Nutr* 2004; 134: 246S–250S.

5. Palozza P, Serini S, Maggiano N, Angelini M, Boninsegna A, Di Nicuolo F, Ranelletti FO, Calviello G. Induction of cell cycle arrest and apoptosis in human colon adenocarcinoma cell lines by β-carotene through down-regulation of cyclin A and Bcl-2 family proteins. *Carcinogenesis* 2002; 23: 11–18.

6. Zhang LX, Cooney RV, Bertram JS. Carotenoids enhance gap junctional communication and inhibit lipid peroxidation in C3H/10T1/2 cells: Relationship to their chemopreventive action. *Carcinogenesis* 1991; 12: 2109–2114.

7. Stahl W, Nicolai S, Briviba K, Hanusch M, Broszeit G, Peters M, Martin HD, Sies H. Biological activities of natural and synthetic carotenoids: Induction of gap junctional communication and singlet oxygen quenching. *Carcinogenesis* 1997; 18: 89–92.

8. Elliott R. Mechanisms of genomic and non-genomic actions of carotenoids. *Biochim Biophys Acta* 2005; 1740: 147–154.

9. Burton GW. Antioxidant action of carotenoids. *J Nutr* 1989; 119: 109–111.

10. Sies H, Stahl W, Sundquist AR. Antioxidant functions of vitamins. Vitamins E and C, beta- and other carotenoids. *Ann N Y Acad Sci* 1992; 669: 7–20.

11. Mortensen A, Skibsted LH, Sampson J, Rice-Evans C, Everett SA. Comparative mechanisms and rates of free radical scavenging by carotenoid antioxidants. *FEBS Lett* 1997; 418: 91–97.

12. Foote CS, Denny RW. Chemistry of singlet oxygen: VII. Quenching by beta-carotene. *J Am Chem Soc* 1968; 90: 6233–6235.

13. Palozza P, Krinsky NI. The inhibition of radical-initiated peroxidation of microsomal lipids by both alpha-tocopherol and beta-carotene. *Free Radic Biol Med* 1991; 11: 407–414.

14. Tsuchihashi H, Kigoshi M, Iwatsuki M, Niki E. Action of β-carotene as an antioxidant against lipid peroxidation. *Arch Biochem Biophys* 1995; 323: 137–147.

15. Sarada SKS, Dipti B, Anju B, Pauline T, Kain AK, Sairam M, Sharma SK, Ilavazhagan G, Kumar D, Selvamurthy W. Antioxidant effect of beta-carotene on hypoxia induced oxidative stress in male albino rats. *J Ethnopharmacol* 2002; 79: 149–153.

16. El-Habit OHM, Saada HN, Azab KS, Abdel-Rahman M, El-Malah DF. The modifying effect of β-carotene on gamma radiation-induced elevation of oxidative reactions and genotoxicity in male rats. *Mutat Res* 2000; 466: 179–186.

17. Salvadori DMF, Ribeiro LR, Xiao Y, Boei JJ, Natarajan AT. Radioprotection of p-carotene evaluated on mouse somatic and germ cells. *Mutat Res* 1996; 356: 163–170.

18. Xue K-X, Wu J-Z, Ma GJ, Juan S, Qin H-L. Comparative studies on genotoxicity and antigenotoxicity of natural and synthetic β-carotene stereoisomers. *Mutat Res* 1998; 418: 73–78.

19. Reed MD, Monske ML, Lauer FT, Meserole SP, Born JL, Burchiel SW. Benzo[*a*]pyrene diones are produced by photochemical and enzymatic oxidation and induce concentration-dependent decreases in the proliferative state of human pulmonary epithelial cells. *J Toxicol Environ Health Part A* 2003; 66: 1189–1205.

20. Flowers L, Ohnishi ST, Penning TM. DNA strand scission by polycyclic aromatic hydrocarbon *o*-quinones: Role of reactive oxygen species, Cu(II)/Cu(I) redox cycling, and *o*-semiquinone anion radicals. *Biochemistry* 1997; 36: 8640–8648.

21. Burdick AD, Davis JW, Liu KJ, Hudson LG, Shi H, Monske ML, Burchiel SW. Benzo(*a*)pyrene quinones increase cell proliferation, generate reactive oxygen species, and transactivate the epidermal growth factor receptor in breast epithelial cells. *Cancer Res* 2003; 63: 7825–7833.

22. Raj AS, Katz M. Beta-carotene as an inhibitor of benzo(a)pyrene and mitomycin C induced chromosomal breaks in the bone marrow of mice. *Can J Genet Cytol* 1985; 27: 598–602.

23. Mukherjee A, Agarwal K, Aguilar MA, Sharma A. Anticlastogenic activity of beta-carotene against cyclophosphamide in mice in vivo. *Mutat Res* 1991; 263: 41–46.

24. Salvadori DMF, Ribeiro LR, Oliveira MDM, Pereira CAB, Becak W. Beta-carotene as a modulator of chromosomal aberrations induced in mouse bone marrow cells. *Environ Mol Mutagen* 1992; 20: 206–210.

25. Salvadori DM, Ribeiro LR, Natarajan AT. The anticlastogenicity of beta-carotene evaluated on human hepatoma cells. *Mutat Res* 1993; 303: 151–156.

26. Aidoo A, Lyn-Cook LE, Lensing S, Bishop ME, Wamer W. *In vivo* antimutagenic activity of beta-carotene in rat spleen lymphocytes. *Carcinogenesis* 1996; 16: 2237–2241.

27. Durnev AD, Tjurina LS, Guseva NV, Oreshchenko AV, Volgareva GM, Seredenin SB. The influence of two carotenoid food dyes on clastogenic activities of cyclophosphamide and dioxidine in mice. *Food Chem Toxicol* 1998; 36: 1–5.

28. Rauscher R, Edenharder R, Platt KL. *In vitro* antimutagenic and *in vivo* anticlastogenic effects of carotenoids and solvent extracts from fruits and vegetables rich in carotenoids. *Mutat Res* 1998; 413: 129–142.

29. Stich HF, Rosin MP, Vallejera MO. Reduction with vitamin A and beta-carotene administration of proportion of micronucleated buccal mucosal cells in Asian betal nut and tobacco chewers. *Lancet* 1984; 323: 1204–1206.

30. Stich HF, Stich W, Rosin MP, Vallejera MO. Use of the micronucleus test to monitor the effect of vitamin A, beta-carotene and canthaxanthin on the buccal mucosa of betel nut/tobacco chewers. *Int J Cancer* 1984; 34: 745–750.

31. Astorg PO, Gradelet S, Leclerc J, Canivenc M-C, Siess M-H. Effects of β-carotene and canthaxanthin on liver xenobiotic-metabolizing enzymes in the rat. *Food Chem Toxicol* 1994; 32: 735–742.

32. Gradelet S, Astorg P, Leclerc J, Siess MH. β-Apo-8′-carotenal, but not β-carotene, is a strong inducer of liver CYP1A1 and 1A2 in the rat. *Xenobiotica* 1996; 26: 909–919.

33. Gradelet S, Astorg P, Leclerc J, Chevalier J, Vemevaut MF, Siess MH. Effects of anthaxanthin, astaxanthin, lycopene and lutein on liver xenobiotic-metabolizing enzymes in the rat. *Xenobiotica* 1996; 26: 49–63.

34. Jewell C, O'Brien MN. Effect of dietary supplementation with carotenoids on xenobiotic metabolizing enzymes in the liver, lung, kidney and small intestine of the rat. *Br J Nutr* 1999; 81: 235–242.

35. Gradelet S, Astorg P, Pineau T, Canivenc M-C, Siess M-H, Leclerc J, Lesca P. Ah receptor-dependent CYP1A induction by two carotenoids, canthaxanthin and β-apo-8′-carotenal, with no affinity for the TCDD binding site. *Biochem Pharmacol* 1997; 54: 307–315.

36. Gradelet S, Le Bon A-M, Berges R, Suschetet M, Astorg P. Dietary carotenoids inhibit aflatoxin B1-induced liver preneoplastic foci and DNA damage in the rat: Role of the modulation of aflatoxin B1 metabolism. *Carcinogenesis* 1998; 19: 403–411.

37. Rühl R, Sczech R, Landes N, Pfluger P, Kluth D, Schweigert FJ. Carotenoids and their metabolites are naturally occurring activators of gene expression via the pregnane X receptor. *Eur J Nutr* 2004; 3: 692–703.

38. Rühl R. Induction of PXR-mediated metabolism by β-carotene. *Biochim Biophys Acta* 2005; 1740: 162–169.

39. Kliewer SA, Willson TM. Regulation of xenobiotic and bile acid metabolism by the nuclear pregnane X receptor. *J Lipid Res* 2002; 43: 359–364.

40. Lehmann JM, McKee DD, Watson MA, Willson TM, Moore JT, Kliewer SA. The human orphan nuclear receptor PXR is activated by compounds that regulate CYP3A4 gene expression and cause drug interactions. *J Clin Invest* 1998; 102: 1016–1023.

41. Falkner KC, Pinaire JA, Xiao GH, Geoghegan TE, Prough RA. Regulation of the rat glutathione S-transferase A2 gene by glucocorticoids: Involvement of both the glucocorticoid and pregnane X receptors. *Mol Pharmacol* 2001; 60: 611–619.

42. Geick A, Eichelbaum M, Burk O. Nuclear receptor response elements mediate induction of intestinal MDR1 by rifampin. *J Biol Chem* 2001; 276: 14581–14587.

43. Payen L, Sparfel L, Courtois A, Vernhet L, Guillouzo A, Fardel O. The drug efflux pump MRP2: Regulation of expression in physiopathological situations and by endogenous and exogenous compounds. *Cell Biol Toxicol* 2002; 18: 221–233.

44. Cui Y, Lu Z, Bai L, Shi Z, Zhao W, Zhao B. β-Carotene induces apoptosis and up-regulates peroxisome proliferator-activated receptor c expression and reactive oxygen species production in MCF-7 cancer cells. *Eur J Cancer* 2007; 43: 2590–2601.

45. Gaziana JM, Manson JE, Branch LG, Colditz GA, Willett WC, During JE. A prospective study of consumption of carotenoids in fruits and vegetables and decreased cardiovascular mortality in the elderly. *Ann Epidemiol* 1995; 5: 255–260.

46. Gey KF, Moser UK, Jordan P, Stähelin HB, Eichholzer M, Lüdin E. Increased risk of cardiovascular disease at suboptimal plasma concentrations of essential antioxidants: An epidemiological update with special attention to carotene and vitamin C. *Am J Clin Nutr* 1993; 57: 787S–797S.

47. Ito Y, Kurata M, Suzuki K, Hamajima M, Hishida H, Aoki K. Cardiovascular disease mortality and serum carotenoid levels: A Japanese population-based follow-up study. *J Epidemiol* 2006; 16: 154–160.

48. Seddon JM, Ajani UA, Sperduto RD, Hiller R, Blair N, Burton DC, Farber MD, Gragoudas ES, Haller J, Miller DT, Yannuzzi LA, Willett W. Dietary carotenoids, vitamins A, C and E, and advanced age-related macular degeneration. *JAMA* 1994; 272(18): 1413–1420.

49. Gale CR, Hall NF, Phillips DIW, Martyn CN. Plasma antioxidant vitamins and carotenoids and age-related cataract. *Ophthalmology* 2001; 108(11): 1992–1998.

50. Jacques PF, Chylack LT. Epidemiologic evidence of a role for the antioxidant vitamins and carotenoids in cataract prevention. *Am J Clin Nutr* 1991; 53: 352S–355S.

51. Ziegler RG. Vegetables, fruits and carotenoids and the risk of cancer. *Am J Clin Nutr* 1991; 53: 251S–259S.

52. Colditz GA, Branch LG, Lipnick RJ, Willett WC, Rosner B, Posner BM, Hennekens CH. Increased green and yellow vegetable intake and lowered cancer deaths in an elderly population. *Am J Clin Nutr* 1985; 41: 32–36.

53. Stähelin HB, Gey KF, Eichholzer M, Lüdin E. β-Carotene and cancer prevention: The Basel Study. *Am J Clin Nutr* 1991; 53: 265S–269S.

54. Giovannucci E, Ascherio A, Rimm EB, Stampfer MJ, Colditz GA, Willett WC. Intake of carotenoids and retinol in relation to risk of prostate cancer. *J Natl Cancer Inst* 1995; 87(23): 1767–1776.

55. Batieha AM, Armenian HK, Norkus EP, Morris JS, Spate VE, Comstock GW. Serum micronutrients and the subsequent risk of cervical cancer in a population-based nested case-control study. *Cancer Epidemiol Biomarkers Prev* 1993; 2: 335–339.

56. Blumberg J, Block G. The alpha-tocopherol, beta-carotene cancer prevention study in Finland. *Nutr Rev* 1994; 52(7): 242–245.

57. Omenn GS, Goodman GE, Thornquist MD, Balmes J, Cullen MR, Glass A, Keogh JP, Meyskens FL, Valanis B, Williams JH, Barnhart S, Cherniack MG, Brodkin CA, Hammar S. Risk factors for lung cancer and for intervention effects in CARET, the Beta-Carotene and Retinol Efficacy Trial. *J Natl Cancer Inst* 1996; 88(21): 1550–1559.

58. Omenn GS, Goodman GE, Thornquist MD, Balmes J, Cullen MR, Glass A, Keogh JP, Meyskens FL, Valanis B, Williams JH, Barnhart S, Hammar S. Effects of a combination of beta carotene and vitamin A on lung cancer and cardiovascular disease. *N Engl J Med* 1996; 334: 1150–1155.

59. Leo MA, Lieber CS. Alcohol, vitamin A and beta-carotene: Adverse interactions, including hepatotoxicity and carcinogenicity. *Am J Clin Nutr* 1999; 69: 1071–1085.

60. van Poppel G, Kok FJ, Duijzings P, de Vogel N. No influence of beta-carotene on smoking-induced DNA damage as reflected by sister chromatid exchanges. *Int J Cancer* 1992; 51: 355–358.

61. Wang XD, Russell RM. Procarcinogenic and anticarcinogenic effects of β-carotene. *Nutr Rev* 1999; 57(9Pt1): 263–272.
62. Sommerburg O, Langhans CD, Arnhold J, Leichsenring M, Salerno C, Crifo C, Hoffmann GF, Debatin KM; Siems WG. β-Carotene cleavage products after oxidation mediated by hypochlorous acid—A model for neutrophil derived degradation. *Free Radic Biol Med* 2003; 35: 1480–1490.
63. Siems W, Capuozzo E, Crifo C, Sommerburg O, Langhans CD, Schlipalius L, Wiswedel I, Kraemer K, Salerno C. Carotenoid cleavage products modify respiratory burst and induce apoptosis of human neutrophils. *Biochim Biophys Acta* 2003; 1639: 27–33.
64. Siems W, Sommerburg O, Schild L, Augustin W, Langhans CD, Wiswedel I. Beta-carotene cleavage products induce oxidative stress in vitro by impairing mitochondrial respiration. *FASEB J* 2002; 16(10): 1289–1291.
65. Siems W, Wiswedel I, Salerno C, Crifo C, Augustin W, Schild L, Langhans CD, Sommerburg O. β-Carotene breakdown products may impair mitochondrial functions-potential side effects of high-dose β-carotene supplementation. *J Nutr Biochem* 2005; 16: 385–397.
66. Alija AJ, Bresgen N, Sommerburg O, Siems W, Eckl PM. Cytotoxic and genotoxic effects of beta-carotene breakdown products on primary rat hepatocytes. *Carcinogenesis* 2004; 25(5): 827–831.
67. Singh NP, McCoy MT, Tice RR, Schneider EL. A simple technique for quantitation of low levels of DNA damage in individual cells. *Exp Cell Res* 1988; 175: 184–191.
68. Hurst JS, Contreras JE, Siems WG, van Kuijk FJ. Oxidation of carotenoids by heat and tobacco smoke. *Biofactors* 2004; 20: 23–35.
69. Kawanishi S, Hiraku Y, Oikawa S. Mechanism of guanine-specific DNA damage by oxidative stress and its role in carcinogenesis and aging. *Mutat Res* 2001; 488: 65–76.
70. Rahman I, MacNee W. Lung glutathione and oxidative stress: Implications in cigarette smoke-induced airway disease. *Am J Physiol Lung Cell Mol Physiol* 1999; 277: 1067–1088.
71. Pryor WA, Hale BJ, Premovic PI, Church DF. The radicals in cigarette tar: Their nature and suggested physiological implications. *Science* 1983; 220: 425–427.
72. Alija AJ, Bresgen N, Sommerburg O, Langhans CD, Siems W, Eckl PM. Beta-carotene breakdown products enhance genotoxic effects of oxidative stress in primary rat hepatocytes. *Carcinogenesis* 2006; 27(6): 1128–1133.
73. Alija AJ, Bresgen N, Sommerburg O, Langhans CD, Siems W, Eckl PM. Cyto- and genotoxic potential of β-carotene and cleavage products under oxidative stress. *Biofactors* 2005; 24: 159–164.
74. Paolini M, Antelli A, Pozzetti L, Spetlova D, Perocco P, Valgimigli L, Pedulli GF, Cantelli-Forti G. Induction of cytochrome P450 enzymes and over-generation of oxygen radicals in beta-carotene supplemented rats. *Carcinogenesis* 2001; 22: 1483–1495.
75. Marques SA, Loureiro APM, Gomes OF, Garcia CCM, Di Mascio P, Medeiros MHG. Induction of 1,N2-etheno-2′-deoxyguanosine in DNA exposed to β-carotene oxidation products. *FEBS Lett* 2004; 560: 125–130.
76. Yeh SL, Wu SH. Effects of quercetin on β-apo-8′-carotenal-induced DNA damage and cytochrome P1A2 expression in A549 cells. *Chem Biol Interact* 2006; 163: 199–206.
77. Woods JA, Bilton RF, Young AJ. β-Carotene enhances hydrogen peroxide-induced DNA damage in human hepatocellular HepG2 Cells. *FEBS Lett* 1999; 449: 255–258.
78. Palozza P, Serini S, Di Nicuolo F, Boninsegna A, Torsello A, Maggiano N, Ranelletti FO, Wolf FI, Calviello G, Cittadini A. β-Carotene exacerbates DNA oxidative damage and modifies p53-related pathways of cell proliferation and apoptosis in cultured cells exposed to tobacco smoke condensate. *Carcinogenesis* 2004; 25: 1315–1325.
79. Cozzi R, Ricordy R, Aglitti T, Gatta V, Perticone P, De Salvia R. Ascorbic acid and β-carotene as modulators of oxidative damage. *Carcinogenesis* 1997; 18: 223–228.

80. Hurst JS, Saini MK, Jin GF, Awasthi YC, van Kuijk FJGM. Toxicity of oxidized β-carotene to cultured human cells. *Exp Eye Res* 2005; 81: 239–243.

81. Cheng JZ, Singhal SS, Saini M, Singhal J, Piper JT, van Kuijk FJGM, Zimniak P, Awasthi YC, Awasthi S. Effects of mGST A4 transfection on 4-hydroxynonenal-mediated apoptosis and differentiation of K562 human erythroleukemia cells. *Arch Biochem Biophys* 1999; 372: 29–36.

82. Eckl PM, Ortner A, Esterbauer H. Genotoxic properties of 4-hydroxyalkenals and analogous aldehydes. *Mutat Res* 1993; 290: 183–192.

83. Alija AJ, Bresgen N, Bojaxhi E, Vogl C, Siems W, Eckl PM. Cytotoxicity of β-carotene cleavage products and its prevention by antioxidants. *Acta Biochim Pol* 2010; 57: 217–221.

84. Matthes-Roth MM. Beta-carotene therapy for erythropoietic protoporphyria and other photosensitivity diseases. *Biochimie* 1986; 68: 875–884.

85. Harper P, Wahlin S. Treatment options in acute porphyria, porphyria cutanea tarda, and erythropoietic protoporphyria. *Curr Treat Options Gastroenterol* 2007; 10: 444–455.

86. Parrish JA, LeVine MJ, Morison WL, Gonzalez E, Fitzpatrick TB. Comparison of PUVA and beta-carotene in the treatment of polymorphous light eruption. *Br J Dermatol* 1979; 100: 187–191.

87. Tronnier H. Protective effect of beta-carotene and canthaxanthin against UV reactions of the skin. *Z Hautkr* 1984; 59: 859–870.

88. Newbold PC. Beta-carotene in the treatment of discoid lupus erythematosus. *Br J Dermatol* 1976; 95: 100–101.

89. Raab W. Photoprotektive Wirkung von Betacarotin. *TW Dermatologie* 1991; 21: 187–201.

90. Bayerl C. Beta-carotene in dermatology: Does it help? *Acta Dermatovenerol Alp Panonica Adriat* 2008; 17: 160–166.

91. Stahl W, Sies H. Carotenoids and flavonoids contribute to nutritional protection against skin damage from sunlight. *Mol Biotechnol* 2007; 37: 26–30.

92. Neuman I, Nahum H, Ben-Amotz A. Prevention of exercise-induced asthma by a natural isomer mixture of beta-carotene. *Ann Allergy Asthma Immunol* 1999; 82: 549–553.

93. van Leeuwen R, Boekhoorn S, Vingerling JR, Witteman JCM, Klaver CCW, Hofman A, de Jong PTVM. Dietary intake of antioxidants and risk of age-related macular degeneration. *JAMA* 2005; 294: 3101–3107.

94. Grodstein F, Kang JH, Glynn RJ, Cook NR, Gaziano JM. A randomized trial of beta carotene supplementation and cognitive function in men. The Physicians' Health Study II. *Arch Intern Med* 2007; 167: 2184–2190.

95. Darlington LG, Stone TW. Antioxidants and fatty acids in the amelioration of rheumatoid arthritis and related disorders. *Br J Nutr* 2001; 85: 251–269.

96. Bae SC, Kim SJ, Sung MK. Inadequate antioxidant nutrient intake and altered plasma antioxidant status of rheumatoid arthritis patients. *J Am Coll Nutr* 2003; 22: 311–315.

97. Naldi L, Parazzini F, Peli L, Chatenoud L, Cainelli T. Dietary factors and the risk of psoriasis. Results of an Italian case-control study. *Br J Dermatol* 1996; 134: 101–106.

98. Sayyah M, Yousefi-Pour M, Narenjkar J. Anti-epileptogenic effect of β-carotene and vitamin A in pentylenetetrazole-kindling model of epilepsy in mice. *Epilepsy Res* 2005; 63: 11–16.

99. Mares-Perlman JA, Brady WE, Klein BE, Klein R, Haus GJ, Palta M, Ritter LL, Shoff SM. Diet and nuclear lens opacities. *Am J Epidemiol* 1995; 141: 322–334.

100. Mares-Perlman JA, Brady WE, Klein BE, Klein R, Palta M, Bowen P, Stacewicz-Sapuntzakis M. Serum carotenoids and tocopherols and severity of nuclear and cortical opacities. *Invest Ophthalmol Vis Sci* 1995; 36: 276–288.

101. Age-Related Eye Disease Study Research Group. A randomized, placebo-controlled, clinical trial of high-dose supplementation with vitamins C and E and beta carotene for age-related cataract and vision loss: AREDS report no. 9. *Arch Ophthalmol* 2001; 119: 1439–1452.

102. Chiu CJ, Taylor A. Nutritional antioxidants and age-related cataract and maculopathy. *Exp Eye Res* 2007; 84: 229–245.

103. Mayne ST. β-Carotene, carotenoids and disease prevention in humans. *FASEB J* 1996; 10: 690–701.

104. Kelly FJ. Oxidative stress: Its role in air pollution and adverse health effects. *Occup Environ Med* 2003; 60: 612–616.

105. Donaldson K, Stone V, Borm PJA, Jimenez LA, Gilmour PS, Schins RPF, Knaapen AM, Rahman I, Faux SP, Brown DM, Macnee W. Oxidative stress and calcium signaling in the adverse effects of environmental particles (PM10). *Free Radic Biol Med* 2003; 34: 1369–1382.

106. Li N, Sioutas C, Cho A, Schmitz D, Misra C, Sempf J, Wang M, Oberley T, Froines J, Nel A. Ultrafine particulate pollutants induce oxidative stress and mitochondrial damage. *Environ Health Perspect* 2003; 111: 455–460.

107. Shvedova AA, Kisin E, Murray AR, Johnson VJ, Gorelik O, Arepalli S, Hubbs AF, Mercer RR, Keohavong P, Sussman N, Jin J, Yin J, Stone S, Chen BT, Deye G, Maynard A, Castranova V, Baron PA, Kagan VE. Inhalation vs. aspiration of single-walled carbon nanotubes in C57BL/6 mice: Inflammation, fibrosis, oxidative stress, and mutagenesis. *Am J Physiol Lung Cell Mol Physiol* 2008; 295: L552–L565.

108. Oberdörster E. Manufactured nanomaterials (fullerenes, C60) induce oxidative stress in the brain of juvenile largemouth bass. *Environ Health Perspect* 2004; 112: 1058–1062.

109. Valko M, Rhodes CJ, Moncol J, Izakovic M, Mazur M. Free radicals, metals and anti-oxidants in oxidative stress-induced cancer. *Chem Biol Interact* 2006; 160: 1–40.

110. Li GJ, Zhang LL, Lu L, Wu P, Zheng W. Occupational exposure to welding fume among welders: Alterations of manganese, iron, zinc, copper, and lead in body fluids and the oxidative stress status. *J Occup Environ Med* 2004; 46: 241–248.

111. Han SG, Kim Y, Kashon ML, Pack DL, Castranova V, Vallyathan V. Correlates of oxidative stress and free-radical activity in serum from asymptomatic shipyard welders. *Am J Respir Crit Care Med* 2005; 172: 1541–1548.

112. Lieber CS. Role of oxidative stress and antioxidant therapy in alcoholic and nonalcoholic liver diseases. *Adv Pharmacol* 1997; 38: 601–628.

113. Leo MA, Lieber CS. Hepatic vitamin A depletion in alcoholic liver injury. *N Engl J Med* 1982; 307: 597–601.

114. Lee M, Lucia SP. Effect of ethanol on the mobilization of vitamin A in the dog and in the rat. *Q J Stud Alcohol* 1965; 26: 1–9.

115. Leo MA, Kim C, Lieber CS. Increased vitamin A in esophagus and other extrahepatic tissues after chronic ethanol consumption in the rat. *Alcohol Clin Exp Res* 1986; 10: 487–492.

116. Leo MA, Aleynik S, Aleynik M, Lieber CS. β-Carotene beadlets potentiate hepatotoxicity of alcohol. *Am J Clin Nutr* 1997; 66: 1461–1469.

117. Murata M, Kawanishi S. Oxidative DNA damage by vitamin A and its derivative via superoxide generation. *J Biol Chem* 2000; 275: 2003–2008.

118. Liu C, Russell RM, Wang XD. α-Tocopherol and ascorbic acid decrease the production of β-apo-carotenals and increase the formation of retinoids from β-carotene in the lung tissues of cigarette smoke–exposed ferrets in vitro. *J Nutr* 2004; 134: 426–430.

119. Salgo MG, Pryor WA. Trolox inhibits peroxynitrite-mediated oxidative stress and apoptosis in rat thymocytes. *Arch Biochem Biophys* 1996; 333: 482–488.

120. Sandoval M, Zhang X-J, Liu X, Mannick EE, Clark DA, Miller MJS. Peroxynitrite-induced apoptosis in T84 and RAW 264.7 cells: Attenuation by L-ascorbic acid. *Free Rad Biol Med* 1997; 22: 489–495.

121. Atkuri KR, Mantovani JJ, Herzenberg LA, Herzenberg LA. *N*-acetylcysteine—A safe antidote for cysteine/glutathione deficiency. *Curr Opin Pharmacol* 2007; 7: 355–359.

122. Watanabe M, Sugimoto M, Ito K. The acrolein cytotoxicity and cytoprotective action of α-tocopherol. *J Gastroenterol* 1992; 27: 199–205.

123. Sprince H, Parker CM, Smith GG, Gonzales LJ. Protective action of ascorbic acid and sulphur compounds against acetaldehyde toxicity: Implication in alcoholism and smoking. *Inflamm Res* 1975; 5: 164–173.

124. Miranda CL, Reed RL, Kuiper HC, Alber S, Stevens JF. Ascorbic acid promotes detoxification and elimination of 4-hydroxy-2(E)-nonenal in human monocytic THP-1 cells. *Chem Res Toxicol* 2009; 22: 863–874.

125. Urso ML, Priscilla M. Clarkson PM. Oxidative stress, exercise, and antioxidant supplementation. *Toxicology* 2003; 189: 41–54.

126. Ji LL. Antioxidants and oxidative stress in exercise. *Proc Soc Exp Biol Med* 1999; 222: 283–292.

127. Radak Z, Chung HY, Koltai E, Taylor AW, Goto S. Exercise, oxidative stress and hormesis. *Ageing Res Rev* 2008; 7: 34–42.

128. Poulsen HE, Loft S, Vistisen K. Extreme exercise and oxidative DNA modification. *J Sports Sci* 1996; 14: 343–346.

129. Umegaki K, Daohua P, Sugisawa A, Kimura M, Higuchi M. Influence of one bout of vigorous exercise on ascorbic acid in plasma and oxidative damage to DNA in blood cells and muscle in untrained rats. *J Nutr Biochem* 2000; 11: 401–407.

130. Tsai K, Hsu TG, Hsu KM, Cheng H, Liu TY, Hsu CF, Kong CW. Oxidative DNA damage in human peripheral leukocytes induced by massive aerobic exercise. *Free Radic Biol Med* 2001; 31: 1465–1472.

131. Mastaloudis A, Yu TW, O'Donnell RP, Frei B, Dashwood RH, Traber MG. Endurance exercise results in DNA damage as detected by the comet assay. *Free Radic Biol Med* 2004; 36: 966–975.

132. Bloomer RJ, Goldfarb AH, McKenzie MJ. Oxidative stress response to aerobic exercise: Comparison of antioxidant supplements. *Med Sci Sports Exerc* 2006; 38: 1098–1105.

133. Baskin CR, Hinchcliff KW, DiSilvestro RA, Reinhart GA, Hayek MG, Chew BP, Burr JR, Swenson RA. Effects of dietary antioxidant supplementation on oxidative damage and resistance to oxidative damage during prolonged exercise in sled dogs. *Am J Vet Res* 2000; 61: 886–891.

134. Teixeira VH, Valente HF, Casal SI, Marques AF, Moreira PA. Antioxidants do not prevent postexercise peroxidation and may delay muscle recovery. *Med Sci Sports Exerc* 2009; 41: 1752–1760.

135. Niess AM, Hartmann A, Grünert-Fuchs M, Poch B, Speit G. DNA damage after exhaustive treadmill running in trained and untrained men. *Int J Sports Med* 1996; 17: 397–403.

136. Radák Z, Kaneko T, Tahara S, Nakamoto H, Ohno H, Sasvári M, Nyakas C, Goto S. The effect of exercise training on oxidative damage of lipids, proteins, and DNA in rat skeletal muscle: Evidence for beneficial outcomes. *Free Radic Biol Med* 1999; 27: 69–74.

137. Salo DC, Donovan CM, Davies KJ. HSP70 and other possible heat shock or oxidative stress proteins are induced in skeletal muscle, heart, and liver during exercise. *Free Radic Biol Med* 1991; 11: 239–246.

138. Radak Z, Kumagai S, Nakamoto H, Goto S. 8-Oxoguanosine and uracil repair of nuclear and mitochondrial DNA in red and white skeletal muscle of exercise-trained old rats. *J Appl Physiol* 2007; 102: 1696–1701.

# 10 Carotenoids and Vitamin A in Lung Cancer Prevention

*Anita Ratnasari Iskandar and Xiang-Dong Wang*

## CONTENTS

## INTRODUCTION

Lung cancer is the leading cause of cancer death worldwide [1]. Despite increased efforts toward prevention, early detection, and myriad of antismoking campaigns, lung cancer accounts for 15% of all new cancers [2]. The addictive power of nicotine and the risk of lung cancer persist for many years after smoking cessation [1]. About 14%–58% of smoking lung cancer patients continue to smoke after the diagnosis [3]. Passive smokers (environmental cigarette smoke-exposed individuals, smoker's spouses, and children) are also at an increased risk of lung cancer [4]. The incidence of lung cancer among nonsmoking women has recently increased, due to unknown factors [5]. Besides smoking, diet and nutrition are other strong determinants of lung cancer risk [6]. Unfortunately, current lung cancer treatments do not offer much benefit [5]. Further, the compounding frustration is that the 5 year survival rate has not improved in three decades [7]. This emphasizes the importance of understanding the molecular mechanism of lung carcinogenesis and the development of an effective dietary chemopreventive agent against lung cancer risk.

Since the 1980s, numerous human studies have pointed out the potential protective effects from the consumption of fruits and vegetables against certain type of cancers [8]. Particularly, the beneficial effects of carotenoid-rich fruits and vegetables on lung cancer risk have been found in many epidemiological and experimental studies [9–19]. In addition, studies have indicated the beneficial association between the intake/serum levels of carotenoids and the risk of lung cancer [13,17,20–29]. In contrast, the two randomized controlled trials (Alpha-Tocopherol, Beta-Carotene Cancer Prevention [ATBC] Study [30], and Beta-Carotene and Retinol Efficacy Trial [CARET] [31]) reported that β-carotene supplementation had detrimental effects on lung cancer incidence and mortality. Although the discrepancies between the observational and trial data have been discussed [32], the biological mechanisms underlying these unexpected adverse effects have not been adequately elucidated. In particular, it is not clear whether any carotenoids other than β-carotene may account for the protective effects of fruits and vegetables on lung cancer risk seen in the observational studies.

This chapter focuses on the human and animal studies on the chemoprevention of retinoids and carotenoids on lung carcinogenesis, including discussions about polymorphisms and the possible mechanisms of the two compounds in lung carcinogenesis specifically. We refer readers to the recent outstanding reviews discussing the general mechanisms of carotenoids and retinoids in cancer prevention [33,34].

## LUNG CANCER

Lung cancer is the number one cause of cancer death in men and women worldwide [35]. By gender, lung cancer is the second most common cancer after prostate cancer in men, and after breast cancer in women [7]. Smoking is the primary risk factor of lung cancer [35]; about 90% of lung cancer cases are caused by cigarette smoke [7]. Both current and former smokers are at high risk of developing lung cancer [35]. As the duration of smoking and the amount of cigarette smoked increase, the risk of lung cancer increases [7]. Other risk factors include environmental exposure of radon, asbestos, air pollution, chromium, nickel, polycyclic aromatic hydrocarbons, and arsenic [7]. Chronic obstructive pulmonary disease is also associated with an increased risk of lung cancer [7].

Worldwide, there are about 1.3 billion people who still smoke [35]. About 70% of smokers attempt to quit each year, but only less than 5% succeed [35]. The prevalence of smoking has decreased in the high-income countries, including the United States, United Kingdom, Canada, and France [1]. On the other hand, the prevalence of smoking in males has risen sharply in low- and middle-income countries, such as China and Indonesia [1]. It was reported that socioeconomic status is inversely correlated with the risk of lung cancer after adjustment for the prevalence of smoking [7]. The 5 year survival rate of lung cancer after diagnosis is approximately only 15%, as the majority of patients already have advanced or metastasized tumors when the disease is discovered [7,35].

Lung cancer is generally categorized as non-small cell lung cancer and small cell lung cancer [7]. Non-small cell lung cancer can be further divided into adenocarcinoma, squamous cell carcinoma, and large cell carcinoma [7]. Small cell lung cancer, squamous cell carcinoma, and about 20% of the adenocarcinoma cases arise in the central compartment, mainly in bronchial airways and basal bronchial cells [36]. The rest of adenocarcinoma cases arise from the peripheral compartment of bronchioles and alveoli [36]. Squamous cell carcinoma was the most common lung cancer subtype before the 1980s; but thereafter, adenocarcinoma started to increase and became the most common subtype [7]. The reason of this alteration has been unclear; however, a few contributing factors have been proposed, including the introduction of filtered-cigarettes, low-tar and low-nicotine cigarettes; and the changes in histological classification, diagnostic and pathological techniques [5,7]. The addition of filters may alter the composition of cigarettes, such that there are increased contents of tobacco-specific nitrosamines and decreased polycyclic aromatic hydrocarbons [5]. Moreover, smokers seem to inhale more deeply when smoking the low-nicotine cigarettes, thus causing a deposition of carcinogens in the periphery of the lung and inciting the development of adenocarcinoma among the other subtypes [5].

The tobacco nitrosamines 4-[methylnitrosamino]-1-[3-pyridyl]-1-butanone (NNK) and $N$-nitrosonornicotine (NNN), and polycyclic aromatic hydrocarbons are metabolized by the phase I P450 cytochrome (CYP) enzymes [36,37] (Figure 10.1). The cytochrome P450 enzymes are part of the normal mammalian system to counter any foreign compound [37]. They convert the tobacco nitrosamines into active metabolites [37], which react with DNA and form DNA adducts [36,37]. If the adducts are not repaired, they cause mutations that often occur in genes encoding the tumor suppressor p53 and the oncogene Ras [36,37]. The consequence can be abnormal cellular growth and tumor development [37]. Recently, studies suggest that there are additional mechanisms of lung carcinogenesis involving the activation of cell-surface receptors, such as the nicotinic acetylcholine receptors (nAChRs), the epidermal growth factor receptors (EGFRs), and the insulin-like growth factor receptors (IGFRs) [37,38]. Nicotine and the tobacco nitrosamines NNK and NNN can exert their actions by binding to the nAChRs or by increasing the expression of both nAChRs and EGFRs [37,39]. Stimulations of these receptors lead to the activation of downstream signaling pathways, such as the phosphatidylinositol 3-kinase (PI3K)/AKT and Ras/mitogen-activated kinase-like protein (MAPK) pathways [36–39]. Persistent stimulation of these pathways can lead to the deregulation of cell growth (proliferation) and cell death (apoptosis), thus contributing to the progression from normal tissues to hyperplasia, dysplasia, and eventually to carcinoma [36–39].

Nonetheless, the development of lung cancer involves multiple factors, such as environmental (diet, smoking, carcinogenic exposures, etc.) and genetic factors [7]. Recently, studies have reported the role of single nucleotide polymorphisms (SNPs) in the pathogenesis of lung cancer, such as those in the genes of the DNA repair systems and phase II detoxification enzymes adding complexity to the perplexing nature of the interaction between nutrients and diseases [40,41].

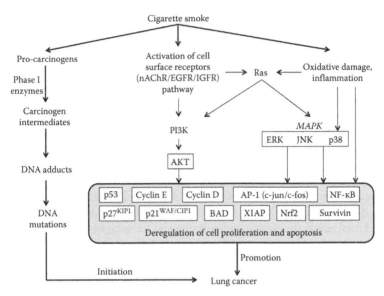

**FIGURE 10.1** Schematic of the proposed mechanisms of the tobacco-smoke contributing lung carcinogenesis. Tobacco smoke carcinogens can be metabolized by the phase I oxidation enzymes into carcinogen intermediates capable of covalently binding to DNA and forming DNA adducts. During DNA replication, DNA adducts cause miscodings of DNA, which subsequently result in permanent mutations if not repaired. The mutations often occur in genes important for cell growth and survival (e.g., Ras, p53). In addition, the high content of radicals and oxidants in cigarette smoke can contribute to oxidative damage and inflammation in the lung, both of which can activate NF-κB leading to the increased expression of cell cycle regulators (e.g., cyclin D1, p21$^{WAF/CIP1}$). Tobacco carcinogens can also bind and/or activate cell surface receptors, such as the nAChRs, the EGFRs, and the IGFRs. Stimulation of these receptors activates the MAPK and PI3K/AKT pathways and several signaling molecules important for the promotion of lung carcinogenesis (e.g., AP-1, p53, P21$^{WAF/CIP1}$, p27$^{KIP1}$, Nrf2, BAD, XIAP, survivin, etc.). AP-1, activator protein 1; MAPK, mitogen-activated protein kinase; Nrf2, nuclear factor E2-related factor 2; PI3K, phosphoinositide 3-kinase; XIAP, X-linked inhibitor of apoptosis protein.

## VITAMIN A AND RETINOIDS

Retinoids are a family of signaling compounds that are structurally related to vitamin A/retinol [34,42]. A retinoid molecule has (1) a cyclic β-ionone group in one end, (2) a polyene side chain, and (3) a polar group in the other end [34]. Retinol is a fat-soluble vitamin important for embryonic growth and organogenesis, tissue homeostasis, cell proliferation, differentiation, and apoptosis [34]. Retinol exerts its role via the conversion to the active acid form (i.e., retinoic acid) by two consecutive oxidation steps, retinol to retinal and retinal to retinoic acid facilitated by retinol dehydrogenase and retinal dehydrogenase, respectively [42]. The classic parent compound of the active acid form is the all-*trans* retinoic acid, which is also the predominant physiological form [43]. The other known biologically active isoforms include 11-*cis*, 13-*cis*, 9,13-di-*cis*, 9-*cis*, and 11,13-di-*cis* retinoic acids [44]. Retinoids are stored in the form of retinyl esters in animal tissues, such as retinyl oleate, retinyl linoleate, retinyl stearate, and the predominate form retinyl palmitate [44,45].

The active retinoids physiologically act via the binding to the nuclear retinoic acid receptors (RARs) and retinoic X receptors (RXRs) [42,44]. RARs and RXRs regulate gene transcription by binding to the response elements (the retinoic acid response element [RARE]) in the promoter region of their target genes [34]. In general, in the absence of ligands, RARs/RXRs repress gene transcription as they interact more with corepressors, whereas in the presence of ligands, they interact with coactivators [42]. Each of the retinoic receptors is further divided to the $\alpha$, $\beta$, and $\gamma$ isoforms; each isoform is encoded by different genes [34]. RXRs regulate gene transcription as homodimers or heterodimers with other nuclear receptors, including the RARs, peroxisome proliferator-activated receptors, vitamin D receptors, and thyroid hormone receptors [42]. Studies have shown that 9-*cis* retinoic acid can bind to both RARs and RXRs, while all-*trans* retinoic acid can bind only to RARs [42,44]. There was a suggestion that all-*trans* retinoic acid can serve as a ligand for peroxisome proliferator-activated receptors $\beta/\gamma$ [42]. Furthermore, in many cancers, the expression of RAR-$\beta$2 isoform is frequently reduced [34].

## CAROTENOIDS

Carotenoids are groups of fat-soluble chemicals responsible for pigmentation in plants and found in many fruits and vegetables [46]. They also contribute to the fall coloration of the leaf after the chlorophyll has been destroyed [47]. Carotenoids are comprised 8 units of isoprene and frequently have a symmetrical skeleton of 40 carbon atoms with a long chain of conjugated carbon–carbon double bonds, which are important for the yellow to red color and the antioxidant characteristics [48]. Based on their structures, carotenoids are divided into xanthophylls (the oxidized derivatives of carotenoids, containing hydroxyl group(s)) and carotenes (the unoxygenated carotenoids including cyclic and linear carotenoids) [49,50].

More than 700 carotenoids have been found in nature [46]. Among the 700 carotenoids, only 40 are commonly consumed by humans [51]. Upon central cleavage, the provitamin A carotenoids can serve as precursors for retinol and exert their actions via the activation of RARs or RXRs [52]. The $\beta$-carotene 15,15′-oxygenase (BCMO) 1 enzyme is the enzyme responsible for the central cleavage of provitamin A carotenoids: It cleaves carotenoids at their 15–15′ carbon–carbon double bond [52,53]. The six carotenoids often detected in the human circulation [46,47,51] are the provitamin A carotenoids ($\alpha$-carotene, $\beta$-carotene, $\beta$-cryptoxanthin), and the non-provitamin A carotenoids (lycopene, lutein, and zeaxanthin) [46,47,51]. Studies in cancer chemoprevention have mainly focused on these six carotenoids.

## CAROTENOIDS AND LUNG CANCER

### HUMAN STUDIES OF CAROTENOIDS

Since the early 1980s, questions were raised whether dietary carotenoids, $\beta$-carotene specifically, can affect cancer risks [54]: some of the early experiments have reported that $\beta$-carotene and carrot extracts had antitumor effects [54]. Peto et al. in 1981 proposed few theoretical mechanisms by which carotenoids may affect cancer risks, such as their convertibility to retinoid-like molecules, their potential to exert retinoid-like functions, and their retinoid-independent actions to enhance immune functions or to quench reactive oxygen species [54].

Unexpectedly, from the ATBC study involving 29,133 men, it was reported that an 18% increase in lung cancer incidence was found in those supplemented with β-carotene (20 mg/day) compared to those in the placebo group [30]. A subsequent analysis of the ATBC trial revealed that this deleterious effect was observed only in heavy smokers [55]. Subsequent to the publication of the ATBC results in 1995, the other randomized, double-blind, placebo-controlled CARET study involving 18,314 smokers, former smokers, and workers exposed to asbestos found a 28% increase in lung cancer risk in subjects supplemented with the combination of β-carotene (30 mg/day) and retinyl palmitate (25,000 UI/day) [31]. Moreover, another randomized, double-blind, placebo-controlled primary prevention Physicians' Health Study examined the effect of β-carotene supplementation (50 mg/day) over 12 years of follow-up in 22,071 male physicians [56]. In this study, β-carotene supplementation did not affect lung cancer incidence in the study subjects who were mostly nonsmokers [56]. Subsequently, in the Women's Health Study, β-carotene supplementation (30 mg/day) also did not affect lung cancer incidence among the apparently healthy female professionals [57]. It is thought that β-carotene may exert its harmful effect only in those with high risk of cancer (e.g., smokers, alcohol drinkers) but not in those with average risk (e.g., nonsmokers, nondrinkers) [58].

After the results of the ATBC study were reported, Holick et al. in 2002 examined the baseline characteristics of the male subjects of the ATBC study [21]. They reported that the baseline serum β-carotene level was associated with a lower risk of lung cancer; this is inconsistent with the detrimental end result of the ATBC trial. Additionally, following the ATBC and CARET studies, other groups also reported the inverse association between the level of serum carotenoids and the risk of lung cancer: In men but not in women, serum levels of total carotenoids [20], α-carotene [22], β-carotene [21,22], and lycopene [22] were inversely correlated with the risk of lung cancer. Moreover, Yuan et al. reported that only serum β-cryptoxanthin level was associated with a lower risk of lung cancer after adjustment with self-report smoking status in a prospective cohort study of Chinese men [59]. However, after readjustment with the urinary cotinine level-adjusted smoking status, this protective association was no longer significant [60].

Nonetheless, several studies have found harmful associations between serum levels of carotenoids and risk of lung cancer confirming the results of the ATBC and CARET trials. Serum levels of β-carotene [61,62] and β-cryptoxanthin [62] were associated with an increased risk of lung cancer in men and women. Ratnasinghe et al. reported that serum α-carotene, β-carotene, β-cryptoxanthin, and lutein/zeaxanthin concentrations were associated with an increased risk of lung cancer only in alcohol drinkers but not in nondrinkers [62].

Whether specific carotenoids can reduce the risk of lung cancer is still unclear. One of the reasons for the inconsistency among the studies may be gene polymorphisms since polymorphisms could affect the metabolism of carotenoids in the body, for example, the polymorphisms in genes related to β-carotene metabolism, DNA repair system, and phase II detoxification enzymes. Recently, a review was published discussing the polymorphisms important for the metabolism of β-carotene, including the genes of the apolipoprotein B, β-carotene 15–15′-oxygenase, hepatic lipase, lipoprotein lipase, and scavenger receptor class B type I [52]. The effects of carrying the minor allele of these genes could either increase or decrease the concentrations of β-carotene in the circulation [52]. Wang et al. reported that smokers having the phase II detoxification enzyme gene glutathione-S-transferase class T (GST-class T, GSTT) null genotype and high plasma levels of β-carotene tended to have higher DNA adduct levels than smokers having the GSTT-present genotype and high plasma β-carotene levels [63]. On the contrary, Mooney et al. reported that plasma level of β-carotene was inversely correlated with the level of DNA adduct in the GST-class M1 (GSTM1) null genotype but not in the GSTM1 present (+/+, +/0) [64]. Furthermore, serum levels of β-carotene and β-cryptoxanthin (highest vs. lowest tertile) but not of lutein/zeaxanthin were directly associated with lung cancer risk in individuals carrying the wild-type genotype of the nucleotide excision repair system XRCC1 (Arg/Arg) [65]. In addition, the discrepancy among the studies may be affected by the variability of the carotenoids' absorption; some individuals may absorb carotenoids better than others [66], and carotenoids themselves may have competitive absorptions [67]. It has been reported that canthaxanthin, zeaxanthin, or lutein may compete with β-carotene for their absorptions [67].

Following the ATBC and CARET studies, examinations on carotenoids besides β-carotene have been conducted. Several case-control studies reported that the intake of total carotenoids [26], α-carotene [23], and β-carotene [13,23,28,29] were significantly associated with a lower risk of lung cancer. The intake of lycopene was suggested to be only protective against lung cancer (highest vs. lowest tertile, odds ratio [OR] = 0.56, 95% confidence interval [CI] = 0.26–1.24) [68]. It should be noted that unbalanced recalls may occur in case–control studies even though both groups were interviewed about their past intakes or other potential risk factors; often the cases are better at recalling their past routines [32,69].

In addition, a prospective cohort study is considered to be superior since recall bias is minimized [32,69]. In this type of study, a large group of people is screened for specific behaviors/risk factors and is subsequently followed for a period of time until they get the specific disease [32,69]. A number of prospective cohort studies observed the inverse association between the risk of lung cancer and the intake of specific carotenoids: α-carotene [24,25], β-cryptoxanthin [17,21,27], lutein/zeaxanthin [17,21], and lycopene [21]. None of the before-mentioned prospective cohort studies found inverse associations between β-carotene intake and lung cancer risk. In 2002, a prospective cohort study of Canadian women found no significant effects of any carotenoids on the risk of lung cancer [70]. Subsequently, in a prospective cohort of French females, Touvier et al. in 2005 reported that β-carotene intake was associated with a lower risk of lung cancer only in nonsmokers, whereas an increased risk was found in smokers [71].

Furthermore, pooled analyses and meta-analyses of the epidemiological studies have been reported after the ATBC and CARET studies. In 2000, Michaud et al. conducted a pooled analysis of two prospective cohort studies and reported a significant 20%–25% lower risk of lung cancer in participants who consumed high intakes of α-carotene and lycopene compared to those who consumed low intake of the carotenoids after adjustment for smoking status [25]. No significant association was found in regard to the intake of β-carotene, β-cryptoxanthin, and lutein [25]. In 2004, another pooled analysis was reported; Mannisto et al. analyzed the primary data from the seven cohort studies in North America and Europe [72]. They observed that only the intake of β-cryptoxanthin was inversely associated with the risk of lung cancer (highest vs. lowest quintile;

pooled RR=0.77; 95% CI: 0.68–0.88), whereas the intake of α-carotene, β-carotene, lutein/zeaxanthin, and lycopene was not inversely associated [72].

Subsequently, in 2008, a systematic review from 6 randomized clinical trials and 25 prospective observational studies was published. The authors conducted a meta-analysis examining the association between carotenoids and lung cancer risk [73]. The intake of α-carotene, β-carotene, β-cryptoxanthin, lycopene, and lutein + zeaxanthin was inversely associated with lung cancer risk after smoking adjustment except for lutein alone. Interestingly, in current smokers, the intake of α-carotene and β-cryptoxanthin was associated with a lower risk of lung cancer (RR = 0.88; 95% CI: 0.78–0.997 and RR 0.78; 95% CI: 0.58–0.96, respectively), but not the intake of β-carotene, lutein/zeaxanthin, lutein, and lycopene [73]. Furthermore, when serum carotenoids were analyzed, they found that only serum levels of lycopene were inversely associated with the risk of lung cancer after smoking adjustment [73].

The more recent systematic review and meta-analysis of randomized controlled trials were published in 2010, reporting the increased incidence of lung cancer in individuals supplemented with β-carotene compared to placebo group [74]. This publication supported the published VITamins And Lifestyle (VITAL) Cohort Study in 2009, which found that the longer duration of use (number of years used over 10 years) of single β-carotene supplements (not classified as multivitamins) was associated with an increased risk of small cell lung cancer and that the longer duration of use of lutein supplements was associated with an increased risk of non-small cell lung cancer and total lung cancer (all subtypes) [75]. Upon stratification by smoking status, only current smokers had a significant direct association between risk of lung cancer and duration of lutein use [75]. So far, this is the only human study reporting a significant effect of lutein on cancer.

Nonetheless, the antitumor effects of the combination of carotenoids and chemotherapeutic agents have been studied for the possible treatments of lung cancer: in 136 patients of stage IIB and IV non-small cell lung cancer, the combination of β-carotene, vitamins C and E, and the chemotherapeutic agents (paclitaxel and carboplatin) did not significantly affect the tumor [76]. The overall survival rate after 1 year of those receiving the combination treatment was 39.1%, which was not significantly different compared to that of those receiving the chemotherapeutic agents (paclitaxel and carboplatin) without β-carotene, and vitamins C and E [76]. The addition of these antioxidants to the chemotherapeutic agents did not alter the toxicity profile [76].

## ANIMAL STUDIES OF CAROTENOIDS

One of the primary advantages of conducting animal studies is to observe the direct effects of dietary intervention with carotenoids on lung carcinogenesis. However, high doses of carotenoids tend to be administered when using animal models [8]. Although a mathematical formula is available for the interspecies dose conversion [77], a careful extrapolation of the dose is necessary since the dose extrapolation is influenced by multiple factors including the size of the organisms, the metabolic rate (see detailed review [78]), and particularly, the differences in the absorptions and metabolisms of carotenoids among species [8].

Majority of the animal studies on carotenoids and lung cancer chemoprevention have focused on β-carotene; the results are summarized in Table 10.1. Despite

**TABLE 10.1**

**Carotenoids Chemoprevention Studies on Lung Carcinogenesis in Animal Model**

| Carotenoid | Dose | Effect on Tumor | Tumor Inducer | Species | N/group | Investigator | Year | References |
|---|---|---|---|---|---|---|---|---|
| β-Carotene | 50 mg/100 g drinking water/day | No effect | 4-NQO | DdY strain mice (m) | 16 | Murakoshi | 1992 | [88] |
| β-Carotene | 50 mg/kg diet/day | No effect | B(a)P | A/J mice (m,f) | 40–41 | Yun | 1995 | [92] |
| β-Carotene | 50 mg/100 g drinking water/day | No effect | 4-NQO | DdY strain mice (m) | 15–16 | Nishino | 1995 | [89] |
| β-Carotene | 16; 32 mg/kg bw s.c. 2×/week | No effect | NNK | A/J mice (f) | 29–30 | Conaway | 1998 | [85] |
| β-Carotene | 2.4 mg/kg bw p.o./day | Induction | Smoke | Ferrets (m) | 6 | Wang | 1999 | [84] |
| β-Carotene | 0.43 mg/kg bw p.o./day | No effect | Smoke | Ferrets (m) | 6 | Liu | 2000 | [87] |
| β-Carotene | 50; 500; 5000 mg/kg diet/day | No effect | Smoke | A/J mice (m) | 10–26 | Obermueller-Jevic | 2002 | [90] |
| β-Carotene | 107.38 mg/kg bw i.p./day | Reduction | Transplant | C57BL/6 mice (na) | 14 | Pradeep | 2003 | [79] |
| β-Carotene | 0.4 mg/mouse by gavage intragastrically 3×/week | No effect | Urethane | BALB/c mice (m,f) | 16–20 | Uleckiene | 2003 | [91] |
| β-Carotene | 5000 mg/kg diet/day | Reduction | Smoke | A/J mice (m) | 13–26 | Witschi | 2005 | [82] |
| β-Carotene | 120; 360; 600; 1800; 3000 mg/kg diet/day | No effect | NNK | A/J mice (m) | 15–40 | Goralczyk | 2006 | [86] |
| β-Carotene[a] | 0.85 mg/kg bw p.o./day | Reduction | NNK | Ferrets (m) | 9–14 | Kim | 2006 | [81] |
| β-Carotene | 20 mg/kg bw p.o./day | Reduction | Transplant | Athymic nude mice (m) | 7–8 | Huang | 2008 | [80] |
| β-Carotene | 0.0005 mg/mouse s.c. 3×/week | Induction | NNK | Syrian golden hamsters (m) | 20 | Al-Wadei | 2009 | [83] |
| α-Carotene | 50 mg/100 g drinking water/day | Reduction | 4-NQO | DdY strain mice (m) | 16 | Murakoshi | 1992 | [88] |
| α-Carotene | 50 mg/100 g drinking water/day | Reduction | 3-NQO | DdY strain mice (m) | 15–16 | Nishino | 1995 | [89] |
| β-Cryptoxanthin[b] | 3.9 mg/100 g mandarin juice/day | Reduction | NNK | A/J mice (m) | 10–23 | Kohno | 2001 | [149] |

*(continued)*

**TABLE 10.1 (continued)**
**Carotenoids Chemoprevention Studies on Lung Carcinogenesis in Animal Model**

| Carotenoid | Dose | Effect on Tumor | Tumor Inducer | Species | N/group | Investigator | Year | References |
|---|---|---|---|---|---|---|---|---|
| β-Cryptoxanthin | 7; 5; 37.5 μg/kg bw p.o./day | Reduction | Smoke | Ferrets (m) | 6 | Liu | 2011 | [140] |
| β-Cryptoxanthin | 1; 10 mg/kg diet/day | Reduction | NNK | A/J mice (m) | 14–15 | Iskandar | 2011 | [150] |
| Lycopene | 250; 500 mg/100 g drinking water/day | Reduction | DMD | B6C3F1 mice (m,f) | 5–16 | Kim | 1997 | [151] |
| Lycopene | 11; 109; 546 mg/kg diet/day | No effect | B(a)P + NNK | A/J mice (f) | 16–20 | Hecht | 1999 | [152] |
| Lycopene | 7.0; 14 g/kg diet/day | Induction | B(a)P | Muta™ mice (m) | 6 | Guttenplan | 2001 | [153] |
| Lycopene | 1.1; 4.3 mg/kg bw p.o./day | Reduction | Smoke | Ferrets (m) | 6 | Liu | 2003 | [135] |
| Lycopene | 1; 20 mg/kg bw p.o./day | Reduction | Transplant | Athymic nude mice (m) | 7–8 | Huang | 2008 | [80] |
| Apo-10'-lycopenoic acid | 10; 40; 120 mg/kg diet/day | Reduction | NNK | A/J mice (m) | 12–14 | Lian | 2007 | [123] |

bw, body weight; f, female; m, males; na, not available; i.p., intraperitoneal injection; p.o., *per os* (oral administration); s.c., subcutaneous injection; B(a)P, benzo(a)pyrene; DMD, diethylnitrosamine + N-methyl-N-nitrosourea + 1,2-dimethylhydrazine; NNK, 4-(methylnitrosamino)-1-(3-pyridyl)-1-butanone; NQO, nitroquinoline 1-oxide.

a   Supplementation was given as a mixture of β-carotene, α-tocopherol (vitamin E), and ascorbic acid (vitamin C).

b   Supplementation was given as mandarin juice rich in the combination of β-cryptoxanthin + the flavonoid hesperidin.

the varieties of the tumor inducers, there are inconsistent effects of β-carotene on lung tumorigenesis: few studies reported a significant tumor reduction [79–82] or tumor induction [83,84]; however, the rest of the studies did not find any effect on tumorigenesis [85–92]. Interestingly, strong proliferative responses in lung tissue and precancerous lesion squamous metaplasia were seen in ferrets treated with high (pharmacological) dose of β-carotene with or without smoke exposure, and this was associated with lower concentrations of retinoic acid in the lung tissue [84]. In contrast, the combined treatment of β-carotene and vitamins C and E did not affect squamous metaplasia and lung concentration of retinoic acid [81]. Therefore, it is possible that the combination of antioxidants partially restored the cigarette smoke-reduced level of retinoic acid.

Furthermore, a small number of studies were conducted to examine the effects of the other carotenoids besides β-carotene, such as α-carotene, β-cryptoxanthin, lycopene, and the metabolite of lycopene (apo-10′-lycopenoic acid) (Table 10.1). Protective effects on lung cancer were observed for the intervention with α-carotene and β-cryptoxanthin, but the data on lycopene are less conclusive. Future studies should examine other carotenoids beyond β-carotene for possible chemopreventive agents against lung cancer.

# RETINOIDS AND LUNG CANCER

## Human Studies of Retinoids

Studies have shown that retinoic acid directs the transcription of genes regulating cell differentiation, cell cycle arrest, and apoptosis [34]. For that reason, retinoids have been clinically utilized for possible treatments for malignancies [42]. However, until today, the use of retinoids for cancer chemotherapy has not been effective (Table 10.2). Nonetheless, alterations of biomarkers were observed in patients who received retinoid treatments: While 13-*cis* retinoic acid did not have any effect on cancer incidence and mortality among smokers [93], it could downregulate the proliferative indexes (proliferating cell nuclear antigen (PCNA) [94] and Ki-67 [95]) and induce the expression of RAR-β [96]. In contrast, retinol did not affect the RAR-β expression in bronchial dysplasia among currents/former smokers [97]. Interestingly, Satia et al. reported that the duration of the use of vitamin A/retinol supplement was associated with an increased risk of non-small cell lung cancer and total lung cancer (all subtypes) [75].

An important finding was reported recently demonstrating the beneficial effect of 13-*cis* retinoic acid on the risk of head-and-neck cancer depending on the patients' genetic background [98]. Lee et al. did an analysis [98] based on the original study in which 1190 stage-I and stage-II head-and-neck cancer patients were assigned to receive 13-*cis* retinoic acid or placebo [99]. In the original parent study, 13-*cis* retinoic acid treatment did not affect the risk of the occurrence of second primary tumors and the recurrence of head-and-neck cancer [99]. Subsequently, the same group conducted a stratified analysis based on the patients' SNPs and found that 13-*cis* retinoic acid was associated with a 38% reduction in the risk of the second primary tumor occurrence and the head-and-neck cancer recurrence in patients carrying the wild-type *RXRA* (SNP rs3118570 located in the intron of the RXR-α gene) [98], whereas

**TABLE 10.2**
**Human Trials with Retinoids on Lung Cancer**

| Agent | Dose | Result | Subject | N | Investigator | Year | References |
|---|---|---|---|---|---|---|---|
| All-*trans* RA | 1.8 mg/kg bw/day[a] | Minimal effect on metastatic | NSCLC patients | 28 | Treat | 1996 | [154] |
| 13-*cis* RA | 1–2.5 mg/kg bw/day | No effect on sputum cytology | Patients with cytologic abnormalities in sputum atypical metaplasia to carcinoma | 26 | Saccomanno | 1982 | [155] |
| 13-*cis* RA | 1 mg/kg bw/day | No effect on squamous metaplasia | Smokers | 152 | Lee | 1994 | [156] |
| 13-*cis* RA | 0.5 mg/kg bw/day[b] | Upregulation of RAR-β | Smokers | 188 | Ayoub | 1999 | [96] |
| 13-*cis* RA | 1 mg/kg bw/day | Reduction of PCNA | Smokers | 86 | Khuri | 2001 | [94] |
| 13-*cis* RA | 0.5 mg/kg bw/day[b] | No effect on second primary tumors and mortality | Stage I NSCLC patients | 1166 | Lippman | 2001 | [157] |
| | | No effect on cancer incidence and mortality | | | Lee | 2010 | [93] |
| 13-*cis* RA | 1 mg/kg bw/day | Reduction of Ki-67 | Former smokers | 151 | Hittelman | 2007 | [95] |
| 13-*cis* RA | 0.83 mg/kg bw/day[b] | No effect on bronchial histology or Ki-67 | Persons with high risk/with prior surgery for NSCLC | 55–86 | Kelly | 2009 | [158] |
| 9-*cis* RA | 1.67 mg/kg bw/day[b] | No effect on Ki-67 | Former smokers | 148 | Hittelman | 2007 | [95] |

| Retinol | 0.25 mg/kg bw/day[b] | No effect on the RAR-β in lesions with bronchial dysplasia | Current or former smokers | 81 | Lam | 2003 | [97] |
|---|---|---|---|---|---|---|---|
| Retinyl palmitate | 1.52 mg/kg bw/day[b] | Reduction of new primary tumors | Stage-I NSCLC patients | 307 | Pastorino | 1993 | [159] |
| Retinyl palmitate | 0.76; 1.52 mg/kg bw/day[b] | No effect on survival and second primary tumors | Patients (60% with head-and-neck cancer and 40% with lung cancer) | 2592 | van Zandwijk | 2000 | [160] |
| 4-HPR | 3.33 mg/kg bw/day[b] | No effect on squamous metaplasia, dysplasia | Smokers | 139 | Kurie | 2000 | [161] |
| 4-HPR | 3.33 mg/kg bw/day[b] | Reduction of hTERT mRNA | Chronic/former smokers | 57 | Soria | 2001 | [162] |

bw, body weight; HPR, N-(4-hydroxyphenyl)retinamide; hTERT, catalytic subunit of telomerase; NSCLC, non-small cell lung cancer; PCNA, proliferating cell nuclear antigen; RA, retinoic acid; RAR, retinoid acid receptor.

a Dose was reported in the original paper as 175 mg/m²/day and was converted to mg/kg bw/day under the assumption that the body weight is 60 kg and the body surface area is 1.62 m² [163].

b Doses were reported in the original publications as daily intake (unit of mass/day) and were converted to (mg/kg bw/day) under the assumption that the body weight is 60 kg. 1 mg Retinol Activity Equivalence = 3300 IU [164].

a 333% induction of the risk of occurrence/recurrence of head-and-neck cancer was found in the placebo group [98]. In addition, significant beneficial effects of the 13-*cis* retinoic acid were also observed in the patients carrying the common genotypes of the JAK/STAT signaling gene (*JAK2*), the matrix metallopeptidase (*MMP3* and *MMP21*), the DNA repair genes (*RAD54L* and *BCCIP*), the membrane transporter gene (*SLC31A1*), the phase II metabolizing enzyme (*GSTM2*), the kinase regulating hematopoiesis gene (*FLT3*), the telomeric poly (ADP-ribose) polymerase gene (*TNKS1BP1*), and the AKT/mTOR signaling gene (*TSC1*) [98].

Recently, concerns were raised about retinoid resistance in cancer therapy [34]. Even though retinoids were able to halt the growth and differentiation of normal cells, the same response was not observed in tumor cells [34]. Loss of RAR-β expression by epigenetic silencing due to hypermethylation, histone deacetylation, or mutation in the ligand-binding domain of RAR has been found and was hypothesized to be responsible for the retinoid resistance in cancer cells [34,100]. Furthermore, hypermethylation of RAR-β was found in heavy smokers [101]. Interestingly, the aberrant methylation at the promoter of RAR-β can be reversed by retinoic acid [102].

Future uses of retinoids for cancer therapy would likely involve a combination therapy with chemotherapeutical agents or with inhibitors of DNA methyltransferase and histone deacetylase [34]. In fact, more pronounced antitumor effects were observed when retinoids 13-*cis* retinoic acid, all-*trans* retinoic acid, retinyl palmitate, or arotinoid (the synthetic retinoid Ro 40-8757) were administered in combination with the chemotherapeutic agent cisplatin to non-small cell lung cancer patients [103–106]. In contrast, in a phase II clinical trial, the combination of cisplatin and all-*trans* retinoic acid did not affect response and survival rates in small cell lung cancer patients [107]. Less effects were seen in either non-small or small cell lung cancer patients when retinoids (13-*cis* retinoic acid or all-*trans* retinoic acid) were administered in combination with the chemotherapeutic agent interferon-α (IFN-α) [108–111]. Only one study reported an increased median survival of small cell lung cancer patients after treatment with the combination of 13-*cis* retinoic acid and INF-α2a [112].

## ANIMAL STUDIES OF RETINOIDS

The first experiment demonstrating the ability of retinoids in suppressing malignancy was reported in 1965 by Chu and Malmgren [113]. They showed that retinol protected the hydrocarbon-induced carcinoma in hamster [113]. Subsequent studies have confirmed the ability of retinoids to prevent epithelial cancer growth in carcinogen-exposed animals. Retinoids do seem to hold promise for the prevention of lung carcinogenesis in animal models despite routes of administration and doses administered (see Table 10.3). However, there is inconsistency between the outcomes of animal studies and human studies. For example, 9-*cis* retinoic acid supplementation at 7.5 and 15.0 mg/kg diet was effective in suppressing the tobacco-nitrosamine NNK-induced lung tumor multiplicity [114]; however, it did not significantly affect the bronchial epithelial cell proliferation in former smokers [95]. The reason for the inconsistency is unclear, but polymorphisms in the human genome, which can

**TABLE 10.3**

**Retinoid Chemoprevention Studies on Lung Carcinogenesis in Animal Model**

| Retinoid | Dose | Antitumor Effect | Tumor Inducer | Species | Investigator | Year | References |
|---|---|---|---|---|---|---|---|
| All-*trans* RA (Ro 1-5488) | 10 mg/kg bw/d i.p. 5 day/week or p.o. daily | Yes | Transplant | BALB/c mice (na) | Hubert | 1983 | [165] |
| All-*trans* RA | 1 μM to cells | Yes | Transplant | C57BL mice (na) | Edward | 1992 | [166] |
| All-*trans* RA | 5 μM to cells | Yes | Transplant | C57BL/6J mice (na) | Gaetano | 1994 | [167] |
| All-*trans* RA | 40 mg/mouse i.p./day | Yes | Transplant | BALBc/byJ (na) | Andela | 2003 | [168] |
| All-*trans* RA | 0.585 mg/kg bw i.v./day | Yes | Transplant | CDF1 mice (na) | Suzuki | 2006 | [169] |
| All-*trans* RA | 20 mg/kg bw i.v./day | Yes | Transplant | C57BL/6J mice (m) | Liu | 2008 | [170] |
| 13-*cis* RA (Ro 4-3780) | 10 mg/kg bw i.p. 5day/week or p.o. daily | No | Transplant | BALB/c mice (na) | Hubert | 1983 | [165] |
| 13-*cis* RA | 20 mg/kg bw i.v./day | Yes | Transplant | C57BL/6J mice (m) | Liu | 2008 | [170] |
| 13-*cis* RA | 1.3; 20.7; 481 μg/L (aerosol)[a] | Yes | B($a$)P, NNK | A/J mice (m) | Dahl | 2000 | [171] |
| 9-*cis* RA | 1.5; 3.0; 6.0 mg/kg bw p.o./day | Yes | NNK | A/J mice (m) | Mernitz | 2007 | [114] |
| 9-*cis* RA | 20 mg/kg bw i.v./day | No | Transplant | C57BL/6J mice (m) | Liu | 2008 | [170] |
| Retinoid (MX-3350-1) | 0.1 and 1–2 μM to cells | Yes | Transplant | Nu/nu mice (m.f) | Lu | 1997 | [172] |

*(continued)*

**TABLE 10.3 (continued)**
**Retinoid Chemoprevention Studies on Lung Carcinogenesis in Animal Model**

| Retinoid | Dose | Antitumor Effect | Tumor Inducer | Species | Investigator | Year | References |
|---|---|---|---|---|---|---|---|
| Retinyl palmitate | 2000 mg/kg bw i.p. 3 times/ week | Yes | Fiberglass | Strain A mice (m) | Morrison | 1981 | [173] |
| Retinyl palmitate | 45 mg/kg bw i.p.[b]/day | No | Transplant | C57BL/6J mice (f) | Pavelic | 1980 | [174] |
| 4-HPR | 782; 1564 mg/kg diet | No | NNK | A/J mice (f) | Conaway | 1998 | [85] |
| N-HTR | 90–1800 mg/kg i.p./week | Yes | Ethyl carbamate, transplant | A/J mice (f) and C57BL/6N mice (m) | McCully | 1987 | [175] |

bw, body weight;f, female; m, male ; na, not available; i.p., intraperitoneal; i.v., intravenous; p.o., per os (oral administration); B(a)P, benzo(a)pyrene; 4-HPR, N-(4-hydroxyphenyl)retinamide; N-HRT, N-homocysteine thiolactonyl retinamide; NNK, 4-(methylnitrosamino)-1-(3-pyridyl)-1-butanone; NQO, nitroquinoline 1-oxide; RA, retinoic acid.

[a] Aerosol 13-*cis* RA was administered 45 min/day and the weekly total deposited doses of 13-*cis* RA in the lung were 0.24, 1.6, and 24.9 mg/kg body weight (bw) as reported in the original publication.

[b] Dose was reported in the original publication as 3000 IU/mouse/day and was converted to mg/kg bw/day under the assumption that the mice weigh 20 g. 1 mg Retinol Activity Equivalence = 3300 IU [164].

influence how one responds to medicine/chemotherapy [115], could partially explain why retinoids are less effective when given to lung cancer patients.

A review paper by Lotan [116] has summarized the early studies on retinoids and the chemoprevention of skin, breast, oral cavity, lung, hepatic, gastrointestinal, prostatic, and bladder cancers in a variety of animal models. The author pointed out that retinoids may target different organs and stages of carcinogenesis; in addition, the choice of animal models/strains is critical since some retinoids were effective in certain animal models/strains but not in others [116].

## POTENTIAL MECHANISMS OF CAROTENOIDS AND RETINOIDS ON LUNG CARCINOGENESIS

### ACTING ON THE RETINOID RECEPTORS

RARs and RXRs can interact only with all-*trans* and 9-*cis* retinoic acid [42]. When compared to the adjacent normal tissues, RAR-β levels were reduced in many solid tumors including non-small cell lung cancer [117]. It has been reported that treatment with 9-*cis* retinoic acid can restore RAR-β expression in the bronchial epithelium of former smokers [118].

Moreover, the provitamin A carotenoids may regulate gene expression via the activation of RARs [52]. In an in vitro study, a physiological dose of β-carotene inhibited the growth of the normal human bronchial epithelial (NHBE) cells and completely reversed the benzo(*a*)pyrene-suppressed RAR-β [119]. In a ferret model, low dose of β-carotene supplementation (0.42 mg/kg bw/day) did not affect or only slightly decreased the smoke-induced precancerous lesion squamous metaplasia and increased RAR-β expression in the lung [84,87]. These in vitro and in vivo observations are supported by the later publication [120] reporting no significant difference between the lung RAR-β expression in the tumor regions of the Physicians' Health Study patients receiving β-carotene and that of the patients receiving placebo (this confirms the main finding of the Physicians' Health Study [56], in which no effect of β-carotene was observed). In contrast, high-dose supplementation of β-carotene (2.4 mg/kg bw/day) increased squamous metaplasia in ferrets, which was associated with a decreased expression of lung RAR-β [84,87].

Interestingly, the provitamin A carotenoid β-cryptoxanthin was also reported to be a direct ligand for RARs [121] and that it can induce the RAR-β mRNA in the immortalized human bronchial epithelial cell line BEAS-2B [122].

There has been increased interest on the role of the cleavage products of carotenoids. The metabolite of β-carotene (β-apo-14′-carotenoic acid) was reported to suppress the growth of NHBE cells, which was associated with an induction of RAR-β [119]. There is also evidence that β-apo-14′-carotenoic acid can transactivate the RAR-β promoter primarily via its conversion to retinoic acid [119]. In addition, the metabolite of the non-provitamin A carotenoid lycopene (apo-10′-lycopenoic acid) was able to inhibit cell growth and to induce RAR-β expression in NHBE, BEAS-2B, and the non-small cell lung cancer cell line A549 [123].

Nonetheless, more studies are increasingly supporting the notion that the different isomers of RAR-β seem to have different effects in lung carcinogenesis. Even though the transcriptions are driven by the same retinoic-acid-responsive promoter, RAR-β2 is thought to function as a tumor suppressor while RAR-β4 as a tumor inducer [124]. The expression of RAR-β2 isoform is frequently lost in the early process of carcinogenesis [34]. Cigarette smoke and the tobacco nitrosamine NNK can induce RAR-β2 promoter methylation, which may contribute to the loss of the tumor suppressor RAR-β2 expression in the lung [124]. Furthermore, it has been reported that the expression of the tumor suppressor RAR-β2 is inducible by retinoic acid [34]. Nonetheless, in the A/J mouse model of lung cancer, β-carotene did not affect lung tumor multiplicity while induced the mRNA of both the tumor suppressor RAR-β2 and the tumor inducer RAR-β4 [86]. Further studies on the specific functions of the different isoforms of RARs and RXRs, and the effects of carotenoids and retinoids on the expression of RARs and RXRs would be of interest.

## ACTING AS ANTIOXIDANTS AND PRO-OXIDANTS

Carotenoids are known for their ability to protect against oxidative damage due to their antioxidant properties [33,47]. Cigarette smoke can contribute to oxidative stress as it contains high levels of radicals and oxidants ($10^{14}$–$10^{16}$ oxidants per puff) [125]. In human lung cancer cells, β-carotene alone could reduce lipid peroxidation, one of the biomarkers for oxidative stress [126,127]. However, the combination of β-carotene and cigarette smoke condensate (tar), which contain high concentration of radicals, increased the oxidative DNA damage marker 8-hydroxyl-2'-deoxyguanosine [128] and the lipid peroxidation marker malondialdehyde [127]. In contrast, the combined β-carotene and vitamins C and E treatment could reduce the smoke-induced DNA damage in ferret lungs [129]. Furthermore, β-cryptoxanthin supplementation was able to reduce the oxidative DNA damage marker 8-hydroxyl-2'-deoxyguanosine in the smoke-exposed ferret lungs [130].

## MODULATING PATHWAYS IN CELL PROLIFERATION AND APOPTOSIS

Tobacco smoke can interfere with the normal regulation of cell proliferation and apoptosis, thereby promoting lung cancer (see Figure 10.2). As mentioned before, high-dose supplementation of β-carotene (2.4 mg/kg bw/day) to ferrets induced lung squamous metaplasia [87,130]. Further analysis indicated that this effect was associated with (1) an increased expressions of AP-1 (c-jun/c-fos), cyclin D1, and PCNA [87,130]; (2) an increased activation (phosphorylation) of c-jun and the MAPK Jun N-terminal kinase (JNK) and p38, and their downstream effector p53 [130]; and (3) a decreased expression of MAPK phosphatase-1 (MKP-1), which is responsible for the dephosphorylation of JNK and p38 in ferret lungs [130]. These changes are thought to play a part in the detrimental effects of high-dose β-carotene.

In contrast, supporting the hypothesis that the low-dose β-carotene did not exert harmful effects, a study reported that low-dose β-carotene supplementation (0.43 mg/kg bw/day) reduced the smoke-induced phosphorylation of p53, MAPK (JNK and p38), and reversed the suppression of MKP-1, all of which contribute to the harmless effect on the smoke-induced squamous metaplasia in the ferret model [130]. Furthermore, in the presence of other antioxidants such as vitamins C and E,

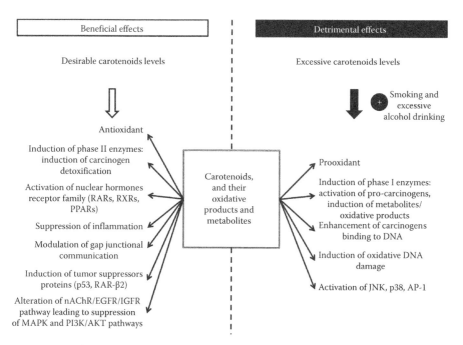

**FIGURE 10.2** Schematic illustration of potential biological effects, both beneficial and harmful, attributed to carotenoids and their metabolites to lung carcinogenesis. Excessive amounts of carotenoids and their metabolites may be harmful, especially when coupled with a highly oxidative environment (e.g., the lungs of cigarette smokers or liver of excessive alcohol drinkers). On the other hand, smaller amounts of carotenoids have been shown to offer beneficial effects contributing to their chemopreventive effects against lung cancer.

β-carotene (0.85 mg/kgbw/day) supplementation did not exacerbate the NNK + smoke-induced squamous metaplasia in ferrets while suppressing the NNK + smoke-induced PCNA, phosphorylation of p53, and phosphorylation of MAPK (JNK and ERK) in the ferret lungs [81]. The nontoxic effect of the combined β-carotene and vitamins C and E supplementation was also confirmed in another ferret model treated with the polycyclic aromatic hydrocarbon benzo(*a*)pyrene [131].

Previously, the influence of retinoids on the activation of JNK and MKP-1 has been reported by Lee et al. [132]. They reported that all-*trans* retinoic acid inhibited the growth of the NHBE cells that was associated with the suppression of the JNK-dependent c-fos expression [133]. They also found that all-*trans* retinoic acid increased MKP-1, which contributed to the inhibition of JNK activation in vitro [132].

Interestingly, two recent studies [120,134] have attempted to address the underlining mechanisms by which β-carotene supplementation was harmful in the ATBC study whereas it was harmless in the Physicians' Health Study. Wright et al. found some evidence that the harmful effect of β-carotene in the ATBC study was associated with aberrant cell growth and significant increase in cyclin D1 in the respiratory epithelium of smokers receiving β-carotene supplementation compared to those of the placebo group [134]. On the other hand, Liu et al. [120]

found no significant difference in the cyclin D1 and the PCNA between the lung tumor tissue from a small subset of the $\beta$-carotene group vs. the placebo group within the Physicians' Health Study, in which $\beta$-carotene had no effect on lung cancer risk [56].

Furthermore, lycopene supplementation could reduce the smoke-induced squamous metaplasia in the ferret lungs [135]. This beneficial effect of lycopene was associated with the reduced plasma concentrations of IGF binding protein-3 (IGF-BP3) and the increased ratio of circulating IGF-I/IGF-BP3 in the cigarette smoke-exposed ferrets [135] (this effect on IGF-I/IGF-BP3 seems to be specific for lycopene since the combined $\beta$-carotene and vitamins C and E did not affect the IGF-I/IGF-BP3 ratio [129]). IGF-BP3 acts by sequestering IGF-I away from the receptor IGFR, thereby IGF-BP3 inhibits the action of IGF-I and interrupts the IGF-I signal transduction pathway [135]. Moreover, independent from the interaction with IGF-I and IGFR, the IGF-BP3 is an inducer of apoptosis and is known to be a potent inhibitor of cell growth, PI3K/AKT and MAPK signaling pathways [135]. The authors also reported that this beneficial effect of lycopene was associated with an induction of the apoptotic marker cleaved-caspase 3 and a reduction of the pro-apoptotic BAD phosphorylation—when phosphorylated, BAD cannot enter the mitochondria, where it exerts its apoptotic effect [135]. In addition, the lycopene metabolite (apo-10-lycopenoic acid) could inhibit the growth of the non-small lung cancer cell line A549, which was associated with an increase in the cyclin inhibitors p21$^{\text{WAF/CIP1}}$ and p27$^{\text{KIP1}}$ proteins [123].

Furthermore, $\beta$-cryptoxanthin has been reported to reduce the proliferation of A549 and BEAS-2B cell lines in a dose-dependent manner [122]. This inhibition was associated with the reduction of cyclin D1 and E, and with the upregulation of p21$^{\text{WAF/CIP1}}$ protein [122]. Cyclin D1 and p21$^{\text{WAF/CIP1}}$ are often upregulated in many cancers [33]. Since cyclin D1 regulates the G1-S phase progression of the cell cycle [136], it frequently becomes a target prevention of lung carcinogenesis [137]. Cyclin D1 can be stimulated by PI3K/AKT pathway via the activation of nuclear factor kappa-B (NF-$\kappa$B) and be repressed by p53 [136].

Aside from the mechanistic studies of carotenoids, more research on the mechanisms of retinoids in the chemoprevention of lung carcinogenesis needs further examination. Recently, a report was published supporting the potential of the rexinoid bexarotene (RXR agonist) bexarotene in combination with the EGFR erlotinib for lung cancer prevention via the repression of cyclin D1 by proteasomal degradation [138]. The role of retinoids on the regulators of cell growth and apoptosis such as p21$^{\text{WAF/CIP1}}$, p27$^{\text{KIP1}}$, p53, IGF-BP3, and AP-1 has been reported in melanoma, breast, and ovarian carcinoma cells (see reviews [34,139]).

## SUPPRESSING INFLAMMATION

Recently, $\beta$-cryptoxanthin was reported to reduce the smoke-induced squamous metaplasia in the ferret lungs, which was associated with the reduction of the pro-inflammatory cytokine TNF-$\alpha$ and the transcription factor NF-$\kappa$B (p65) [140]. NF-$\kappa$B can be activated upon the stimulation of the TNF-$\alpha$ receptor and by cigarette smoke [140,141]. High levels of NF-$\kappa$B, which is important for the regulation of cell proliferation and apoptosis, were found in tumor samples of SCLC and NSCLC patients [141]. A recent overview of the NF-κB signaling in lung cancer has been published [141].

## INDUCING PHASE I AND II ENZYMES

There is evidence that β-carotene and its eccentric cleavage product (β-apo-8′-carotenal) could induce several of the cytochrome P450 phase I oxidizing enzymes (i.e., CYP1A1, CYP1A2, CYP2B, and CYP2E1) that convert the pro-carcinogens to active carcinogens [142]. Moreover, β-carotene could reduce the binding of the P450-catalyzed benzo(*a*)pyrene to DNA thereby preventing the formation of DNA adducts [143]. In contrast, the oxidation products of β-carotene increased this benzo(*a*)pyrene binding to DNA [143].

As mentioned before, the increased squamous metaplasia in the ferret lungs after smoke exposure, high-dose β-carotene supplementation, or both are caused partly by the reduced level of retinoic acid in the lung [84]. Our lab further demonstrated that this is due to the induction of the catabolism of retinoic acid via the enhancement of the cytochrome P450 enzymes CYP1A1 and CYP1A2 in particular. A dose-dependent increase in the polar metabolites (4-oxo-retinoic acid and 18-hydroxy retinoic acid) was observed after retinoic acid was incubated with the lung microsomal fractions of ferrets exposed to cigarette smoke and/or given a pharmacological dose of β-carotene [144].

Furthermore, the apo-10′-lycopenoic acid treatment was shown to increase the nuclear accumulation of the transcription factor nuclear factor E2-related factor 2 (Nrf2) in a time- and a dose-dependent manner in BEAS-2B cells [145]. The accumulation of Nrf2 in the nucleus regulates the expression of phase II detoxification/antioxidant enzymes, which are responsible for the detoxification of carcinogens, including heme oxygenase-1, NAD(P)H:quinone oxidoreductase 1, glutathione S-transferase, glutathione reductase, glutamate-cysteine ligase, microsomal epoxide hydrolase 1, and UDP glucuronosyltransferase 1 family, polypeptide A6 [145]. The apo-10′-lycopenoic acid treatment to the BEAS-2B cells not only increased the mRNA of these phase II enzymes but also suppressed the generation of reactive oxygen species [145].

## MODULATING CELL-TO-CELL COMMUNICATION

β-Carotene could alter the expression of connexin 43 [146], the best-known connexin [147]. Connexin is the major component of connexons (the channels of the gap junctional communication, which has been used for a general marker of malignancy) [147]. β-Carotene modulation of connexin 43 may be cell-specific as β-carotene reduced connexin 43 expressions in the murine fibroblast but not in the lung epithelial cells [146].

Injections of all-*trans* retinoic acid-treated melanoma cells to mice resulted in the reduction of the tumor cell colonization in the lung, compared to the colonization of the untreated cells [148]. The authors reported that the discrepancy was due to the apparent difference of the glycosaminoglycan profile [148].

Glycosaminoglycan side chains of the surface macromolecules of tumor cells are important for tumor cells adhesion to target organs [148].

## CONCLUSION

Understanding of the actions of carotenoids at the molecular level is critically needed for future human studies involving carotenoids for the prevention of lung and other cancers. Although the original hypothesis prompting the ATBC and CARET intervention trials was the potential beneficial effects of β-carotene through its anti-oxidant activity, the unexpected harmful effects of β-carotene supplementation was observed in the two trials. This led to subsequent animal studies supporting the presumption that the free radical-rich environment of the lung of smokers decreased the stability of the β-carotene molecule and/or retinoic acid thereby promoting lung carcinogenesis rather than preventing. In contrast, when given at physiological dose, β-carotene supplementation with other antioxidants (such as vitamins E and C) was able to inhibit the tobacco smoke-induced activation of several signaling molecules important for cell proliferation and survival, leading to the prevention of the smoke-induced carcinogenesis in animal studies. The in vivo studies have indicated that the beneficial vs. detrimental effects of carotenoids are related to the dose administered, the accumulation, and the metabolites/decomposition products in the lung, as well as the specific target molecules of the cellular signaling pathways (Figure 10.2).

Recent studies have supported the chemopreventive action of carotenoids beyond β-carotene. There have been reports supporting the action of lycopene for the prevention of the smoke-induced lung carcinogenesis via upregulation of IGFBP-3 and induction of apoptotic signals. The metabolite of lycopene (the apo-10′-lycopenoic acid) was able to alter RAR-β expression and to increase the nuclear Nrf2 and the cell cycle inhibitors (p21$^{WAF/CIP1}$ and p27$^{KIP1}$). Hence, it has the potential to inhibit cancer cell growth. In addition, β-cryptoxanthin itself was reported to be a ligand for RARs and could increase p21$^{WAF/CIP1}$, decrease cyclin D1, cyclin E, AP-1, TNF-α, and NF-κB expressions thus preventing lung carcinogenesis. So far, little attention has been given to lutein in regard to lung cancer prevention. Future chemoprevention studies should examine the specific roles and mechanisms of other carotenoids besides β-carotene while taking into account the possible detrimental effects from the supra-physiological-dose carotenoid administrations.

There is no compelling evidence that retinoids are effective for the chemoprevention of human lung cancer. Interestingly, the recent report about the effectiveness of 13-*cis* retinoic acid on the risk of head-and-neck cancer in certain patients carrying a specific SNP in the region of the RXR-α gene has emphasized the importance of incorporating patients' genetic background for future studies on cancer. This opens up the possibility of personalized chemoprevention using carotenoids or retinoids. Still, there is limited understanding on the specific molecular targets of retinoids.

Carotenoids and retinoids have preventive and therapeutic potential on lung carcinogenesis, but there seems to be apparent differences in the efficacy of the different compounds. The effective doses of retinoids seem to be relatively high with apparent toxicity. In addition, both carotenoids and retinoids have distinctive effects on current smokers, former smokers, and never smokers. More studies need to address the intricate interactions among carotenoids (and their cleavage products), retinoids, nuclear receptors (RARs and RXRs), and cellular signaling molecules. This knowledge will hopefully provide a better foundation for pursuing carotenoids and retinoids as chemopreventive agents against lung carcinogenesis.

## REFERENCES

1. Jha P. Avoidable global cancer deaths and total deaths from smoking. *Nat Rev Cancer* 2009; 9: 655–664.
2. Society AC. Learn about cancer: Lung cancer-non-small cell. Detailed guide: What are the key statistics about lung cancer? Internet: http://www.cancer.org/Cancer/LungCancer-Non-SmallCell/DetailedGuide/index (accessed June 3, 2011).
3. Cox LS, Africano NL, Tercyak KP, Taylor KL. Nicotine dependence treatment for patients with cancer. *Cancer* 2003; 98: 632–644.
4. Sun S, Schiller JH, Gazdar AF. Lung cancer in never smokers—A different disease. *Nat Rev Cancer* 2007; 7: 778–790.
5. Toh CK. The changing epidemiology of lung cancer. *Methods Mol Biol* 2009; 472: 397–411.
6. Forman MR, Hursting SD, Umar A, Barrett JC. Nutrition and cancer prevention: A multidisciplinary perspective on human trials. *Annu Rev Nutr* 2004; 24: 223–254.
7. Gadgeel S, Kalemkerian GP. Lung cancer: Overview. In: Keshamouni VG, Kalemkerian GP, Arenberg DA, eds. *Lung Cancer Metastasis*. New York: Springer Science+Business Media, 2009.
8. Ruano-Ravina A, Figueiras A, Freire-Garabal M, Barros-Dios JM. Antioxidant vitamins and risk of lung cancer. *Curr Pharm Des* 2006; 12: 599–613.
9. Buchner FL, Bueno-de-Mesquita HB, Ros MM et al. Variety in fruit and vegetable consumption and the risk of lung cancer in the European prospective investigation into cancer and nutrition. *Cancer Epidemiol Biomarkers Prev* 2010; 19: 2278–2286.
10. Darby S, Whitley E, Doll R, Key T, Silcocks P. Diet, smoking and lung cancer: A case–control study of 1000 cases and 1500 controls in south-west England. *Br J Cancer* 2001; 84: 728–735.
11. Galeone C, Negri E, Pelucchi C, La Vecchia C, Bosetti C, Hu J. Dietary intake of fruit and vegetable and lung cancer risk: A case–control study in Harbin, northeast China. *Ann Oncol* 2007; 18: 388–392.
12. Kubik A, Zatloukal P, Tomasek L et al. Interactions between smoking and other exposures associated with lung cancer risk in women: Diet and physical activity. *Neoplasma* 2007; 54: 83–88.
13. Mayne ST, Janerich DT, Greenwald P et al. Dietary beta carotene and lung cancer risk in U.S. nonsmokers. *J Natl Cancer Inst* 1994; 86: 33–38.
14. Pawlega J, Rachtan J, Dyba T. Evaluation of certain risk factors for lung cancer in Cracow (Poland)—A case–control study. *Acta Oncol* 1997; 36: 471–476.
15. Skuladottir H, Tjoenneland A, Overvad K et al. Does insufficient adjustment for smoking explain the preventive effects of fruit and vegetables on lung cancer? *Lung Cancer* 2004; 45: 1–10.
16. Skuladottir H, Tjoenneland A, Overvad K, Stripp C, Olsen JH. Does high intake of fruit and vegetables improve lung cancer survival? *Lung Cancer* 2006; 51: 267–273.
17. Voorrips LE, Goldbohm RA, Verhoeven DT et al. Vegetable and fruit consumption and lung cancer risk in the Netherlands Cohort Study on diet and cancer. *Cancer Causes Control* 2000; 11: 101–115.
18. Yong LC, Brown CC, Schatzkin A et al. Intake of vitamins E, C, and A and risk of lung cancer. The NHANES I epidemiologic followup study. First National Health and Nutrition Examination Survey. *Am J Epidemiol* 1997; 146: 231–243.

19. Ziegler RG, Colavito EA, Hartge P et al. Importance of alpha-carotene, beta-carotene, and other phytochemicals in the etiology of lung cancer. *J Natl Cancer Inst* 1996; 88: 612–615.

20. Epplein M, Franke AA, Cooney RV et al. Association of plasma micronutrient levels and urinary isoprostane with risk of lung cancer: The multiethnic cohort study. *Cancer Epidemiol Biomarkers Prev* 2009; 18: 1962–1970.

21. Holick CN, Michaud DS, Stolzenberg-Solomon R et al. Dietary carotenoids, serum beta-carotene, and retinol and risk of lung cancer in the alpha-tocopherol, beta-carotene cohort study. *Am J Epidemiol* 2002; 156: 536–547.

22. Ito Y, Wakai K, Suzuki K et al. Lung cancer mortality and serum levels of carotenoids, retinol, tocopherols, and folic acid in men and women: A case–control study nested in the JACC Study. *J Epidemiol* 2005; 15(Suppl 2): S140–S149.

23. Jin YR, Lee MS, Lee JH et al. Intake of vitamin A-rich foods and lung cancer risk in Taiwan: With special reference to garland chrysanthemum and sweet potato leaf consumption. *Asia Pac J Clin Nutr* 2007; 16: 477–488.

24. Knekt P, Jarvinen R, Teppo L, Aromaa A, Seppanen R. Role of various carotenoids in lung cancer prevention. *J Natl Cancer Inst* 1999; 91: 182–184.

25. Michaud DS, Feskanich D, Rimm EB et al. Intake of specific carotenoids and risk of lung cancer in 2 prospective US cohorts. *Am J Clin Nutr* 2000; 72: 990–997.

26. Stefani ED, Boffetta P, Deneo-Pellegrini H et al. Dietary antioxidants and lung cancer risk: A case–control study in Uruguay. *Nutr Cancer* 1999; 34: 100–110.

27. Yuan JM, Stram DO, Arakawa K, Lee HP, Yu MC. Dietary cryptoxanthin and reduced risk of lung cancer: The Singapore Chinese Health Study. *Cancer Epidemiol Biomarkers Prev* 2003; 12: 890–898.

28. Zhou B, Wang T, Sun G, Guan P, Wu JM. A case–control study of the relationship between dietary factors and risk of lung cancer in women of Shenyang, China. *Oncol Rep* 1999; 6: 139–143.

29. Zhou BS, Wang TJ, Guan P, Wu JM. Indoor air pollution and pulmonary adenocarcinoma among females: A case–control study in Shenyang, China. *Oncol Rep* 2000; 7: 1253–1259.

30. ATBC Group. The effect of vitamin E and beta carotene on the incidence of lung cancer and other cancers in male smokers. The Alpha-Tocopherol, Beta Carotene Cancer Prevention Study Group. *N Engl J Med* 1994; 330: 1029–1035.

31. Omenn GS, Goodman GE, Thornquist MD et al. Effects of a combination of beta carotene and vitamin A on lung cancer and cardiovascular disease. *N Engl J Med* 1996; 334: 1150–1155.

32. Epstein KR. The role of carotenoids on the risk of lung cancer. *Semin Oncol* 2003; 30: 86–93.

33. Palozza P, Simone R, Mele MC. Interplay of carotenoids with cigarette smoking: Implications in lung cancer. *Curr Med Chem* 2008; 15: 844–854.

34. Tang XH, Gudas LJ. Retinoids, retinoic acid receptors, and cancer. *Annu Rev Pathol* 2011; 6: 345–364.

35. Hecht SS, Kassie F, Hatsukami DK. Chemoprevention of lung carcinogenesis in addicted smokers and ex-smokers. *Nat Rev Cancer* 2009; 9: 476–488.

36. Brambilla E, Gazdar A. Pathogenesis of lung cancer signalling pathways: Roadmap for therapies. *Eur Respir J* 2009; 33: 1485–1497.

37. Hecht SS. Tobacco carcinogens, their biomarkers and tobacco-induced cancer. *Nat Rev Cancer* 2003; 3: 733–744.

38. Schuller HM. Is cancer triggered by altered signalling of nicotinic acetylcholine receptors? *Nat Rev Cancer* 2009; 9: 195–205.

39. Memmott RM, Dennis PA. The role of the Akt/mTOR pathway in tobacco carcinogen-induced lung tumorigenesis. *Clin Cancer Res* 2009; 16: 4–10.

40. Di Pietro G, Magno LA, Rios-Santos F. Glutathione *S*-transferases: An overview in cancer research. *Expert Opin Drug Metab Toxicol* 2010; 6: 153–170.

41. Kiyohara C, Yoshimasu K. Genetic polymorphisms in the nucleotide excision repair pathway and lung cancer risk: A meta-analysis. *Int J Med Sci* 2007; 4: 59–71.

42. Noy N. Between death and survival: Retinoic acid in regulation of apoptosis. *Annu Rev Nutr* 2010; 30: 201–217.

43. Poulain S, Evenou F, Carre MC, Corbel S, Vignaud JM, Martinet N. Vitamin A/retinoids signalling in the human lung. *Lung Cancer* 2009; 66: 1–7.

44. Theodosiou M, Laudet V, Schubert M. From carrot to clinic: An overview of the retinoic acid signaling pathway. *Cell Mol Life Sci* 2010; 67: 1423–1445.

45. Schaffer MW, Roy SS, Mukherjee S et al. Qualitative and quantitative analysis of retinol, retinyl esters, tocopherols and selected carotenoids out of various internal organs from different species by HPLC. *Anal Methods* 2010; 2: 1320–1332.

46. Arscott SA, Simon PW, Tanumihardjo SA. Anthocyanins in purple-orange carrots (*Daucus carota* L.) do not influence the bioavailability of beta-carotene in young women. *J Agric Food Chem* 2010; 58: 2877–2881.

47. Krinsky NI, Johnson EJ. Carotenoid actions and their relation to health and disease. *Mol Aspects Med* 2005; 26: 459–516.

48. Nagao A. Absorption and function of dietary carotenoids. *Forum Nutr* 2009; 61: 55–63.

49. Goodwin TW. Metabolism, nutrition, and function of carotenoids. *Annu Rev Nutr* 1986; 6: 273–297.

50. Khoo HE, Prasad KN, Kong KW, Jiang Y, Ismail A. Carotenoids and their isomers: Color pigments in fruits and vegetables. *Molecules* 2011; 16: 1710–1738.

51. Nishino H, Murakoshi M, Tokuda H, Satomi Y. Cancer prevention by carotenoids. *Arch Biochem Biophys* 2008; 483: 165–168.

52. von Lintig J. Colors with functions: Elucidating the biochemical and molecular basis of carotenoid metabolism. *Annu Rev Nutr* 2010; 30: 35–56.

53. Wyss A. Carotene oxygenases: A new family of double bond cleavage enzymes. *J Nutr* 2004; 134: 246S–250S.

54. Peto R, Doll R, Buckley JD, Sporn MB. Can dietary beta-carotene materially reduce human cancer rates? *Nature* 1981; 290: 201–208.

55. Albanes D, Heinonen OP, Taylor PR et al. Alpha-tocopherol and beta-carotene supplements and lung cancer incidence in the alpha-tocopherol, beta-carotene cancer prevention study: Effects of base-line characteristics and study compliance. *J Natl Cancer Inst* 1996; 88: 1560–1570.

56. Hennekens CH, Buring JE, Manson JE et al. Lack of effect of long-term supplementation with beta carotene on the incidence of malignant neoplasms and cardiovascular disease. *N Engl J Med* 1996; 334: 1145–1149.

57. Lee IM, Cook NR, Manson JE, Buring JE, Hennekens CH. Beta-carotene supplementation and incidence of cancer and cardiovascular disease: The Women's Health Study. *J Natl Cancer Inst* 1999; 91: 2102–2106.

58. Pastorino U. beta-Carotene and the risk of lung cancer. *J Natl Cancer Inst* 1997; 89: 456–457; author reply 457–458.

59. Yuan JM, Ross RK, Chu XD, Gao YT, Yu MC. Prediagnostic levels of serum beta-cryptoxanthin and retinol predict smoking-related lung cancer risk in Shanghai, China. *Cancer Epidemiol Biomarkers Prev* 2001; 10: 767–773.

60. Stram DO, Yuan JM, Chan KK, Gao YT, Ross RK, Yu MC. Beta-cryptoxanthin and lung cancer in Shanghai, China—An examination of potential confounding with cigarette smoking using urinary cotinine as a biomarker for true tobacco exposure. *Nutr Cancer* 2007; 57: 123–129.

61. Kumagai Y, Pi JB, Lee S et al. Serum antioxidant vitamins and risk of lung and stomach cancers in Shenyang, China. *Cancer Lett* 1998; 129: 145–149.

62. Ratnasinghe D, Forman MR, Tangrea JA et al. Serum carotenoids are associated with increased lung cancer risk among alcohol drinkers, but not among non-drinkers in a cohort of tin miners. *Alcohol Alcohol* 2000; 35: 355–360.

63. Wang Y, Ichiba M, Iyadomi M, Zhang J, Tomokuni K. Effects of genetic polymorphism of metabolic enzymes, nutrition, and lifestyle factors on DNA adduct formation in lymphocytes. *Ind Health* 1998; 36: 337–346.

64. Mooney LA, Bell DA, Santella RM et al. Contribution of genetic and nutritional factors to DNA damage in heavy smokers. *Carcinogenesis* 1997; 18: 503–509.

65. Ratnasinghe DL, Yao SX, Forman M et al. Gene-environment interactions between the codon 194 polymorphism of XRCC1 and antioxidants influence lung cancer risk. *Anticancer Res* 2003; 23: 627–632.

66. Maiani G, Caston MJ, Catasta G et al. Carotenoids: Actual knowledge on food sources, intakes, stability and bioavailability and their protective role in humans. *Mol Nutr Food Res* 2009; 53(Suppl 2): S194–S218.

67. Lietz G, Lange J, Rimbach G. Molecular and dietary regulation of beta,beta-carotene 15,15′-monooxygenase 1 (BCMO1). *Arch Biochem Biophys* 2010; 502: 8–16.

68. Garcia-Closas R, Agudo A, Gonzalez CA, Riboli E. Intake of specific carotenoids and flavonoids and the risk of lung cancer in women in Barcelona, Spain. *Nutr Cancer* 1998; 32: 154–158.

69. Bruemmer B, Harris J, Gleason P et al. Publishing nutrition research: A review of epidemiologic methods. *J Am Diet Assoc* 2009; 109: 1728–1737.

70. Rohan TE, Jain M, Howe GR, Miller AB. A cohort study of dietary carotenoids and lung cancer risk in women (Canada). *Cancer Causes Control* 2002; 13: 231–237.

71. Touvier M, Kesse E, Clavel-Chapelon F, Boutron-Ruault MC. Dual Association of beta-carotene with risk of tobacco-related cancers in a cohort of French women. *J Natl Cancer Inst* 2005; 97: 1338–1344.

72. Mannisto S, Smith-Warner SA, Spiegelman D et al. Dietary carotenoids and risk of lung cancer in a pooled analysis of seven cohort studies. *Cancer Epidemiol Biomarkers Prev* 2004; 13: 40–48.

73. Gallicchio L, Boyd K, Matanoski G et al. Carotenoids and the risk of developing lung cancer: A systematic review. *Am J Clin Nutr* 2008; 88: 372–383.

74. Druesne-Pecollo N, Latino-Martel P, Norat T et al. Beta-carotene supplementation and cancer risk: A systematic review and metaanalysis of randomized controlled trials. *Int J Cancer* 2010; 127: 172–184.

75. Satia JA, Littman A, Slatore CG, Galanko JA, White E. Long-term use of beta-carotene, retinol, lycopene, and lutein supplements and lung cancer risk: Results from the VITamins And Lifestyle (VITAL) study. *Am J Epidemiol* 2009; 169: 815–828.

76. Pathak AK, Bhutani M, Guleria R et al. Chemotherapy alone vs. chemotherapy plus high dose multiple antioxidants in patients with advanced non small cell lung cancer. *J Am Coll Nutr* 2005; 24: 16–21.

77. Reagan-Shaw S, Nihal M, Ahmad N. Dose translation from animal to human studies revisited. *FASEB J* 2008; 22: 659–661.

78. Sharma V, McNeill JH. To scale or not to scale: The principles of dose extrapolation. *Br J Pharmacol* 2009; 157: 907–921.

79. Pradeep CR, Kuttan G. Effect of beta-carotene on the inhibition of lung metastasis in mice. *Phytomedicine* 2003; 10: 159–164.

80. Huang CS, Liao JW, Hu ML. Lycopene inhibits experimental metastasis of human hepatoma SK-Hep-1 cells in athymic nude mice. *J Nutr* 2008; 138: 538–543.

81. Kim Y, Chongviriyaphan N, Liu C, Russell RM, Wang XD. Combined antioxidant (beta-carotene, alpha-tocopherol and ascorbic acid) supplementation increases the levels of lung retinoic acid and inhibits the activation of mitogen-activated protein kinase in the ferret lung cancer model. *Carcinogenesis* 2006; 27: 1410–1419.

82. Witschi H. Carcinogenic activity of cigarette smoke gas phase and its modulation by beta-carotene and *N*-acetylcysteine. *Toxicol Sci* 2005; 84: 81–87.

83. Al-Wadei HA, Schuller HM. Beta-carotene promotes the development of NNK-induced small airway-derived lung adenocarcinoma. *Eur J Cancer* 2009; 45: 1257–1264.

84. Wang XD, Liu C, Bronson RT, Smith DE, Krinsky NI, Russell M. Retinoid signaling and activator protein-1 expression in ferrets given beta-carotene supplements and exposed to tobacco smoke. *J Natl Cancer Inst* 1999; 91: 60–66.

85. Conaway CC, Jiao D, Kelloff GJ, Steele VE, Rivenson A, Chung FL. Chemopreventive potential of fumaric acid, *N*-acetylcysteine, *N*-(4-hydroxyphenyl) retinamide and beta-carotene for tobacco-nitrosamine-induced lung tumors in A/J mice. *Cancer Lett* 1998; 124: 85–93.

86. Goralczyk R, Bachmann H, Wertz K et al. beta-carotene-induced changes in RARbeta isoform mRNA expression patterns do not influence lung adenoma multiplicity in the NNK-initiated A/J mouse model. *Nutr Cancer* 2006; 54: 252–262.

87. Liu C, Wang XD, Bronson RT, Smith DE, Krinsky NI, Russell RM. Effects of physiological versus pharmacological beta-carotene supplementation on cell proliferation and histopathological changes in the lungs of cigarette smoke-exposed ferrets. *Carcinogenesis* 2000; 21: 2245–2253.

88. Murakoshi M, Nishino H, Satomi Y et al. Potent preventive action of alpha-carotene against carcinogenesis: Spontaneous liver carcinogenesis and promoting stage of lung and skin carcinogenesis in mice are suppressed more effectively by alpha-carotene than by beta-carotene. *Cancer Res* 1992; 52: 6583–6587.

89. Nishino H. Cancer chemoprevention by natural carotenoids and their related compounds. *J Cell Biochem Suppl* 1995; 22: 231–235.

90. Obermueller-Jevic UC, Espiritu I, Corbacho AM, Cross CE, Witschi H. Lung tumor development in mice exposed to tobacco smoke and fed beta-carotene diets. *Toxicol Sci* 2002; 69: 23–29.

91. Uleckiene S, Domkiene V. Investigation of ethyl alcohol and beta-carotene effect on two models of carcinogenesis. *Acta Biol Hung* 2003; 54: 89–93.

92. Yun TK, Kim SH, Lee YS. Trial of a new medium-term model using benzo(*a*)pyrene induced lung tumor in newborn mice. *Anticancer Res* 1995; 15: 839–845.

93. Lee JJ, Feng L, Reshef DS et al. Mortality in the randomized, controlled lung intergroup trial of isotretinoin. *Cancer Prev Res (Phila)* 2010; 3: 738–744.

94. Khuri FR, Lee JS, Lippman SM et al. Modulation of proliferating cell nuclear antigen in the bronchial epithelium of smokers. *Cancer Epidemiol Biomarkers Prev* 2001; 10: 311–318.

95. Hittelman WN, Liu DD, Kurie JM et al. Proliferative changes in the bronchial epithelium of former smokers treated with retinoids. *J Natl Cancer Inst* 2007; 99: 1603–1612.

96. Ayoub J, Jean-Francois R, Cormier Y et al. Placebo-controlled trial of 13-*cis*-retinoic acid activity on retinoic acid receptor-beta expression in a population at high risk: Implications for chemoprevention of lung cancer. *J Clin Oncol* 1999; 17: 3546–3552.

97. Lam S, Xu X, Parker-Klein H et al. Surrogate end-point biomarker analysis in a retinol chemoprevention trial in current and former smokers with bronchial dysplasia. *Int J Oncol* 2003; 23: 1607–1613.

98. Lee JJ, Wu X, Hildebrandt MA et al. Global assessment of genetic variation influencing response to retinoid chemoprevention in head and neck cancer patients. *Cancer Prev Res (Phila)* 2011; 4: 185–193.

99. Khuri FR, Lee JJ, Lippman SM et al. Randomized phase III trial of low-dose isotretinoin for prevention of second primary tumors in stage I and II head and neck cancer patients. *J Natl Cancer Inst* 2006; 98: 441–450.

100. Digel W, Lubbert M. DNA methylation disturbances as novel therapeutic target in lung cancer: Preclinical and clinical results. *Crit Rev Oncol Hematol* 2005; 55: 1–11.

101. Jin Y, Xu H, Zhang C et al. Combined effects of cigarette smoking, gene polymorphisms and methylations of tumor suppressor genes on non small cell lung cancer: A hospital-based case–control study in China. *BMC Cancer* 2010; 10: 422.
102. Stefanska B, Rudnicka K, Bednarek A, Fabianowska-Majewska K. Hypomethylation and induction of retinoic acid receptor beta 2 by concurrent action of adenosine analogues and natural compounds in breast cancer cells. *Eur J Pharmacol* 2010; 638: 47–53.
103. Arrieta O, Gonzalez-De la Rosa CH, Arechaga-Ocampo E et al. Randomized phase II trial of All-*trans*-retinoic acid with chemotherapy based on paclitaxel and cisplatin as first-line treatment in patients with advanced non-small-cell lung cancer. *J Clin Oncol* 2010; 28: 3463–3471.
104. Recchia F, Sica G, De Filippis S et al. Cisplatin, vindesine, mitomycin-C and 13-*cis* retinoic acid in the treatment of advanced non small cell lung cancer. A phase II pilot study. *Anticancer Res* 2000; 20: 1985–1990.
105. Thiruvengadam R, Atiba JO, Azawi SH. A phase II trial of a differentiating agent (tRA) with cisplatin-VP 16 chemotherapy in advanced non-small cell lung cancer. *Invest New Drugs* 1996; 14: 395–401.
106. van Zuylen L, Schellens JH, Goey SH et al. Phase I and pharmacologic study of the arotinoid Ro 40-8757 in combination with cisplatin and etoposide in patients with non-small cell lung cancer. *Anticancer Drugs* 1999; 10: 361–368.
107. Kalemkerian GP, Jiroutek M, Ettinger DS, Dorighi JA, Johnson DH, Mabry M. A phase II study of all-*trans*-retinoic acid plus cisplatin and etoposide in patients with extensive stage small cell lung carcinoma: An Eastern Cooperative Oncology Group Study. *Cancer* 1998; 83: 1102–1108.
108. Arnold A, Ayoub J, Douglas L et al. Phase II trial of 13-*cis*-retinoic acid plus interferon alpha in non-small-cell lung cancer. The National Cancer Institute of Canada Clinical Trials Group. *J Natl Cancer Inst* 1994; 86: 306–309.
109. Athanasiadis I, Kies MS, Miller M et al. Phase II study of all-*trans*-retinoic acid and alpha-interferon in patients with advanced non small cell lung cancer. *Clin Cancer Res* 1995; 1: 973–979.
110. Rinaldi DA, Lippman SM, Burris HA, 3rd, Chou C, Von Hoff DD, Hong WK. Phase II study of 13-*cis*-retinoic acid and interferon-alpha 2a in patients with advanced squamous cell lung cancer. *Anticancer Drugs* 1993; 4: 33–36.
111. Roth AD, Abele R, Alberto P. 13-*cis*-retinoic acid plus interferon-alpha: A phase II clinical study in squamous cell carcinoma of the lung and the head and neck. *Oncology* 1994; 51: 84–86.
112. Ruotsalainen T, Halme M, Isokangas OP et al. Interferon-alpha and 13-*cis*-retinoic acid as maintenance therapy after high-dose combination chemotherapy with growth factor support for small cell lung cancer—A feasibility study. *Anticancer Drugs* 2000; 11: 101–108.
113. Costa A, Pastorino U, Andreoli C, Barbieri A, Marubini E, Veronesi U. Vitamin A and retinoids: A hypothesis of tumour chemoprevention. *Int Adv Surg Oncol* 1984; 7: 271–295.
114. Mernitz H, Smith DE, Wood RJ, Russell RM, Wang XD. Inhibition of lung carcinogenesis by 1alpha,25-dihydroxyvitamin D3 and 9-*cis* retinoic acid in the A/J mouse model: Evidence of retinoid mitigation of vitamin D toxicity. *Int J Cancer* 2007; 120: 1402–1409.
115. Sauna ZE, Kimchi-Sarfaty C, Ambudkar SV, Gottesman MM. Silent polymorphisms speak: How they affect pharmacogenomics and the treatment of cancer. *Cancer Res* 2007; 67: 9609–9612.
116. Lotan R. Retinoids in cancer chemoprevention. *FASEB J* 1996; 10: 1031–1039.
117. Cras C, Guidez F, Chomienne C. Retinoic acid receptor. In: Bunce, CM, Campbell MK, eds. *Nuclear Receptor: Current Concepts and Future Challenges, Proteins and Cell Regulation*. London, U.K.: Springer Science+Business Media, 2010; pp. 237–258.

118. Kurie JM, Lotan R, Lee JJ et al. Treatment of former smokers with 9-*cis*-retinoic acid reverses loss of retinoic acid receptor-beta expression in the bronchial epithelium: Results from a randomized placebo-controlled trial. *J Natl Cancer Inst* 2003; 95: 206–214.
119. Prakash P, Liu C, Hu KQ, Krinsky NI, Russell RM, Wang XD. Beta-carotene and beta-apo-14'-carotenoic acid prevent the reduction of retinoic acid receptor beta in benzo[*a*] pyrene-treated normal human bronchial epithelial cells. *J Nutr* 2004; 134: 667–673.
120. Liu C, Wang XD, Mucci L, Gaziano JM, Zhang SM. Modulation of lung molecular biomarkers by beta-carotene in the Physicians' Health Study. *Cancer* 2009; 115: 1049–1058.
121. Matsumoto A, Mizukami H, Mizuno S et al. beta-Cryptoxanthin, a novel natural RAR ligand, induces ATP-binding cassette transporters in macrophages. *Biochem Pharmacol* 2007; 74: 256–264.
122. Lian F, Hu KQ, Russell RM, Wang XD. Beta-cryptoxanthin suppresses the growth of immortalized human bronchial epithelial cells and non-small-cell lung cancer cells and up-regulates retinoic acid receptor beta expression. *Int J Cancer* 2006; 119: 2084–2089.
123. Lian F, Smith DE, Ernst H, Russell RM, Wang XD. Apo-10'-lycopenoic acid inhibits lung cancer cell growth in vitro, and suppresses lung tumorigenesis in the A/J mouse model in vivo. *Carcinogenesis* 2007; 28: 1567–1574.
124. Xu XC. Tumor-suppressive activity of retinoic acid receptor-beta in cancer. *Cancer Lett* 2007; 253: 14–24.
125. Faux SP, Tai T, Thorne D, Xu Y, Breheny D, Gaca M. The role of oxidative stress in the biological responses of lung epithelial cells to cigarette smoke. *Biomarkers* 2009; 14(Suppl 1): 90–96.
126. Baker DL, Krol ES, Jacobsen N, Liebler DC. Reactions of beta-carotene with cigarette smoke oxidants. Identification of carotenoid oxidation products and evaluation of the prooxidant/antioxidant effect. *Chem Res Toxicol* 1999; 12: 535–543.
127. Palozza P, Serini S, Trombino S, Lauriola L, Ranelletti FO, Calviello G. Dual role of beta-carotene in combination with cigarette smoke aqueous extract on the formation of mutagenic lipid peroxidation products in lung membranes: Dependence on pO2. *Carcinogenesis* 2006; 27: 2383–2391.
128. Palozza P, Serini S, Di Nicuolo F et al. beta-Carotene exacerbates DNA oxidative damage and modifies p53-related pathways of cell proliferation and apoptosis in cultured cells exposed to tobacco smoke condensate. *Carcinogenesis* 2004; 25: 1315–1325.
129. Kim Y, Lian F, Yeum KJ et al. The effects of combined antioxidant (beta-carotene, alpha-tocopherol and ascorbic acid) supplementation on antioxidant capacity, DNA single-strand breaks and levels of insulin-like growth factor-1/IGF-binding protein 3 in the ferret model of lung cancer. *Int J Cancer* 2007; 120: 1847–1854.
130. Liu C, Russell RM, Wang XD. Low dose beta-carotene supplementation of ferrets attenuates smoke-induced lung phosphorylation of JNK, p38 MAPK, and p53 proteins. *J Nutr* 2004; 134: 2705–2710.
131. Fuster A, Pico C, Sanchez J et al. Effects of 6-month daily supplementation with oral beta-carotene in combination or not with benzo[a]pyrene on cell-cycle markers in the lung of ferrets. *J Nutr Biochem* 2008; 19: 295–304.
132. Lee HY, Sueoka N, Hong WK, Mangelsdorf DJ, Claret FX, Kurie JM. All-*trans*-retinoic acid inhibits Jun N-terminal kinase by increasing dual-specificity phosphatase activity. *Mol Cell Biol* 1999; 19: 1973–1980.
133. Lee HY, Walsh GL, Dawson MI, Hong WK, Kurie JM. All-*trans*-retinoic acid inhibits Jun N-terminal kinase-dependent signaling pathways. *J Biol Chem* 1998; 273: 7066–7071.
134. Wright ME, Groshong SD, Husgafvel-Pursiainen K et al. Effects of beta-carotene supplementation on molecular markers of lung carcinogenesis in male smokers. *Cancer Prev Res (Phila)* 2010; 3: 745–752.

135. Liu C, Lian F, Smith DE, Russell RM, Wang XD. Lycopene supplementation inhibits lung squamous metaplasia and induces apoptosis via up-regulating insulin-like growth factor-binding protein 3 in cigarette smoke-exposed ferrets. *Cancer Res* 2003; 63: 3138–3144.

136. Gautschi O, Ratschiller D, Gugger M, Betticher DC, Heighway J. Cyclin D1 in non-small cell lung cancer: A key driver of malignant transformation. *Lung Cancer* 2007; 55: 1–14.

137. Kim ES, Lee JJ, Wistuba, II. Cotargeting cyclin d1 starts a new chapter in lung cancer prevention and therapy. *Cancer Prev Res (Phila)* 2011; 4: 779–782.

138. Dragnev KH, Ma T, Cyrus J et al. Bexarotene plus erlotinib suppress lung carcinogenesis independent of KRAS mutations in two clinical trials and transgenic models. *Cancer Prev Res (Phila)* 2011; 4: 818–828.

139. Evans TR, Kaye SB. Retinoids: Present role and future potential. *Br J Cancer* 1999; 80: 1–8.

140. Liu C, Bronson RT, Russell RM, Wang XD. {beta}-Cryptoxanthin supplementation prevents cigarette smoke-induced lung inflammation, oxidative damage and squamous metaplasia in ferrets. *Cancer Prev Res (Phila)* 2011; 4(8): 1255–1266.

141. Chen W, Li Z, Bai L, Lin Y. NF-kappaB in lung cancer, a carcinogenesis mediator and a prevention and therapy target. *Front Biosci* 2011; 16: 1172–1185.

142. Paolini M, Cantelli-Forti G, Perocco P, Pedulli GF, Abdel-Rahman SZ, Legator MS. Co-carcinogenic effect of beta-carotene. *Nature* 1999; 398: 760–761.

143. Salgo MG, Cueto R, Winston GW, Pryor WA. Beta carotene and its oxidation products have different effects on microsome mediated binding of benzo[a]pyrene to DNA. *Free Radic Biol Med* 1999; 26: 162–173.

144. Liu C, Russell RM, Wang XD. Exposing ferrets to cigarette smoke and a pharmacological dose of beta-carotene supplementation enhance in vitro retinoic acid catabolism in lungs via induction of cytochrome P450 enzymes. *J Nutr* 2003; 133: 173–179.

145. Lian F, Wang XD. Enzymatic metabolites of lycopene induce Nrf2-mediated expression of phase II detoxifying/antioxidant enzymes in human bronchial epithelial cells. *Int J Cancer* 2008; 123: 1262–1268.

146. Banoub RW, Fernstrom M, Ruch RJ. Lack of growth inhibition or enhancement of gap junctional intercellular communication and connexin43 expression by beta-carotene in murine lung epithelial cells in vitro. *Cancer Lett* 1996; 108: 35–40.

147. Kandouz M, Batist G. Gap junctions and connexins as therapeutic targets in cancer. *Expert Opin Ther Targets* 2010; 14: 681–692.

148. Edward M, MacKie RM. Retinoic acid-induced inhibition of lung colonization and changes in the synthesis and properties of glycosaminoglycans of metastatic B16 melanoma cells. *J Cell Sci* 1989; 94(Pt 3): 537–543.

149. Kohno H, Taima M, Sumida T, Azuma Y, Ogawa H, Tanaka T. Inhibitory effect of mandarin juice rich in beta-cryptoxanthin and hesperidin on 4-(methylnitrosamino)-1-(3-pyridyl)-1-butanone-induced pulmonary tumorigenesis in mice. *Cancer Lett* 2001; 174: 141–150.

150. Iskandar AR, Wang XD. Beta-cryptoxanthin supplementation inhibits NNK-induced lung tumor multiplicity via inhibition of alpha 7 nicotinic receptor expression and AKT activation in A/J mice. *FASEB J* 2011; 25: 344.3.

151. Kim YH, Dohi DF, Han GR et al. Retinoid refractoriness occurs during lung carcinogenesis despite functional retinoid receptors. *Cancer Res* 1995; 55: 5603–5310.

152. Hecht SS, Kenney PM, Wang M et al. Evaluation of butylated hydroxyanisole, myo-inositol, curcumin, esculetin, resveratrol and lycopene as inhibitors of benzo[*a*]pyrene plus 4-(methylnitrosamino)-1-(3-pyridyl)-1-butanone-induced lung tumorigenesis in A/J mice. *Cancer Lett* 1999; 137: 123–130.

153. Guttenplan JB, Chen M, Kosinska W, Thompson S, Zhao Z, Cohen LA. Effects of a lycopene-rich diet on spontaneous and benzo[a]pyrene-induced mutagenesis in prostate, colon and lungs of the lacZ mouse. *Cancer Lett* 2001; 164: 1–6.

154. Treat J, Friedland D, Luginbuhl W et al. Phase II trial of all-*trans* retinoic acid in metastatic non-small cell lung cancer. *Cancer Invest* 1996; 14: 415–420.

155. Saccomanno G, Moran PG, Schmidt R et al. Effects of 13-CIS retinoids on premalignant and malignant cells of lung origin. *Acta Cytol* 1982; 26: 78–85.

156. Lee JS, Lippman SM, Benner SE et al. Randomized placebo-controlled trial of isotretinoin in chemoprevention of bronchial squamous metaplasia. *J Clin Oncol* 1994; 12: 937–945.

157. Lippman SM, Lee JJ, Karp DD et al. Randomized phase III intergroup trial of isotretinoin to prevent second primary tumors in stage I non-small-cell lung cancer. *J Natl Cancer Inst* 2001; 93: 605–618.

158. Kelly K, Kittelson J, Franklin WA et al. A randomized phase II chemoprevention trial of 13-CIS retinoic acid with or without alpha tocopherol or observation in subjects at high risk for lung cancer. *Cancer Prev Res (Phila)* 2009; 2: 440–449.

159. Pastorino U, Infante M, Maioli M et al. Adjuvant treatment of stage I lung cancer with high-dose vitamin A. *J Clin Oncol* 1993; 11: 1216–1222.

160. van Zandwijk N, Dalesio O, Pastorino U, de Vries N, van Tinteren H. EUROSCAN, a randomized trial of vitamin A and *N*-acetylcysteine in patients with head and neck cancer or lung cancer. For the EUropean Organization for Research and Treatment of Cancer Head and Neck and Lung Cancer Cooperative Groups. *J Natl Cancer Inst* 2000; 92: 977–986.

161. Kurie JM, Lee JS, Khuri FR et al. *N*-(4-hydroxyphenyl)retinamide in the chemoprevention of squamous metaplasia and dysplasia of the bronchial epithelium. *Clin Cancer Res* 2000; 6: 2973–2979.

162. Soria JC, Moon C, Wang L et al. Effects of *N*-(4-hydroxyphenyl)retinamide on hTERT expression in the bronchial epithelium of cigarette smokers. *J Natl Cancer Inst* 2001; 93: 1257–1263.

163. Administration. FaD. Guidance for industry: Estimating the maximum safe starting dose in initial clinical trials for therapeutics in adult healthy volunteers: Internet: http://www. fda.gov/downloads/Drugs/GuidanceComplianceRegulatoryInformation/Guidances/ ucm078932.pdf (accessed June 2, 2011).

164. Office of Dietary Supplements NIoH. Dietary supplement fact sheet: Vitamin A and carotenoids. Internet: http://ods.od.nih.gov/factsheets/vitamina/ (accessed June 2, 2011).

165. Hubert DD, Holiat SM, Smith WE, Baylouny RA. Inhibition of transplanted carcinomas in mice by retinoids but not by vitamin C. *Cancer Treat Rep* 1983; 67: 1061–1065.

166. Edward M, Gold JA, Mackie RM. Retinoic acid-induced inhibition of metastatic melanoma cell lung colonization and adhesion to endothelium and subendothelial extracellular matrix. *Clin Exp Metastasis* 1992; 10: 61–67.

167. Gaetano C, Melchiori A, Albini A et al. Retinoic acid negatively regulates beta 4 integrin expression and suppresses the malignant phenotype in a Lewis lung carcinoma cell line. *Clin Exp Metastasis* 1994; 12: 63–72.

168. Andela VB, Gingold BI, Souza MD et al. Clinical relevance of increased retinoid and cAMP transcriptional programs in tumor cells rendered non-malignant by dominant negative inhibition of NFkappaB. *Cancer Lett* 2003; 194: 37–43.

169. Suzuki S, Kawakami S, Chansri N, Yamashita F, Hashida M. Inhibition of pulmonary metastasis in mice by all-*trans* retinoic acid incorporated in cationic liposomes. *J Control Release* 2006; 116: 58–63.

170. Liu X, Chan SY, Ho PC. Comparison of the in vitro and in vivo effects of retinoids either alone or in combination with cisplatin and 5-fluorouracil on tumor development and metastasis of melanoma. *Cancer Chemother Pharmacol* 2008; 63: 167–174.

171. Dahl AR, Grossi IM, Houchens DP et al. Inhaled isotretinoin (13-*cis* retinoic acid) is an effective lung cancer chemopreventive agent in A/J mice at low doses: A pilot study. *Clin Cancer Res* 2000; 6: 3015–3024.

172. Lu XP, Fanjul A, Picard N et al. Novel retinoid-related molecules as apoptosis inducers and effective inhibitors of human lung cancer cells in vivo. *Nat Med* 1997; 3: 686–690.

173. Morrison DG, Daniel J, Lynd FT et al. Retinyl palmitate and ascorbic acid inhibit pulmonary neoplasms in mice exposed to fiberglass dust. *Nutr Cancer* 1981; 3: 81–85.

174. Pavelic ZP, Dave S, Bialkowski S, Priore RL, Greco WR. Antitumor activity of *Corynebacterium parvum* and retinyl palmitate used in combination on the Lewis lung carcinoma. *Cancer Res* 1980; 40: 4617–4621.

175. McCully KS, Vezeridis MP. Chemopreventive and antineoplastic activity of *N*-homocysteine thiolactonyl retinamide. *Carcinogenesis* 1987; 8: 1559–1562.

# 11 Is the Effect of β-Carotene on Prostate Cancer Cells Dependent on Their Androgen Sensitivity?*

*Joanna Dulińska-Litewka, Gerd Schmitz, Aldona Kieć-Dembińska, and Piotr Laidler*

## CONTENTS

## INTRODUCTION

Prostate cancer is the second leading cause of male deaths in Western countries. As androgen is essential for normal prostate function and has also been implicated in the pathogenesis of prostate cancer, inoperable malignant prostate cancer is treated in

---

* Parts of this chapter were presented at the *16th International Symposium on Carotenoids towards a Brighter Side of Life*, Polish Academy of Sciences—Cracow Branch, Jagiellonian University, Cracow, July 17–22, 2011, Cracow, Poland.

general with androgen deprivation therapy (ADT) [1]. Unfortunately, it often returns as more invasive androgen-insensitive form, offering little hope for successful treatment. The mechanisms responsible for that phenomenon are not clear. Androgens act mainly through the androgen receptor (AR). The regulation of AR function involves a cross talk with other signaling pathways (PI3K/Akt) [2], transcription factors, and coregulatory proteins, for example, β-catenin, which seems to play a critical role in tumorigenesis.

Carotenoids have distinct antioxidative properties, protecting against free radicals that can damage DNA and other important biomolecules [3]. Because oxidative stress increases with androgen exposure and age—factors related to prostate cancer risk—carotenoids may be particularly relevant in the prevention of this disease [4,5]. Despite the extensive knowledge on the role of β-carotene (BC) in major cellular processes, there are some controversies with respect to its potential effect on cancer progression [6–9]. Observational data indicate that diets high in cruciferous vegetables, vitamins E and C, lycopene, selenium, and green tea are associated with a lower risk of prostate cancer [10]. However, a recent randomized Prostate, Lung, Colorectal, Ovarian (PLCO) Trial did not confirm their positive effects, and data suggest that the benefit of lycopene is small and that BC, an antioxidant related to lycopene, even increases the risk of aggressive prostate cancer [6].

It was often shown that more than 30 μM BC decreased proliferation of prostate cancer cells in vitro by conversion of BC to retinol and related compounds by prostate cells, what caused changes in the concentration of intracellular ligands for retinoid receptors, subsequently influencing gene expression and biological effects regulated by the steroid receptor superfamily [11]. However, we recently reported that—surprisingly—3 and 10 μM BC increased proliferation of LNCaP cells, whereas it had a weaker effect on PC-3 cells. Global microarray analysis showed remarkable differences between BC-treated and untreated cells with respect to the number of responsive genes. Slight induction of c-myc, c-jun genes affecting proliferation and apoptosis (bcl-2) with no significant effect on major cell cycle control genes (cdk2, RB, E2F-1) in LNCaP cells and the lack of effect on PC-3 cells on any of the studied genes was observed [12].

It was also shown that ligands of peroxisome proliferator-activated receptor gamma (PPARγ), which require for their transcription the activity of the signaling retinoid receptor (RXR), were more effective as anticancer agents in androgen-sensitive LNCaP cells than in androgen-insensitive PC-3 cells. The results suggested that inhibition of β-catenin and in effect c-myc expression through activation of PPARγ might help prostate cancer cells to restore several characteristics of normal prostate cell phenotype [13]. The initial analysis of those results allows for a careful conclusion that differences among prostate cancer cells in response to both BC and PPARγ ligand treatment are due to various androgen sensitivities of LNCaP and PC-3 cells.

To find out if the hypothesis was justified, we decided to follow the effect of 20 μM BC on prostate cells with various sensitivity to androgen, and the respective results are presented in this report.

## MATERIALS AND METHODS

### CELL CULTURES

The studies were carried out on LNCaP-FGC-10 (human prostate cancer, lymph node metastasis) and PC-3 (human prostate adenocarcinoma, bone metastasis) (American Type Culture Collection). Treated and untreated cells were maintained in RPMI-1640 medium (Sigma) supplemented with 10% fetal bovine serum and 100 U/mL penicillin and 100 µg/mL streptomycin (Sigma). FCS was substituted with charcoal-stripped fetal calf serum (SF—steroid-free IMMUNIQ), if DHT (MERCK) was to be added to the medium. Semi-confluent cell cultures (initially seeded with $10^4$ cells/well on 96-well plates or $1–5 \times 10^5$ cells/75 cm$^2$ culture dish) were supplemented with (a) BC (all-*trans*-BC—the chemical laboratories of DSM Nutritional Products Ltd., Kaiseraugst, Switzerland) (10, 20 µM in 0.25% tetrahydrofuran (THF)/Et 1:1) and (b) 5nM DHT (Fluka); 5 nM DHT and 10 or 20 µM BC were maintained at 37°C in a humidified atmosphere of 5% $CO_2$ for 24, 48, and 72 h. Every 24 h, the culture medium was replaced by the fresh one. We used untreated cells as the control group (CON).

### UPTAKE

The methods that were used to determine the uptake of BC were described previously by Duliñska et al. [12]. Data were obtained from three experiments, each with three replications ($P < 0.005$). Mean $\pm$ SD (all such values).

### CELL PROLIFERATION (ELISA, BrdU)

The cells were seeded in triplicates into 96-microwell plates at a density of $1.5 \times 10^3$ cells per well and cultured in the medium without or with various concentration of BCs (10 or 20 µM), 5 nM DHT, and BC with DHT (from 1 to 3 days). The ELISA BrdU (Boehringer–Roche) colorimetric immunoassay test was used as described previously [12]. The absorbance was measured at 450 nm using an ELISA reader (BIO-TEK Synergy HT-R&D; Microplate Date Analysis Software KC4). Data were obtained from four experiments, each with three replications ($P < 0.0001$). Mean $\pm$ SD (all such values).

### CYTOTOXICITY ASSAY (ELISA—LDH)

The cells were seeded in triplicates into 96-microwell plates at a density of $1.5 \times 10^3$ cells per well and incubated without or with BC (10 or 20 µM), 5 nM DHT, and BC + DHT for 24, 48, and 72 h. Afterward, the culture growth media were mixed with the reaction mixture from Cytotoxicity Detection Kit (Boehringer–Roche); the method was described previously [14]. Data were obtained from three experiments, each with four replications ($P < 0.005$).

### RNA EXTRACTION, cDNA SYNTHESIS, AND RT-PCR ANALYSIS

RNA extraction, preparation of cDNA, and RT-PCR reaction were carried out as previously described [14]. The temperature profile of RT-PCR amplification consisted

**TABLE 11.1**

**Primers Used for RT-PCR Analysis**

| Protein | Primer Sequences | Conditions |
|---------|------------------|------------|
| GAPDH | 5' CACCGCCTCGGCTTGTCACAT 3'<br>5' CTGCTGTCTTGGGTGCATTGC 3' | 59°C, 30 s |
| PSA | 5' TTGTCTTCCTCACCCTGTCC 3'<br>5' TGTCCTTGATCCACTTCCGGTA 3' | 59°C, 40 s |
| AR | 5' TGTCAACTCCAGGATGCTCTACTT 3'<br>5' ATTCGGACACCACTGGCTGTACA 3' | 55°C, 45 s |
| β-CATENIN | 5' GCTGATTTGATGGAGTTGGA 3'<br>5' TTCACTCAAGAACAAGTAGC 3' | 55°C, 120 s |
| BCL-2 | 5' GGTGCCACCTGGGTCCACCT 3'<br>5' CTTCACTTGTGGCCCAGATAGG 3' | 60°C, 60 s |
| BAX | 5' CCGGAATTCCGGATGGACGGGTCCGGGGAGCAG 3'<br>5' TGCTCTAGAGCATCAGCCCATCTTCTTCCAG 3' | 60°C, 60 s |
| C-MYC | 5' CAAGAGGCGAACACACAACGTCT 3'<br>5' AACTGTTCTCGTCGTTTCCGCAA 3' | 55°C, 60 s |
| MMP2 | 5' GTGCTGAAGGACACACTAAAGAAGA 3'<br>5' TTGCCATCCTTCTCAAGTTGTAGG 3' | 59°C, 60 s |
| MMP9 | 5' AGATTTCGACTCTCCACGCA 3'<br>5' CTGAGAACCAATCTCACCGACA 3' | 60°C, 60 s |

of activation of Taq polymerase at 95°C for 4 min, denaturation of cDNA at 95°C for 30–45 s, applying primer annealing temperature for 30–40 s, elongation at 72°C for 40–60 s for the following 35 cycles, and completion of the process by the extension step for 10 min. PCR protocols were optimized for each studied gene according to the specific annealing temperature (Table 11.1). PCR data were analyzed by the Relative Quantification Software (Roche) and expressed as target to reference ratios. A standard protocol was used for expression analysis in eukaryotic cells, as described in the manufacturer's manual, including hybridization of fragment biotin-labeled cRNA molecules and double phycoerythrin staining. The starting material used per one chip was 10 ng total RNA. The detection of the fluorescent signal was performed by the GeneChip® Scanner 3000 System with the GeneChip® Operating Software Version 1.0 (Affymetrix, Inc., Santa Clara, CA).

## RQ-PCR

The gene expression analysis was done using the two-step RT-PCR with real-time quantitative amplification performed in the continuous fluorescence detection system—DNA Engine Opticon (MJ Research Inc., St. Bruno, Quebec, Canada). Equal amounts of total RNA (4 μg) were reverse-transcribed using the Superscript II Kit (QIAGEN). The amplification was performed with the QuantiTect™ SYBR® Green PCR Kit (QIAGEN) using SYBR Green as the fluorescent dye. The amplification conditions were as follows: initial incubation at 95°C for 15 min, followed by 40 cycles of denaturation at 94°C for 15 s, annealing at 59°C for 30 s, and extension at 72°C for 30 s. Subsequently, melting curve analysis was performed to verify the specificity of the PCR products. RNA extraction, preparation of cDNA, and RT-PCR reaction were carried out as previously described [12,15]. The sequences of the primers for MMPs used in this study were as follows: forward, AAT CTC TTC TAG AGA CTG GGA AGG AG; reverse, AGC TGA TTG ACT AAA GTA GCT GGA. The detection of the fluorescent signal was performed by the GeneChip® Scanner 3000 System with the GeneChip® Operating Software Version 1.0 (Affymetrix, Inc., Santa Clara, CA). Each sample was tested in triplicate using quantitative RT-PCR, and samples obtained from three independent experiments were used for analysis of relative gene expression data using the $2^{-\Delta\Delta CT}$ method as described previously [12].

## CELL MIGRATION AND INVASION ASSAY

Cell migration and invasion assays were performed using conventional Boyden chamber transfer methods according to the manufacturer's protocol (BD BioCoat-Tumor Invasion System No. 354166). Quantitation of invading cell was achieved by post-invasion cell labeling with calcein and measurement of the fluorescence of invading cells.

## CASPASE ACTIVITY TEST

All the studied cell lines were tested with the ApoFluor®Green Caspase Activity Assay (ICN Biomedical Inc., Warrendale, PA). The cells ($5 \times 10^5$; triplicate) were cultured for 24–48 h in the presence of BC, DHT, or BC+DHT. Staurosporine (1 µM) was used as a positive control for apoptosis (12 and 24 h incubation with Staurosporine (SIGMA)). After the treatment without or with the reagents, the cells were incubated with the ApoFluor®Green reagent according to the manufacturer's recommendation. The cells were harvested, washed to remove any unbound dye, and placed in black microtiter plates (NUNK). The fluorescence was detected at 520 nm emission wavelength after excitation at 488 nm (ELISA reader).

## ZYMOGRAPHY

Gelatinolytic activities of metalloproteinases MMP-2 and MMP-9 were evaluated from the conditioned medium. The proteins were separated on 10% polyacrylamide gel containing 0.1% gelatin (Sigma) in nonreducing conditions. After electrophoresis, the gel was washed two times for 30 min in 2.5% TritonX-100. After 48 h incubation at 37°C in buffer (50 mM Tris pH 7.5; 10 mM $CaCl_2$; 0.15 mM NaCl), the gel was stained with 1% Coomassie blue R250 for 1 h, and its excess was washed out with methanol (50%)/acetic acid (10%) solvent. Gelatinolytic activity was observed as clear areas in the gel.

## WESTERN BLOT ANALYSIS

The cells were lysed in sample buffer (0.0625 M Tris/HCl pH 6.8, 2% SDS, 10% glycerol, 5% b-mercaptoethanol). Cell lysates containing equal amounts of protein were separated on 10% SDS-PAGE gels and subsequently transferred onto a PVDF membrane. Antibodies against AKT, phospho-AKT(S473), AR, β-catenin, phospho-β-catenin(S552), c-myc, caspase 3, −8, −9 (all Cell Signaling Technology), β-actin (Sigma), PSA (DAKO), and MMP −2, −9 (Santa Cruz) were used to detect the indicated proteins. Total protein loading was determined by probing the membranes for β-actin. Bands were visualized using alkaline phosphate-coupled secondary anti-mouse or anti-rabbit antibody (Sigma). Finally, immunoreactions were visualized by NBT/BCIP staining (Roche). After incubation with the secondary antibody, the immune complexes were visualized by the enhanced chemiluminescence detection system and quantified by densitometric scanning. The PCR reaction products were separated electrophoretically on 2% agarose gel and visualized with ethidium bromide.

## STATISTICAL ANALYSIS

Statistical analysis was performed using the Microsoft EXCEL 5 program and by one-way ANOVA. All the results are expressed as mean values ± standard error or (SE). Before statistical analysis, normal distributions and homogeneity of the variables were tested. The parameters that did not fulfill these tests were logarithmically transformed. Statistical comparisons were made by the unpaired $t$-test for comparisons of quantitative variables. $P \leq 0.05$ was considered significant. All the data from the proliferation and cytotoxicity experiments represent the average of five wells in each experiment ($n = 4$). The variables were summarized as mean ± standard deviation of the mean. In case of RT-PCR and western blotting, each analysis was performed in triplicate.

## RESULTS

Our recent study indicated that a lower concentration of BC (3–10 μM) was more effective as a growth-stimulating agent in androgen-sensitive LNCaP prostate cancer cells than in androgen-insensitive PC-3 cells. In contrast, after treatment of PC-3 cells with 20 μM BC, a significant inhibition of PC-3 cell growth was observed, but no major effect on the growth of LNCaP cells was noted [12]. Therefore, in view of the observed significant differences in the uptake of BC and in the proliferation of various prostate cancer cells, we decided to try to explain the effect of androgens as well as the function of AR in the regulation of proliferation of prostate cancer cells after treatment with 10 and 20 μM BC. The analysis of uptake of BC alone or in the presence of DHT showed successful uptake of BC by both androgen-sensitive and androgen-insensitive cells. They absorbed BC from media in a line-, dose-, and time-dependent manner, although some kind of saturation effect could have been observed for LNCaP cells that absorbed significantly more BC than PC-3 cells. The significant effect of DHT on the uptake of BC by cells of both lines suggested that absorption of BC by prostate cells might be steroid dependent and that it may require AR as well, which would explain a much higher uptake of BC by LNCaP androgen-sensitive than PC-3 androgen-insensitive cells. Elimination of the steroids from the serum

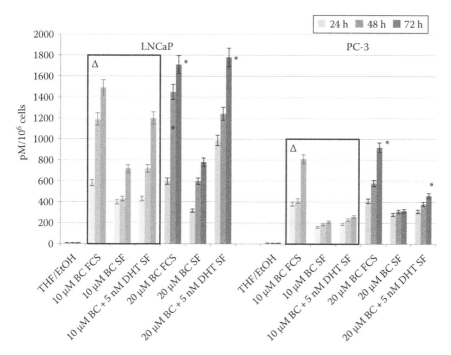

**FIGURE 11.1** Uptake of BC by LNCaP (androgen-sensitive) and PC-3 (androgen-insensitive) prostate cancer cells in RPMI 1640 media supplemented with 10 or 20 μM BC in 0.25% tetrahydrofuran (THF)/ethanol 1:1 with or without 5 nM DHT. The experiments were carried out as described under M&M. FCS—fetal calf serum, SF—steroid-free serum or THF/ethanol were used as vehicle—control. The cells were kept untreated or treated for 24–72 h. The uptake of BC is expressed as pmole per $10^6$ cell. All the procedures were carried out under dimmed light and cooled conditions to prevent photodamage to the compounds. Some results concerning the treatment of the cells with 10 μM BE, placed in a frame (Δ), have been presented previously [12]. Data represent mean values ± SE of three independent experiments done in triplicates. Significantly different from the corresponding control cells *$P<0.05$.

(SF-condition) evidently changed the uptake in both types of cells, but to a higher degree in the androgen-dependent cell line (Figure 11.1). We did not observe any cytotoxic effect of the concentration of both BC and DHT in effect of 24–72 h treatment of the studied cells.

As was formerly presented, 10 μM BC had no effect on the expression of genes coding for proteins involved in BC metabolism as retinol-binding protein (RBP or CREBP), retinol dehydrogenase (RDH11, RDH14), or retinoic receptors (RARA, RARB, RXRA) in PC-3 cells and influenced the expression of the same genes in LNCaP cells [12]. LNCaP cells treated with 5 nM DHT as well as 10 μM BC+5 nM DHT showed significant androgen-induced changes in comparison to those cells treated with BC alone. The microarray analysis confirmed the influence of DHT on the expression of genes responsible for the transport of retinals—RBP and cellular retinol binding protein (CREBP) and also some genes involved in BC

**TABLE 11.2**

**Effect of β-Carotene on the Expression of Genes Coding for Proteins Involved in Its Intracellular Metabolism in LNCaP Cell Line**

TARGET—Transport of Retinal

| Title | Symbol | No UG | LNCaP 5 nM DHT/SF | LNCaP 10 μM BC/SF | LNCaP 10 uM BC/SF C+5 nM DHT/SF | PC-3 10 μM BC/SF | PC-3 10uM BC+5 nM DHT/SF |
|---|---|---|---|---|---|---|---|
| | | | $n = 3$ | $n = 3$ | $n = 3$ | $n = 3$ | $n = 3$ |
| Retinol binding protein | RBP | Hs.857 | 1.6 | 1.1 | 2.4 | NC | NC |
| Cellular retinal binding protein | CREBP | Hs.270804 | NC | 1.3 | 1.2 | NC | NC |
| Retinol dehydrogenase | RDH11 | Hs.226007 | 1.4 | 1.2 | 1.6 | NC | NC |
| Retinol dehydrogenase | RDH14 | Hs.288880 | 1.3 | 1.2 | 1.6 | NC | NC |
| Retinoic acid receptor alpha | RARA | Hs.361071 | NC | 2.4 | −0.5 | NC | NC |
| Retinoic acid receptor beta | RARB | Hs.436538 | 1.6 | 1.3 | NC | NC | NC |
| Retinoid X receptor | RXRA | Hs.20084 | −1.3 | 1 | −1.5 | NC | NC |

[a] Treated and untreated cells were maintained in RPMI-1640 medium (Sigma) supplemented with 10% SF—Steroid-free fetal bovine serum (IMMUNIQ). Data are presented as mean ± SD of at least three independent experiments performed in triplicates.

metabolism in the androgen-sensitive cell line. No such effect was observed in the PC-3 androgen independent (Table 11.2).

Prostate cancer cells reacted differently to BC treatment in respect to proliferation. An increase in proliferation of LNCaP cells was observed for 10 μM BC, while there was a slight decrease or no effect for 20 μM BC. Incubation of PC-3 cells with 10 μM BC did not significantly influence the growth of prostate cancer cells, but after the treatment with 20 μM BC, a significant growth inhibition was observed. DHT added to the medium led to a marked increase in proliferation of LNCaP cells and had no effect on PC-3 cells. However, if present in the medium of cells treated with BC, it caused a very effective reduction of their growth, in particular when a higher 20 μM BC concentration was used, and surprisingly, the effect was stronger in the case of PC-3 cells (Figure 11.2).

Androgen-dependent LNCaP prostate cancer cells stimulated with DHT in addition to proliferation (Figure 11.2) increased migration, while there was no such effect on androgen-insensitive PC-3 prostate cancer cells (Table 11.3). LNCaP cells treated with 5 nM DHT or 10 μM BC increased invasiveness about 1.5–2.5 times, treated with 10 μM BC + 5 nM DHT—less than 50%, but treatment with 20 μM BC + 5 nM DHT decreased migration through Matrigel in both cell lines, but this effect was

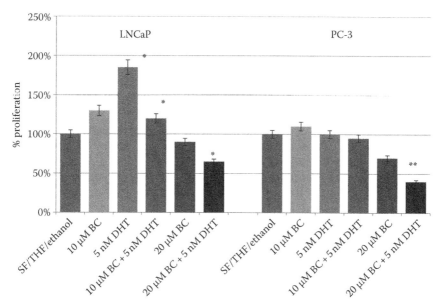

**FIGURE 11.2** The effect of BC and DHT or vehicle alone on proliferation of the LNCaP and PC-3 prostate cancer cell lines. Prostate cells cultured in RPMI containing 10% SF treated in culture growth medium with vehicle 0.25% THF/ethanol (1:1), 10 μM BC, 20 μM BC, 5 nM DHT, or a combination of both (BC/DHT) were labeled after 24–72 h with BrDU. BrDU incorporation was determined by colorimetric immunoassay. The results are given as the percent of the control proliferation. Data represent mean values ± SE of three independent experiments performed in triplicates. Significantly different from the corresponding control, *P < 0.05, **P < 0.005.

**TABLE 11.3**
**Cell Invasion Assay through Matrigel-Coated Boyden Chamber**

| Cell Line | LNCaP | PC-3 |
|---|---|---|
| | $n^a = 9$ | $n = 9$ |
| **Factor** | **% of Migrated Cells[b]** | **% of Migrated Cells[c]** |
| 5 nM DHT | 189–247 | 99–100 |
| 10 μBC | 130–165 | 105–135 |
| 10 μM BC + 5 nM DHT | 110–146 | 95–97 |
| 20 μM BC | 94–99 | 78–85 |
| 20 μM BC + 5 nM DHT | 62–81 | 42–58 |

Data are presented as mean ± SD of at least three independent experiments.

[a]  $n$—the experiments were performed in triplicates and repeated at least trice.

[b,c] Number of the migrated cells was calculated as a % versus 100% migrated cells as control THF/ethanol/SF.

more visible in the androgen-insensitive cell line (Table 11.3). LNCaP cells showed 5% and 20%–30% reduction of the ability to migrate through Matrigel-coated Boyden chambers when compared with the control cells after treatment with 20 μM BC or 20 μM BC + 5 nM DHT, respectively, while PC-3 cells reduced the ability to migrate by about 17% and 40% when treated with 20 μM BC or 20 μM BC + 5 nM DHT, respectively (Table 11.3). We also observed changes in the activity of MMPs upon treatment of cells with BC and/or DHT.

The invasive potential of cancer cells could be regulated through the expression of MMPs. The activity of MMP9 and MMP2 were very low in the LNCaP cell line. After treatment with 5 nM DHT, the cells significantly increased the activity of MMP2 and slightly increased the activity of MMP9. Both MMP9 and MMP2 activities were decreased in LNCaP cells after treatment with 20 μM BC + 5 nM DHT (Figure 11.3). No significant differences in the activity of MMPs were noted when PC-3 cells were treated with 5 nM DHT, but an increased activity of MMP9 was observed after treatment with 10 μM BC, while a decreased activity of both MMPs was seen after treatment with 20 μM BC + 5 nM DHT (Figure 11.3).

We also studied the effect of BC and DHT on the expression of genes/proteins related to proliferation and apoptosis of human prostate carcinoma cells (Table 11.4). Global microarray analysis of genes' expression in LNCaP and PC-3 cells treated for 48 h with 10 μM BC alone or in presence of 5 nM DHT revealed significant differences between both lines in the overall number of up- and downregulated genes. The response of both lines to BC treatment was not very profound, but in the case of androgen-sensitive LNCaP cells, the number of genes—the expression of which changed upon incubation with BC—was much higher than that in PC-3 cells. A detailed analysis of microarray data in respect to LNCaP cells treated with BC showed that numerous genes were involved in replication, transcription, and translation processes, cholesterol metabolism, and expression after treatment with 10 μM BC alone or in the presence of 5 nM DHT, but, surprisingly, LNCaP cells treated with BC + DHT were not affected or rather decreased steroid metabolism as well as eicosanoid metabolism. In general, the number of responsive genes after BC treatment of LNCaP cells was relatively low (165) in comparison to the number of responsive genes in the same cells after BC + DHT treatment (3900) or 5 nM DHT treatment (2850 genes) (Table 11.4). The alterations in the expression of AR or

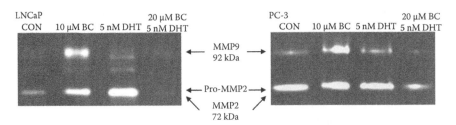

**FIGURE 11.3** Gelatinolytic activity of prostate cancer cells treated with 10 μM BC, 5 nM DHT, and 20 μM BC + 5 nM DHT in the LNCaP and PC-3 cell lines. Gelatinolytic activities of MMP-2 and MMP-9 in prostate LNCaP and PC-3 cells appear as clear zones in the gel. The results are mean SD from three independent experiments.

**TABLE 11.4**

**Group of Genes with the Strongest Response to β-Carotene in the Microarray Experiments According to the Major Processes in the Prostate Cancer Cell Lines**

| Cell Line | LNCaP | | LNCaP | | LNCaP | | PC-3 | | PC-3 | |
|---|---|---|---|---|---|---|---|---|---|---|
| Factor[a] | 5 nM DHT | | 10 µM BC | | 5 nM DHT + 10 µM BC | | 10 µM BC | | 5 nM DHT + 10 µM BC | |
| Number of | 2850 | | 165 | | 3900 | | 45 | | 64 | |
| Changing | Up | Down | Up | Down | Up | Down | Up | Down | Up | Down |
| Genes | 1250 | 1600 | 102 | 63 | 1890 | 2010 | 10 | 25 | 15 | 39 |
| Process | n = 3 | n = 3 | n = 3 | n = 3 | n = 3 | n = 3 | n = 3 | n = 3 | n = 3 | n = 3 |
| Signal transduction | 32 | 22 | 18 | 13 | 48 | 78 | 1 | 2 | 2 | 3 |
| Cell adhesion | 56 | 21 | 8 | 2 | 93 | 64 | 0 | 2 | 0 | 1 |
| Transcription and RNA processing | 33 | 42 | 12 | 3 | 45 | 102 | 1 | 0 | 1 | 3 |
| Protein synthesis and modification | 567 | 356 | 6 | 3 | 321 | 167 | 1 | 4 | 1 | 4 |
| Cell growth and division | 15 | 32 | 4 | 1 | 23 | 63 | 1 | 0 | 1 | 0 |
| Cell cycle | 25 | 13 | 6 | 2 | 67 | 187 | 1 | 2 | 1 | 3 |
| Intracellular trafficking | 145 | 67 | 2 | 0 | 145 | 97 | 5 | 10 | 8 | 20 |
| Steroid metabolism | 32 | 12 | 6 | 1 | 42 | 23 | 0 | 1 | 0 | 0 |
| Lipid metabolism | 18 | 23 | 4 | 6 | 21 | 43 | 0 | 1 | 0 | 1 |

[a] Treated and untreated cells were maintained in RPMI-1640 medium (Sigma) supplemented with 10% SF—steroid free (IMMUNIQ)—fetal bovine serum and 100 U/mL penicillin and 100 µg/mL streptomycin (Sigma). Data are presented as mean ± SD of at least three independent experiments performed in triplicates.

kallikreins—typical markers that have been implicated in the development of prostate cancer—were clearly noted (Table 11.5).

A further careful analysis of microarray data allowed for selecting a limited number of genes, the expression of which was affected by 10 µM BC in LNCaP cells, while it remained without any major changes in PC-3 cells. The increase expression of AR, prostate specific antigen-KLK2 and KLK3, prostatic acid phosphatase, as well c-myc or α- and β-catenin expression of the key molecules participating in prostate cancer progression, were observed.

**TABLE 11.5**

**Microarray Analysis**

| Factor[a] | 5 nM DHT | | 10 µM BC | | 5 nM DHT + 10 µM BC | |
|---|---|---|---|---|---|---|
| Cell Line | LNCaP | PC-3 | LNCaP | PC-3 | LNCaP | PC-3 |
| AR | 1.8 | ND | 1.2 | ND | −0.5 | ND |
| ACPP (PAP) | 1.5 | ND | 1.2 | ND | +0.5 | ND |
| KLK2 | 18.4 | ND | 10.4 | ND | −1.5 | ND |
| KLK3 | 5.7 | ND | 3.2 | ND | −0.5 | ND |
| MAZ | 2.3 | NC | 2.3 | NC | NC | 0.5 |
| MYB | −2.1 | 0.5 | −2.1 | NC | NC | NC |
| MYC | 2.8 | NC | 1.44 | NC | NC | −0.5 |
| JUN | −1.9 | 1 | −1.9 | −0.5 | NC | −0.5 |
| CTNNA | 10.5 | 1 | 10.5 | 4 | −0.5 | NC |
| CDH1 | NC | ND | NC | ND | 1.5 | 1 |
| CTNNB | 2.1 | NC | 1.6 | 1.4 | NC | −0.5 |
| ADAM2 | 6 | NC | 2 | NC | −2 | −1 |
| ADAM7 | 8 | 1 | 2 | 3 | −1.4 | −4 |
| ADAM 9 | 2 | NC | 6 | 1 | −2.6 | −3 |

List of genes related to proliferation and markers of prostate cancer regulated by DHT or BC in LNCaP cell line. ND, not detected; NC, no change.

[a] Treated and untreated cells were maintained in RPMI-1640 medium (Sigma) supplemented with 10% SF—steroid free (IMMUNIQ)—fetal bovine serum and 100 U/mL penicillin and 100 µg/mL streptomycin (Sigma). Data are presented as mean ± SD of at least three independent experiments performed in triplicates.

The changes in prostate cell proliferation after their treatment with 10 µM BC were reported earlier [12]. Due to the observed differences in cell proliferation after treatment with 20 µM BC when compared to 10 µM BC, particularly in PC-3 cells, attempts were made to detect these genes at the mRNA level (RT-PCR) and to identify the proteins in extracts of LNCaP and PC-3 cells (western blot) to confirm the observed changes in gene expression revealed by microarray analysis. Decreased expression of AR, PSA, and c-myc without significant changes in the expression of total Akt and β-catenin at the mRNA level was observed in androgen-sensitive LNCaP cells, in particular after their treatment with 20 µM BC and 5 nM DHT (Figure 11.4A). In PC-3 cells, among the analyzed proteins, a slightly decreased expression of total Akt, β-catenin, and c-myc at the mRNA level was found. We did not detect the expression of AR and PSA in the PC-3 cell line (Figure 11.4A).

However, despite the lack of changes in the expression of total Akt and β-catenin at the mRNA level in LNCaP cells treated with 20 µM BC + DHT, a significant decrease in P-Akt and β-catenin as well as c-myc and AR expression at the protein

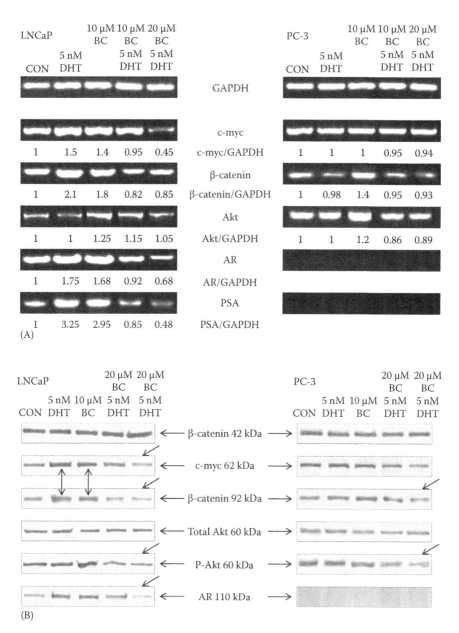

**FIGURE 11.4** Expression of c-myc, β-catenin, Akt, AR, PSA by BC, and/or DHT in LNCaP and PC-3 cells. (A) Representative mRNA level and (B) Western blot analyses. Specific antibody probes were used to localize c-myc, β-catenin, Akt, p-Akt after 72 h of incubation in the absence or in the presence of BC. Total protein loading was determined by probing the membranes for β-actin. The values are mean ± SE, $n = 6$ per group. Significantly different from the corresponding control *$P < 0.05$.

level was discovered (Figure 11.4B). Similar effects—except AR expression—were observed for PC-3 cells treated with 20 μM BC+DHT (Figure 11.4B). To determine if the decreased proliferation was due to the increase in apoptosis, we studied the effect of BC on the expression of caspases as well as antiapoptotic proteins. In general, no effect on global cellular caspase activity following 72 h incubation of both LNCaP and PC-3 cells with 10 μM BC alone or in the presence of DHT was noted (Figure 11.5A).

However, the apoptosis of both, LNCaP and PC-3, cells markedly increased after their treatment with 20 μM BC, in particular in the presence of DHT (Figure 11.5A and B). An additional analysis of various caspases in BC-treated and untreated cells showed that the cells of both lines did not respond in the same manner. The highest increase in caspase-3 and—9 activity was observed in PC-3 cells. We did not observe any change in casp-8 expression (Figure 11.5B). In addition, decreased expression of the Bcl-2 gene after treatment with 20 μM BC, in particular in the presence of DHT, was observed in both LNCaP and PC-3 cells. We did not note significant changes of Bax expression (Figure 11.5B).

However, a quantitative analysis of BAX and BCL-2 gene expression in LNCaP and PC-3 cells confirmed differences in the expression of these genes after treatment with 20 μM BC or BC+DHT (Figure 11.6). The RQ-PCR analysis established that DHT changed the expression of BAX, BCL-2, and MMP9 only in the androgen-sensitive LNCaP cell line (Figure 11.6).

The higher (20 μM) BC concentration showed an opposite effect on the expression of BAX, BCL-2, and MMP9 as compared to that of 10 μM BC. Increased BAX and decreased BCL-2 and MMP expressions in both LNCaP and PC-3 cells were

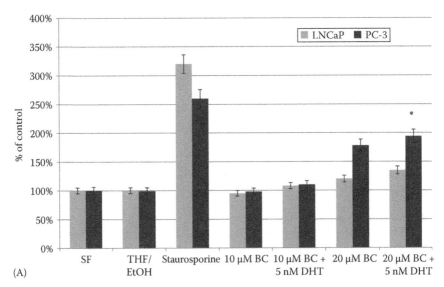

(A)

**FIGURE 11.5** Effects of DHT or BC on apoptosis and apoptosis-related proteins in LNCaP and PC-3 cells. Apoptosis was measured as caspase activity test (A).

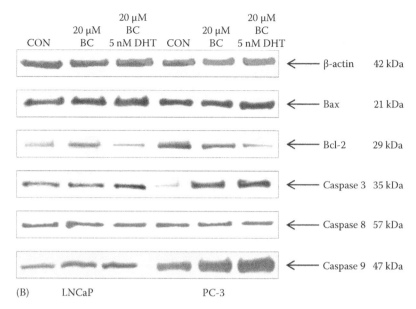

(B)        LNCaP                        PC-3

**FIGURE 11.5 (continued)** Effects of DHT or BC on apoptosis and apoptosis-related proteins in LNCaP and PC-3 cells. Representative western blot analyses of Bcl-2, Bax, caspase-3, caspase-8, caspase-9 (B). The cells were treated with 10 or 20 µLM BC, 5 nM DHT, or both for 72 h. Significantly different from the corresponding control *$P < 0.05$. Protein levels of Bcl-2, Bax, caspase-3, -8, -9 were measured using western blot analysis, and β-actin was used to normalize the quantity of protein. The results are representative of four independent experiments. The values were the means ± SE.

observed (Figure 11.6A). A profound effect was seen when the cells were treated with 20 µM BC in the presence of 5 nM DHT.

The higher concentration of BC alone as well as the presence of DHT stimulated changes in the expression of proteins involved in apoptosis that remained in agreement with the activity of caspases, indicating that apoptosis might be responsible for the observed decrease in the number of cells grown in the higher BC concentration, in particular in the presence of DHT (Figure 11.6B).

## DISCUSSION

The regulation of gene expression by steroid hormones plays an important role in the normal development and strongly influences the pathogenesis of steroid-dependent prostate cancers. In the course of tumor progression, fundamental changes in the expression of cell signaling proteins that control principal functions of the prostate gland are observed. In light of ambiguous results of numerous studies on the effect of dietary factors, including broadly understood carotenoids, on prostate cancer development and progression, we sought to determine if and how one of the most common carotenoids, BC, influences important signaling pathways and respective proteins that likely contribute to prostate physiology and pathology. Cook et al. [16] suggested that BC supplementation might reduce the risk of

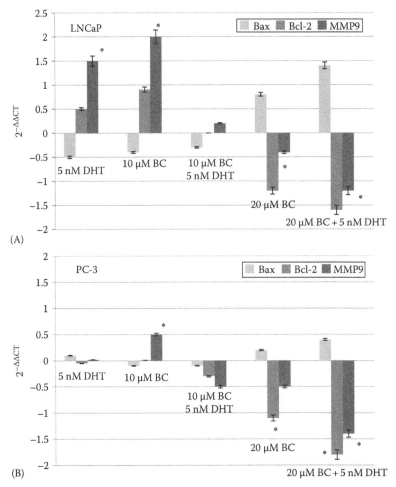

**FIGURE 11.6** Real-time PCR analysis of Bax, Bcl-2, and MMP9 transcripts in the (A) LNCaP and (B) PC-3 prostate cancer cell lines. The total RNA samples were prepared from untreated cell lines and from treated cells with 5 nM DHT, 10 μM BC (72 h), or both. Equal amounts of total RNA (400 ng) were reverse transcribed. The amplification conditions were as follows: initial incubation at 95°C for 15 min., followed by 40 or 45 cycles of denaturation at 94°C for 15 s, annealing at 59°C for 30 s, and extension at 72°C for 30 s. Subsequently, melting curve analysis was performed to verify the specificity of the PCR products. The mean values ± SE. The measurements were done in $n = 3$ (each three times) according to the standard $2^{-\Delta\Delta Ct}$ method with GAPDH used as a calibrator. All primer pairs were initially validated passing the test for equal amplification efficiencies. Using the presented conditions, the amplification efficiency was close to 2. All primer pairs delivered single gene products, which proved to be specific in sequencing. Significantly different from the corresponding control *$P < 0.05$.

prostate carcinoma among men with low baseline levels of plasma BC. Once in the cells, BC is most likely metabolized to retinol and possibility further. The results presented here clearly indicate a link between androgens (DHT as well as testosterone—which, although not as effective, acted analogously—data not shown) and the uptake of BC by both the androgen-sensitive (LNCaP) and androgen-insensitive (PC-3) cells. One has to agree that lack of androgen sensitivity/dependence in case of PC-3 is in light of the presented data rather theoretical and may reflect the effect of androgens on cell behavior with respect to the role of AR. The comparison of the genes, the expression of which was down- or upregulated in the LNCaP and PC-3 cells as an effect of DHT and/or BC treatment, indicates that the observed changes in response between the two prostate cancer cell lines are most likely due to a different status of AR expressed in LNCaP or missing in PC-3. How DHT increases BC uptake in the latter remains to be elucidated, but there is little doubt that in addition to the main AR contribution to the uptake of BC by LNCaP cell, such an unknown mechanism is also likely involved. Taken together, this would explain the effective increase in intracellular BC in LNCaP cells. To our knowledge, this is the first report on the effect of androgens on the uptake of BC by prostate cancer cells. Our results confirm the positive effect of BC on prostate cancer cells, but only in higher concentration. Williams et al. [11] showed that retinol was produced in vitro by the incubation of BC with prostate cancer cells. The mechanism by which BC is converted to retinol in vitro in the presence of prostate cells is unknown. Prostate cells may express a specific enzyme that cleaves BC as has been suggested for hepatocytes [17], human colon cancer cells [18], human lung fibroblasts [19], and human skin fibroblasts [20]. It may be equally intriguing to consider the impact dietary carotenoids could have on prostate biology and disease risk if local cleavage to retinol or other metabolites, such as retinoic acid, does occur in vivo. There is some evidence of androgen–retinoid cross talk and ligand antagonism in the prostate and other hormone-responsive tissues [21,22]; however, interactions between androgens and retinoids in coordinating the expression and regulation of genes involved in RA metabolism and signaling remain unclear. RA has been found to repress androgen-stimulated cell growth of LNCaP cells [20]. Decreased cellular replication has been also observed in PC-3 prostate carcinoma cells incubated with 4-HPR (10 μmol/L), a synthetic retinoid [23]. Since serum concentrations of vitamin A are tightly regulated in vivo, the conversion of provitamin A carotenoids to retinol in the prostate could serve as a "by-pass" mechanism to increase local retinol concentrations. Kotake-Nara et al. [24] reported that alpha-carotene and BC at 20 μmol/L concentration significantly reduced the viability of PC-3 and Du 145 cells, while it had no effect on LNCaP cells. Numerous studies have suggested that retinoids and RA can inhibit carcinogenesis of the prostate and that loss of normal RA metabolism is involved in prostate cancer development [25–29]. It was shown that RA in low concentration increased proliferation of human squamous cell carcinoma lines, but higher concentration of RA inhibited cell division [30]. Interestingly, all-*trans* BC enhanced the proliferation and DNA synthesis of the BALB/c 3T3 fibroblast cell line induced by a tumor promoter (TPA) [31], which is in accordance with our results showing that low concentrations of BC as well as its derivative RA significantly increase proliferation of prostate cancer cells [12]. It was demonstrated that RXR can form a heterodimer with the AR under in vitro and in vivo conditions. Functional analyses further demonstrated that AR, in the presence or absence of androgen, can function as a repressor to suppress RXR target genes, thereby preventing RXR binding to the RXR DNA response element. In contrast, RXR can function as a repressor to suppress AR target genes in the presence of RA, but unliganded RXR can function as a weak coactivator to moderately enhance AR transactivation. Together, these results not only reveal a unique interaction between members of the two nuclear receptor subfamilies, but also show that a nuclear receptor (RXR) may function as either a repressor or a coactivator based on the ligand binding status. [32]. Dulińska et al. observed increased expression of RARα and RARβ upon the treatment of LNCaP cells with 10 μM BC. This concentration of BC rather led to a decreased caspase activity in LNCaP cells, indicating that at relatively low concentration of BC in the medium, rather pro-proliferative and antiapoptotic signals prevailed [12]. Low doses of BC induced proliferation, and the effect was more profound in the LNCaP cell line, since it absorbed BC more efficiently, leading to its higher concentration in the cells, but still remaining within a disadvantageous range of cellular BC concentration. However, the higher concentration of BC (20 μM) as well as a higher uptake of BC led to a relative decrease in proliferation in the case of LNCaP cells and quite remarkable growth inhibition of PC-3 cells. A surprising effect, proliferation decreased by about 35% and 60% in LNCaP and PC-3, respectively, was observed when prostate cells were treated with 20 μM BC + 5 nM DHT. Carotenoids act not only as antioxidants but also as prooxidants, and the prooxidant action of carotenoids has been suggested to induce apoptosis in tumor cells [33]. BC of 20 μM increased the level of caspase 3 and 9 to a higher degree in androgen-insensitive as compared to androgen-sensitive cells.

To find out if the changes in expression of the cadherin/catenin complex might be somehow linked to c-myc, the expression of the latter was examined in different primary and cancer human prostate cell lines. The most significant effect, a decreased expression of c-myc, was observed upon treatment of the LNCaP cell line with ciglitazone or linoleic acid. In addition, the most profound changes of c-myc expression were observed in LNCaP and in CA-K primary cells isolated from cancer tissue, while no changes in its expression in primary hyperplasia BPH-K cells were noted [13]. Androgens have strong and dose-dependent effects both on proliferation and on differentiation of LNCaP cells [34,35]. However, LNCaP cell proliferation is modulated not only by androgens but also by a number of other ligands acting via nuclear receptors such as VD3, T3, and retinoids. VD3 or PPARγ provokes both stimulatory and inhibitory effects [13,36]. The inhibitory effect of 20 μM BC on proliferation in both cell lines may be caused by its—as yet not identified—oxidized forms. Perhaps higher doses of BC lead to DNA damage and in consequence cell cycle arrest. Williams et al. [11] observed cell cycle arrest and apoptosis process in prostatic cancer cell lines independently with p53 and p21 activity. In case of LNCaP, the expression of many genes of steroid metabolism was affected, while no such effect was noted in PC-3, androgen-insensitive cells. Growth factors and steroid hormones activate and may quickly and coordinately modulate the expression of nuclear proto-oncogenes, such as c-myc, c-fos, c-jun, which play a crucial role in the steps that can lead to proliferation and/or differentiation of cells [37]. Several mechanisms have been proposed to explain the putative role of BC in the modulation of cell growth, for example, its ability to act as a redox agent, to enhance immune response and gap junction communication [38]; at the moment, no investigations have directly probed into a possible influence of BC on the androgen pathway and the possible consequent effects on tumor cell growth, particularly in prostate cells. The higher concentration of BC in the presence of DHT increased the activity of caspase 3 and 9 in both the studied cell lines, but a more profound effect was observed in the androgen-insensitive cell line. Proapoptotic effects of BC and other carotenoids have been reported previously in cancer cells, and different mechanisms have been implicated, including carotenoid's ability to induce the caspase cascade [39,40], affect mitochondrial functions [39], modulate the expression of apoptosis-related proteins, including Bcl-2 and Bcl-xL [40,41], Bad [42], Bid [42], and Bax [41,43], and the levels of transcription factors involved in apoptosis induction [8,9,44]. Palozza et al. showed that the cell growth–inhibitory effects of BC may occur via a mechanism involving increased production of ROS and activation of NF-kB. The delicate balance between oxidants and antioxidants, such as glutathione and thioredoxin, ultimately determines the activity profile for many transcription factors involved in cell proliferation and apoptosis [45]. These authors demonstrated that one of the possible mechanisms responsible for the control of cell growth by BC may be the redox regulation of NF-kB and the consequent expression of proteins, such as c-myc, involved in apoptosis induction [33]. Palozza et al. [7] showed that BC acted differently in the cells with different androgen sensitivity. We noted decreased pAkt and AR expression in LNCaP cells after treatment with 20 μM BC and consequently a decreased β-catenin level, c-myc and increased caspase 3- and 9- activity. Yang et al. [46] showed that BC was likely associated with the reduction of proliferation (attenuation of PCNA expression and IGFBP3 levels) and interfered with IGF-1 signaling (reduction of IGFBP3 levels) in PC-3-bearing nude mice. Alimiraha et al. [47] reported that the DU-145 and PC-3 cell lines expressed AR mRNA. They found that the relative levels of AR mRNA and protein that were detected in the DU-145 and PC-3 cell lines were lower than that in LNCaP, an AR-positive cell line. Notably, treatment of these cell lines with DHT resulted in measurable increases in the AR protein levels and considerable nuclear accumulation. Although treatment of DU-145 and PC-3 cells with DHT did not result in stimulation of the activity of an AR-responsive reporter, knockdown of AR expression in PC-3 cells resulted in decreases in p21CIP1 protein levels and a measurable decrease in the activity of the p21-luc-reporter. However, we did not observe the expression of AR in the PC-3 cell line, but the influence of DHT on the uptake of BC and proliferation as well as apoptosis effect in androgen-insensitive PC-3 cells was observed.

The reported results indicate links between BC and steroid signaling not only in androgen-sensitive cells. In view of the recent interest in the prognostic significance of lipogenic enzymes and their potential role as targets for antineoplastic therapy, our findings on the regulation of these enzymes by BC may provide a novel insight into the complex mechanisms, by which androgens affect prostate cancer cells.

## ACKNOWLEDGMENTS

The authors' responsibilities were as follows—JDL: designed and conducted the research; JDL: provided essential reagents or materials; GS and JDL: conducted and analyzed microarray data; JDL and PL: analysis and interpretation of data; JDL: performed statistical analyses; JDL and PL: wrote the chapter; AKD: critical revision of the manuscript for important intellectual content. JDL, PL, AKD, GS had no conflicts of interest. This chapter was supported by grants MNiSW through the Jagiellonian University Medical College K/ZDS/001003 and K/PBW/ 000561-NN401 125638 (1256/B/P01/2010/38) (to JDL); microarray data (to GS).

## REFERENCES

1. World Health Statistics 2010 *WHO Library Cataloguing-in-Publication Data*, ISBN 978 92 4 156398 7 (NLM classification: WA 900.1)
2. Dulińska-Litewka J, Gil D, Ciołczyk-Wierzbicka D, Laidler P. Modulation of androgen receptor transcriptional activity in human prostate. *FEBS J* 2011; 278(suppl. 1): 356–357.
3. Krinsky NI. The antioxidant and biological properties of the carotenoids. *Ann NY Acad Sci* 1998; 854: 443–447.
4. Iynem AH, Alademir AZ, Obek C, Kural AR, Konukoglu D, Akcay T. The effect of prostate cancer and antiandrogenic therapy on lipid peroxidation and antioxidant systems. *Int Urol Nephrol* 2004; 36: 57–62. PMID 15338676.
5. Tam NN, Gao Y, Leung YK, Ho SM. Androgenic regulation of oxidative stress in the rat prostate: Involvement of NAD(P)H oxidases and antioxidant defense machinery during prostatic involution and regrowth. *Am J Pathol* 2003; 163: 2513–2522.
6. Peters U, Leitzmann MF, Chatterjee N, Wang Y, Albanes D, Gelmann EP, Friesen MD, Riboli E, Hayes RB. Serum lycopene, other carotenoids, and prostate cancer risk: A nested case-control study in the prostate, lung, colorectal, and ovarian cancer screening trial. *Cancer Epidemiol Biomarkers Prev* 2007; 16: 962–968. DOI: 10.1158/1055–9965. EPI-06-0861.
7. Palozza P, Sestito R, Picci N, Lanza P, Monego G, Ranelletti FO. The sensitivity to b-carotene growth-inhibitory and proapoptotic effects is regulated by caveolin-1 expression in human colon and prostate cancer cells. *Carcinogenesis* 2008; 29(11): 2153–2161. DOI: 10.1093/carcin/bgn018.
8. Kieć-Wilk B, Śliwa A, Mikołajczyk M, Małecki MT, Mathers JC. The CpG island methylation regulated expression of endothelial proangiogenic genes in response to β-carotene and arachidonic acid. *Nutr Cancer* 201; 63(7): 1053–1063.
9. Dembińska-Kieć A, Polus A, Kieć-Wilk B, Grzybowska J, Stachura J, Dyduch G, Pryjma J, Skrzeczyńska J, Langman T, Bodzioch M, Partyka Ł, Schmitz G. Beta-carotene and angiogenesis. *Pure Appl Chem* 2006; 78: 1519–1537.
10. Brosman SA, Prostate cancer and nutrition. *Drug, Diseases & Procedures* 2008; Updated: October 27, 2011.
11. Williams AW, Boileau TW, Zhou JR, Clinton SK, Erdman JW. β-carotene modulates human prostate cancer cell growth and may undergo intracellular metabolism to retinol. *J Nutr* 2000; 130: 728–732.
12. Dulińska J, Gil D, Zagajewski J, Hartwich J, Bodzioch M, Dembińska-Kieć A, Langmann T, Schmitz G, Laidler P. Different effect of beta-carotene on proliferation of prostate cancer cells. *Biochim Biophys Acta* 2005; 1740(2): 189–201. PMID: 15949686.
13. Laidler P, Dulińska J, Mrozicki S. Does the inhibition of c-myc expression mediate the anti-tumor activity of PPAR's ligands in prostate cancer cell lines? *Arch Biochem Biophys* 2007; 462(1): 1–12. PMID: 17466258.

14. Jach R, Dulińska-Litewka J, Laidler P, Szczudrawa A, Kopera A, Szczudlik Ł, Pawlik M, Zając K, Mak M, Basta A. Expression of VEGF, VEGF-C and VEGFR-2 in in situ and invasive SCC of cervix. *Front Biosci* 2010; 2: 411–423. PMID 20036889.

15. Dudzik P, Dulińska-Litewka J, Wyszko E, Jędrychowska P, Barciszewski J, Laidler. P. Effects of kinetin riboside on proliferation and proapoptotic activities in human normal and cancer cell lines. *JCB* 2011; 112: 2115–2124. DOI: 10.1002/jcb.23132.

16. Cook NR, Stampfe MJ, Manson JE, Sacks FM, Buring J, Hennekens CE. Beta-carotene supplementation of low baseline levels and decreased risk of total and prostate cancer. *Cancer* 1999; 86: 1783–1792.

17. Blaner WS, Olson JA. Retinol and retinoic acid metabolism, In: *The Retinoids: Biology, Chemistry and Medicine* (Sporn, M. B., Roberts, A. B. and Goodwin, D. S. eds.) 2nd edn., Raven Press Ltd., New York, 1994, pp. 229–255.

18. During A, Albaugh G, Smith JC. Characterization of β-carotene 15,15′-dioxygenase activity in TC7 clone of human intestinal cell line Caco-2. *Biochem Biophys Res Commun* 1998; 249: 467–474.

19. Scita G, Aponte GW, Wolf G. Uptake and cleavage of b-carotene by cultures of rat small intestinal cells and human lung fibroblasts. *J Nutr Biochem* 1992; 3: 118–123.

20. Wei RR, Wamer WG, Lambert LA, Kornhauser A. β-carotene uptake and effects on intracellular levels of retinol in vitro. *Nutr Cancer* 1998; 30: 53–58.

21. Kuang-Hsiang C, Yi-Fen L, Wen-Jye L, Chin-Yi C, Saleh A, Yu-Jui YW, Chawnshang C. 9-cis-retinoic acid inhibits androgen receptor activity through activation of retinoid X receptor. *Mol Endocrinol* 2005; 19: 1200–1212.

22. Young CY, Murtha PE, Andrews PE, Lindzey JK, Tindall DJ. Antagonism of androgen action in prostate tumor cells by retinoic acid. *Prostate* 1994; 25: 39–45.

23. Igawa M, Tanabe T, Chodak GW, Rukstalis DB. N-(4-hydroxyphenyl) retinamide induces cell cycle specific growth inhibition in PC3 cells. *Prostate* 1994; 24: 299–305.

24. Kotake-Nara E, Kushiro M, Zhang H, Sugawara T, Miyashita K, Nagao A. Carotenoids affect proliferation of human prostate cancer cells. *J Nutr* 2001; 131: 3303–3306.

25. Chopra DP, Wilkoff LJ. Inhibition and reversal by b-retinoic acid of hyperplasia induced in cultured mouse prostate tissue by 3-methylcholanthrene or N-methyl-N9-nitro-N-nitrosoguanidine. *J Natl Cancer Inst* 1976; 56: 583–589.

26. Pasquali D, Thaller C, Eichele G. Abnormal level of retinoic acid in prostate cancer tissues. *J Clin Endocrinol Metab* 1996; 81: 2186–2191.

27. Guo X, Knudsen BS, Peehl DM, Ruiz A, Bok D, Rando RR, Rhim JS, Nanus DM, Gudas LJ. Retinol metabolism and lecithin:retinol acyltransferase levels are reduced in cultured human prostate cancer cells and tissue specimens. *Cancer Res* 2002; 62: 1654–1661.

28. Reichman ME, Hayes RB, Ziegler RG, Schatzkin A, Taylor PR, Kahle LL, Fraumeni JF. Serum vitamin A and subsequent development of prostate cancer in the first national health and nutrition examination survey epidemiologic follow-up study. *Cancer Res* 1990; 50: 2311–2315.

29. Ren MQ, Pozzi S, Bistulfi G, Somenzi G, Rossetti S, Sacchi N. Impaired RA-signal leads to RAR-beta 2 epigenetic silencing and RA resistance. *Mol Cell Biol* 2005; 25: 10591–10603.

30. Crowe DL, Kim R, Chandrarratna AS. Retinoic acid differentially regulates cancer cell proliferation via does-dependent modulation of the mitogen-activated protein kinase pathway. *Mol Cancer Res* 2003; 1: 532–540.

31. Okai Y, Higashi-Okai K, Yano Y, Otani S. All-tran beta carotene enhances mitogenic responses and ornithine decarboxylase activity of BALB/c 3T3 fibroblast cells induced by tumor promoter and fetal bovine serum but suppressed mutagen dependent umu Cgene expression in salmonella typhimurium (TA 1535/pSK 1002). *Cancer Lett* 1996; 99: 15–21.

32. Chuang KH, Lee YF, Lin WJ, Chu CY, Altuwaijri S, Wan YJY, Chang C. 9-cis-retinoic acid inhibits androgen receptor activity through activation of retinoid X receptor. *Mol Endocrinol* 2005; 19: 1200–1212.

33. Nakamura H, Nakamura K, Yodoi J. Redox regulation of cellular activation. *Annu Rev Immunol* 1997; 15: 351–369.

34. Lin MF, DaVolio J, Garcia-Arenas R. Expression of human prostatic acid phosphatase activity and the growth of prostate carcinoma cells. *Cancer Res* 1992; 52: 4600–4607.

35. Swinnen JV, Ulrix W, Heyns W, Verhoeven G. Coordinate regulation of lipogenic gene expression by androgens: Evidence for a cascade mechanism involving sterol regulatory element binding proteins. *Proc Natl Acad Sci* 1997; 94: 12975–12980.

36. Esquenet M, Swinnen JV, Veldhoven PP, Van Deneff C, Heyns W, Verhoeven G. Retinoids stimulate lipid synthesis and accumulation in LNCaP prostatic adenocarcinoma cells. *Mol Cell Endocrinol* 1997; 136: 37–46.

37. Bernard D, Pourtier-Manzanedo A, Gil J, Beach DH. Myc confers androgen-independent prostate cancer cell growth. *J Clin Invest* 2003; 112(11): 1724–1731.

38. Thompson TC, Tahir SA, Li L, Watanabe M, Naruishi K, Yang G, Kadmon D et al. The role of caveolin-1 in prostate cancer: Clinical implications. *Prostate Cancer Prostatic Dis* 2010; 13: 6–11. DOI: 10.1038/pcan.2009.29.

39. Palozza P, Serini S, Torsello A, Nicuolo FD, Maggiano N, Ranelletti FO, Wolf F, Calviello G. Mechanism of activation of caspase cascade during beta carotene- induced apoptosis in human tumor cells. *Nutr Cancer* 2003; 47: 76–87.

40. Prasad V, Chandele A, Jagtap JC, Kumar S, Shastry P. ROS-triggered caspase 2 activation and feedback amplification loop in β-carotene-induced apoptosis. *Free Radic Biol Med* 2006; 41: 431–442.

41. Terao J, Yamauchi R, Murakami H, Matsushita S. Inhibitory effects of tocopherols and β-carotene on singlet oxygen-initiated photooxidation of methyl linoleate and soybean oil. *J Food Process Preserv* 1980; 4: 79–93.

42. Liu C, Lian F, Smith DE, Russell RM, Wang XD. Lycopene supplementation inhibits lung squamous metaplasia and induces apoptosis via up-regulating insulin-like growth factor binding protein 3 in cigarette smoke-exposed ferrets. *Cancer Res* 2003; 63: 3138–3144.

43. Sacha T, Zawada M, Hartwich J, Lach Z, Polus A, Szostek M, Zdziowska E et al. The effect of beta-carotene and its derivatives on cytotoxicity, differentiation, proliferative potential and apoptosis on the three human acute leukemia cell lines: U-937, HL-60 and TF-1. *Biochim Biophys Acta* 2005; 1740(2): 206–214.

44. Bodzioch M, Dembinska-Kiec A, Hartwich J, Lapicka-Bodzioch K, Banas A, Polus A, Grzybowska J et al. The microarray expression analysis identifies Bax as a mediator of beta-carotene effects on apoptosis. *Nutr Cancer* 2005; 51: 226–235. DOI: 10.1207/s15327914nc5102_13.

45. Palozza P, Serini S, Torsello A, Di Nicuolo F, Piccioni E, Ubaldi V, Pioli C, Wolf FI, Calviello G. β-carotene regulates NF-kappaB DNA-binding activity by a redox mechanism in human leukaemia and colon adenocarcinoma cells. *J Nutr* 2003; 133: 381–388.

46. Yang CM, Yen YT, Huang CS, Hu ML. Growth inhibitory efficacy of lycopene and b-carotene against androgen-independent prostate tumor cells xenografted in nude mice. *Mol Nutr Food Res* 2011; 55: 606–612. DOI: 10.1002/mnfr.201000308.

47. Alimiraha F, Chena J, Basrawalab Z, Xina H, Choubey D. DU-145 and PC-3 human prostate cancer cell lines express androgen receptor: Implications for the androgen receptor functions and regulation. *FEBS Lett* 2006; 580: 200.

# 12 Carotenoid Intake and Supplementation in Cancer
## Pro and Con

*Werner G. Siems, Olaf Sommerburg,
Ingrid Wiswedel, Peter M. Eckl, Nikolaus Bresgen,
Frederik J.G.M. van Kuijk, Carlo Crifò,
and Costantino Salerno*

## CONTENTS

## INTRODUCTION

The aim of this review is to summarize the literature data concerning statistical correlations between carotenoid intake and cancer and to provide evidence, if any, that can justify carotenoid supplementation in protocols designed for the prevention and the treatment of cancer diseases. Particular attention will be drawn to lung cancer because of its surprisingly high incidence in cigarette smokers and asbestos workers who received high doses of β-carotene. The efficacy of carotenoids and retinol as adjuvants in the therapeutic procedures against cancer and the clinical indications for carotenoid supplementation will be discussed.

## RELATIONSHIP BETWEEN CAROTENOID
## INTAKE AND CANCER INCIDENCE

Of all known carotenoids (about 600 different types are found in nature), about 40 are regularly consumed by humans. Carotenoids are known to be biologically important micronutrients with many varied functions. Around 50 carotenoids display provitamin A activity [1–3]. Carotenoids are also precursors of retinoids. Retinoids are a class of over 4000 natural and synthetic molecules structurally and/or functionally related to fat-soluble vitamin A [4]. It has been suggested that the antioxidant potency of β-carotene is carried out by scavenging oxygen radicals [5,6]. The intake of β-carotene has been recommended, and carotenoid supplementation has been used for the prevention and treatment of diseases related to oxidative stress [7,8], such as cancer [9–11], UV-mediated skin diseases, neurodegenerative diseases, and cystic fibrosis (CF).

The association between food intake and cancer is still under intense debate [12,13]. Although there are inconsistencies across studies that have investigated the relationship between diet and cancer, it is still widely accepted that dietary factors influence cancer risk and that diet plays a major role in cancer etiology and prevention [14,15]. Several epidemiological studies consistently showed that increased consumption of food rich in β-carotene (i.e., vegetables and fruits) is associated with a reduced risk of lung and some other types of cancer [9,16]. A similar relationship has been found between plasma levels of β-carotene and risk of cancer [9,17].

In the "Third National Health and Nutrition Examination Survey Follow-up Study" (with an average follow-up period of 13.9 years), investigating the relationship between α-carotene concentrations and risk of death among 15,318 U.S. adults 20 years and older, it was found that serum α-carotene concentration was inversely associated with the risk of death from all causes including cardiovascular disease and cancer [18]. The cancer cohort consisted of 834 persons who suffered mainly from cancer of aerodigestive system (i.e., cancers of lip, oral cavity, pharynx, esophagus, stomach, colon, rectum, anus, liver, intrahepatic bile ducts, pancreas, and larynx) and of respiratory tract (lung, trachea, or bronchus). The authors found that serum α-carotene concentration was inversely associated with the unadjusted risk of death from cancer (P < 0.001 for linear trend). After adjusting for the demographic characteristics, lifestyle habits, and health risk factors, they found that serum α-carotene was inversely associated with adjusted risk of death from cancer with P = 0.02 for linear trend. In the subcategories for cause of death, they found a significant inverse association between serum α-carotene concentration and risk of death from cancer of the aerodigestive system with P < 0.001 for linear trend and of the respiratory tract with P = 0.002. The results were consistent with findings from previous studies including those conducted among 6832 American men of Japanese ancestry in Hawaii [19], which showed that low serum α-carotene concentrations were associated with increased risk of cancers of the upper aerodigestive system including esophageal cancer, laryngeal cancer, and oral-pharyngeal cancer.

Two meta-analyses clearly indicate that the intake of vitamin A and related compounds is inversely associated with the risk of cervical cancer. In the first work [20],

11 articles on dietary vitamin A and 4 articles on blood vitamin A were selected according to the eligibility criteria and were included in the meta-analysis, for a total of 12,136 participants. Obviously, the number of included participants was high enough to exclude in causal influence of HPV and increased risk of cervical cancer with increasing duration of oral contraceptive use. The pooled odds ratios (ORs) of cervical cancer were 0.59 (95% CI 0.49–0.72) for total vitamin A intake and 0.60 (95% CI 0.41–0.89) for blood vitamin A levels. The combined ORs of cervical cancer were 0.80 (95% CI 0.64–1.00), 0.51 (95% CI 0.35–0.73), and 0.60 (95% CI 0.43–0.84) for retinol, carotene, and other carotenoid intake, and 1.14 (95% CI 0.83–1.56) and 0.48 (95% CI 0.30–0.77) for blood retinol and carotene. In the second investigation [21], of 274 articles meeting the initial criteria, the authors included 22 case–control studies involving a total of 10,073 participants. In meta-analyses by the type of vitamin or antioxidant, a significant preventive effect on cervical neoplasm was found in intakes of vitamin $B_{12}$ (OR 0.35, 95% CI 0.19–0.63; n=2), vitamin C (OR 0.67, 95% CI 0.55–0.82; n=8), vitamin E (OR 0.56, 95% CI 0.35–0.88; n=10), and β-carotene (OR 0.68, 95% CI 0.55–0.84; n=9).

Studies on prostate cancer gave rise to contradictory results. Indeed, it has been reported that consumption of lycopene, an important red pigment and the predominant carotenoid of tomatoes and tomato products [22], is associated with a reduced incidence of prostate cancer [23–32] and that three to five servings of tomato products a week reduces the risk of prostate cancer significantly [23,30]. Additionally, blood serum levels of lycopene are inversely associated with prostate cancer risk [12,31,33]. Moreover, it has been found that, among young men, diets rich in β-carotene may play a protective role in prostate carcinogenesis [33]. In vitro evidence suggested that lycopene is metabolized by carotenoid monooxygenase II [34] resulting in an asymmetric chain cleavage of lycopene and other carotenoids to form aldehyde metabolites. In vivo studies on the plasma from subjects consuming tomato juice for 8 weeks [35] led to the identification of tomato carotenoid intermediates that are partially pro-oxidative compounds [34,36] such as apo-6′-lycopenal, apo-8′-lycopenal, apo-10′-lycopenal, and apo-12′-lycopenal. It was reported that lycopene alone is not responsible for in vitro and in vivo medicinal effects, but lycopene and apo-12′-lycopenal reduce the proliferation of prostate cancer cells by inhibiting normal cell cycle progression [32]. Arguments against a role for lycopene in prostate cancer prevention came from the work of Kristal et al. [37], who studied 1683 subjects and offered a prostate biopsy at the trial end to all participants not previously diagnosed with prostate cancer. The data appeared to show no association between serum lycopene and prostate cancer risk, but this conclusion was criticized by Giovannucci [38], who suggested that lycopene or related compounds could inhibit growth of extant prostate cancers and that biopsy in apparently healthy subjects could reflect the absence of progression factors rather than the presence of them. The puzzle is unsolved since a study on 29,133 smokers from Finland indicates that higher serum retinol is associated with elevated prostate cancer risk (instead of a lower risk) in subjects under conditions of heavy oxidative stress [39]. Randomized controlled trials in this field appear hitherto still scanty, taking into account that, within 64 citations from electronic databases, only 3 studies with a total of 154 participants met the inclusion criteria in a Cochrane Review in 2011 [40]. Given the high risk of bias in two of the three studies, there was insufficient evidence to either support, or refute, the use of lycopene for the prevention of prostate cancer. Similarly, there was no robust evidence to identify the impact of lycopene consumption upon the incidence of prostate cancer, prostate symptoms, prostate-specific antigen levels, or adverse events. Further information on possible correlations between carotenoid levels and risk of prostate cancer is reported in other chapters of this book (Omer Kucuk: Lower risk of prostate cancer at higher lycopene levels; JM Schenk: Lower risk of prostate cancer at higher circulating retinol).

Studies on lung cancers gave rise to ambiguous results as well. Low intake of vegetables and fruits was consistently associated for a long time with increased risk of lung cancer [41,42]. The interest was focused on the potential protective role of β-carotene and evidence from observational and epidemiological studies rapidly accumulated, which supported an inverse relationship between cancer incidence and β-carotene intake and serum concentration. This conclusion was criticized by Gallicchio et al. [43] in a systematic review of the literature data, who suggested that the inverse relationship might be the result of carotenoid measurements' function as a marker of a healthier lifestyle with higher fruit and vegetable consumption. The problem was reexamined in a recent meta-analysis [44], which reviewed 248 studies and came to the conclusion that, although some reports demonstrated benefits, there was insufficient evidence overall to support the use of vitamin A or related retinoids for the treatment or prevention of lung cancers. The efficacy and safety of a daily combination of 30 mg of β-carotene and 25,000 IU of retinyl palmitate (vitamin A) was assessed in a multicenter, randomized, double-blind, placebo-controlled primary prevention trial—the Beta-Carotene and Retinol Efficacy Trial (CARET)—involving a total of 18,314 smokers, former smokers, and workers exposed to asbestos [45]. CARET was stopped ahead of schedule in January 1996 because participants, who were randomly assigned to receive the active intervention, were found to have a 28% increase in the incidence of lung cancer, a 17% increase in the incidence of death, and a higher rate of cardiovascular disease mortality compared with participants in the placebo group. These results were in line with those obtained in the "Alpha-Tocopherol, Beta-Carotene Cancer Prevention Study" (ATBC) pertaining to supplementation of 20 mg β-carotene per day to 29,133 male smokers of 50–69 years of age from southwestern Finland [46]. It was found that total mortality after 5–8 years of dietary supplementation was 8% higher (95% confidence interval, 1%–16%) among the participants who received β-carotene than that among those who did not, primarily because there were more deaths from lung cancer and ischemic heart disease.

Epidemiologic studies on vitamin A and carotenoids in relation to breast cancer risk have been inconclusive [47]. There is no evidence on significant influence of carotenoid or retinol intake on the prevention of breast cancer, but fenretinide, a pro-apoptotic and pro-oxidant vitamin A derivative, has shown promise in several trials, and its preventive potential is being assessed in young women at very high risk for breast cancer [48]. The available data, especially from prospective studies, do not support an association between vitamins E and C and risk of breast cancer. By contrast, several studies have shown [48] that a high proportion of women at risk for breast cancer or affected by the disease have deficient vitamin D levels,

**TABLE 12.1**

**Pro and Con Arguments for Positive Influence of Carotenoid on Cancer Prevention**

|  | Methods of Analysis | Pro | Con |
|---|---|---|---|
| Aerodigestive track | Epidemiologic studies | [18] |  |
| Respiratory system | Epidemiologic studies | [9,11,16–18] |  |
|  | Clinical trials |  | [45,46] |
|  | Meta-analyses |  | [44] |
| Breast | Epidemiologic studies |  | [47] |
|  | Clinical trials | [48] |  |
| Cervix | Meta-analyses | [20,21] |  |
| Prostate | Epidemiologic studies | [23,33] | [39] |
|  | Clinical trials |  | [37] |
|  | Meta-analyses | [30,31] | [40] |

that is, 250 H-D < 20 ng/mL or 50 nmol/L. While the association between vitamin D levels and breast cancer risk/prognosis is still controversial, the U-shaped relationship between 250 H-D levels and breast cancer observed in different studies suggests the need to avoid both deficient and too high levels of vitamin D. Further trials using an optimal dose range are needed to assess the preventive and therapeutic effect of vitamin D.

Arguments for and against a significant influence of carotenoid intake on the prevention of cancer diseases are summarized in Table 12.1. The results do not allow unequivocal conclusions at least in part because of the great differences among various types of cancer, such as differences in cellular metabolism, host interaction, and aggressiveness. Other discrepancies could be due to differences in project design and participants included. On the whole, the efficacy of carotenoids in cancer prevention was more easily demonstrated in nutrition examination surveys and epidemiological studies than in randomized placebo-controlled trials. Indeed, epidemiological studies attempting to clarify the relationships between diet and cancer face several obstacles such as determining whether cancer correlates with diet quality or with socioeconomic conditions and/or other lifestyle choices. An unbalanced diet is not necessarily the direct cause of cancer, but only the hallmark of unhealthy lifestyle. Thus, diet could be connected to cancer through a complex network of cause–effect relationships that, if misunderstood, could result in thorough misinterpretation of the data.

## HARMFUL EFFECT OF HIGH DOSES OF CAROTENOIDS

Since CARET and ATBC trials demonstrated [45,46] a remarkably increased incidence of lung cancer in cigarette smokers and asbestos workers who received high doses of β-carotene and/or vitamin A, several efforts have been made [49–57] to correlate this unexpected effect to the pro-oxidant activity of β-carotene cleavage products (CCP) in order to justify the pro-carcinogenic action in the case of preexisting

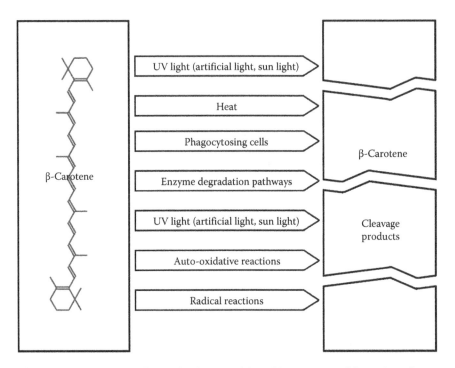

**FIGURE 12.1** Factors leading to the decomposition of β-carotene and formation of potentially toxic cleavage products. β-Carotene is characterized by extensive delocalization of polyene π-electrons enabling to absorb UV and visible light. This electron-rich system makes β-carotene highly susceptible to attacks of electrophilic reagents.

premalignant lesions. Factors that may lead to the decomposition of β-carotene and the formation of potentially toxic cleavage products are summarized in Figure 12.1.

Edge and Truscott [58] discussed the switch from antioxidant to pro-oxidant behavior of carotenoids using results based on pulsed radiation techniques. They suggested a synergistic antioxidant protection by mixtures of carotenoids and vitamins E and C and a switch from antioxidant to pro-oxidant behavior as a function of oxygen concentration. Moreover, they speculated that the key pro-oxidant step involves oxygen being "carried" from peroxyl radical to lipid via carotenoids.

Murata and Kawanishi [59] reported that low concentrations of retinal and retinol cause cellular DNA cleavage and induction of 8-oxo-7,8-dihydro-2-deoxyguanosine formation in HL-60 and HP100 cells. They concluded that superoxide radicals generated by autoxidation of carotenoid derivatives are dismutated to hydrogen peroxide, which is responsible for DNA damage. They also concluded that retinal has a pro-oxidant capability, which might lead to carcinogenesis.

It can be hypothesized that aldehydic and epoxidic cleavage products of carotenoids exert pro-oxidative effects, which are responsible for the damages observed in patients undergoing extreme oxidative stress while taking carotenoid supplements. In a previous paper, we provided evidence that CCP inhibit $Na^+$-$K^+$-ATPase activity [60]. Interestingly, CCP were even stronger inhibitors of $Na^+$-$K^+$-ATPase activity

**TABLE 12.2**
**Potential Biological Effects of β-Carotene Cleavage Products**

| | |
|---|---|
| Protein structure | Reaction with –SH and –NH |
| | Protein cross-linking |
| | Protein-DNA cross-linking |
| Enzyme activity | Inhibition |
| | Activation |
| Membrane effects | Destabilization |
| | Inhibition of transporters |
| Chemotaxis | Inhibition |
| | Activation |
| Cell viability | Cytotoxicity |
| | Genotoxicity |
| | Mutagenesis |
| | Carcinogenesis |

than the endogenous major lipid peroxidation product 4-hydroxynonenal (HNE) [60], which is known to be extremely reactive. An overview of potential biological effects of β-carotene cleavage products is reported in Table 12.2.

In pathophysiological situations, the mitochondria play a special role for multiplicating potential damage. The mitochondria are the main producers of superoxide radicals and $H_2O_2$ within the cell [61]. Impairment of mitochondrial function including changes in calcium homeostasis [62] can cause the increase in the formation of superoxide [63], thus promoting oxidative stress resulting in the oxidation of lipids, proteins, and DNA molecules. Oxidative DNA damage is a hallmark of cancerogenesis. We have found that CCP inhibit state-3 respiration in isolated rat liver mitochondria. Furthermore, we provided evidence that CCPs increase oxidative stress in mitochondria, which is characterized by decreases in mitochondrial GSH and protein-SH and increased formation of malonic dialdehyde (MDA) [64].

Oxidation of β-carotene and lycopene and the consecutive degradation of these molecules have been intensively studied under different conditions mimicking more or less pathophysiologically relevant situations. There are various possibilities to degrade β-carotene and other carotenoids. One important source for the destruction of carotenoids is UV light exposure that has been associated with carotenoid oxidation in eye tissues, skin, or food. For the corresponding in vitro model, Rose Bengal was used to study the photosensitized oxidation of β-carotene [60,65]. Other photosensitizers, such as methylene blue or 12-(pyrene)dodecanoic acid, were reported as well [66]. In other studies, initiators of radical reactions such as 2,2′-azo-bis-isobutyronitril (AIBN) and 2,2′-azo-bis(2,4-dimethylvaleronitril) (AMVN) were used for carotenoid degradation in the presence of oxygen. These substances were used to model the radical trapping properties of carotenoids [60,67–71]. In another study, the effect of ozone and molecular oxygen on carotenoids in an aqueous model system was investigated [72].

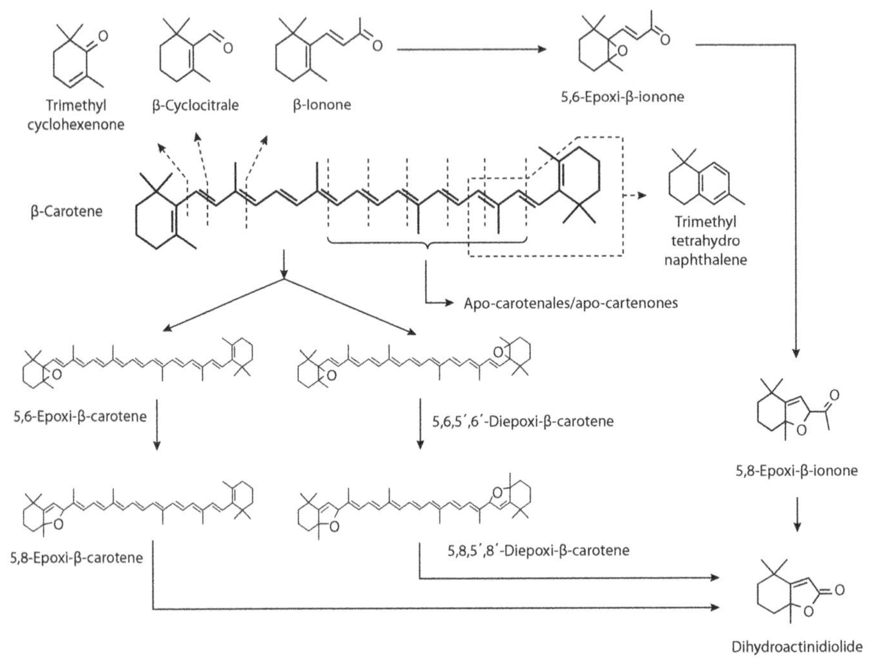

**FIGURE 12.2** Possible pathways of formation and structural formulas of products after HOCl/OCl⁻ mediated degradation of β-carotene. Oxidation and breakdown products of β-carotene include high-molecular-weight and short-chain aldehydes and epoxides that show high reactivity. (Modified according to Sommerburg, O. et al., *Free Radic. Biol. Med.*, 35, 1480, 2003.)

Hypochlorous acid is another pathophysiologically relevant oxidant that is able to degrade β-carotene [60,71]. This compound is used by phagocytosing cells for antibacterial defense, and it can be assumed that hypochlorous acid plays a relevant role in β-carotene degradation mediated by neutrophils in vivo [73]. Figure 12.2 shows a schematic representation of the HOCl/OCl⁻ mediated oxidative degradation of β-carotene.

In a number of studies, successful attempts were made to identify β-carotene cleavage products. It became obvious that during mild oxidative stress, independent of its source, mostly high-molecular-weight products, such as apo-carotenals, are formed. Enzymatic formation of apo-carotenoids was described intensively [74]. However, further oxidation of these high-molecular-weight products to so-called short-chain cleavage products occurs. As shown in a number of studies, different oxidants lead to the formation of a variety β-carotene cleavage products. Those products are, for example, the high-molecular-weight products β-apo-8′-carotenal, β-apo-10′-carotenal, β-apo-12′-carotenal, β-apo-14′-carotenal, and β-apo-15′-carotenal [65,68,71,73–78]. However, also short-chain products, such as β-cyclocitral, β-ionone, ionene, 5,6-epoxi-β-ionone, dihydroactinidiolide, and 4-oxo-ionone, are found constantly after the oxidation of β-carotene with different oxidants [65,68,73–75,79–83]. In contrast, other products are formed only under certain conditions. This is true, for example, for β-ionylidenacetaldehyde, 4-oxo-β-ionylidenacetaldehyde, and 5,8-epoxi-β-ionone, which were found only after reaction of β-carotene with molecular oxygen [74,75].

The most important structural feature of carotenoids is the extensive delocalization of the polyene π-electrons enabling carotenoids to absorb UV and visible light. This electron-rich system makes carotenoids highly susceptible to attacks of electrophilic reagents, which are produced by activated neutrophils and serve in host defense against invading micro-organisms. Indeed, β-carotene is degraded to colorless compound(s) in culture medium of human neutrophils [73] or rat peritoneal macrophages [84], which have been activated in the presence of phorbol myristate acetate (PMA), but not in the medium of nonactivated cells. The degradation rate is about 0.7 nmol/(h × $10^5$ cells) after addition of 1 μM β-carotene to human neutrophil suspension in phosphate-buffered saline. In our opinion, the carotenoid degradation by stimulated neutrophils is the dominant source of CCPs at high-dose carotenoid supplementation in patients suffering from several type of inflammation—what we have in the lung of cigarette smokers and asbestos workers and in patients suffering from bronchitis, pneumonia, and other inflammatory lung disease.

Carotenoids may scavenge a variety of toxic oxygen metabolites released by macrophages during respiratory burst. Quenching activities of carotenoids against singlet oxygen from a hydrogen peroxide/hypochlorite system have been widely studied [85–88]. The ability of carotenoids to quench singlet oxygen decreases in the order of lycopene, γ-carotene, astaxanthin, canthaxanthin, α-carotene, β-carotene, bixin, zeaxanthin, and lutein. However, it has been observed [89] that β-carotene readily reacts with hypochlorite with a rate constant of $2.3 \times 10^4$ mol/(L·s), and this suggests that carotenoids may protect living cells from hypochlorite, rather than from singlet oxygen. Several papers report that carotenoids react with hydrogen peroxide, nitric oxide (NO), and peroxynitrite in cell-free systems [84,90]. On the whole, the available data indicate that carotenoids may scavenge a variety of toxic oxygen metabolites released by macrophages during respiratory burst. It remains to elucidate the contribution of the various pathways in the degradation of carotenoids under different physiopathological conditions.

It was already stressed that in CARET and ATBC studies, harmful effects of carotenoids have been described [45,46], for example, a higher incidence of lung cancer in individuals exposed to extraordinary oxidative stress (heavy smoking or influence of asbestos). They may be caused by oxidatively formed carotene cleavage products including highly reactive aldehydes and epoxides, which are generated during oxidative attacks in the course of antioxidative action. Carotenoids are split by oxidative enzymes, such as dioxygenases, epoxidases, hydroxylases, dehydrogenases, and aldehyde oxidases [91]. Retinal is the primary cleavage product of β-carotene, which is formed by 15,15′-dioxygenase [92,93]. Furthermore, CCP are formed nonenzymatically by attack of different free radical species under conditions of enhanced oxidative stress [74,94–96]. Handelman et al. [71] performed CCP generation in the presence of hypochlorite in order to mimic the in vivo formation in inflammatory regions following activation of neutrophils. Many of these products are aldehydes, carbonyls, and epoxides [64,65,97–99]. Our research group has intensively studied chemical reactions and biological effects of aldehydic lipid peroxidation products such as malondialdehyde (MDA) and HNE. If many of the carotenoid cleavage products (CCPs) are aldehydes, one should assume that a major part of their chemical reactions and biological effects will be similar to the reactions and effects of other aldehydic lipid oxidation products such as MDA and HNE due to the aldehydic function. CCPs that are aldehydes, carbonyls, and epoxides, of course, are able to exert the typical chemical reactions of aldehydes, carbonyls, and epoxides with biomolecules such as proteins, peptides, nucleic acids, and lipids. Those reactions are the chemical basis of the potential toxicity of CCP.

It has been demonstrated [64] that exogenously added CCP (caroteneCP, retinalCP, iononeCP, retinal, and β-ionone) disturbed important mitochondrial functions. They inhibited ADP-stimulated respiration in the micromolar concentration range (between 0.5 and 20 µM) in isolated mitochondria or various tissues.

**TABLE 12.3**

**Biological Effects of CCPs in Mitochondria**

| Toxic Effect by CCP | Tissue/Organ | References |
|---|---|---|
| Decrease in ADP-stimulated respiration | Rat liver, brain, lung | [56,64,197] |
| Decrease in protein-SH, GSH | Rat liver | [56,64,197] |
| MDA formation | Rat liver | [64] |
| Impairment of ANT activity | Rat liver, heart | [64,101] |
| Association between mitochondrial function and all-*trans* retinoic acid-induced apoptosis | Acute myeloblastic leukemia cells | [198] |
| Role for mitochondrial respiration in apoptosis induction by synthetic retinoid CD437 | Human cutaneous squamous cell carcinoma COLO 16 | [199] |
| Decreased mitochondrial functions by CCP | Human K562 erythroleukemic cells, 28 SV4 retinal pigment epithelium cells | [200] |
| Oxidative stress in submitochondrial particles after vitamin A supplementation | Rat cerebral cortex and cerebellum | [201] |
| Distortion of redox and bioenergetic state homeostasis | Rat liver | [202] |
| Caspase activation and mitochondria-associated cell death in light-induced retinal degeneration | Mice retina | [203] |
| Postnatal oxidative damage of respiratory complex I by embryonal exposure to all-*trans* RA | Rat cerebellum | [204] |

In previous experiments, we found that under almost identical incubation conditions, lung mitochondria were more susceptible than brain mitochondria, and brain mitochondria were more susceptible than liver mitochondria. During incubation with CCP, mitochondrial protein SH and glutathione levels were diminished, whereas oxidized glutathione and thiobarbituric acid reactive substances (TBA-RS) were accumulated [64]. As one of the reliable mechanisms for the perturbation of mitochondrial functions by CCP, the adenine nucleotide translocator (ANT) was discussed [64]. At first, ANT is rich in SH groups in the active center, and oxidation of SH impairs ANT activity [100]. Second, the ATP-synthetase could be mostly excluded by measurements of the dissipation of membrane potential [64]. Chen et al. showed that the transporting activity of ANT was also inhibited by two unsaturated aldehydes, HNE and 4-hydroxyhexenal [100]. Concerning the underlying mechanism of this inhibition, they concluded from reconstitution experiments that the loss of sulfhydryl groups and/or the alteration of the physiochemical status of the lipid environment, in which the ANT is embedded, may be responsible for the impairment of membrane functions by aldehydic products. Notario et al. demonstrated that carotenoids as retinoic acids inhibit ANT activity in mitochondria of heart and liver [101]. Table 12.3 summarizes biological effects of CCP in mitochondria.

## BALANCE BETWEEN ANTIOXIDATIVE AND PRO-OXIDATIVE MECHANISMS WITHIN THE GAME BETWEEN CANCER PREVENTION AND CANCER INDUCTION

Depending on their chemical nature, retinoids can either ameliorate or exacerbate stress-related cellular damages [102]. Under heavy oxidative stress, supplemented carotenoids are oxidatively degraded leading to the formation of high amounts of CCP [103]. These compounds (shortened carbonyls, aldehydes, epoxides, and others) are very reactive and may exert harmful effects. A scheme on antioxidative and pro-oxidative activities of carotenoids and derived products is reported in Figure 12.3.

Under conditions of moderate oxidative stress (see right), the antioxidant effects of β-carotene and its analogs predominate. Under heavy oxidative stress (see left), the bulk of supplemented carotenoids is degraded in cleavage products exerting pro-oxidative actions that overcome the antioxidant activity of precursor carotenoid. This results in the impairment of ANT, accompanied by an increase in oxidative stress, which is indicated by reduction of protein sulfhydryl content and glutathione levels and accumulation of MDA and TBA-RS. The increased levels of reactive oxygen species (ROS), in particular superoxide anion radicals and hydrogen peroxide, inhibit energy metabolism and damage the surrounding macromolecules, such as proteins, lipids, and nucleic acids, thereby increasing the risk of cancer. This could be a conceivable mechanism for the increased incidence of lung cancer in smokers with high-dosage carotenoid supplementation as observed in the ATBC and CARET studies. The hypothesis underlines the opinion that in

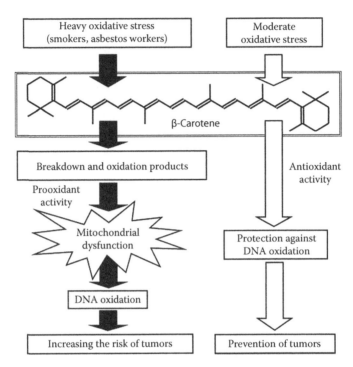

**FIGURE 12.3** Hypothetic scheme of the mechanism of antioxidant and pro-oxidant effects of β-carotene. The switch from antioxidant to pro-oxidant behavior of β-carotene is dependent on several variables, including oxygen partial tension, carotenoid concentration, and macrophage activation. Under heavy oxidative stress, the bulk of supplemented carotenoids is degraded in breakdown and oxidation products exerting pro-oxidative actions that overcome the antioxidant activity of precursor carotenoid.

all other conditions except heavy oxidative stress, such as in smokers or asbestos workers, the antioxidative reactions of carotenoids will lead to beneficial biological and medical effects.

Indeed, at low oxygen concentrations, β-carotene is a powerful scavenger of singlet oxygen and acts as chain-breaking antioxidant toward lipid peroxyl radicals. β-Carotene scavenge other free radicals as nitrogen dioxide, thiyl, and sulfonyl radicals, as well. Moreover, it has been shown that carotenoids efficiently inhibit lipid peroxidation, and the interaction of β-carotene with lipid peroxyl radical results in the formation of a radical adduct [6].

When β-carotene and α-tocopherol are present together in homogeneous solution, α-tocopherol is consumed predominantly and β-carotene is spared. On the contrary, β-carotene is consumed faster than α-tocopherol, when the radicals are generated within the lipophilic compartment of the membranes (scavenging lipophilic radicals within the membrane by β-carotene). β-Carotene was found to be less potent as antioxidant than α-tocopherol. The stable β-carotene radical reacts with oxygen to give β-carotene-peroxyl radical, which is not stable but able to attack lipid to continue chain oxidation [104].

It has been reported [102] that retinal should exert a protective action, since retinal deprivation enhances oxidative damage, as indicated by rapid loss of mitochondrial membrane potential. Supplementation with a physiological concentration of retinal reversed this effect. Anhydroretinol, a known antagonist, working by displacing retinal from the common binding sites on serine/threonine kinases, also caused mitochondrial membrane depolarization. This effect is $Ca^{2+}$-dependent and cyclosporin-sensitive, suggesting an upstream signaling mechanism rather than direct membrane effect. Retinoids are known to act as survival factors of cardiac cells, because retinal reduced the incidence of heart disease and protected the heart from inflammation, oxidative damage, and degeneration. Conversely, chronic retinol deficiency in rats decreased the respiratory activity of their heart mitochondria [105]. Retinol is able to protect, at least in part, mitochondria from oxidative damage. It is known that retinoids modulate—independently of nuclear receptors—cell proliferation, differentiation, and apoptosis [106], which is also dependent on ROS and reactive nitrogen species. Retinol and appropriate derivatives can be expected to find clinical application for stabilization of cellular redox state after ischemia/reperfusion and other conditions of oxidative stress.

## LUNG AS A CRITICAL ORGAN

Supplemented β-carotene is distributed via lipoproteins throughout the human body into different tissues. As known from previous studies, β-carotene accumulates in lung tissue in comparable amounts compared to other organs [107]. In CARET and ATBC trials, high doses of β-carotene were given as supplements. This caused an increase in carotenoid blood level to 2.1 and 3.0 µg/mL, respectively, as compared to the average blood level for the U.S. population (0.05–0.5 µg/mL) [108]. It might be assumed that a similar effect of increase in the β-carotene concentration is also true for the lung tissue of other supplemented subjects.

In the respiratory system, oxygen enters the lung with a pressure of about 150 mmHg as present in normal air. In the alveoli, an extremely thin barrier between air and capillaries allows oxygen to move from the alveoli into the blood and allows carbon dioxide to move from the blood in the capillaries into the alveoli. In this situation, β-carotene present in lung tissue is exposed to a relatively high partial pressure of oxygen. Burton and Ingold demonstrated already in 1984 that the antioxidant behavior of β-carotene was dependent in part upon the partial pressure of oxygen [6]. As they could show, β-carotene exhibits good radical-trapping antioxidant activity only at partial pressures of oxygen significantly less than 150 mmHg. At higher oxygen pressure and in in vitro situations, the carotenoid loses its antioxidant activity and shows an autocatalytic pro-oxidant effect, particularly at relatively high concentrations. The consequences of this effect in vivo are still under discussion [49]. In this discussion, it has to be considered that the lung is the inner organ with the highest partial oxygen pressure (about 100 mmHg in the alveoli) compared to other tissues in the human body. In the case of high-dosage supplementation, tissue β-carotene concentration might be close to the in vitro conditions showing the pro-oxidative behavior of β-carotene.

Another specific aspect of the lung is its powerful defense system. An average subject who is moderately active breathes about 20,000 L of air every 24 h. The immune system protects the lung from foreign and possibly pathogenic microorganisms, such as bacteria, viruses, fungi, and yeasts, which infiltrate the lung with every breath. However, before coming into contact with the alveolar lining layer, most invading microorganisms are removed by mucociliary clearance. The organisms that reach the alveolar compartment via the pulmonary airway first are brought into contact with the pulmonary epithelium and the resident macrophages. These cells are able to release cytokines and chemokines to recruit white blood cells in the circulation, such as neutrophils. In this way, alveolar macrophages initiate and coordinate the host response to infection, including adaptive immunity.

Phagocytic cells are able to liberate a number of oxidants to kill microorganism. One of these oxidants is hypochlorite that is able to damage a high number of different biologically important macromolecules [109–111]. As revealed in vitro, the reaction rate of hypochlorite with β-carotene is slow as compared to other reducing agents, for example, reduced glutathione [89]. However, myeloperoxidase, the source of hypochlorite, is a highly cationic protein that is attached nearly to the membranes after its release from stimulated neutrophils [112]. Thus, an interaction of hypochlorite with more lipophilic constituents such as carotenoids seems to be realistic under conditions where neutrophils accumulate extensively. Hypochlorite is able to destroy carotenoids in low-density lipoproteins [113]. Calculations revealed that $2-4 \times 10^6$ leukocytes/mL produce 100–140 μM hypochlorite in 1 h [114] leading to realistic β-carotene/hypochlorite ratios between 1:100 and 1:10 [115]. If so, inflammatory cells present in the lung should be able to degrade β-carotene. In vitro experiments with primary cultures of PML clearly indicated that the activated cells are able to degrade substantial amounts of carotenoids.

Under normal conditions, carotenoid degradation by activated macrophages might not be of relevance. However, several factors may alter the physiological balance regarding recruitment and activity of inflammatory cells in lung tissue. One of these factors is the smoking behavior of people. Amin et al. investigated bronchial biopsies of asymptomatic smokers (smokers without respiratory problems) and never-smokers [116]. In their study, smokers show an increased thickness of the laminin and tenascin layers in the epithelial tissue as a sign for structural changes and for disturbance of the epithelial integrity. Furthermore, the number of inflammatory cells was significantly increased in smokers compared to never-smokers. Neutrophils, detected by immunostaining with monoclonal antibodies to IL-8, were found in the biopsy specimen of smokers in the concentration of about 12 cells/mm² (median of n=29). By contrast, never-smokers showed only about 2 IL-8 positive cells/mm² (median of n=16) [116]. These results are in agreement with the work of Eidelman et al., who described a fivefold increased number of neutrophils in lung tissue of smokers compared to nonsmokers [117]. Other authors [118,119] reported significantly increased numbers of inflammatory cells in lung tissue of smokers as well.

Smokers are not the only group showing increased numbers of inflammatory cells in the lung. In atopic and non-atopic asthmatics, eosinophils and mast cells—which are able to liberate oxidants as well—were shown to be increased compared to controls [120]. However, the number of neutrophils was found increased only in non-atopic asthmatics. A more detailed overview about oxidative events in the patho-genesis of asthma and, in particular, about the role of inflammatory cells are given in reviews published by Andreadis et al. [121] and Maddox and Schwartz [122], who describe how increased amounts of eosinophils and neutrophils have the potential to harm host tissue and contribute to inflammatory injury. Primary Sjogren's syndrome is another disease in which increased numbers of neutrophils was observed in lung tissue [123].

Patients with the mentioned disorders do not have necessarily the need for a β-carotene supplementation. However, most patients with CF show pancreatic insuf-ficiency leading to a malabsorption of carotenoids. Since the patients may benefit from an additional β-carotene intake, it is interesting to have a look on the conditions in the lungs in these subjects. Hubeau et al. published in 2001 a comprehensive quan-titative analysis of inflammatory cells of the airway mucosa of CF patients using immunohistochemistry. They found that in CF patients, elastase$^+$ neutrophils were particularly numerous at the level of the segmental bronchi (CF: $199 \pm 54$ cells/mm$^2$ vs. non-CF: $27 \pm 7$ cells/mm$^2$) [124]. This means that the number of neutrophils in lung tissue of CF patients seems to be even seven times higher than that in non-CF people. This shows that risk of oxidative stress mediated by neutrophils is in CF patients at the same level or even higher compared to smokers.

## ROLE OF NEUTROPHILS IN INFLAMMATORY DISEASES

In inflammatory response, neutrophil activation is characterized by a burst of oxygen metabolism and the release of soluble antimicrobials (including granule proteins) and small cell-signaling molecules, which in turn amplify the inflammatory reac-tion. Reduction of molecular oxygen through a series of electron transfers gener-ates highly reactive oxygen products that not only degrade retinoids to pro-oxidative breakdown products, as discussed earlier, but also play an important role in the killing of microbial pathogens [125,126] and contribute per se to tissue damage. Pulmonary diseases that are associated with greatest risk for lung cancer are characterized by abundant and deregulated inflammation [127–129]. Pulmonary disorders such as chronic obstructive pulmonary disease, which can be induced by tobacco smoking, are characterized by profound abnormalities in inflammatory pathway [130]. For example, among the cytokines, growth factors, and mediators released in these lung diseases and in the developing tumor microenvironment, interleukin (IL)-1β, prosta-glandin (PG)E2, and transforming growth factor-β have been found to have deleteri-ous properties that simultaneously pave the way for both epithelial–mesenchymal transition and destruction of specific host cell-mediated immuneresponses against tumor antigens [131,132].

Oxidant production begins with a cytoplasmic membrane-associated NADPH oxidase, which reduces molecular oxygen to superoxide ($O_2^-$). As shown in Figure 12.4, most of this $O_2^-$ undergoes to dismutation reaction (either spontaneously or more rapidly in the presence of superoxide dismutase) to form hydrogen peroxide ($H_2O_2$), the second product of the respiratory burst. In the presence of chloride ($Cl^-$), $H_2O_2$ is used in a large extent (25%–40%) to form hypochlorous acid (HOCl) and its anion, hypochlorite ($OCl^-$), through the reaction catalyzed by the heme protein

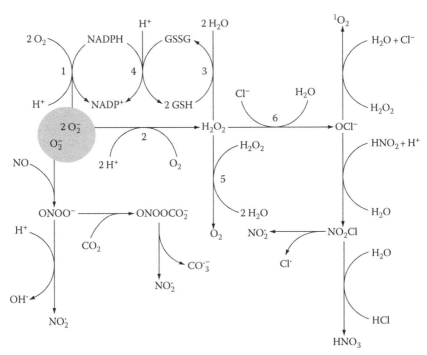

**FIGURE 12.4** Simplified pathways leading to oxygen radical production by activated neutrophils. Superoxide anion ($O_2^-$) is formed through the NADPH-oxidase-catalyzed reaction. This compound undergoes dismutation to hydrogen peroxide ($H_2O_2$) that is used in a large extent to form hypochlorite ($OCl^-$). Superoxide anion reacts also with NO to form peroxynitrite ($ONOO^-$) and nitrosoperoxocarbonate ($ONOOCO_2^-$). These compounds lead to the formation of strongly reactive radicals. 1, NADPH oxidase; 2, glutathione reductase; 3, glutathione peroxidase; 4, superoxide dismutase; 5, myeloperoxidase; 6, catalase.

myeloperoxidase, which constitutes about 5% of the cellular proteins in activated neutrophils [133]. Both superoxide and hydrogen peroxide are relatively nontoxic to bacteria. By contrast, HOCl is capable of oxidizing many important biomolecules, such as plasma membrane components, proteins, and ascorbate. Moreover, HOCl depletes intracellular ATP and glutathione, and causes cell death [134]. Bactericidal action is mainly due to respiration injury by singlet oxygen ($^1O_2$) that is formed non-enzymatically from HOCl and $H_2O_2$ [135].

Treatment of proteins with excess of HOCl leads to preferential depletion of cysteine and methionine residues, yielding stable sulfinic/sulfonic acids and/or cross-linked sulfinamides and methionine sulfoxide, respectively [136]. Other major products of HOCl oxidation of proteins are chloramines, which are formed on lysine and histidine side chains and other free amino groups [137]. Chloramines are moderately stable species, but they are also oxidants (although less potent than HOCl itself) and are capable of oxidizing cysteine and methionine residues, as well as some antioxidants. Depletion of tryptophan residues has been attributed to the formation of 2-oxoindolone derivatives [138]. Oxidation of the fatty acid side chains of lipids results in the formation of chlorohydrins, probably via addition of HOCl across the double bonds [139]. It is possible that chloramines formed on the head groups of phospholipids play a role in this process. Peroxidation may occur mainly via reaction of HOCl with low levels of peroxides that are already present within the lipids [140]. In vitro studies of LDL oxidation suggest that apo B-100 is the major target of HOCl oxidation and that lipid peroxidation occurs via a further formation of radicals through the decomposition of chloramines on the protein lysine side chains [141]. As far as the nucleosomes are concerned, it has been suggested that histones are the major target for HOCl [142]. The resulting protein chloramines and the radicals derived from them might act as contributing agents in HOCl-mediated DNA oxidation. These radicals can react with plasmid DNA to cause strand breaks as well as with pyrimidine nucleosides to give nucleobase radicals.

An additional reaction of HOCl is with nitrite ($NO_2^-$) to form nitryl chloride ($NO_2Cl$), which then hydrolyzes to nitrate ($NO_3^-$) [143] or decomposes to give strong oxidants such as nitrogen dioxide ($NO_2^{\bullet}$) and $Cl^{\bullet}$ radicals. Levels of $NO_2^-$ in plasma from healthy human volunteers range between 0.5 and 21.0 $\mu M$ [144] and reach millimolar concentrations in patients with rheumatoid arthritis [145], systemic sclerosis [146], Sjogren's syndrome [147], and systemic lupus erythematosus [148]. The reaction of HOCl with $NO_2^-$ is favored with decreasing pH, such as that may occur during chronic inflammation [149]. $NO_2Cl$ is capable of nitrating, chlorinating, and dimerizing phenolic compounds such as tyrosine [150], and exposure of isolated human low-density lipoprotein to $NO_2Cl$ results in the depletion of $\beta$-carotene and $\alpha$-tocopherol as well as in protein modification [149]. Moreover, $NO_2Cl$ causes cytosine chlorination and DNA oxidation in intact cells and in cell-free systems [151]. Nevertheless, it has been reported that $NO_2^-$ substantially decreases HOCl-dependent cellular toxicity even if added at low (micromolar) concentrations. It has been suggested [152] that the most probable mechanism for the inhibition of HOCl toxicity is that $NO_2^-$ removes HOCl via a two-electron oxidation to $NO_3^-$ in a 1:1 molar ratio. This may represent an endogenous defense mechanism to protect ascorbate, $\alpha_1$-antiproteinase, and DNA against HOCl in vivo.

The production of NO by oxidizing L-arginine to L-citrulline is another oxygen-dependent pathway that may contribute to human host defense [153,154]. There are three different NO synthetases, each encoded by a different gene. Two are constitutively expressed in several tissues, including endothelium, brain, and neutrophils. Expression of a third form, known as inducible NO synthetase, can be induced by inflammatory stimuli in a variety of cells. Human neutrophils in resting state generate modest amount of NO [155], while a dramatic increase in inducible NO synthetase activity is observed in these cells during bacterial infection [156].

The direct toxicity of NO is poor, but is greatly enhanced by reacting with $O_2^-$ to give peroxynitrite ($ONOO^-$) [157]. Upon protonation to peroxynitrous acid ($ONOOH$, $pK_a = 6.8$), this compound rapidly decomposes to form nitrate ($NO_3^-$) as well as free hydroxyl radical ($OH^{\bullet}$) and nitrogen dioxide radical ($NO_2^{\bullet}$). The yield of $OH^{\bullet}$ and $NO_2^{\bullet}$ formation approximates 25%–35% in the absence of competitive reactions [158]. In physiological settings where bicarbonate abounds, the formation of $OH^{\bullet}$ and $NO_2^{\bullet}$ is limited by the reaction of peroxynitrite with carbon dioxide ($CO_2$) to form nitrosoperoxocarbonate anions ($ONOOCO_2^-$), one-third of them decomposes to carbonate radical ($CO_3^{\cdot-}$) and nitrogen dioxide ($NO_2^\cdot$) [159]. Because of its fast reaction with $CO_2$, peroxynitrite reacts in vivo with a limited number of targets, which include heme-containing proteins such as hemoglobin, peroxidases such as myeloperoxidase, seleno-proteins such as glutathione peroxidase, and proteins containing zinc-thiolate centers such as DNA-binding transcription factors [160]. Reactive intermediates responsible for reactions with a variety of substrates may be the free radicals, $NO_2^\cdot$ and $CO_3^{\cdot-}$. Biologically important reactions of these free radicals include the nitration of tyrosine residues, which may interfere with tyrosine phosphorylation, a generalized means of controlling enzymatic activity [161]. Irrespective of the mechanisms involved, $NO_2^-$ is a major oxidation product derived from NO, and increased $NO_2^-$ levels can often be detected in situations where NO production is elevated [162].

NO may also undergo nonenzymatic autoxidation in the presence of $O_2$ to form $NO_2^-$, but at physiological concentrations of NO and $O_2$, this reaction may be too slow to be of major importance in vivo [157]. Anyhow, $NO_2^-$ does not accumulate in vivo, but is rapidly oxidized to nitrate ($NO_3^-$) [163]. In the vascular system, $NO_2^-$ can be oxidized by oxyhemoglobin to form methemoglobin and $NO_3^-$ [164]. Catalase has been demonstrated to contribute to $NO_2^-$ oxidation. Furthermore, $NO_2^-$ can also be oxidized by myeloperoxidase in the presence of $H_2O_2$ [162]. It has been suggested that reactive nitrogen intermediates are produced during this process, but the physiological relevance of these reactions is not yet known.

Carotenoid breakdown products may affect neutrophil response in different ways that depend on the concentration that is reached by these products in the medium (Figure 12.5). Experiments with purified suspensions of human neutrophils [165] show that stimulation of superoxide production can be induced by nanomolar and micromolar concentrations of carotenoid derivatives with aliphatic chains of

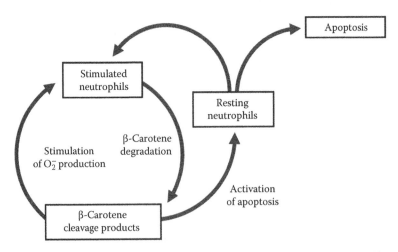

**FIGURE 12.5** Mutual effects exerted by β-carotene cleavage products and neutrophils. Activated neutrophils degrade β-carotene generating harmful CCPs. These compounds in turn stimulate superoxide production by neutrophils, while at higher concentrations, they inhibit neutrophil function by promoting cell apoptosis.

different length (retinal, β-ionone, mixture of CCPs), but not by carotenoids lacking the carbonyl moiety. It is noteworthy that stimulatory effect of carotenoids can be observed only with cells activated by PMA, while a slight inhibition of superoxide production is noticed if the chemotactic tripeptide *N*-formyl-Met-Leu-Phe (f-MLP) is employed to trigger cell response. At slightly higher concentrations, carotenoids inhibit superoxide production in the presence of both PMA and f-MLP. Under these conditions, carotenoid oxidation products and derivatives lacking the carbonyl moiety exert similar inhibitory effects.

It has been reported [54,165] that addition of 5–20 μM retinal or mixture of β-carotene breakdown products to unstimulated neutrophil suspensions causes a remarkable increase in intracellular caspase-3 activity, which is regarded as a biochemical marker of cell apoptosis [166]. Smaller effect on the intracellular caspase-3 level is exerted by β-apo-8′-carotenal and β-ionone. Electrophoretic analysis of neutrophil chromatin extracts shows that activation of caspase-3 activity is accompanied by an increase in DNA fragmentation, while a significant decrease in DNA fragmentation can be obtained by adding z-VAD, a broad-spectrum caspase inhibitor. These results are in good agreement with previous reports [167], indicating that carotenoids induce apoptosis in T-lymphoblast cell line in a time- and concentration-dependent manner with the lowest effective concentration of about 3 μM. T-Lymphoblast apoptosis is clearly evident in the presence of 20 μM carotenoids and after 24 h incubation, β-carotene and its cleavage products are more cytotoxic than lycopene. In contrast, β-carotene breakdown products did not show pro-apoptotic activity on hepatocytes in primary culture, suggesting that there might be a cell- and tissue-dependent response [50].

It is noteworthy [54] that the pro-apoptotic activity of carotenoid breakdown products on neutrophils is affected neither by uric acid, an efficient antioxidant that behaves as peroxynitrite scavenger [168,169], nor by ascorbic acid that has been reported to exert synergistic protective effect from oxidative stress in model systems [170]. Moreover, a pronounced increase in neutrophil apoptosis (instead of an increase in cell viability, as expected) is observed for combinations of α-tocopherol and carotenoid breakdown products. Up-regulating effect of α-tocopherol was not observed in the presence of retinol that markedly stimulates apoptosis by itself, whereas increase in caspase-3 activity was induced by concomitant addition of α-tocopherol and β-ionone, a cyclohexenyl breakdown product of β-carotene with shorter aliphatic chain. The effect of CCPs on neutrophil viability was partially influenced by the addition of methanol. This suggests that solvents, which may influence the repartition of lipophilic vitamins between different compartments in the cells, should be considered to explain, at least in part, the effects exerted by carotenoid derivatives in combination with other compounds.

The mechanisms that cause cell apoptosis in the presence of β-carotene breakdown products and α-tocopherol are still unclear. It has been reported [167] that α-tocopherol did not protect T-lymphoblast cell line from apoptosis induced by carotenoids. Moreover, the antioxidant BO-653 led to a slight but statistically significant increase in cell apoptosis when it was used alone and to a dramatic increase in apoptosis when it was used in combination with β-carotene. It was suggested that a second pathway by which carotenoids might induce apoptosis is via the formation of retinoic acid or retinoic acid analogues [171,172]. Other mechanisms, which were taken into account to explain cell apoptosis, include direct DNA damage [171] and down-regulation of cyclooxigenase-2 activity [173] and epidermal growth-factor receptors [174].

Activation of apoptosis may contribute to reduce neutrophil responsiveness that has been observed incubating the cells in the presence of relatively high concentrations of carotenoid breakdown products. It is quite hard to predict the effect of abnormal activation of neutrophil apoptosis on lung cancerogenesis. Neutrophil apoptosis results in the loss of expression of adhesion molecules and responsiveness to external stimuli, so that the cells become functionally isolated from the environment. Since specific changes in plasma membrane allow neutrophil recognition and ingestion by macrophages, apoptosis ends up in neutrophil clearance, which prevents the release of cytotoxic agents that could both damage DNA and promote abnormal cell proliferation. On the other hand, premature activation of neutrophil apoptosis not only may greatly increase the potential risk for host tissue injury by bacteria but also may interfere with the clearance of arrested neoplastic cells that determine the metastatic behavior of cancer [175,176].

## CLINICAL INDICATIONS FOR CAROTENOID SUPPLEMENTATION

Carotenoids and vitamin A are indicated in physiological and pathophysiological conditions that clearly require supplementation of these micronutrients, that is, vitamin deficiency, newborn feeding, and chronic diseases. Severe vitamin A deficiency requires an intake of 600–1500 µg retinal or β-carotene daily. At birth, the liver has low vitamin A content that rapidly increases, because colostrum, breast milk, and formula preparations furnish large amounts of the vitamin. States of deficiency are rare in healthy children with a balanced diet. However, vitamin A deficiency is still a problem in underdeveloped countries, especially in those in which rice is the staple food. A new approach to prevent the problem in these countries is the distribution of so-called golden rice. The rice is genetically manipulated so that β-carotene is produced in the grains [177–179]. However, since absorption of β-carotene depends also on the amount of fat in the diet, this factor should be considered in the children's diet. Whether or not the delivery of β-carotene via rice may establish sufficient plasma levels of carotenoids and prevent the problem of vitamin A deficiency in these countries will be seen in the future.

Lipid-soluble vitamins were found markedly lower in cord blood of neonates in comparison with the serum levels of their mothers [180–184]. Since carotenoids in plasma are transported by lipoproteins, this might be due to the lower amount of body lipids in neonates. Carotenoid concentrations in cord blood plasma were reported to be approximately 9%–25% of plasma levels of the mothers [182–184]. For newborns, breast milk or formula preparations are the only way to obtain carotenoids. Mothers represent a broad variety of carotenoids in their milk [185]. Like other nutrients, carotenoids are higher concentrated in colostrum (by factor 3.5–4.5) than in mature mother's milk [186]. Furthermore, concentration of β-carotene in the milk of lactating women might be influenced by the daily diet [187,188]. Ostrea et al. reported for β-carotene an increase in plasma levels within 5 days in breast-fed infants but not in infants fed formula preparation that did not contain supplemented β-carotene [182]. In a newer study, a total of eight formula preparations available on the European market were investigated for their content of different carotenoids [186]. The results of this study showed that still four out of eight formula preparations did not contain β-carotene. Moreover, none of these formula preparations met the profile of the main carotenoids found in mother's milk. It was reported [186] that plasma concentration of β-carotene at birth was 24 (19–31) μg/L (median and interquartile ranges). After 3 weeks, β-carotene was higher in breastfed infants, 32 (22–63) μg/L, P<0.05. However, this was not the case for infants fed mother's milk and formula preparations, 20 (14–27) μg/L, and in the group of formula fed infants, 14 (0–32) μg/L. In the latter, β-carotene was found even significantly lower than that in infants after birth (P<0.05) and breastfed babies (P<0.01). Since a fast replenishment of plasma carotenoids as seen during feeding with mother's milk is desired, supplementation of formula preparations with β-carotene and other carotenoids is clearly necessary.

Diseases such as chronic intestinal disorders, celiac disease, hepatic and pancreatic disease, iron-deficiency anemia, or chronic infectious diseases may also lead to disturbances of absorption and consumption of β-carotene and consequently to a state of deficiency of carotenoids and vitamin A. In these cases, absorption of lipids has to be improved as well as β-carotene content in the diet to reach adequate β-carotene levels in the plasma. CF is an example for a disease with true carotenoid deficiency. CF is an inherited multisystem disorder affecting most critically lungs, and also pancreas, liver, and intestine. Exocrine pancreas insufficiency leads to a malabsorption of lipids and lipid-soluble vitamins. Supplementation of vitamin A, D, E, and K is already widely accepted, but this is not the case for carotenoids. Therefore, most CF patients suffer a distinctive carotenoid deficiency [189,190], and in some cases, carotenoids are not detectable at all in plasma of CF patients. To correct carotenoid plasma levels in these patients, pharmacological doses of β-carotene (approximately 0.5–1.0 mg/kg body weight) are required [191,192]. When β-carotene was supplemented, CF patients showed a markedly increase in the carotenoid in the plasma and a decrease in plasma levels of lipid peroxidation parameters such as malondialdehyde and TNF-α [191,193,194]. Interestingly, patients under β-carotene supplementation seem to have significant clinical benefit. It has been reported that these CF patients show a decrease in pulmonary exacerbations and an improvement of lung function parameters [194,195] suggesting that β-carotene deficiency in CF is clinically relevant and the correction of carotenoid levels should be required as regular treatment of this disease.

## WHAT WE CAN LEARN FROM STUDIES ON THE RELATIONSHIP BETWEEN DIET AND CANCER

All the epidemiological studies on diet and cancer prevention—measuring either the intake level or the serum concentration of carotenoids—argue for a "sufficient" rather than "optimal" amount of these micronutrients in human nutrition. That is to say, if the intake of carotenoids is too low, the state of health and well-being is expected to decrease. On the other hand, a lifestyle in which there is a sufficient or surplus consumption of carotenoids will allow you to benefit from all of the beneficial effects of these compounds. Most of the observational data showing an inverse relationship between cancer risk and serum level of carotenoids or retinol were interpreted in that light [44]. Thus, considerable effort was made to restructure nutrition guidelines and promote the intake of vegetables and fruits. Daily ingestion of vitamin supplements was often recommended because they could assure the intake of a sufficient amount of carotenoids in case of unexpected oxidative stresses. However, the schematic dietary rule for carotenoid intake reported earlier proves inadequate, since trials involving high-dose supplementation of β-carotene have demonstrated that a surplus of carotenoids might be harmful as well, at least in heavy smokers and asbestos-exposed workers. In vitro experiments indicate that either carotenoid depletion or excessive supplementation of carotenoids can increase the oxidative stress and lead to mitochondrial dysfunction [105,196]. It is well known from several in vitro studies that, at nearly physiological conditions, carotenoid may scavenge a variety of toxic oxygen metabolites and exert a protective action from oxidative damage of cell structures. Unfortunately, the switch from antioxidant to pro-oxidant behavior of carotenoids is dependent on several variables, including carotenoid concentration, oxygen partial tension, and macrophage activation, that is, factors that are particularly altered in conditions of severe oxidative stress and tissue inflammation. In conclusion, carotenoids are a powerful medical resource, but they could fail in some clinical situations where their help would be expected. Pollution and smoking-attributable morbidity cannot be reduced by taking high doses of β-carotene or other carotenoids, and trusting in their antioxidant properties, neither vitamin assumption can be used as a justification for cigarette smoking and other unhealthy behaviors. The primary medical aim in these cases is to prevent smoking and guarantee safe environments, instead of prescribing antioxidants.

## REFERENCES

1. Olson JA, Krinsky NI. The colorful, fascinating world of the carotenoids: Important physiologic modulators. *FASEB J* 1995; 9:1547–1550.
2. Parker RS. Absorption, metabolism, and transport of carotenoids. *FASEB J* 1996; 10:542–551.
3. Rock CL. Carotenoids: Biology and treatment. *Pharmacol Ther* 1997; 75:185–197.
4. Smith W, Saba N. Retinoids as chemoprevention for head and neck cancer: Where do we go from here? *Crit Rev Oncol Hematol* 2005; 55:143–152.
5. Slaga TJ. Inhibition of the induction of cancer by antioxidants. *Adv Exp Med Biol* 1995; 369:167–174.
6. Burton GW, Ingold KU. β-Carotene: An unusual type of lipid antioxidant. *Science* 1984; 224:569–573.
7. Packer L. Antioxidant action of carotenoids in vitro and in vivo and protection against oxidation of human low-density lipoproteins. *Ann NY Acad Sci* 1993; 691:48–60.
8. Sies H, Stahl W, Sundquist AR. Antioxidant functions of vitamins. Vitamins E and C, β-carotene and other carotenoids. *Ann NY Acad Sci* 1992; 669:7–20.
9. Ziegler RG, Mayne ST, Swanson CA. Nutrition and lung cancer. *Cancer Causes Control* 1996; 7:157–177.
10. Van Poppel G, Goldbohm RA. Epidemiologic evidence for β-carotene and cancer prevention. *Am J Clin Nutr* 1995; 62:1393S–1402S.

11. Cooper DA, Eldridge AL, Peters JC. Dietary carotenoids and lung cancer: A review of recent research. *Nutr Rev* 1999; 57:133–145.
12. Lippi G, Tarther G. Tomatoes, lycopene-containing foods and cancer risk. *Br J Cancer* 2011; 104:1234–1235.
13. Mamede AC, Tavares SD, Abrantes AM, Trindade J, Maia JM, Botelho MF. The role of vitamins in cancer: A review. *Nutr Cancer* 2011; 63:479–494.
14. Greenwald P, Clifford CK, Milner JA. Diet and cancer prevention. *Eur J Cancer* 2001; 37:948–965.
15. Greenwald P, McDonald SS. The β-carotene story. *Adv Exp Med Biol* 2001; 492:219–231.
16. Wakai K, Matsuo K, Nagata C, Mizoue T, Tanaka K, Tsuji I, Sasazuki S et al. (Research Group for the Development and Evaluation of Cancer Prevention Strategies in Japan). Lung cancer risk and consumption of vegetables and fruit: An evaluation based on a systematic review of epidemiological evidence from Japan. *Jpn J Clin Oncol* 2011; 441:693–708.
17. Peto R, Doll R, Buckley JD, Sporn MB. Can dietary β-carotene materially reduce human cancer rates? *Nature* 1981; 290:201–208.
18. Li C, Ford ES, Zhao G, Balluz LS, Giles WH, Liu S. Serum alpha-carotene concentrations and risk of death among US adults. The Third National Health and Nutrition Examination Survey Follow-up Study. *Arch Intern Med* 2011; 171:507–515.
19. Nomura AM, Ziegler RG, Stemmermann GN, Chyou PH, Craft NE. Serum micronutrients and upper aerodigestive tract cancer. *Cancer Epidemiol Biomarkers Prev* 1997; 6:407–412.
20. Zhang X, Dai B, Zhang B, Wang Z. Vitamin A and risk of cervical cancer: A meta-analysis. *Gynecol Oncol* 2012; 124(2):366–373.
21. Myung SK, Ju W, Kim SC, Kim H, Korean Meta-analysis (KORMA) Study Group. Vitamin or antioxidant intake (or serum level) and risk of cervical neoplasm: A meta-analysis. *BJOG* 2011; 118:1285–1291.
22. Story EN, Kopec RE, Schwartz SJ, Harris GK. An update on the health effects of tomato lycopene. *Annu Rev Food Sci Technol* 2010; 1:189–210.
23. Giovannucci E, Ascherio A, Rimm EB, Stampfer MJ, Colditz GA, Willett WC. Intake of carotenoids and retinol in relation to risk of prostate cancer. *J Natl Cancer Inst* 1995; 87:1767–1776.
24. Hsing QW, Comstock GW, Abbey H, Polk BF. Serologic precursors of cancer. Retinol, carotenoids, and tocopherol and risk of prostate cancer. *J Natl Cancer Inst* 1990; 82:941–946.
25. Gann PH, Ma J, Giovannucci E, Willett W, Sacks FM, Hennekens CH, Stampfer MJ. Lower prostate cancer risk in men with elevated plasma lycopene levels: Results of a prospective analysis. *Cancer Res* 1999; 59:1225–1230.
26. Tzonou A, Signorello LB, Lagiou P, Wuu J, Trichopoulos D, Trichopoulou A. Diet and cancer of the prostate: A case–control study in Greece. *Int J Cancer* 1999; 80:704–708.
27. Mills PK, Beeson WL, Philips RL, Fraser GE. Cohort study of diet, lifestyle, and prostate cancer in Adventist men. *Cancer* 1989; 64:598–604.
28. Vogt TM, Mayne ST, Graubard BI, Swanson CA, Sowell AL, Schoenberg JB, Swanson GM et al. Serum lycopene, other serum carotenoids, and risk of prostate cancer in US blacks and whites. *Am J Epidemiol* 2002; 155:1023–1032.
29. Key TJ, Appleby PN, Allen NE, Travis RC, Roddam AW, Jenab M, Egevad L et al. Plasma carotenoids, retinol, and tocopherols and the risk of prostate cancer in the European prospective investigation into cancer and nutrition study. *Am J Clin Nutr* 2007; 86:672–681.
30. Etminan M, Takkouche B, Caamano-Isorna F. The role of tomato products and lycopene in the prevention of prostate cancer: A meta-analysis of observational studies. *Cancer Epidemiol Biomarkers Prev* 2004; 13:340–345.

31. Van Patten CL, de Boer JG, Tomlinson Guns ES. Diet and dietary supplement intervention trials for the prevention of prostate cancer recurrence: A review of the randomized controlled trial evidence. *J Urol* 2008; 180:2314–2321 and (discussion) 2721–2722.

32. Ford NA, Elsen AC, Yuniga K, Lindshield BL, Erdman JW Jr. Lycopene and apo-12′-lycopenal reduce cell proliferation and alter cell cycle progression in human prostate cancer cells. *Nutr Cancer* 2011; 63:256–263.

33. Wu K, Erdman JW, Schwartz SJ, Platz EA, Leitzmann M, Clinton SK, DeGroff V, Willett WC, Giovannucci E. Plasma and dietary carotenoids, and the risk of prostate cancer: A nested case–control study. *Cancer Epidemiol Biomarkers Prev* 2004; 13:260–269.

34. Hu KQ, Liu C, Ernst H, Krinsky NI, Russell RM, Wang XD. The biochemical characterization of ferret carotene-9′,10′-monooxygenase catalyzing cleavage of carotenoids in vitro and in vivo. *J Biol Chem* 2006; 281:19327–19338.

35. Kopec RE, Riedl KM, Harrison EH, Curley RW Jr., Hruszkewycz DP, Clinton SK, Schwartz SJ. Identification and quantification of apo-lycopenals in fruits, vegetables, and human plasma. *J Agric Food Chem* 2010; 58:3290–3296.

36. Gajic M, Zaripheh S, Sun F, Erdman JW. Apo-8′-lycopenal and apo-12′-lycopenal are metabolic products of lycopene in rat liver. *J Nutr* 2006; 136:1552–1557.

37. Kristal AR, Till C, Platz EA, Song X, King IB, Neuhouser ML, Ambrosone CB, Thompson IM. Serum lycopene concentration and prostate cancer risk: Results from the prostate cancer prevention trial. *Cancer Epidemiol Biomarkers Prev* 2011; 20:638–646.

38. Giovannucci E. Commentary: Serum lycopene and prostate cancer progression: A re-consideration of findings from the prostate cancer prevention trial. *Cancer Causes Control* 2011; 22:1055–1059.

39. Mondul AM, Watters JL, Mannisto S, Weinstein SJ, Knyder K, Virtamo J, Albanes D. Serum retinol and risk of prostate cancer. *Am J Epidemiol* 2011; 173:813–821.

40. Ilic D, Forbes KM, Hassed C. Lycopene for the prevention of prostate cancer. *Cochrane Database Syst Rev* 2011; 11:CD008007.

41. Bjelke E. Dietary vitamin A and human lung cancer. *Int J Cancer* 1975; 15:561–565.

42. Ziegler RG. A review of epidemiologic evidence that carotenoids reduce the risk of cancer. *J Nutr* 1989; 119:116–122.

43. Gallicchio L, Boyd K, Matanoski G, Tao XG, Chen L, Lam TK, Shiels M et al. Carotenoids and the risk of developing lung cancer: A systematic review. *Am J Clin Nutr* 2008; 88:372–383.

44. Fritz H, Kennedy D, Fergusson D, Fernandez R, Doucette S, Cooley K, Seely A, Sagar S, Wong R, Seely D. Vitamin A and retinoid derivatives for lung cancer: A systematic review and meta analysis. *PLoS One* 2011; 6:e21107; 1–11.

45. Virtamo J, Pietinen P, Huttunen JK, Korhonen P, Malila N, Virtanen MJ, Albanes D, Taylor PR, Albert P; ATBC Study Group. Incidence of cancer and mortality following alpha-tocopherol and β-carotene supplementation: A postintervention follow-up. *JAMA* 2003; 290:476–485.

46. The Alpha-Tocopherol, Beta-Carotene Cancer Prevention Study Group. The effect of vitamin E and β-carotene on the incidence of lung cancer and other cancers in male smokers. *N Engl J Med* 1994; 330:1029–1035.

47. Zhang SM. Role of vitamins in the risk, prevention, and treatment of breast cancer. *Curr Opin Obstet Gynecol* 2004; 16:19–25.

48. Lazzeroni M, Gandini S, Puntoni M, Bonanni B, Gennari A, DeCensi A. The science behind vitamins and natural compounds for breast cancer prevention. Getting the most prevention out of it. *Breast* 2011; 20(Suppl 3):S36–S41.

49. Young AJ, Lowe GM. Antioxidant and prooxidant properties of carotenoids. *Arch Biochem Biophys* 2001; 385:20–27.

50. Alija AJ, Bresgen N, Sommerburg O, Siems W, Eckl PM. Cytotoxic and genotoxic effects of β-carotene breakdown products on primary rat hepatocytes. *Carcinogenesis* 2004; 25:827–831.

51. Siems W, Wiswedel I, Sommerburg O, Langhans CD, Salerno C, Crifò C, Capuozzo C, Alija A, Bresgen N, Eckl P. Clinical use of carotenoids—Antioxidative protection versus prooxidative side effects. In: *Free Radicals and Diseases: Gene Expression, Cellular Metabolism and Pathophysiology. NATO Science Series* (T Grune, ed.). IOS Press, Amsterdam, the Netherlands, 2005; pp. 177–192.

52. Alija AJ, Bresgen N, Sommerburg O, Langhans CD, Siems W, Eckl PM. Cyto- and genotoxic potential of â-carotene and cleavage products under oxidative stress. *BioFactors* 2005; 24:159–164.

53. Alija AJ, Bresgen N, Sommerburg O, Langhans CD, Siems W, Eckl PM. β-Carotene breakdown products enhance genotoxic effects of oxidative stress in primary rat hepatocytes. *Carcinogenesis* 2006; 27:1128–1133.

54. Salerno C, Capuozzo E, Crifò C, Siems W. Alpha-tocopherol increases caspase-3 up-regulation and apoptosis by β-carotene cleavage products in human neutrophils. *BBA* 2007; 1772:1052–1056.

55. Palozza P, Simone R, Mele MC. Interplay of carotenoids with cigarette smoking: Implications in lung cancer. *Curr Med Chem* 2008; 15:844–854.

56. Siems W, Salerno C, Crifò C, Sommerburg O, Wiswedel I. β-Carotene degradation products: Formation, toxicity and prevention of toxicity. In: *Forum of Nutrition 61: Food Factors for Health Promotion* (T Yoshikawa, Y Naito, eds.). Karger Publ. Co., Basel, Switzerland, 2009; pp. 75–86.

57. Alija A, Bresgen N, Bojaxhi E, Vogl C, Siems W, Eckl PM. Cytotoxicity of β-carotene cleavage products and its prevention by antioxidants. *Acta Biochim Pol* 2010; 57:217–221.

58. Edge R, Truscott TG. Prooxidant and antioxidant reaction mechanisms of carotene and radical interactions with vitamins E and C. *Nutrition* 1997; 13:992–994.

59. Murata M, Kawanishi S. Oxidative DNA damage by Vitamin A and its derivative via superoxide generation. *J Biol Chem* 2000; 275:2003–2008.

60. Siems WG, Sommerburg O, Hurst JS, van Kuijk FJGM. Carotenoid oxidative degradation products inhibit $Na^+$-$K^+$-ATPase. *Free Radic Res* 2000; 33:427–435.

61. Kowaltowski AJ, Vercesi AE. Mitochondrial damage induced by conditions of oxidative stress. *Free Radic Biol Med* 1999; 26:463–471.

62. Schild L, Keilhoff G, Augustin W, Reiser G, Striggow F. Distinct $Ca^{2+}$ thresholds determine cytochrome c release or permeability transition pore opening in brain mitochondria. *FASEB J* 2001; 15:565–567.

63. Turrens JF. Superoxide production by the mitochondrial respiratory chain. *Biosci Rep* 1997; 17:3–8.

64. Siems W, Sommerburg O, Schild L, Augustin W, Langhans CD, Wiswedel I. β-Carotene cleavage products induce oxidative stress in vitro by impairing mitochondrial respiration. *FASEB J* 2002; 16:1289–1291.

65. Stratton SP, Schaefer WH, Liebler DC. Isolation and identification of singlet oxygen oxidation products of β-carotene. *Chem Res Toxicol* 1993; 6:542–547.

66. Ojima F, Sakamoto H, Ishiguro Y, Terao J. Consumption of carotenoids in photosensitized oxidation of human plasma and plasma low-density lipoprotein. *Free Radic Biol Med* 1993; 15:377–384.

67. Liebler DC, McClure TD. Antioxidant reactions of β-carotene: Identification of carotenoid-radical adducts. *Chem Res Toxicol* 1996; 9:8–11.

68. McClure TD, Liebler DC. A rapid method for profiling the products of antioxidant reactions by negative ion chemical ionization mass spectrometry. *Chem Res Toxicol* 1995; 8:128–135.

69. Kennedy TA, Liebler DC. Peroxyl radical oxidation of β-carotene: Formation of β-carotene epoxides. *Chem Res Toxicol* 1991; 4:290–295.

70. El-Tinay AH, Chichester CO. Oxidation of β-carotene. Site of initial attack. *J Org Chem* 1970; 35:2290–2293.

71. Handelman GJ, van Kuijk FJ, Chatterjee A, Krinsky NI. Characterization of products formed during the autoxidation of β-carotene. *Free Radic Biol Med* 1991; 10:427–437.

72. Henry LK, Puspitasari-Nienaber NL, Jaren-Galan M, van Breemen RB, Catignani GL, Schwartz SJ. Effects of ozone and oxygen on the degradation of carotenoids in an aqueous model system. *J Agric Food Chem* 2000; 48:5008–5013.

73. Sommerburg O, Langhans CD, Arnhold J, Leichsenring M, Salerno C, Crifò C, Hoffmann GF, Debatin KM, Siems WG. β-Carotene cleavage products after oxidation mediated by hypochlorous acid. A model for neutrophil-derived degradation. *Free Radic Biol Med* 2003; 35:1480–1490.

74. Mein JR, Dolnikowski GG, Ernst H, Russell RM, Wang XD. Enzymatic formation of apo-carotenoids from the xanthophyll carotenoids lutein, zeaxanthin and β-cryptoxanthin by ferret carotene-9′,10′-monooxygenase. *Arch Biochem Biophys* 2011; 506:108–121.

75. Mordi RC, Walton JC, Burton GW, Hughes L, Ingold KU, Lindsay DA, Moffatt D. Oxidative degradation of β-carotene and β-apo-8′-carotenal. *Tetrahedron* 1993; 49:911–928.

76. Mordi RC, Walton JC, Burton GW, Hughes L, Ingold KU, Lindsay DA. Exploratory study of β-carotene auto-oxidation. *Tetrahedron Lett* 1991; 32:4203–4206.

77. Ouyang JM, Daun H, Chang SS, Ho CT. Formation of carbonyl compounds from β-carotene during palm oil deodorization. *J Food Sci* 1980; 45:1214–1222.

78. Marty D, Berset C. Degradation products of trans-β-carotene produced during extrusion cooking. *J Food Sci* 1988; 53:1880–1886.

79. Gloria MB, Grulke EA, Gray JI. Effect of type of oxidation on β-carotene loss and volatile products formation in model systems. *Food Chem* 1993; 46:401–406.

80. Onyewu PN, Ho CT, Daun H. Characterization of β-carotene thermal degradation products in a model food system. *J Am Oil Chem Soc* 1986; 63:1437–1441.

81. Schreier P, Drawert F, Bhiwapurkar S. Volatile compounds formed by thermal degradation of β-carotene. *Chem Mikrobiol Technol Lebensm* 1979; 6:90–91.

82. Mader I. β-Carotene: Thermal degradation. *Science* 1964; 144:533–534.

83. Day WC, Erdman JG. Ionene: A thermal degradation product of β-carotene. *Science* 1963; 141:808.

84. Zhao W, Han Y, Zhao B, Hirota S, Hou J, Xin W. Effect of carotenoids on the respiratory burst of rat peritoneal macrophages. *Biochim Biophys Acta* 1998; 1381:77–88.

85. Foote CS, Chang YC, Denny RW. Chemistry of singlet oxygen: Carotenoid quenching parallels biological protection. *J Am Chem Soc* 1970; 92:5216–5218.

86. Di Mascio P, Devasagayam TP, Kaiser S, Sies H. Carotenoids, tocopherols and thiols as biological singlet molecular oxygen quenchers. *Biochem Soc Trans* 1990; 18:1054–1056.

87. Conn PF, Schalch W, Truscott TG. The singlet oxygen and carotenoid interaction. *J Photochem Photobiol B* 1991; 11:41–47.

88. Sundquist AR, Briviba K, Sies H. Singlet oxygen quenching by carotenoids. *Methods Enzymol* 1994; 234:384–388.

89. Albrich JM, McCarthy CA, Hurst JK. Biological reactivity of hypochlorous acid: Implications for microbicidal mechanisms of leukocyte myeloperoxidase. *Proc Natl Acad Sci U S A* 1981; 78:210–214.

90. Panasenko OM, Sharov VS, Briviba K, Sies H. Interaction of peroxynitrite with carotenoids in human low density lipoproteins. *Arch Biochem Biophys* 2000; 373:302–305.

91. Olson JA. Molecular actions of carotenoids. *Ann NY Acad Sci* 1993; 691:156–166.

92. Von Lintig J, Wyss A. Molecular analysis of vitamin A formation: Cloning and characterization of ß-carotene 15,15′-dioxygenases. *Arch Biochem Biophys* 2001; 385:47–52.

93. Von Lintig J, Vogt K. Filling the gap in vitamin A research. Molecular identification of an enzyme cleaving β-carotene to retinal. *J Biol Chem* 2000; 275:11915–11920.

94. Krinsky NI. The antioxidant and biological properties of the carotenoids. *Ann NY Acad Sci* 1998; 854:443–447.

95. Krinsky NI. Actions of carotenoids in biological systems. *Annu Rev Nutr* 1993; 13:561–587.

96. Krinsky NI. Antioxidant functions of carotenoids. *Free Radic Biol Med* 1989; 7:617–635.

97. Liebler DC. Antioxidant reactions of carotenoids. *Ann NY Acad Sci* 1993; 691:20–31.

98. Siems W, Sommerburg O, van Kuijk F. Oxidative breakdown of carotenoids and biological effects of their metabolites. In: *Handbook of Antioxidants* (E Cadenas, L Packer, eds.). Marcel Dekker, Inc., New York, 2001; pp. 235–249.

99. Salerno C, Crifò C, Siems W. Carotenoids and lung cancer: Biochemical aspects. *Cent Eur J Chem* 2011; 9:1–6.

100. Chen JJ, Bertrand H, Yu BP. Inhibition of adenine nucleotide translocator by lipid peroxidation products. *Free Radic Biol Med* 1995; 19:583–590.

101. Notario B, Zamora M, Vinas O, Mampel T. All-trans-retinoic acid binds to and inhibits adenine nucleotide translocase and induces mitochondrial permeability transition. *Mol Pharmacol* 2003; 63:224–231.

102. Korichneva I, Waka J, Hammerling U. Regulation of the cardiac mitochondrial membrane potential by retinoids. *J Pharmacol Exp Ther* 2003; 305:426–433.

103. Britton G. Structure and properties of carotenoids in relation to function. *FASEB J* 1995; 9:1551–1558.

104. Tsuchihashi H, Kigoshi M, Iwatsuki M, Niki E. Action of β-carotene as an antioxidant against lipid peroxidation. *Arch Biochem Biophys* 1995; 323:137–147.

105. Estornell E, Tormo JR, Marin P, Renau-Piqueras J, Timoneda J, Barber T. Effects of vitamin A deficiency on mitochondrial function in rat liver and heart. *Br J Nutr* 2000; 84:927–934.

106. Lotan R. Receptor-independent induction of apoptosis by synthetic retinoids. *J Biol Regul Homeost Agents* 2003; 17:13–28.

107. Schmitz HH, Poor CL, Wellman RB, Erdman JJ. Concentrations of selected carotenoids and vitamin A in human liver, kidney and lung tissue. *J Nutr* 1991; 121:1613–1621.

108. Mayne ST. β-Carotene, carotenoids, and disease prevention in humans. *FASEB J* 1996; 10:690–701.

109. Winterbourn CC. Comparative reactivities of various biological compounds with myeloperoxidase-hydrogen peroxide-chloride, and similarity of the oxidant to hypochlorite. *Biochim Biophys Acta* 1985; 840:204–210.

110. Schiller J, Arnhold J, Grunder W, Arnold K. The action of hypochlorous acid on polymeric components of cartilage. *Biol Chem Hoppe Seyler* 1994; 375:167–172.

111. Schaur RJ, Jerlich A, Stelmaszynska T. Hypochlorous acid as reactive oxygen species. *Curr Top Biophys* 1998; 22:176–185.

112. Nauseef WM. Insights into myeloperoxidase biosynthesis from its inherited deficiency. *J Mol Med* 1998; 76:661–668.

113. Panasenko OM, Panasenko OO, Briviba K, Sies H. Hypochlorite destroys carotenoids in low density lipoproteins thus decreasing their resistance to peroxidative modification. *Biochemistry (Mosc)* 1997; 62:1140–1145.

114. Vogt W, Hesse D. Oxidants generated by the myeloperoxidase-halide system activate the fifth component of human complement, C5. *Immunobiology* 1994; 192:1–9.

115. Weiss SJ, Klein R, Slivka A, Wei M. Chlorination of taurine by human neutrophils. Evidence for hypochlorous acid generation. *J Clin Invest* 1982; 70:598–607.

116. Amin K, Ekberg-Jansson A, Lofdahl CG, Venge P. Relationship between inflammatory cells and structural changes in the lungs of asymptomatic and never smokers: A biopsy study. *Thorax* 2003; 58:135–142.

117. Eidelman D, Saetta MP, Ghezzo H, Wang NS, Hoidal JR, King M, Cosio MG. Cellularity of the alveolar walls in smokers and its relation to alveolar destruction. Functional implications. *Am Rev Respir Dis* 1990; 141:1547–1552.

118. Saetta M, Turato G, Baraldo S, Zanin A, Braccioni F, Mapp CE, Maestrelli P, Cavallesco G, Papi A, Fabbri LM. Goblet cell hyperplasia and epithelial inflammation in peripheral airways of smokers with both symptoms of chronic bronchitis and chronic airflow limitation. *Am J Respir Crit Care Med* 2000; 161:1016–1021.

119. Di Stefano A, Capelli A, Lusuardi M, Balbo P, Vecchio C, Maestrelli P, Mapp CE, Fabbri LM, Donner CF, Saetta M. Severity of airflow limitation is associated with severity of airway inflammation in smokers. *Am J Respir Crit Care Med* 1998; 158:1277–1285.

120. Amin K, Ludviksdottir D, Janson C, Nettelbladt O, Bjornsson E, Roomans GM, Boman G, Seveus L, Venge P. Inflammation and structural changes in the airways of patients with atopic and nonatopic asthma. BHR Group. *Am J Respir Crit Care Med* 2000; 162:2295–2301.

121. Andreadis AA, Hazen SL, Comhair SA, Erzurum SC. Oxidative and nitrosative events in asthma. *Free Radic Biol Med* 2003; 35:213–221.

122. Maddox L, Schwartz DA. The pathophysiology of asthma. *Annu Rev Med* 2002; 53:477–498.

123. Amin K, Ludviksdottir D, Janson C, Nettelbladt O, Gudbjornsson B, Valtysdottir S, Bjornsson E et al. Inflammation and structural changes in the airways of patients with primary Sjogren's syndrome. *Respir Med* 2001; 95:904–910.

124. Hubeau C, Lorenzato M, Couetil JP, Hubert D, Dusser D, Puchelle E, Gaillard D. Quantitative analysis of inflammatory cells infiltrating the cystic fibrosis airway mucosa. *Clin Exp Immunol* 2001; 124:69–76.

125. Hallett MB, Cole C, Dewitt S. Detection and visualization of oxidase activity in phagocytes. *Methods Mol Biol* 2003; 225:61–67.

126. Rojkind M, Dominguez-Rosales JA, Nieto N, Greenwel P. Role of hydrogen peroxide and oxidative stress in healing responses. *Cell Mol Life Sci* 2002; 59:1872–1891.

127. Reynolds PR, Cosio MG, Hoidal JR. Cigarette smoke-induced Egr-1 upregulates pro-inflammatory cytokines in pulmonary epithelial cells. *Am J Respir Cell Mol Biol* 2006; 35:314–319.

128. Smith CJ, Perfetti TA, King JA. Perspectives on pulmonary inflammation and lung cancer risk in cigarette smokers. *Inhal Toxicol* 2006; 18:667–677.

129. Walser T, Cui X, Yanagawa J, Lee JM, Heinrich E, Lee G, Sharma S, Dubinett SM. Smoking and lung cancer: The role of inflammation. *Proc Am Thorac Soc* 2008; 5:811–815.

130. Kim V, Rogers TJ, Criner GJ. New concepts in the pathobiology of chronic obstructive pulmonary disease. *Proc Am Thorac Soc* 2008; 5:478–485.

131. Baratelli F, Lin Y, Zhu L, Yang SC, Heuzé-Vourc'h N, Zeng G, Reckamp K, Dohadwala M, Sharma S, Dubinett SM. Prostaglandin E2 induces FOXP3 gene expression and T regulatory cell function in human CD4+ T cells. *J Immunol* 2005; 175:1483–1490.

132. Leng Q, Bentwich Z, Borkow G. Increased TGF-β, Cbl-b and CTLA-4 levels and immunosuppression in association with chronic immune activation. *Int Immunol* 2006; 18:637–644.

133. Bainton DF, Ullyot JL, Farquhar MG. The development of neutrophilic polymorphonuclear leukocytes in human bone marrow. *J Exp Med* 1971; 134:907–934.

134. Jenner AM, Ruiz EJ, Dunster C, Halliwell B, Mann GE, Siow RCM. Vitamin C protects against hypochlorous acid-induced glutathione depletion and DNA base and protein damage in human vascular smooth muscle cells. *Arterioscler Thromb Vasc Biol* 2002; 22:574–580.

135. Arisawa F, Tatsuzawa H, Kambayashi Y, Kuwano H, Fujimori K, Nakano M. MCLA-dependent chemiluminescence suggests that singlet oxygen plays a pivotal role in myeloperoxidase-catalysed bactericidal action in neutrophil phagosomes. *Luminescence* 2003; 18:229–238.

136. Fu X, Mueller DM, Heinecke JW. Generation of intramolecular and intermolecular sulfenamides, sulfinamides, and sulfonamides by hypochlorous acid: A potential pathway for oxidative cross-linking of low-density lipoprotein by myeloperoxidase. *Biochemistry* 2002; 41:1293–1301.

137. Peskin AV, Winterbourn CC. Kinetics of the reactions of hypochlorous acid and amino acid chloramines with thiols, methionine, and ascorbate. *Free Radic Biol Med* 2001; 30:572–579.

138. Naskalski JW. Oxidative modification of protein structures under the action of myeloperoxidase and the hydrogen peroxide and chloride system. *Ann Biol Clin* 1994; 52:541–456.

139. Carr AC, Van den Berg JJM, Winterbourn CC. Differential reactivities of hypochlorous and hypobromous acids with purified *Escherichia coli* phospholipids: Formation of haloamines and halohydrins. *Biochim Biophys Acta* 1998; 1392:254–264.

140. Pattison DI, Hawkins CL, Davies MJ. Hypochlorous acid-mediated oxidation of lipid components and antioxidants present in low-density lipoproteins: Absolute rate constants, product analysis, and computational modelling. *Chem Res Toxicol* 2003; 16:439–449.

141. Hazel LJ, Davies MJ, Stocker R. Secondary radicals derived from chloramines of apolipoprotein B-100 contribute to HOCl-induced LDL lipid peroxidation. *Biochem J* 1999; 339:489–495.

142. Hawkins CL, Pattison DI, Davies MJ. Reaction of protein chloramines with DNA and nucleotides: Evidence for the formation of radicals, protein-DNA cross-links and DNA fragmentation. *Biochem J* 2002; 365:805–615.

143. Whiteman M, Hooper DC, Scott GS, Koprowski H, Halliwell B. Inhibition of hypochlorous acid-induced cellular toxicity by nitrite. *Proc Nat Acad Sci U S A* 2002; 99:12061–12066.

144. Ueda T, Maekawa T, Sadamitsu D, Ohshita S, Ogino K, Nakamura K. The determination of nitrite and nitrate in human blood plasma by capillary zone. *Electrophoresis* 1995; 16:1002–1004.

145. Wanchu A, Agnihotri N, Deodhar SD, Ganguly NK. Plasma reactive nitrogen intermediate levels in patients with clinically active rheumatoid arthritis. *Indian J Med Res* 1996; 104:263–268.

146. Sud A, Khullar M, Wanchu A, Bambery P. Increased nitric oxide production in patients with systemic sclerosis. *Nitric Oxide* 2000; 4:615–619.

147. Wanchu A, Khullar M, Sud A, Bambery P. Elevated nitric oxide production in patients with primary Sjogren's syndrome. *Clin Rheumatol* 2000; 19:360–364.

148. Wanchu A, Khullar M, Deodhar SD, Bambery P, Sud A. Nitric oxide synthesis is increased in patients with systemic lupus erythematosus. *Rheumatol Int* 1998; 18:41–43.

149. Panasenko OM, Briviba K, Klotz LO, Sies H. Oxidative modification and nitration of human low-density lipoproteins by the reaction of hypochlorous acid with nitrite. *Arch Biochem Biophys* 1997; 343:254–259.

150. Eiserich JP, Hristova M, Cross CE, Jones AD, Freeman BA, Halliwell B, van der Vliet A. Formation of nitric oxide-derived inflammatory oxidants by myeloperoxidase in neutrophils. *Nature* 1998; 391:393–397.

151. Whiteman M, Spencer JP, Jenner A, Halliwell B. Hypochlorous acid-induced DNA base modification: Potentiation by nitrite: Biomarkers of DNA damage by reactive oxygen species. *Biochem Biophys Res Commun* 1999; 257:572–576.

152. Whiteman M, Rose P, Halliwell B. Inhibition of hypochlorous acid-induced oxidative reactions by nitrite: Is nitrite an antioxidant? *Biochem Biophys Res Commun* 2003; 303:1217–1224.

153. Davies CA, Rocks SA, O'Shaughnessy MC, Perrett D, Winyard PG. Analysis of nitrite and nitrate in the study of inflammation. *Methods Mol Biol* 2003; 225:305–320.

154. Pagliaro P. Differential biological effects of products of nitric oxide (NO) synthase: It is not enough to say NO. *Life Sci* 2003; 73:2137–2149.

155. Albina JE. On the expression of nitric oxide synthase by human macrophages. Why no NO? *J Leukoc Biol* 1995; 58:643–649.

156. Wheeler MA, Smith SD, Garcia-Cardena G, Nathan CF, Weiss RM, Sessa WC. Bacterial infection induces nitric oxide synthase in human neutrophils. *J Clin Invest* 1997; 99:110–116.

157. Beckman JS, Koppenol WH. Nitric oxide, superoxide, and peroxynitrite: The good, the bad, and ugly. *Am J Physiol* 1996; 271:C1424–1437.

158. Kissner R, Koppenol WH. Product distribution of peroxynitrite decay as a function of pH, temperature, and concentration. *J Am Chem Soc* 2002; 124:234–239.

159. Jourd'heuil D, Jourd'heuil FL, Kutchukian PS, Musah RA, Wink DA, Grisham MB. Reaction of superoxide and nitric oxide with peroxynitrite. Implications for peroxynitrite-mediated oxidation reactions in vivo. *J Biol Chem* 2001; 276:28799–28805.

160. Squadrito GL, Pryor WA. Oxidative chemistry of nitric oxide: The roles of superoxide, peroxynitrite, and carbon dioxide. *Free Radic Biol Med* 1998; 25:392–403.

161. Beckman JS. Oxidative damage and tyrosine nitration from peroxynitrite. *Chem Res Toxicol* 1996; 9:836–844.

162. Van der Vliet A, Eiserich JP, Halliwell B, Cross CE. Formation of reactive nitrogen species during peroxidase-catalyzed oxidation of nitrite: A potential additional mechanism of nitric oxide-dependent toxicity. *J Biol Chem* 1997; 272:7617–7625.

163. Parks NJ, Krohn KJ, Mathis CA, Chasko JH, Geiger KR, Gregor ME, Peek NF. Nitrogen-13-labeled nitrite and nitrate: Distribution and metabolism after intratracheal administration. *Science* 1981; 212:58–60.

164. Radi R. Reactions of nitric oxide with metalloproteins. *Chem Res Toxicol* 1996; 9:828–835.

165. Siems W, Capuozzo E, Crifò C, Sommerburg O, Langhans CD, Schlipalius L, Wiswedel I, Kraemer K, Salerno C. Carotenoid cleavage products modify respiratory burst and induce apoptosis of human neutrophils. *Biochim Biophys Acta* 2003; 1639:27–33.

166. Hengartner MO. The biochemistry of apoptosis. *Nature* 2000; 407:770–776.

167. Müller K, Carpenter KL, Challis IR, Skepper JN, Arends MJ. Carotenoids induce apoptosis in the T-lymphoblast cell line Jurkat E6.1. *Free Radic Res* 2002; 36:791–802.

168. Robinson KM, Morre JT, Beckman JS. Triuret: A novel product of peroxynitrite-mediated oxidation of urate. *Arch Biochem Biophys* 2004; 423:213–217.

169. Spitsin SV, Scott GS, Mikheeva T, Zborek A, Kean RB, Brimer CM, Koprowski H, Hooper DC. Comparison of uric acid and ascorbic acid in protection against EAE. *Free Radic Biol Med* 2002; 33:1363–1371.

170. Böhm F, Edge R, McGarvey DJ, Truscott TG. β-Carotene with vitamins E and C offers synergistic cell protection against NOx. *FEBS Lett* 1998; 436:387–389.

171. Lowe GM, Booth LA, Young AJ, Bilton RF. Lycopene and β-carotene protect against oxidative damage in HT29 cells at low concentrations but rapidly lose this capacity at higher doses. *Free Radic Res* 1999; 30:141–151.

172. Konta T, Xu Q, Furusu A, Nakayama K, Kitamura M. Selective roles of retinoic acid receptor and retinoid X receptor in the suppression of apoptosis by all-trans-retinoic acid. *J Biol Chem* 2001; 276:12697–12701.

173. Cacciola SA, Cohen LA, Kashfi K. Lycopene inhibits proliferation and regulates cyclooxygenase-2 gene expression in neoplastic rat mammary epithelial cells. *FASEB J* 1999; 13:A551.

174. Muto Y, Fujii J, Shidoji Y, Moriwaki H, Kawaguchi T, Noda T. Growth-retardation in human cervical dysplasia-derived cell lines by β-carotene through down-regulation of epidermal growth-factor receptor. *Am J Clin Nutr* 1995; 62:S1535–S1540.

175. Glaves D. Role of polymorphonuclear leukocytes in the pulmonary clearance of arrested cancer cells. *Invas Metast* 1983; 3:160–173.

176. Chen YL, Wang JY, Chen SH, Yang BC. Granulocytes mediates the Fas-L-associated apoptosis during lung metastasis of melanoma that determines the metastatic behaviour. *Br J Cancer* 2002; 87:359–365

177. Beyer P, Al Babili S, Ye X, Lucca P, Schaub P, Welsch R, Potrykus I. Golden *Rice*: Introducing the β-carotene biosynthesis pathway into rice endosperm by genetic engineering to defeat vitamin A deficiency. *J Nutr* 2002; 132:506S–510S.

178. Potrykus I. Golden rice and beyond. *Plant Physiol* 2001; 125:1157–1161.

179. Hoa TT, Al Babili S, Schaub P, Potrykus I, Beyer P. Golden indica and japonica rice lines amenable to deregulation. *Plant Physiol* 2003; 133:161–169.

180. Leonard PJ, Doyle E, Harrington W. Levels of vitamin E in the plasma of newborn infants and of the mothers. *Am J Clin Nutr* 1972; 25:480–484.

181. Mino M, Nishino H. Fetal and maternal relationship in serum vitamin E level. *J Nutr Sci Vitaminol (Tokyo)* 1973; 19:475–482.

182. Ostrea EM Jr., Balun JE, Winkler R, Porter T. Influence of breast-feeding on the restoration of the low serum concentration of vitamin E and β-carotene in the newborn infant. *Am J Obstet Gynecol* 1986; 154:1014–1017.

183. Oostenbrug GS, Mensink RP, Al MD, van Houwelingen A, Hornstra G. Maternal and neonatal plasma antioxidant levels in normal pregnancy, and the relationship with fatty acid unsaturation. *Br J Nutr* 1998; 80:67–73.

184. Kiely M, Cogan PF, Kearney PJ, Morrissey PA. Concentrations of tocopherols and carotenoids in maternal and cord blood plasma. *Eur J Clin Nutr* 1999; 53:711–715.

185. Khachik F, Spangler CJ, Smith JJ, Canfield LM, Steck A, Pfander H. Identification, quantification, and relative concentrations of carotenoids and their metabolites in human milk and serum. *Anal Chem* 1997; 69:1873–1881.

186. Sommerburg O, Meissner K, Nelle M, Lenhartz H, Leichsenring M. Carotenoid supply in breast-fed and formula-fed neonates. *Eur J Pediatr* 2000; 159:86–90.

187. Johnson EJ, Qin J, Krinsky NI, Russell RM. β-Carotene isomers in human serum, breast milk and buccal mucosa cells after continuous oral doses of all-trans and 9-cis β-carotene. *J Nutr* 1997; 127:1993–1999.

188. Canfield LM, Giuliano AR, Neilson EM, Blashil BM, Graver EJ, Yap HH. Kinetics of the response of milk and serum β-carotene to daily β-carotene supplementation in healthy, lactating women. *Am J Clin Nutr* 1998; 67:276–283.

189. Benabdeslam H, Abidi H, Garcia I, Bellon G, Gilly R, Revol A. Lipid peroxidation and antioxidant defenses in cystic fibrosis patients. *Clin Chem Lab Med* 1999; 37:511–516.

190. Portal BC, Richard MJ, Faure HS, Hadjian AJ, Favier AE. Altered antioxidant status and increased lipid peroxidation in children with cystic fibrosis. *Am J Clin Nutr* 1995; 61:843–847.

191. Rust P, Eichler I, Renner S, Elmadfa I. Effects of long-term oral β-carotene supplementation on lipid peroxidation in patients with cystic fibrosis. *Int J Vitam Nutr Res* 1998; 68:83–87.

192. Winklhofer-Roob BM, van't Hof MA, Shmerling DH. Response to oral β-carotene supplementation in patients with cystic fibrosis: A 16-month follow-up study [published erratum in *Acta Paediatr* 1996; 85:124]. *Acta Paediatr* 1995; 84:1132–1136.

193. Cobanoglu N, Ozcelik U, Gocmen A, Kiper N, Dogru D. Antioxidant effect of β-carotene in cystic fibrosis and bronchiectasis: Clinical and laboratory parameters of a pilot study. *Acta Paediatr* 2002; 91:793–798.

194. Wood LG, Fitzgerald DA, Lee AK, Garg ML. Improved antioxidant and fatty acid status of patients with cystic fibrosis after antioxidant supplementation is linked to improved lung function. *Am J Clin Nutr* 2003; 77:150–159.

195. Renner S, Rath R, Rust P, Lehr S, Frischer T, Elmadfa I, Eichler I. Effects of β-carotene supplementation for six months on clinical and laboratory parameters in patients with cystic fibrosis. *Thorax* 2001; 56:48–52.

196. Barber T, Borras E, Torres L, Garcia C, Cabezuelo F, Lloret A, Pallordo FV, Vina JR. Vitamin A deficiency causes oxidative damage to liver mitochondria in rats. *Free Radic Biol Med* 2000; 29:1–7.

197. Siems W, Wiswedel I, Salerno C, Crifò C, Augustin W, Schild L, Langhans CD, Sommerburg O. Beta-carotene breakdown products may impair mitochondrial functions—Potential side effects of high-dose beta-carotene supplementation. *J Nutr Biochem* 2005; 16:385–397.

198. Zheng A, Mäntymaa P, Säily M, Siitonen T, Savolainen ER, Koistinen P. An association between mitochondrial function and all-trans retinoic acid-induced apoptosis in acute myeloblastic leukaemia cells. *Br J Haematol* 1999; 105:215–224.

199. Hail N Jr., Youssef EM, Lotan R. Evidence supporting a role for mitochondrial respiration in apoptosis induction by the synthetic retinoid CD437. *Cancer Res* 2001; 61:6698–6702.

200. Hurst JS, Saini MK, Jin GF, Awasthi YC, van Kuijk FJ. Toxicity of oxidized beta-carotene to cultured human cells. *Exp Eye Res* 2005; 81:239–243.

201. De Oliveira MR, Moreira JC. Acute and chronic vitamin A supplementation at therapeutic doses induces oxidative stress in submitochondrial particles isolated from cerebral cortex and cerebellum of adult rats. *Toxicol Lett* 2007; 173:145–150.

202. De Oliveira MR, Oliveira MW, Lorenzi R, Fagundes da Rocha R, Fonseca Moreira JC. Short-term vitamin A supplementation at therapeutic doses induces a pro-oxidative state in the hepatic environment and facilitates calcium-ion-induced oxidative stress in rat liver mitochondria independently from permeability transition pore formation: Detrimental effects of vitamin A supplementation on rat liver redox and bioenergetic states homeostasis. *Cell Biol Toxicol* 2009; 25:545–560.

203. Maeda A, Maeda T, Golczak M, Chou S, Desai A, Hoppel CL, Matsuyama S, Palczewski K. Involvement of all-trans-retinal in acute light-induced retinopathy of mice. *J Biol Chem* 2009; 284:15173–15183.

204. Signorile A, Sardaro N, De Rasmo D, Scacco S, Papa F, Borracci P, Carratu MR, Papa S. Rat embryo exposure to all-trans retinoic acid results in postnatal oxidative damage of respiratory complex I in the cerebellum. *Mol Pharmacol* 2011; 80:704–713.

# Part IV

*Carotenoids and Vitamin A
in Other Diseases*

# 13 Vitamin A in Lung Development and Function

*Hans-Konrad Biesalski*

## CONTENTS

## INTRODUCTION

Vitamin A seems to be a two-sided sword with respect to pregnancy and child outcome. On the one side, it is discussed that a high intake of vitamin A will exert birth defects related to the cranial neural crest; on the other side, there is clear evidence that low vitamin A intake during pregnancy is related to low birth weight and impaired lung function in the newborn. Due to the controversial discussion regarding the questionable toxicity of preformed vitamin A, it is recommended to avoid food with high vitamin A concentrations (beef or chicken liver) or supplements containing vitamin A. In contrast, supplements containing vitamin A show beneficial effects especially in countries with low vitamin A intake on pregnancy outcome. Preformed vitamin A and its provitamin, β-carotene, play a major role with respect to lung development, maturation, and, at least, function.

The majority (90%) of preformed vitamin A in food occurs as retinol esterified with fatty acids, predominantly palmitate and stearate. Retinyl esters (REs) entering the intestine are hydrolyzed to retinol and fatty acid esters by a pancreatic lipase during lipid digestion. Following re-esterification in the intestinal mucosa, vitamin A is incorporated into chylomicrons and transported via the lymphatic route. During the circulation, the chylomicrons undergo a conversion as a result of the delivery of triglycerides to peripheral tissues mediated by the lipoprotein lipase.

The apolipoprotein E (apoE) incorporation into the chylomicrons with subsequent formation of chylomicron remnants ensures an uptake of the remnants and the incorporated RE via the apoE receptor into the hepatic stellate cells. From there, vitamin A is released as retinol bound to RBP to the blood stream under strictly controlled conditions.

Vitamin A can be delivered to extrahepatic target tissues in two different ways:

1. Via circulating retinol, released from the liver stores and bound to its binding protein RBP (retinol binding protein)
2. From chylomicrons via LPL activity during circulation following postprandial uptake

Whereas the first way was claimed to be the most important way of delivery, it has been overlooked that the delivery of REs to tissues is of great importance and may serve as an RBP-independent way.[1] The delivery of REs to peripheral cells and tissues contributes to the formation of local storage sides including the respiratory epithelium.[2] These local stores may serve as a pool for an acute increased demand.[3]

Intracellular hydrolysis of REs and formation of retinol and at least retinoic acid (RA) are strictly controlled to avoid uncontrolled formation of the active compound RA. Retinyl ester hydrolase (REH) is the enzyme that splits REs into retinol and the remaining fatty acid. Activity and at least expression of the REH are controlled via the concentration of "free" cytoplasmatic binding retinol and RA proteins apo-CRBP or apo-CRABP. The concentration of the binding proteins controls the formation of retinol and RA in a way that both metabolites are completely bound to the proteins. The tight metabolic control of RA formation may explain the teratogenic effect of exogenous supplied RA. At least only unbound RA can serve as a ligand to the nuclear receptor and exert its role in gene expression. If RA is delivered as a drug, the concentration of free RA exceeds the binding affinity of CRABP that may result in an overloading of the nuclear RA concentration.[3] The time to increase CRABP or to activate Cyp26, which metabolizes RA, is too long to protect the cells from RA overload (for review see D'Ambrosio[4]).

## IMPORTANCE OF VITAMIN A FOR THE MATURATION AND DIFFERENTIATION OF THE LUNG

The alveolar cells of type II are especially prepared to synthesize and secrete the surfactant.[5] RA is able to stop the expression of the surfactant protein-A (SP-A) concentration dependently[6] in human fetal lung explants. Insulin, TGF-β, and high concentrations of glucocorticoids can also down-regulate the SP-A-mRNA expression,[7] but lower concentrations of glucocorticoids are stimulating the expression of these genes.[8] In contrast, the SP-B-mRNA expression is increased in human fetal lung explants both by hyperoxia (rats) and by dexamethasone (human fetal lung explant).[9] Consequently, the formation of some surfactant proteins is regulated differently and selectively by RA together with glucocorticoids.

Prostaglandins of type $PGE_2$ are able to increase the surfactant synthesis. Under the influence of epidermal growth factor (EGF), the formation of prostaglandin rises, especially of $PGE_2$. On the other hand, the expression of the EGF receptor is increased by RA. EGF increases the proliferation of the lung tissues, and this leads to an amplified formation of surfactant phospholipids.[10] RA and EGF are both leading to an increase (40%, 80%) in the $PGE_2$ secretion in fetal lung cells of the rat in vitro.[11] The combination of RA and EGF though leads to a more than a sixfold increase in the $PGE_2$ secretion. Consequently, RA can interfere in the lung development by its modulating effect on the EGF expression and the subsequent $PGE_2$-induced surfactant formation. A sufficient and continuous availability (either on the blood pathway or by local storage sides) is pivotal, especially for a time-dependent regulation of the lung development and the related formation of the active metabolite RA.

## KINETIC PATTERNS OF RETINYL ESTERS DURING FETAL LUNG MATURATION

In fibroblast-like cells close to the alveolar cells, in type-II cells as well as in the respiratory epithelium RE, local extrahepatic stores are present.[2] The importance of these REs as "acute reserve" during the development of the lung becomes apparent during the late phase of gestation and the beginning of lung maturation. During this period, a rapid emptying of the RE storages in the lung of rat embryos occurs.[12] This depletion is the result of an increased demand in the process of the lung development, because the RA is "instantly" needed for the process of cellular differentiation (e.g., proximalization) and metabolic work (surfactant).

Three days prior delivery, the RE stores of rat fetal lungs decline and the retinol concentration increases.[12] Retinol is the source for an adequate supply with RA as ligand for the nuclear receptors, which control gene expression of different proteins responsible for late lung development and maturation. The prenatal lung development is also influenced by glucocorticoids. The steroid hormones have a similar effect on lung development as vitamin A; the two factors complement each other. This is not surprising, because the nuclear receptors for steroids and retinoids belong to the same multireceptor complex. The mode of action of glucocorticoids does not only come into action on the level of gene expression, but seems to have an impact on a much earlier phase of the vitamin release. The application of dexamethasone leads to an increase in the maternal and fetal retinol-binding protein, leading to an improvement of the vitamin A supply by channeling out via the normal hepatic pathway. Such an increase in the vitamin A concentration in the systemic circulation diminishes obviously the morbidity and mortality of prematures due to bronchopulmonal dysplasia.[13,14]

Dexamethasone and glucocorticoids are not only leading to an improvement of the total vitamin A supply through a change of the release from the liver, but they also influence, as recently described,[12] the metabolization of the vitamin A esters, which are stored in the lung. After administration of dexamethasone, but also without steroid application, a significant reduction of REs in the maturing lung can be detected, as well as a moderate increase in retinol, the hydrolyzation product of RE.

This observation may explain the therapeutic success with steroids and also their failures during the therapy of lung-distress syndrome of prematures. As far as an insufficient supply is concerned, inappropriate RE stores, caused by a shortage of supply of the fetal lung during the late pregnancy, the regulatory effect of glucocorticoids for the vitamin A metabolism of the lung cells cannot take place.

## PULMONARY CONSEQUENCES OF MARGINAL DEFICIENCY

Masuyama et al. demonstrated that a marginal vitamin A deficiency, which is not necessarily detected via low plasma retinol due to homoeostatic control, may have an important impact on late lung development.[15] In a rat study, they documented that RE increased rapidly to a peak on day 17 of gestation and decreased to a minimum on day 21 of gestation; there was a slight increase after birth. These data show that there might be a very short window in which the REs are stored in the lung, shortly before they are needed. If in case of early delivery the stores are not adequately filled, this might have serious consequences for lung maturation. Retinoid acid receptor (RAR)-alpha and -beta mRNA were detected in all samples obtained from perinatal and adult rat lung, and only a trace of RAR-gamma mRNA was detected in the fetuses on days 15, 17, and 19 of gestation and in the adults. After a maternal retinol deficiency of 28 days' duration, fetal body and lung weights were significantly lower than those of controls; the concentrations of retinyl palmitate and phosphatidylcholine (PC) in the lung after a maternal retinol deficiency of 14, 21, or 28 days were significantly lower than those of controls. Expression of RAR-beta mRNA in the group with 28 day retinol deficiency was lower than that in controls, that of RAR-alpha mRNA was increased, and that of RAR-gamma mRNA was not influenced by retinol deficiency. The rate of choline incorporation into PC in fetal lung explants was significantly higher in the group treated with RA than in controls. RA enhanced the effect of EGF on choline incorporation and prevented that of dexamethasone. Taken together, marginal deficiency results in altered expression of nuclear receptors of vitamin A with the consequence of impaired maturation prior to birth and even postnatal. Very recently, it has been shown that maternal vitamin A supplementation before, during, and after pregnancy improved lung function in offspring aged from 9 to 13 years.[16]

## INFLUENCE OF VITAMIN A SUPPLY FOR THE POSTNATAL DEVELOPMENT OF THE LUNG

An adequate vitamin A intake during pregnancy is of great importance for the formation of RE stores in the developing lung. These stores are the basis for RA formation during lung maturation and postnatal function. However, in the case of early delivery or very low birth weight, an insufficient vitamin A supply during pregnancy together with the poor vitamin A status of preterm neonates in general might have serious consequences.

A disease observed recurrently in connection with a poor vitamin A status is broncho-pulmonary dysplasia (BPD). The pathogenesis of BPD certainly depends on a multitude of factors. Some of the observed morphological changes are strongly reminiscent of vitamin A deficiency in the case of humans and animals. Particularly noted should be the focal loss of ciliated cells with keratinizing metaplasia and necrosis of the bronchial mucosa, as well as the increase in mucus-secreting cells.[17]

Focal keratinizing metaplasia, as it may occur after a vitamin A deficiency, especially strengthens the assumption of an impairment of the differentiation on the level of gene expression. Since vitamin A regulates the expression of different cytokeratins and therefore influences terminal differentiation, it seems obvious to suppose the existence of common mechanisms. Consequently, premature neonates are dependent on a sufficient supply of vitamin A to ensure adequate lung maturation. The earlier a child is born before its due date, the lower are its serum retinol levels.[18]

It was shown repeatedly that serum retinol level and RBP level depend on birth weight and are significantly lower in prematures with low birth weight compared to neonates with higher birth weight.[19] In addition, mothers from low-income groups had lower levels of serum vitamin A and a higher incidence of prematurity.[20] In the liver of prematures, significantly lower retinol levels can be found in prematures in comparison to neonates.[21] Plasma values lower than 20 µg/dL are not rare in this case, and they should be taken as an indicator of a relative vitamin A deficit.

Very low plasma vitamin A levels can be found recurrently in preterm compared to term neonates.[18,22] This can, among other things, be attributed to the relative immaturity of the liver for the synthesis of retinol-binding proteins. The neonate is almost exclusively dependent on the mother for its supply: this includes the lung REs, which are either directly absorbed by the cells (from chylomicrons) or else by esterification of retinol after uptake into the cells. These lung RE stores can be sufficiently filled only if the mother guarantees an appropriate vitamin A supply during late pregnancy.

Reduced plasma levels during the first developmental months have a considerable influence on the total development of infants as well as on their susceptibility to infections. In the case of reduced retinol plasma levels, repeated infections are more often described,[23,24] and they are counted among the main complications of a poor vitamin A supply in developing countries. In addition, the serum vitamin A level during infectious diseases, particularly of the respiratory tract, continues to drop.[25,26] This can be explained, on the one hand, by an increased metabolic demand and, on the other hand, by a renal elimination of retinol and of RBP during the process of acute infections.[24]

If the RE stores of the lung are low at delivery, these storage sites can hardly be replenished, and as a consequence lung function may be impaired. Even in developed countries with a wide variety of food containing preformed vitamin A, very low plasma levels (<1.4 µmol) occur and result in low umbilical cord blood levels und subsequently low retinol supply of the newborn.[27]

Despite low plasma retinol levels, most of the women with blood retinol levels <1.4 µmol/L had high plasma β-carotene levels. However, the high β-carotene in plasma may be taken as a sign for low conversion to vitamin A due to the recently described BCO polymorphism.[28] As a consequence, these women are, despite high dietary β-carotene intake, at risk of low vitamin A supply due to a very low intake of preformed vitamin A. This results in very low levels of retinol in blood and breast milk for the newborn. Normal levels of retinol in umbilical cord blood are in a range >0.7–2.3 µmol/L; levels <0.7 as measured in our study are defined as deficient.[29]

There are limited data linking the intake of vitamin A during pregnancy to cord blood levels and fetal development. Shah et al. demonstrated a strong relationship between low socioeconomic status, low cord blood, and low body weight of the newborn.[18]

Taken together, young women are at risk for low vitamin A intake due to misleading information and special dietary habits. As a consequence, the intake of preformed vitamin A from dietary sources (e.g., liver, liver products) is poor, and β-carotene cannot really compensate the higher need for vitamin A during pregnancy. Due to the fact that the human body can store vitamin A for 6–10 months, an inadequate supply might be a minor problem in women practicing a balanced diet. However, in cases of an inadequate diet, low socioeconomic status, short birthrate, or twins, the problem can become more or less serious.

## PREVENTION AND THERAPY

The best prevention of inadequate vitamin A supply of the newborn is an adequate mixed diet of the mother. During pregnancy and the breast-feeding period, the need for vitamin A and β-carotene is increased. On average, intake should be one-third higher during pregnancy, and during the breast-feeding period, it should be 0.7 mg/day higher than that for non-pregnant or non-breast-feeding women. Due to the importance of the vitamin for late lung development and maturation, sufficient intake should be especially ensured during the second and third pregnancy trimesters.

The German Nutrition Society (DGE) recommends a 40% increase in vitamin A intake for pregnant women and a 90% increase for breast-feeding women.[30] However, pregnant women or those considering becoming pregnant are generally advised to avoid the intake of vitamin A–rich liver and liver foods, based upon unsupported scientific findings. As a result, the provitamin A carotenoid β-carotene remains their essential source of vitamin A. Basic sources of provitamin A are orange and dark green vegetables, followed by fortified beverages that represent between 20% and 40% of the daily supply. The average intake of β-carotene in Germany is about 1.5–2 mg a day. Assuming a vitamin A conversion rate for β-carotene for juices of 4:1, and fruit and vegetables between 12:1 and 26:1, the total vitamin A contribution from β-carotene intake represents 10%–15% of the RDA.

The American Pediatrics Association[31] cites vitamin A as one of the most critical vitamins during pregnancy and the breast-feeding period, especially in terms of lung function and maturation. If the vitamin A supply of the mother is inadequate, her supply to the fetus will also be inadequate, as will later be in her milk. These inadequacies cannot be compensated by postnatal supplementation. A clinical study in pregnant women with short birth intervals or multiple births showed that almost one-third of the women showed plasma retinol levels below 1.4 μmol/L, which can be taken as borderline deficiency. Despite the fact that vitamin A– and β-carotene–rich food is generally available, risk groups for low vitamin A supply exist in the Western world.

One of the major findings of a poor vitamin A status of the mother is an increased risk of delivery of a very low-birth-weight infant.[20] The consequences of a poor vitamin A status on respiratory outcome in very low-birth-weight infants was estimated in 350 very low-birth-weight infants. Of that, 192 (55%) had a low vitamin A status.[32] Compared to infants with normal vitamin A status, the adjusted odds ratio for respiratory disability was 2.6. Adjusted odds ratios for BPD with low vitamin A were 3.5.

Doyle et al.[33] examined the nutrition of women who gave birth to children with low birth weight. The examinations focused on folic acid and iron, both micronutrients, which are mainly obtained from animal products (best bioavailability) and again from liver as the main source. In regard to the examined women (n = 55), it became clear that they did not have merely an insufficient folic acid intake but also low serum folic acid, erythrocyte folic acid, and iron values. The authors concluded that especially for women with multiple births, the plasma values of these micronutrients decrease rapidly and therefore fall within the critical area. This assumption was proven in a study with 500 pregnant adolescents aged 14–18 years.[34] Fifty-two percent had iron-deficiency anemia and 30% vitamin D deficiency (serum vitamin D < 25 nmol/L). The incidence of delivering a child small for gestational age was significantly higher in subjects with poorer folate status or low folate or iron intake. Low folate and iron might be a sign for inadequate nutrition. The best dietary source with respect to bioavailability for folate and iron is liver!

Again intake levels are to be seen as critical for young women, and it would be advisable if the corresponding enriched food could be recommended to women with multiple births, to women with a low socio economic status, and to breast-feeding women. In the case of β-carotene, the bioavailability in its isolated form is much higher than that in the food matrix. With regard to the study by Doyle, it is apparent that in the group of inadequate nutrition compared to the group of adequate nutrition, not only were the folic acid and iron values very low but also the vitamin A intake was 20% lower. The overall energy intake was reported to be 1633 kcal in the inadequately fed group compared to 2263 kcal in the adequately fed group, showing that a reduction of food energy below 1800 kcal does not meet requirements in terms of micronutrients. In conclusion, it has to be said that Doyle's and Baker's studies clearly show that young, pregnant, and breast-feeding women have to be seen as a group at risk in terms of the intake of micronutrients in particular of animal origin. In order to avoid a supply bottleneck and to avoid creating a problem that should not exist in industrialized countries, these groups should be advised to consume food enriched not only with folic acid and iron but also with β-carotene or even better with vitamin A and β-carotene. Supplementation with multi-micronutrients, including vitamin A, from the first trimester of pregnancy showed a significant lower number of SGA infants.[35] At recruitment of the participants (402), the prevalence of anemia was 13%, vitamin D insufficiency 72%, thiamine deficiency 12%, and folate deficiency 5%.

Sufficient data regarding a teratogenic effect of preformed vitamin A (retinol or RE) are missing. In particular, there are no data showing any relationship between consumption of liver and malformations. Nevertheless, a daily intake of more than 10,000 IU (3 mg preformed vitamin A) should be avoided to be on the safe site. During the first trimester of pregnancy, vitamin A intake from liver or supplements is not recommended, and in cases of a mixed diet (which may ensure sufficient liver stores) not really needed. However, females practicing a strong vegetarian or even vegan diet over the last years are at risk for insufficient vitamin A status and should be advised to take a supplement. A recent meta-analysis on the effect of multi-micro-nutrient supplementation including vitamin A (800–5000 IU/day) on pregnancy outcome showed a significant reduction of low birth weight when compared with the frequently recommended supplementation of iron and folic acid alone.[36]

Nevertheless, especially during the third trimester, a sufficient dietary vitamin A intake is strongly recommended. This ensures adequate liver stores of the newborn and adequate vitamin A stores in the lung to ensure sufficient maturation.

## THERAPEUTIC DELIVERY OF VITAMIN A TO THE NEWBORN

In case of an insufficient supply with vitamin A, it is controversially discussed whether and if how much vitamin A should be delivered. In the data from two Cochrane analyses[37,38] including 9 trials with 1291 infants, it was documented that vitamin A (intramuscular) supplementation in low-birth-weight infants (<1500 g or <32 week gestation) significantly reduced death or oxygen requirement at 1 month of age.

However, high doses in newborns may be not sufficient to supply cells and tissues with adequate vitamin A for a longer time period. This is due to two different physiological aspects of the newborn:

1. Inadequate protein synthesis of the liver including RBP
2. Inadequate (immature) fat digestion

1. In case of a high dose (50,000 IU) orally, the majority of the vitamin appears as RE in chylomicrons immediately following absorption. These chylomicrons are transported to the liver. During the transport, a part of the REs can be taken up by different target tissues, as shown in a recent published report on siblings without RBP.[39] As shown by Ross et al.,[40,41] administration of REs alone did not result in a sufficient increase in lung tissue RE levels of rats. However, the combination with RA results in an increased storage of REs due to "metabolic priming" of the cells of the lung.[42] In case of lower doses (5000 IU), the short-term increase in REs in the blood is not sufficient to supply much REs to target tissues. Indeed, low levels of vitamin A as usually delivered within breast milk of either supplemented or vitamin A–sufficient mothers may be more effective. Maternal supplementation postpartum improves breast milk retinol and vitamin A status of the newborn.[43,44] Colostral retinol increased significantly compared to an unsupplemented control group whereas the retinol level of mature milk (30 days after delivery) increased only slightly but significantly. The impact on child vitamin A status over a time period of 6 months shows that small doses might be better to improve the status than large doses. That might be explained by the special conditions of fat absorption (including fat soluble vitamins) in newborns and infants in the first 6 months of life.
2. Fat digestion is different in newborn infants compared to adults. Activities of different lipases vary and may contribute to differences in absorption depending on the concentration and at least way of administration via supplement or breast milk. The abrupt transition from carbohydrate to fat as

the main energy source that occurs at birth is not matched by commensurate endogenous fat-digesting capacity in the newborn. The activity of the gastric lipase and the lipase present in human milk may compensate the low activity of the pancreatic lipase and low bile salts in newborn.[45] In addition, the lingual lipase plays a dominant role in the hydrolysis of the milk fat globule, containing vitamin A.[4,53] This hydrolysis compensates the low luminal concentration of bile salts of newborns. Consequently, fat digestion from breast milk derived fat superior than the fat from supplements.

The favor of a lower dose on improving vitamin A status was shown in a recent study[46] with three different regimens in low-birth-weight infants (5,000 IU vitamin A three times per week for 4 weeks, 10,000 IU three times a week, and 15,000 IU once per week). The two higher doses did improve neither vitamin A status (retinol or RBP and RDR) nor outcome.

Thus, the attention should be directed to their supply with vitamin A. On the other hand, either the vitamin A supply of the premature for achieving adequate concentrations in the lung seems not sufficient or the availability of the vitamin for the corresponding cells of the lung is not guaranteed. An alternative solution could be the inhalative application of vitamin A. With this, the lung is directly attained and the inhalatively applied REs can be absorbed into the cells and metabolized in a controlled way, as shown in animal studies.[47]

The effect of inhalation was demonstrated in 25 vitamin A–deficient children compared to 115 untreated, who were randomly assigned to each of two treatment groups: one receiving retinyl palmitate by inhalation of two puffs of an aerosol containing 1 mg (3000 IU) per delivery to give a total of 2 mg (6000 IU); and the other receiving an aerosol without retinyl palmitate. Both treatments were administered every 2 weeks for 3 months. Serum retinol and retinol-binding protein concentrations in the vitamin A–treated group were 0.68 (SD 0.31) μmol/L and 59.4 (SD 24.2) mg/L before and 1.43 (SD 0.46) μmol/L (P<0.01) and 97.3 (SD 31.2) mg/L (P < 0.05) 3 months after supplementation with retinyl palmitate.[48] These data clearly show that the inhalative approach will improve vitamin A status and might have an impact on lung vitamin A concentration.

## HOW FAR IS PREFORMED VITAMIN A A RISK FOR THE DEVELOPING FETUS?

A low intake of vitamin A during pregnancy is a risk factor for low-birth-weight and increased neonatal mortality. Vitamin A deficiency during pregnancy results in typical malformations called the vitamin A deficiency syndrome. Occular defects (coloboma, retinal eversion, and further malformations) are the most frequently observed, followed by defects of the urogenital tract and the cardiovascular system. We described coloboma formation in siblings with genetically impaired RBP synthesis and subsequent low retinol delivery to the developing eye.[49] Low retinol supply to cells may indeed be responsible for this kind of malformation. Coloboma is claimed to be the phenotype of an STRA6 mutation, the membrane protein for retinol transport, in an Irish family.[50]

Because vitamin A is also important for fetal lung development and maturation, sufficient intake should be ensured during the second and third trimesters of pregnancy. Depending on animal feeding practices, liver may contain very excessively high concentrations of REs. This has led the German Federal Institute for Consumer Protection and Veterinary Medicine to advise pregnant women to avoid consuming liver. Most often, this results in an insufficient supply of vitamin A in pregnant and breast-feeding women who are therefore reliant on β-carotene as a source of vitamin A. The low intakes of vitamin A and β-carotene are also of concern in young women especially those with multiple births, of low social economic status, and breastfeeding women. In a recent study,[29] it was documented that a low vitamin A intake from preformed vitamin A resulted in low plasma retinol, low cord blood levels in newborns and colostrum- and breast-milk levels in 28% of the participants (10 out of 36). Based on preformed vitamin A sources, 75% of subjects did not reach the recommended intake level for pregnant women (1.1 mg/day) and 90% did not reach the recommended level for breast-feeding women (1.5 mg/day). Considering carotenoid intake in terms of retinol equivalents using the accepted conversion factor of 12,[51] still 67.8% of women of the German Nutrition Survey 2008[52] do not reach the vitamin A requirement for the breast-feeding period (see Chapter 18).

This situation is based upon two main causes: Young women and those considering pregnancy have been repeatedly advised to avoid the consumption of liver due to the excessive vitamin A content. From the consumption of 100 g of liver, only about 40% of its vitamin A content is absorbed, and it is absorbed more slowly than that from capsules with very little formation of RA, a formation that is tightly controlled. Thus the likelihood that critical vitamin A levels, in particular an increased formation of the teratogenic compound RA, occur from a dietary intake of liver is minimal and without any concern if ingested within the third trimester.

## CONCLUSION

Vitamin A is essential for adequate lung development and function. Inadequate intake during pregnancy might result in an increased risk for impaired lung maturation of the newborn resulting in an increased risk for BPD. Both preformed vitamin A from animal sources (in particular liver) and provitamin A from plant-based resources are needed to cover the increased demand during pregnancy.

## REFERENCES

1. Biesalski HK, Nohr D. New aspects in vitamin A metabolism: The role of retinyl esters as systemic and local sources for retinol in mucous epithelia. *J Nutr.* 2004;134 (12 Suppl):3453S–3457S.
2. Biesalski HK. Separation of retinyl esters and their geometric isomers by isocratic adsorption high-performance liquid chromatography. *Methods Enzymol.* 1990;189:181–189.
3. Gerlach T, Biesalski HK, Weiser H, Haeussermann B, Baessler KH. Vitamin A in parenteral nutrition: Uptake and distribution of retinyl esters after intravenous application. *Am J Clin Nutr.* 1989;50(5):1029–1038.

4. D'Ambrosio DN, Clugston RD, Blaner WS. Vitamin A metabolism: An update. *Nutrients.* 2011;3:63–103.

5. Zachman RD. Retinol (vitamin A) and the neonate: Special problems of the human premature infant. *Am J Clin Nutr.* 1989;50:413–424.

6. Metzler MD, Snyder JM. Retinoic acid differentially regulates expression of surfactant-associated proteins in human fetal lung. *Endocrinology.* 1993;133:1990–1998.

7. Weaver TE, Whitsett JA. Function and regulation of expression of pulmonary surfactant-associated proteins. *Biochem J.* 1991;273:249–264.

8. Odom MJ, Snyder JM, Boggaram V, Mendelson CR. Glucocorticoid regulation of the major surfactant associated protein (SP-A) and its messenger ribonucleic acid and of morphological development of human fetal lung in vitro. *Endocrinology.* 1988;123:1712–1720.

9. Allred TF, Mercer RR, Thomas RF, Deng H, Auten RL. Brief 95% $O_2$ exposure effects on surfactant protein and mRNA in rat alveolar and bronchiolar epithelium. *Am J Physiol.* 1999;276:L999–L1009.

10. Sundell HW, Gray ME, Serenius FS, Escobedo MB, Stahlman MT. Effects of epidermal growth factor on lung maturation in fetal lambs. *Am J Pathol.* 1980;100:707–725.

11. Oberg KC, Carpenter G. EGF-induced PGE2 release is synergistically enhanced in retinoic acid treated fetal rat lung cells. *Biochem Biophys Res Commun.* 1989;162(3):1515–1521.

12. Geevarghese SK, Chytil F. Depletion of retinyl esters in the lungs coincides with lung prenatal morphological maturation. *Biochem Biophys Res Commun.* 1994;200:529–535.

13. Shenai JP, Kennedy KA, Chytil F, Stahlman MT. Clinical trial of vitamin A supplementation in infants susceptible to bronchopulmonary dysplasia. *J Pediatr.* 1987;111:269–277.

14. Shenai JP, Rush MG, Stahlman MT, Chytil F. Plasma retinol-binding protein response to vitamin A administration in infants susceptible to bronchopulmonary dysplasia. *J Pediatr.* 1990;116:607–614.

15. Masuyama H, Hiramatsu Y, Kudo T. Effects of retinoids on fetal lung development in the rat. *Biol Neonate.* 1995;67:264–273.

16. Checkley W, West KP Jr., Wise RA, Baldwin MR, Wu L, LeClerq SC, Christian P, Katz J, Tielsch JM, Khatry S, Sommer A. Maternal vitamin A supplementation and lung function in offspring. *N Engl J Med.* 2010;362:1784–1794.

17. Stofft E, Biesalski HK, Zschaebitz A, Weiser H. Morphological changes in the tracheal epithelium of guinea pigs in conditions of "marginal" vitamin A deficiency. *Int J Nutr Res.* 1992;62:134–142.

18. Shah RS, Rajalekshmi R. Vitamin A status of the newborn in relation to gestational age, body weight, and maternal nutritional status. *Am J Clin Nutr.* 1984;40:794–800.

19. Mupanemunda RH, Lee DSC, Fraher LJ, Koura IR, Chance GW. Postnatal changes in serum retinol status in very low birth weight infants. *Early Hum Dev.* 1994;38:45–54.

20. Radhika MS, Bhaskaram P, Balakrishna N, Ramalakshmi BA, Devi S, Kumar BS. Effects of vitamin A deficiency during pregnancy on maternal and child health. *BJOG.* 2002;109(6):689–693.

21. Shenai JP, Chytil F, Stahlman MT. Liver vitamin A reserves of very low birth weight neonates. *Pediatr Res.* 1985;19:892–893.

22. Coutsoudis A, Adhikari M, Coovadia HM. Serum vitamin A (retinol) concentrations and association with respiratory disease in premature infants. *J Trop Pediatr.* 1995;41(4):230–233.

23. Pinnock CB, Douglas RM, Badcock NR. Vitamin a status in children who are prone to respiratory tract infections. *Aust Paediatr J.* 1986;2:95–99.

24. Filteau SM, Morris SS, Abbott RA et al. Influence of morbidity on serum retinol of children in a community-based study in northern Ghana. *Am J Clin Nutr.* 1993;58:192–197.

25. Neuzil KM, Gruber WC, Chytil F, Stahlman MT, Engelhardt B, Graham BS. Serum vitamin A levels in respiratory syncytial virus infection. *J Pediatr.* 1994;124:433–436.

26. Agarwal DK, Singh SV, Gupta V, Agarwal KN. Vitamin A status in early childhood diarrhoea, respiratory infection and in maternal and cord blood. *J Trop Pediatr.* 1996;42:12–14.

27. Schulz C, Engel U, Kreienberg R, Biesalski HK. Vitamin A and beta-carotene supply of women with gemini or short birth intervals: A pilot study. *Eur J Nutr.* 2007;46:12–20.

28. Leung WC, Hessel S, Méplan C, Flint J, Oberhauser V, Tourniaire F, Hesketh JE, von Lintig J, Lietz G. Two common single nucleotide polymorphisms in the gene encoding beta-carotene 15,15′-monoxygenase alter beta-carotene metabolism in female volunteers. *FASEB J.* 2009;23(4):1041–1045.

29. Godel JC, Basu TK, Pabst HF, Hodges RS, Hodges PE, Ng ML. Perinatal vitamin A (retinol) status of northern Canadian mothers and their infants. *Biol Neonate.* 1996;69(3):133–139.

30. DACH: Deutsche Gesellschaft für Ernährung, Österreichische Gesellschaft für Ernährung, Schweizerische Gesellschaft für Ernährungsforschung, Schweizerische Vereinigung für Ernährung: Referenzwerte für die Nährstoffzufuhr. 1. Auflage, 3. korrigierter Nachdruck, Umschau/Braus, 2008.

31. American Academy of Pediatrics. Committee on Nutrition. Nutritional needs of preterm infants. In: *Pediatric Nutrition Handbook* (Kleinman RE, ed.). 4th edn, pp. 55–87, American Academy of Pediatrics, Elk Grove Village, IL, 1998.

32. Spears K, Cheney C, Zerzan J. Low plasma retinol concentrations increase the risk of developing bronchopulmonary dysplasia and long-term respiratory disability in very-low-birth-weight infants. *Am J Clin Nutr.* 2004;80(6):1589–1594.

33. Doyle W, Srivastava A, Crawford MA, Bhatti R, Brooke Z, Costeloe KL. Inter-pregnancy folate and iron status of women in an inner-city population. *Br J Nutr.* 2001:86: 81–87.

34. Baker PN, Wheeler SJ, Sanders TA. A prospective study of micronutrient status in adolescent pregnancy. *Am J Clin Nutr.* 2009;89:1114–1124.

35. Brough L, Rees GA, Crawford MA. Effect of multiple micronutrient supplementation on maternal nutrient status, infant birth weight and gestational age at birth in low income, multi-ethnic population. *Br J Nutr.* 2010;104:437–445.

36. Shah PS, Ohlsson A. Effects of prenatal multimicronutrient supplementation on pregnancy outcome: A meta analysis. *CMAJ.* 2009;180:99–108.

37. Darlow BA, Graham PJ. Vitamin A supplementation for preventing morbidity and mortality in very low birth weight infants. *Cochrane Database Syst Rev.* 2002;(4):CD000501. Review. Update in: *Cochrane Database Syst Rev.* 2007;(4):CD000501.

38. Darlow BA, Graham PJ. Vitamin A supplementation to prevent mortality and short-long-term morbidity in very low birth weight infants. *Cochrane Database Syst Rev.* 2011;(4):CD 000501.

39. Biesalski HK, Frank J, Beck SC, Heinrich F, Illek B, Reifen R, Gollnick H, Seeliger MW, Wissinger B, Zrenner E. Biochemical but not clinical vitamin A deficiency results from mutations in the gene for retinol binding protein. *Am J Clin Nutr.* 1999;69:931–936. Erratum in: *Am J Clin Nutr.* 2000;71(4):1010.

40. Ross AC, Ambalavanan N. Retinoic acid combined with vitamin A synergizes to increase retinyl ester storage in the lungs of newborn and dexamethasone-treated neonatal rats. *Neonatology.* 2007;92:26–32.

41. Wu L, Ross AC. Acidic retinoids synergize with vitamin A to enhance retinol uptake and STRA6, LRAT, and CYP26B1 expression in neonatal lung. *J Lipid Res.* 2010;51:378–387.

42. Ross AC, Ambalavanan N, Zolfaghari R, Li NQ. Vitamin A combined with retinoic acid increases retinol uptake and lung retinyl ester formation in a synergistic manner in neonatal rats. *J Lipid Res.* 2006;47:1844–1851.

43. Bahl R, Bhandari N, Wahed MA, Kumar GT, Bhan MK; WHO/CHD Immunization-Linked Vitamin A Group. Vitamin A supplementation of women postpartum and of their infants at immunization alters breast milk retinol and infant vitamin A status. *J Nutr.* 2002;132(11):3243–3248.

44. Bezerra DS, Araújo KF, Azevêdo GM, Dimenstein R. Maternal supplementation with retinyl palmitate during immediate postpartum period: Potential consumption by infants. *Rev Saude Publica.* 2009;43(4):572–579.

45. Lindquist S, Hernell O. Lipid digestion and absorption in early life: An update. *Curr Opin Clin Nutr Metab Care.* 2010;13(3):314–312.

46. Ambalavanan N, Wu TJ, Tyson JE, Kennedy KA, Roane C, Carlo WA. A comparison of three vitamin A dosing regimens in extremely-low-birth-weight infants. *J Pediatr.* 2003;142:656–661.

47. Biesalski HK. Effects of intra-tracheal application of vitamin A on concentrations of retinol derivatives in plasma, lungs and selected tissues of rats. *Int J Vitam Nutr Res.* 1996;66(2):106–112.

48. Biesalski H, Reifen R, Fürst P, Edris M. Retinyl palmitate supplementation by inhalation of an aerosol improves vitamin A status of preschool children in Gondar (Ethiopia). *Br J Nutr.* 1999;82:179–182.

49. Seeliger MW, Biesalski HK, Wissinger B, Gollnick H, Gielen S, Frank J, Beck S, Zrenner E. Phenotype in retinol deficiency due to a hereditary defect in retinol binding protein synthesis. *Invest Ophthalmol Vis Sci.* 1999;40(1):3–11.

50. Casey J, Kawaguchi R, Morrissey M, Sun H, McGettigan P, Nielsen JE, Conroy J et al. First implication of STRA6 mutations in isolated anophthalmia, microphthalmia, and coloboma: A new dimension to the STRA6 phenotype. *Hum Mutat.* 2011;32(12):1417–1426.

51. IOM: Institute of Medicine. *Dietary Reference Intakes for Vitamin A, Vitamin K, Arsenic, Boron, Chromium, Copper, Iodine, Iron, Manganese, Molybdenum, Nickel, Silicon, Vanadium, and Zinc.* National Academy Press, Washington, DC, 2001.

52. NVS II. Max-Rubner-Institut. Nationale Verzehrsstudie. Ergebnisbericht Teil II. Bundesforschungsinstitut für Ernährung und Lebensmittel, 2008.

53. Hamosh M. The role of lingual lipase in neonatal fat digestion. *Ciba Found Symp.* 1979;70:69–98.

# 14 Vitamin A and Carotenoids in Lung Diseases

*Olaf Sommerburg and Werner G. Siems*

## CONTENTS

## INTRODUCTION

A number of respiratory defects were described in early vitamin A deficiency syndrome embryos, but were characterized as rare anomalies [1]. It is known today that vitamin A acquired from dietary retinol or carotenoids undergoes oxidative conversion to all-*trans* or 9-*cis* retinoic acid (RA), before exerting its biological effects during embryogenesis and after birth in lung tissue [2]. The RA isomers exert their biological effects by binding to RA receptors and retinoid-X receptors, respectively [3]. The two receptors belong to the steroid hormone receptor superfamily. They are able to form heterodimers, which bind to RA-responsive elements in genes expressed in the epithelium of pulmonary tissue. Rat embryos develop lung hypoplasia when RA in the maternal diet is reduced [4]. Retinoid acid receptors (RARs) are expressed throughout lung development, and in a mouse model (RAR compound null mutant mice), it could be shown that those animals develop left lung agenesis and also lung hypoplasia [5]. In addition, a wide spectrum of RA synthesizing, metabolizing, and binding proteins are found in developing lung tissue [6–8]. It is interesting that besides the liver, the lung is also a large store of retinoids in the body. The retinoids are stored as retinyl esters in lipid-laden fibroblasts that are abundant in the alveolar wall and often in close vicinity to type II pneumocytes [9–11]. Action of RA in lung development is well regulated. For initial budding, RA signaling is required. Thereafter, RA levels are down-regulated by RA-degrading enzymes to enable more distal branching and distal airway formation [8,12 ]. Furthermore, it could be shown that RARs are also required for correct lung alveolar septation [13–15]. A role for

RA in alveoli formation is supported by the finding that exogenous RA can stimulate alveoli formation in immature rat and mouse lung [10,16,17].

These data raise the question whether pharmacological regenerative therapy with retinoids might be a potential approach for human diseases characterized by reduced surface area such as emphysema or other diseases with changes of lung tissue architecture. A recent follow-up study showed that in a region with endemic vitamin A deficiency, children whose mothers had received vitamin A supplementation before, during, and for 6 months after pregnancy had better lung function when they were tested at 9–11 years of age when compared to children whose mothers had received placebo [18]. Furthermore, it was found that the period during which supplementation with vitamin A was most important was from gestation through a postnatal age of 6 months [18]. This suggests conservation of retinoid signaling in human lung development and demonstrates that gas-exchanging surface area can be manipulated by retinoids in man. The chapter reviews the literature on observational and supplementation studies on vitamin A and carotenoid intake (diet and supplements) with regard to different lung diseases. The chapter does not provide information about the role of vitamin A, retinoids, and carotenoids with regard to pulmonary infectious diseases and the topic of lung cancer.

## BRONCHOPULMONARY DYSPLASIA

Bronchopulmonary dysplasia (BPD) is the most common chronic respiratory disease in infants. BPD occurs in infants with immature or damaged lungs and requires often mechanical ventilation and prolonged oxygen therapy. The pathophysiology and the management of BPD have evolved over the past four decades as improved neonatal intensive care unit modalities have increased survival rates. Oxygen toxicity is believed to play the major role in the lung injury process leading to extensive inflammation [19]. However, also other issues such as barotrauma and/or volutrauma resulting from mechanical ventilation are discussed in the process of scarring the infant's lungs [20,21]. The incidence of BPD is about 34% in preterm babies weighing less than 1000 g and 15% in babies weighing between 1000 and 1200 g [22], but it occurs also in term babies with postnatal complications, such as neonatal sepsis, pneumonia, congenital heart defects, or meconium aspiration syndrome requiring prolonged ventilatory support. Infants with BPD have an increased tendency toward pulmonary problems in the following years of childhood, for example, respiratory infections, childhood asthma, and exercise intolerance. The treatment of BPD at present comprises symptomatic therapy with diuretics, bronchodilators, and steroids and involves providing a sufficient nutrition [23]. One of the current recommendations for BPD prevention is administration of vitamin A since retinoids are believed to promote lung healing, to increase the number of alveoli, and to decrease susceptibility to infections [22,24].

Evidence for the positive action of retinoids in the process of lung healing after BPD comes from animal experiments, for example, [25,26]. Albertine et al. [27] were able to show that daily vitamin A supplementation in preterm and mechanically ventilated lambs (125 days' gestation, term approximately 150 days) had a positive effect. They reported increased alveolar secondary septation, decreased thickness of the mesenchymal tissue cores between the distal air space walls, and increased alveolar capillary growth [27]. In postmortem lungs, they were also able to show important changes on the molecular level, such as less tropoelastin mRNA expression, less matrix elastin deposition, and less deposition of proliferating cell nuclear antigen in the mesenchymal tissue core of the distal air space walls [27]. On the other hand, mRNA expression and protein abundance of vascular endothelial growth factor (VEGF), VEGF receptor 2, midkine, and cleaved caspase 3 were increased [27]. This suggests that vitamin A treatment may partially improve lung development in injured lung tissue due to long-term ventilation in preterm neonates as seen for BPD by modulating expression of structural proteins. Also vitamin A given prenatally to pregnant rats has been found to stimulate alveologenesis of hypoplastic fetal lungs via up-regulation of certain transcription factors [28].

After birth, preterm infants have low levels of carotenoids [29] and also low levels of vitamin A [30,31]. Since low levels of vitamin A were shown to lead to an increased risk of BPD [32,33], supplementation of retinol or its precursors was developed as a strategy for prevention of that disease. In the last three decades, a number of clinical trials have been conducted investigating the effect of vitamin A supplementation for prevention of mortality and morbidity in very low-birth-weight infants (e.g., [34–38]). The last Cochrane review on the efficacy of vitamin A supplementation in very low-birth-weight infants (by intramuscular injection or in milk formula) supported the use of vitamin A [24]. Supplementation was associated with a trend toward a reduced number of deaths or oxygen requirement at 1 month of age compared to placebo [24]. Furthermore, in the largest study considered for this review, among infants with a birth weight lower than 1,000 g, death or chronic lung disease was lower in the vitamin A group compared to the placebo group [37]. These data provide evidence that supplementation of vitamin A should be part of the therapeutic management of BPD. Whether clinicians decide to utilize repeated intramuscular doses of vitamin A to prevent BPD may depend upon the local incidence of this disease.

Nowadays, it seems that administration of antenatal vitamin A to the mother in late pregnancy together with neonatal supplementation can prevent development of BPD not only in areas of endemic vitamin A deficiency but also in the developed world [39]. Supplementation in late pregnancy increases the cord blood vitamin A levels proportionately without having harmful effects for the mother and/or the fetus. In contrast, high doses of vitamin A given in early pregnancy are known to cause teratogenicity; however, those effects are not expected after the period of embryogenesis. According to a position paper of the Teratology Society, vitamin A supplementation in late pregnancy is safe and can be recommended [40]. In a recent paper, the impact of providing vitamin A to the routine pulmonary care of extremely low-birth-weight infants was reported. It could be shown that between pre- and post-vitamin A, the incidence of moderate to severe BPD decreased by 11%, from 33% to 22%, although no difference was found in the number of ventilator days or in the incidence of any other neonatal complications [41].

While the benefit of vitamin A with regard to BPD is obvious, this is not the case for carotenoids other than beta-carotene. Vogelsang et al. [42] reported in 2009 about a study in which lipid peroxidation (LPO) products, carotenoids, and other antioxidants in preterm infants with and without BPD were measured. When comparing the BPD group and the control group, infants in the BPD group had lower beta-carotene and vitamin A concentrations at day 7 after birth [42]. Interestingly, this was not the case for the other carotenoids, lycopene, lutein, alpha-carotene, and also not for vitamin E [42]. In a more recent study, Manzoni et al. investigated the effect of supplementation of the non-pro-vitamin-A carotenoids lutein and zeaxanthin in preterm very low-birth-weight neonates on severe neonatal diseases such as prematurity (ROP), necrotizing enterocolitis (NEC), and BPD [43]. The preterm infants received daily carotenoid supplementation (0.14 mg lutein + 0.0006 mg zeaxanthin) or placebo (5% glucose solution) from day of birth until the 36th week of life. Thereafter, ROP, NEC, and BPD incidence did not significantly differ in treated versus nontreated infants [43]. The authors concluded that lutein and zeaxanthin supplementation in preterm infants does not have an effect on BPD outcome.

## ASTHMA

Asthma is one of the leading chronic diseases in children and adults. It affects up to 20% of children in the United States. In Europe, asthma prevalence varies widely, from about 18% in Scotland to 3% in Russia or Switzerland [44]. Most children develop allergic asthma not before the age of 5 years. However, many of these children start with wheezing and show recurrent episodes of obstructive bronchitis already after infancy. Wheezing and viral respiratory tract infections (e.g., caused by respiratory syncytial virus, parainfluenza virus, human rhinovirus) in infants and young children are known to be associated with the development of nonallergic and allergic childhood asthma. The asthma prevalence has increased rapidly over the last decades in both developed and developing countries. Asthma currently affects more that 300 million individuals, and it is responsible for 180,000 deaths each year [45]. The cause of the increasing asthma prevalence is multifactorial. Dietary intake has also been implemented in the etiology of that disease. In particular, lower levels of antioxidant vitamins were associated with a greater risk of asthma in observational studies [46–49], but randomized controlled trials investigating the effect of supplementation of vitamin C and E on asthma risk were inconclusive [50–52].

Very little is known about the causal influence of vitamin A or carotenoids on asthma risk. RA is known to play a key role in lung development and regeneration of lung tissue [1,53], and we learned from animal studies that deficiency of vitamin A leads to changes in the lung architecture [8,14,54,55]. Ten years ago, McGowan et al. showed that vitamin A deficiency might be responsible for airway hyperresponsiveness, which is part of the asthma disease. Vitamin A deficiency increases airway reactivity in rats, and this was associated with a decrease in airway muscarinic receptor-2 (M2R) expression and function [56]. In prejunctional nerves, M2R is known to limit the release of acetylcholine from prejunctional neurons in bronchial smooth muscles [57]. This results in an attenuation of a nervus-vagus-induced bronchoconstriction, which may represent the link to the pathogenesis of allergic and nonallergic asthma [58,59]. However, the mechanisms for a vitamin-A-deficiency-related increased reactivity to cholinergic agents are not completely understood. On the one hand, a decrease in M2R gene expression could be involved; on the other hand, vitamin A deficiency might have an influence on the development of junctional cholinergic ganglia. Alternatively, deficiency of vitamin A could result in structural changes of lung parenchyma or airway walls that may promote responsiveness to a cholinergic stimulus. There is some evidence that altered mechanical properties of pulmonary parenchyma and airway walls influence airway responsiveness to cholinergic agents [60,61]. Therefore, in a follow-up study, McGowan et al. investigated whether vitamin-A-deficiency-related alterations in parenchymal and airway architecture contribute to airway hyperresponsiveness. They showed that RA given to vitamin-A-deficient rats over 12 days was able to reverse hyperresponsiveness by restoring M2R expression in bronchial tissue and prejunctional M2R function [62]. However, administration of RA did not alter the gas-exchanging surface area or airway wall density [62]. Nevertheless, these important results give insights into a possible role of retinol in the pathogenesis of asthma.

Multiple observational studies have found that vitamin A deficiency is associated with a higher risk of asthma; however, others did not. A meta-analysis published in 2008 did not support the hypothesis that dietary intake of antioxidants including vitamins C, E, and beta-carotene influences the risk of asthma [63]. However, a more recent review conducted in accordance with the MOOSE guidelines [64] and using more comprehensive sources and search strategies showed a negative association between overall dietary retinol intake and asthma [65]. Self-reported dietary intake was significantly reduced in patients with asthma by 182 mg/day corresponding to 26% and 30% of the currently recommended daily intake [66] of vitamin A for men and women, respectively [65]. Also pooled serum levels seemed to be decreased in asthmatics diagnosed by physicians, although only the data in two children studies reached statistical significance [67,68]. A significantly lower dietary intake of vitamin A could also be seen comparing patients with severe asthma to patients with mild asthma disease [69,70]. Furthermore, patients of a subgroup with physician-diagnosed severe asthma had also significantly lower serum levels of vitamin A [67,69,70]. However, there was no association between wheeze as outcome and low dietary intake of vitamin A. Nevertheless, low vitamin A serum levels were significantly associated with decreased odds of wheeze [65]. Several studies assessed also the effect of beta-carotene. However, only one study carried out in adolescents showed that lower serum levels were associated with a significantly increased risk of asthma [71], while other studies performed in adults showed no effect [72–74]. No study found an effect for dietary and serum levels of alpha-carotene [65]. Altogether, the meta-analysis showed a consistent negative association between overall dietary vitamin A intake and the odds of asthma and severe asthma, although the findings for wheeze were less consistent [65].

To the best of our knowledge, there are so far no randomized trials to support a causal relationship of vitamin A and/or carotenoids and asthma. However, Checkley et al. undertook a follow-up study in an area with chronic vitamin A deficiency to determine the effects of supplementation early in life with preformed vitamin A on subsequent asthma risk [75]. Their hypothesis was that vitamin A deficiency early in life may predispose to an increased risk of asthma. In the first original trial

(NNIPS-1), vitamin A or placebo was given to children during preschool years. Every 4 months, an oil-based supplement containing placebo or 200,000 IU of vitamin A (60,000 μg retinol equivalents [RE]) was given by trained field staff. Infants aged <6 months received a quarter dose (15,000 RE), children 6–12 months received half-dose (30,000 μg RE), and children 12–60 months of age received 60,000 μg RE. The supplementation intervals among the children were variable; however, more than 70% received four doses over 16 months [76]. The second trial (NNIPS-2) evaluated the effect of weekly supplementation with vitamin A or beta-carotene of females on mortality from all causes related to pregnancy [77]. The women received 7000 μg RE, 42 mg of beta-carotene, or placebo each week. The follow-up study included only a subgroup of children born to mothers who completed the pregnancy-to-postpartum dosing protocol and receiving afterward biannually vitamin A supplementation through a national program. However, the follow-up study did not find any differences between the vitamin A supplemented and placebo groups from either trial in the prevalence of lifetime or current asthma and wheeze or in spirometric indices of obstruction suggesting that vitamin A supplementation early in life is not associated with a decreased risk of asthma in an area with chronic vitamin A deficiency [75]. However, there might be several reasons for these results. One reason might be that asthma exacerbations, pulmonary infections, or airway inflammation may alter the vitamin A metabolism. Decreased serum retinol levels or urinary excretion of retinol was shown by increasing cellular demand [68,78–80]. If true, this would explain why observational studies have found an association between asthma and vitamin A [62,67,68,81,82] and would explain why vitamin A supplementation did not reduce the risk of asthma in either of the trials. However, our major criticism is aimed at the relatively short duration of vitamin A supplementation early in live and the implementation of a national vitamin A supplementation program in preschool-aged children that may have masked the effects of the original trials.

Finally, up to now, only data from epidemiological studies suggest that consumption of food containing vitamin A and beta-carotene such as fresh fruits, salads, and green vegetables has a beneficial effect on lung function and wheeze in children [83,84]. In a more recent paper it could be shown that early introduction of fresh fruit or vegetables in infancy, but not extra vitamins or cod-liver oil supplements might decrease the risk of asthma [85].

## CHRONIC OBSTRUCTIVE PULMONARY DISEASE/EMPHYSEMA

Chronic obstructive pulmonary disease (COPD) is a common disease of older people comprising the related diseases of airways to cigarette smoking such as chronic bronchitis and emphysema [86]. Patients with COPD describe breathlessness, physical examination reveals hyperinflation, decreased breath sounds, wheezes, crackles at the lung bases, and/or distant heart sounds [87]. The symptoms of the disease are related to changes in lung structure including reduction in gas-exchanging surface area, small airway obstruction caused in part by loss of alveolar attachments, chronic bronchitis including goblet cell hyperplasia and pulmonary hypertension [86,88]. These pathologic changes lead to impaired lung mechanics, reduced alveolar ventilation, and recurrent infection associated exacerbations [89].

Data of the animal experiments described in the first part of the chapter (e.g., [7,16,55]) and from the vitamin A supplementation study in women in areas of endemic dietary retinoid deficiency showing increased lung function in their offspring [18] suggest that retinoids may also have therapeutic effects also in COPD and emphysema.

To the best of our knowledge, no randomized trials are reported investigating the effects of vitamin A and/or carotenoid supplementation in patients with COPD or emphysema. However, few studies were published reporting therapeutic approaches with RA. A first study in patients with emphysema was reported in 2002 by Mao et al. [90]. In a 3 month placebo-controlled crossover study, 50 mg/m$^2$ all-*trans*-RA was given orally to 20 patients with severe emphysema [90]. The dose of all-*trans*-RA was well tolerated with only minor side effects. However, computer tomography (CT) imaging, lung function, and quality of life scores remained unchanged in this study. Nevertheless, all-*trans*-RA was able to alter the protease/antiprotease balance as seen for the MMP-9/TIMP-1 molar ratio in those patients [91]. A larger randomized multi-center feasibility study of all-*trans*-RA was undertaken in patients with moderately severe emphysema [92]. In this crossover designed study, patients received all-*trans*-RA either in a low dose (1 mg/kg), a high dose (2 mg/kg), 13-*cis* RA (1 mg/kg), or placebo for 6 months followed by a 3 month crossover period. The side effects were generally mild and self-limiting, but there were no changes in CT densitometry or CT imaging [92]. However, a delayed improvement was found in gas-transfer (DLCO) measurements that correlated with plasma drug levels in those patients receiving the higher dose of all-*trans*-RA [92]. Furthermore, 5 out of the 25 patients that received the higher dose had also delayed improvements in CT densitometry scores that correlated with plasma drug levels [92]. These three studies suggest that orally given retinoids may have positive effects in patients with emphysema or COPD. Therefore, development of RAR-specific agonists, which is currently in progress might be promising.

## CYSTIC FIBROSIS LUNG DISEASE

Cystic fibrosis (CF) is the most common fatal autosomal recessive disease caused by mutations in a single large gene on the long arm of chromosome 7 that encodes the transmembrane conductance regulator (CFTR) protein. CFTR is mainly expressed in the apical membrane of epithelial cells and submucosal glands [93]. The most common mutation in Caucasians Phe508del was found in 1989 [94], but today more than 1900 *CFTR* mutations are described. *CFTR* mutations are classified in five mutation classes determining the more or less pronounced course of CF disease. In the United States, an overall prevalence of roughly 1:4000 is found, but the prevalence varies according to ethnic background. Approximately, 1 in 3,300 Caucasian children, 1 in 15,000 African American children, and 1 in 32,000 Asian children are born with CF in the United States [95]. In the German population the CF prevalence is roughly 1:4500 [96]. CFTR protein dysfunction is associated with chloride efflux failure, which has to be compensated by increased sodium absorption. With active sodium influx, water and chloride will passively follow into the cell leading to changed cell volume, pH, transepithelial transport, and membrane conductance.

As a result, fluid and electrolyte composition of secretions are altered leading to progressive obstruction of airways and glandular ducts. This leads to fibrosis of affected organs [97], such as the lung, pancreas, liver, and the reproductive system of males. Since affection of the lung is life limiting, the understanding of the pathogenesis of CF lung disease is essential. In brief, reduction of periciliary lining fluid in the bronchi of CF patients leads to an increased viscosity of bronchial mucus resulting in an impaired mucociliary clearance. Consequently, the mucus will be permanently colonized by bacteria leading to recurrent infections. Those infections trigger the presence of activated neutrophils and lead to a permanent liberation of reactive oxygen species and proteases as part of their defense mechanism. This results in a state of chronic inflammation with increased oxidative stress and consequently in structural destruction of lung tissue. Affection of the pancreas leads to exocrine pancreatic insufficiency impeding the absorption of lipids, lipid-soluble micronutrients including vitamins A and E, and carotenoids [98]. In a longitudinal study of the oxidative status in 312 CF patients, it could be shown that carotenoid and vitamin E deficiencies occur very early in the course of CF disease [99]. Notably, antioxidants decrease further with bronchial infections [99]. Especially serum vitamin A seems to be associated with inflammation and disease severity and was shown to be negatively associated with C-reactive protein in CF patients [100]. Others have shown before that levels of selected carotenoids correlated negatively with IgG levels, an indirect measure of inflammation [98]. By contrast, lower antioxidant serum levels may have a direct influence on the frequency of respiratory infections in CF patients. Hakim et al. studied the longitudinal effect of serum vitamin A and E levels on the incidence of pulmonary exacerbations in pancreatic-insufficient and pancreatic-sufficient CF patients. He found that incidence of pulmonary exacerbations was directly correlated with lower vitamin A and E levels, even when found in the normal range [101]. Nevertheless, the presence of both, increased oxidative stress due to chronic infection and a defective antioxidant defense system, leads to deterioration of the clinical status of CF patients [102–104]. In view of these aspects, CF patients have a higher demand of antioxidants. This has provided the rationale for systematic studies on supplementation of antioxidants in CF patients, in particular not only of vitamin A and E, but also of carotenoids.

Besides antioxidant deficiency, there is also evidence of increased LPO parameters in blood of patients with CF lung disease [99]. Already 15 years ago, it was demonstrated in a 2 month open trial with 12 CF patients that supplementation of 4.4 mg beta-carotene three times a day led to correction of increased LPO levels in these patients [105]. Rust et al. [106] reported that long-term oral supplementation with 50 mg beta-carotene per day (approximately 1 mg of beta-carotene per kg body weight and day) restored levels of this carotenoid, while suboptimal supplementation done with doses of 10 mg beta-carotene per day or lower did not. The reason to give those high doses of beta-carotene was seen in the need to overcome the limited absorption and thus to achieve serum concentrations of healthy subjects. The high-dose treatment appeared also to be successful in lowering LPO products and in correction of total antioxidant capacity of blood plasma. Winklhofer-Roob et al. showed in another study that beta-carotene supplementation decreased malondialdehyde (MDA) formation and enhanced the resistance to copper(II)ion-induced oxidation of low-density lipoproteins [107]. Lung inflammation in CF is known to be associated with an accumulation of neutrophils in lung tissue and an increased release of neutrophilic elastase and other proteinases from those activated neutrophils. Winklhofer-Roob et al. hypothesized that enhanced antioxidant protection could represent an additional long-term strategy to attenuate host inflammatory response in CF patients [108]. To prove this, they evaluated the effect of oral beta-carotene supplementation on plasma neutrophil elastase/ alpha 1-proteinase inhibitor (NE/alpha1-PI) complex as a marker of lung inflammation and plasma MDA concentrations as LPO marker in 33 CF patients. In the presence of a more than 10-fold increase in serum beta-carotene concentrations, a more than 50% decrease in plasma MDA concentrations was found and plasma NE/alpha1-PI complex decreased about 20% [108]. At the same time, serum retinol concentrations increased significantly due to conversion of beta-carotene to retinol, which could have contributed to the decrease in NE/alpha1–PI complex levels. These results show that efficient antioxidant supplementation could attenuate lung inflammation in CF. Renner et al. evaluated in another double-blind, placebo-controlled study the effect of an oral beta-carotene supplementation for 6 months on clinical parameters in 24 CF patients [109]. The CF patients were randomized to receive beta-carotene 1 mg/kg/day (maximum 50 mg/day) for 3 months (high-dose supplementation) and 10 mg/day for a further 3 months (low-dose supplementation) or placebo. At monthly follow-up visits, they assessed the plasma beta-carotene concentration, total antioxidant capacity, MDA, and clinical parameters such as Shwachmann–Kulczycki score (special score for evaluation of clinical state of CF patients), body mass index (BMI), and lung function as forced expiratory volume in one second ($FEV_1$). No adverse events were observed during the supplementation period. After the 3 months of high-dose supplementation, concentration of beta-carotene increased significantly to the normal range, but decreased again to subnormal levels in the period of low-dose supplementation. MDA fell to normal levels, and the total antioxidant capacity showed a nonsignificant trend toward improvement during high-dose supplementation. However, the Shwachmann–Kulczycki score, lung function, and BMI did not show any changes in either of the treatment groups. Of note, the need of antibiotic treatment decreased significantly in the supplementation group from 14.5 days per patient during the 3 months before the study to 9.8 days per patient during high-dose supplementation and to 10.5 days per patient during low-dose supplementation, but increased in the placebo group [109]. Although those data have to be confirmed in larger cohorts, it gives an idea that orally given beta-carotene in a dose of 1 mg/kg/day might decrease pulmonary exacerbations. These data suggest that CF patients may benefit clinically from supplementation with beta-carotene [109]. Interestingly, after these results, supplementation of beta-carotene in a dose of 1 mg/kg/day was recommended in the European consensus on nutrition in patients with CF published in 2002 [110], but not in the consensus report on nutrition in pediatric CF patients published in the United States in the same year [111]. One reason for that might be that beta-carotene is seen only as precursor of retinol. Of note, vitamin A supplementation is recommended in a dose of 1,500 IU for infants with CF (0–12 month), 5,000 IU for CF patients aged from 1 to 3 years, 5,000–10,000 IU for patients aged from

4 to 8 years, and >10,000 IU in patients older than 8 years [110]. However, following the recommendations, elevated serum retinol levels were reported in CF patients of different age groups, and further studies were requested to evaluate dosing and monitoring care practices of vitamin A to ensure adequacy and avoid toxicity [112,113]. Whether and how much retinol and beta-carotene should be supplemented simultaneously was not scientifically discussed yet. Experiences from the CARET trial raised the question about a potential toxicity of carotenoid supplementation [114–117]. However, in the previous supplementation studies, beta-carotene supplementation seemed to be safe since plasma concentrations of other carotenoids and of retinol, as well as of other fat-soluble vitamins were not affected in CF patients [106]. Recent studies evaluated the clinical efficacy of a CF-tailored multivitamin formulation (AquADEKs®). The supplement contains 18,167 IU of retinol in a formulation containing 92% beta-carotene and 8% palmitate. CF patients aged from 4 to 10 years are recommended to take two tablets daily. The safety of this type of formulation was carefully evaluated, and it could be demonstrated that taking the supplement does not increase vitamin A above the normal serum levels observed in healthy controls [118–120]. However, normalization of beta-carotene plasma levels obtained in these studies was associated only with minor improvements on clinical parameters [118] and only with nonsignificant changes regarding LPO parameters [119,120]. In a pilot study conducted by the authors of that chapter, beta-carotene serum levels of 12 CF patients could be normalized by feeding them over 8 weeks with three table spoons of natural red palm oil containing high amounts of beta-carotene (unpublished data). When this effect can be confirmed in a larger trial, red palm oil simply added to daily food could be another way of safe beta-carotene and retinol supplementation.

There are no data about supplementation of other carotenoids in CF lung disease. However, CF patients show also deficiency of other carotenoids as seen in a number of studies, for example, [98,121].

## CONCLUSIONS

Although more than 600 carotenoids have been identified, the majority of research has focused on beta-carotene as precursor of retinol. Studies that examine fruit and vegetable intake have reported beneficial effects supporting that carotenoids might play a protective role in lung diseases. Published observational studies generally support the association between total carotenoids and some individual carotenoids. However, mechanisms of action in certain lung diseases are reported only for retinol and its derivatives. For few lung diseases, the use of retinol and beta-carotene can be recommended under certain conditions; however, the reported effect of supplement use was often disappointing. Therefore, additional data from interventions of dietary or supplemental approaches focused on increasing total or specific carotenoid intake can provide critical information on the biologic mechanisms supporting observational findings. Supplementation of retinol seems to be more critical and should be limited on states of deficiencies and other special conditions. Nevertheless, natural or designed derivatives of retinol represent pharmaceuticals with the potential to affect positively in a number of pulmonary disorders.

## REFERENCES

1. Wilson JG, Roth CB, Warkany J. An analysis of the syndrome of malformations induced by maternal vitamin A deficiency. Effects of restoration of vitamin A at various times during gestation. *Am J Anat* 1953;92:189–217.
2. Blomhoff R, Green MH, Berg T, Norum KR. Transport and storage of vitamin A. *Science* 1990;250:399–404.
3. Mangelsdorf DJ, Thummel C, Beato M et al. The nuclear receptor superfamily: The second decade. *Cell* 1995;83:835–839.
4. See AW, Kaiser ME, White JC, Clagett-Dame M. A nutritional model of late embryonic vitamin A deficiency produces defects in organogenesis at a high penetrance and reveals new roles for the vitamin in skeletal development. *Dev Biol* 2008;316:171–190.
5. Ghyselinck NB, Dupe V, Dierich A et al. Role of the retinoic acid receptor beta (RARbeta) during mouse development. *Int J Dev Biol* 1997;41:425–447.
6. Mollard R, Viville S, Ward SJ, Decimo D, Chambon P, Dolle P. Tissue-specific expression of retinoic acid receptor isoform transcripts in the mouse embryo. *Mech Dev* 2000;94:223–232.
7. Hind M, Corcoran J, Maden M. Temporal/spatial expression of retinoid binding proteins and RAR isoforms in the postnatal lung. *Am J Physiol Lung Cell Mol Physiol* 2002;282:L468–L476.
8. Malpel S, Mendelsohn C, Cardoso WV. Regulation of retinoic acid signaling during lung morphogenesis. *Development* 2000;127:3057–3067.
9. Hind M, Gilthorpe A, Stinchcombe S, Maden M. Retinoid induction of alveolar regeneration: From mice to man? *Thorax* 2009;64:451–457.
10. Dirami G, Massaro GD, Clerch LB, Ryan US, Reczek PR, Massaro D. Lung retinol storing cells synthesize and secrete retinoic acid, an inducer of alveolus formation. *Am J Physiol Lung Cell Mol Physiol* 2004;286:L249–L256.
11. Okabe T, Yorifuji H, Yamada E, Takaku F. Isolation and characterization of vitamin-A-storing lung cells. *Exp Cell Res* 1984;154:125–135.
12. Wongtrakool C, Malpel S, Gorenstein J et al. Down-regulation of retinoic acid receptor alpha signaling is required for sacculation and type I cell formation in the developing lung. *J Biol Chem* 2003;278:46911–46918.
13. McGowan S, Jackson SK, Jenkins-Moore M, Dai HH, Chambon P, Snyder JM. Mice bearing deletions of retinoic acid receptors demonstrate reduced lung elastin and alveolar numbers. *Am J Respir Cell Mol Biol* 2000;23:162–167.
14. Massaro GD, Massaro D, Chambon P. Retinoic acid receptor-alpha regulates pulmonary alveolus formation in mice after, but not during, perinatal period. *Am J Physiol Lung Cell Mol Physiol* 2003;284:L431–L433.
15. Massaro GD, Massaro D, Chan WY et al. Retinoic acid receptor-beta: An endogenous inhibitor of the perinatal formation of pulmonary alveoli. *Physiol Genomics* 2000;4:51–57.
16. Massaro GD, Massaro D. Postnatal treatment with retinoic acid increases the number of pulmonary alveoli in rats. *Am J Physiol* 1996;270:L305–L310.
17. Massaro GD, Radaeva S, Clerch LB, Massaro D. Lung alveoli: Endogenous programmed destruction and regeneration. *Am J Physiol Lung Cell Mol Physiol* 2002;283:L305–L309.
18. Checkley W, West KP, Jr., Wise RA et al. Maternal vitamin A supplementation and lung function in offspring. *N Engl J Med* 2010;362:1784–1794.
19. deLemos RA, Coalson JJ, Gerstmann DR, Kuehl TJ, Null DM, Jr. Oxygen toxicity in the premature baboon with hyaline membrane disease. *Am Rev Respir Dis* 1987;136:677–682.
20. Tremblay LN, Slutsky AS. Ventilator-induced injury: From barotrauma to biotrauma. *Proc Assoc Am Physicians* 1998;110:482–488.

21. Dreyfuss D, Saumon G. Ventilator-induced lung injury: Lessons from experimental studies. *Am J Respir Crit Care Med* 1998;157:294–323.

22. Bhandari A, Bhandari V. Bronchopulmonary dysplasia: An update. *Indian J Pediatr* 2007;74:73–77.

23. Deakins KM. Bronchopulmonary dysplasia. *Respir Care* 2009;54:1252–1262.

24. Darlow BA, Graham PJ. Vitamin A supplementation to prevent mortality and short- and long-term morbidity in very low birthweight infants. *Cochrane Database Syst Rev* 2011;CD000501.

25. Nadeau K, Jankov RP, Tanswell AK, Sweezey NB, Kaplan F. Lgl1 is suppressed in oxygen toxicity animal models of bronchopulmonary dysplasia and normalizes during recovery in air. *Pediatr Res* 2006;59:389–395.

26. Ruttenstock E, Doi T, Dingemann J, Puri P. Prenatal administration of retinoic acid upregulates insulin-like growth factor receptors in the nitrofen-induced hypoplastic lung. *Birth Defects Res B Dev Reprod Toxicol* 2011;92:148–151.

27. Albertine KH, Dahl MJ, Gonzales LW et al. Chronic lung disease in preterm lambs: Effect of daily vitamin A treatment on alveolarization. *Am J Physiol Lung Cell Mol Physiol* 2010;299:L59–L72.

28. Doi T, Sugimoto K, Puri P. Prenatal retinoic acid up-regulates pulmonary gene expression of COUP-TFII, FOG2, and GATA4 in pulmonary hypoplasia. *J Pediatr Surg* 2009;44:1933–1937.

29. Sommerburg O, Meissner K, Nelle M, Lenhartz H, Leichsenring M. Carotenoid supply in breast-fed and formula-fed neonates. *Eur J Pediatr* 2000;159:86–90.

30. Shenai JP, Chytil F, Jhaveri A, Stahlman MT. Plasma vitamin A and retinol-binding protein in premature and term neonates. *J Pediatr* 1981;99:302–305.

31. Peeples JM, Carlson SE, Werkman SH, Cooke RJ. Vitamin A status of preterm infants during infancy. *Am J Clin Nutr* 1991;53:1455–1459.

32. Hustead VA, Gutcher GR, Anderson SA, Zachman RD. Relationship of vitamin A (retinol) status to lung disease in the preterm infant. *J Pediatr* 1984;105:610–615.

33. Shenai JP, Chytil F, Stahlman MT. Vitamin A status of neonates with bronchopulmonary dysplasia. *Pediatr Res* 1985;19:185–188.

34. Pearson E, Bose C, Snidow T et al. Trial of vitamin A supplementation in very low birth weight infants at risk for bronchopulmonary dysplasia. *J Pediatr* 1992;121:420–427.

35. Ravishankar C, Nafday S, Green RS et al. A trial of vitamin A therapy to facilitate ductal closure in premature infants. *J Pediatr* 2003;143:644–648.

36. Shenai JP, Kennedy KA, Chytil F, Stahlman MT. Clinical trial of vitamin A supplementation in infants susceptible to bronchopulmonary dysplasia. *J Pediatr* 1987;111:269–277.

37. Tyson JE, Wright LL, Oh W et al. Vitamin A supplementation for extremely-low-birth-weight infants. National Institute of Child Health and Human Development Neonatal Research Network. *N Engl J Med* 1999;340:1962–1968.

38. Wardle SP, Hughes A, Chen S, Shaw NJ. Randomised controlled trial of oral vitamin A supplementation in preterm infants to prevent chronic lung disease. *Arch Dis Child Fetal Neonatal Ed* 2001;84:F9–F13.

39. Babu TA, Sharmila V. Vitamin A supplementation in late pregnancy can decrease the incidence of bronchopulmonary dysplasia in newborns. *J Matern Fetal Neonatal Med* 2010;23:1468–1469.

40. Public Affairs Committee of the Teratology Society. Teratology Society position paper: Recommendations for vitamin A use during pregnancy. *Teratology* 1987;35:269–275.

41. Moreira A, Caskey M, Fonseca R, Malloy M, Geary C. Impact of providing vitamin A to the routine pulmonary care of extremely low birth weight infants. *J Matern Fetal Neonatal Med* 2012;25:84–88.

42. Vogelsang A, van Lingen RA, Slootstra J et al. Antioxidant role of plasma carotenoids in bronchopulmonary dysplasia in preterm infants. *Int J Vitam Nutr Res* 2009;79:288–296.

43. Manzoni P, Guardione R, Bonetti P et al. Lutein and zeaxanthin supplementation in preterm very low-birth-weight neonates in neonatal intensive care units: A multicenter randomized controlled trial. *Am J Perinatol* 2012.

44. Masoli M, Fabian D, Holt S, Beasley R. The global burden of asthma: Executive summary of the GINA Dissemination Committee report. *Allergy* 2004;59:469–478.

45. Braman SS. The global burden of asthma. *Chest* 2006;130:4S–12S.

46. Schwartz J, Weiss ST. Dietary factors and their relation to respiratory symptoms. The Second National Health and Nutrition Examination Survey. *Am J Epidemiol* 1990;132:67–76.

47. Kelly FJ, Mudway I, Blomberg A, Frew A, Sandstrom T. Altered lung antioxidant status in patients with mild asthma. *Lancet* 1999;354:482–483.

48. Bodner C, Godden D, Brown K, Little J, Ross S, Seaton A. Antioxidant intake and adult-onset wheeze: A case-control study. Aberdeen WHEASE Study Group. *Eur Respir J* 1999;13:22–30.

49. Schunemann HJ, Grant BJ, Freudenheim JL et al. The relation of serum levels of antioxidant vitamins C and E, retinol and carotenoids with pulmonary function in the general population. *Am J Respir Crit Care Med* 2001;163:1246–1255.

50. Fogarty A, Lewis SA, Scrivener SL et al. Oral magnesium and vitamin C supplements in asthma: A parallel group randomized placebo-controlled trial. *Clin Exp Allergy* 2003;33:1355–1359.

51. Pearson PJ, Lewis SA, Britton J, Fogarty A. Vitamin E supplements in asthma: A parallel group randomised placebo controlled trial. *Thorax* 2004;59:652–656.

52. Romieu I, Sienra-Monge JJ, Ramirez-Aguilar M et al. Antioxidant supplementation and lung functions among children with asthma exposed to high levels of air pollutants. *Am J Respir Crit Care Med* 2002;166:703–709.

53. Chytil F. Retinoids in lung development. *FASEB J* 1996;10:986–992.

54. Desai TJ, Chen F, Lu J et al. Distinct roles for retinoic acid receptors alpha and beta in early lung morphogenesis. *Dev Biol* 2006;291:12–24.

55. Massaro D, Massaro GD. Retinoids, alveolus formation, and alveolar deficiency: Clinical implications. *Am J Respir Cell Mol Biol* 2003;28:271–274.

56. McGowan SE, Smith J, Holmes AJ et al. Vitamin A deficiency promotes bronchial hyperreactivity in rats by altering muscarinic M(2) receptor function. *Am J Physiol Lung Cell Mol Physiol* 2002;282:L1031–L1039.

57. Fryer AD, Jacoby DB. Muscarinic receptors and control of airway smooth muscle. *Am J Respir Crit Care Med* 1998;158:S154–S160.

58. Costello RW, Jacoby DB, Fryer AD. Pulmonary neuronal M2 muscarinic receptor function in asthma and animal models of hyperreactivity. *Thorax* 1998;53:613–616.

59. Adamko DJ, Yost BL, Gleich GJ, Fryer AD, Jacoby DB. Ovalbumin sensitization changes the inflammatory response to subsequent parainfluenza infection. Eosinophils mediate airway hyperresponsiveness, m(2) muscarinic receptor dysfunction, and antiviral effects. *J Exp Med* 1999;190:1465–1478.

60. Adler A, Cowley EA, Bates JH, Eidelman DH. Airway-parenchymal interdependence after airway contraction in rat lung explants. *J Appl Physiol* 1998;85:231–237.

61. Mitchell HW, Turner DJ, Noble PB. Cholinergic responsiveness of the individual airway after allergen instillation in sensitised pigs. *Pulm Pharmacol Ther* 2004;17:81–88.

62. McGowan SE, Holmes AJ, Smith J. Retinoic acid reverses the airway hyperresponsiveness but not the parenchymal defect that is associated with vitamin A deficiency. *Am J Physiol Lung Cell Mol Physiol* 2004;286:L437–L444.

63. Gao J, Gao X, Li W, Zhu Y, Thompson PJ. Observational studies on the effect of dietary antioxidants on asthma: A meta-analysis. *Respirology* 2008;13:528–536.

64. Stroup DF, Berlin JA, Morton SC et al. Meta-analysis of observational studies in epidemiology: A proposal for reporting. Meta-analysis of Observational Studies in Epidemiology (MOOSE) group. *JAMA* 2000;283:2008–2012.

65. Allen S, Britton JR, Leonardi-Bee JA. Association between antioxidant vitamins and asthma outcome measures: Systematic review and meta-analysis. *Thorax* 2009;64:610–619.

66. British Nutrition Foundation. Nutrition basics: Nutrient requirements and recommendations. http://www.nutrition.org.uk/attachments/044_SACN%20-%20Vit%20A%20 -%20update.pdf (accessed June 12, 2012).

67. Arora P, Kumar V, Batra S. Vitamin A status in children with asthma. *Pediatr Allergy Immunol* 2002;13:223–226.

68. Mizuno Y, Furusho T, Yoshida A, Nakamura H, Matsuura T, Eto Y. Serum vitamin A concentrations in asthmatic children in Japan. *Pediatr Int* 2006;48:261–264.

69. Picado C, Deulofeu R, Lleonart R et al. Dietary micronutrients/antioxidants and their relationship with bronchial asthma severity. *Allergy* 2001;56:43–49.

70. Baker JC, Tunnicliffe WS, Duncanson RC, Ayres JG. Dietary antioxidants and magnesium in type 1 brittle asthma: A case control study. *Thorax* 1999;54:115–118.

71. Rubin RN, Navon L, Cassano PA. Relationship of serum antioxidants to asthma prevalence in youth. *Am J Respir Crit Care Med* 2004;169:393–398.

72. Ford ES, Mannino DM, Redd SC. Serum antioxidant concentrations among U.S. adults with self-reported asthma. *J Asthma* 2004;41:179–187.

73. Nagel G, Linseisen J. Dietary intake of fatty acids, antioxidants and selected food groups and asthma in adults. *Eur J Clin Nutr* 2005;59:8–15.

74. Woods RK, Walters EH, Raven JM et al. Food and nutrient intakes and asthma risk in young adults. *Am J Clin Nutr* 2003;78:414–421.

75. Checkley W, West KP, Jr., Wise RA et al. Supplementation with vitamin A early in life and subsequent risk of asthma. *Eur Respir J* 2011;38:1310–1319.

76. West KP, Jr., Pokhrel RP, Katz J et al. Efficacy of vitamin A in reducing preschool child mortality in Nepal. *Lancet* 1991;338:67–71.

77. West KP, Jr., Katz J, Khatry SK et al. Double blind, cluster randomised trial of low dose supplementation with vitamin A or beta carotene on mortality related to pregnancy in Nepal. The NNIPS-2 Study Group. *BMJ* 1999;318:570–575.

78. Filteau SM, Morris SS, Abbott RA et al. Influence of morbidity on serum retinol of children in a community-based study in northern Ghana. *Am J Clin Nutr* 1993;58:192–197.

79. Morabia A, Sorenson A, Kumanyika SK, Abbey H, Cohen BH, Chee E. Vitamin A, cigarette smoking, and airway obstruction. *Am Rev Respir Dis* 1989;140:1312–1316.

80. Riccioni G, Barbara M, Bucciarelli T, Di IC, D'Orazio N. Antioxidant vitamin supplementation in asthma. *Ann Clin Lab Sci* 2007;37:96–101.

81. Riccioni G, Bucciarelli T, Mancini B, Di IC, Della VR, D'Orazio N. Plasma lycopene and antioxidant vitamins in asthma: The PLAVA study. *J Asthma* 2007;44:429–432.

82. Morabia A, Menkes MJ, Comstock GW, Tockman MS. Serum retinol and airway obstruction. *Am J Epidemiol* 1990;132:77–82.

83. Cook DG, Carey IM, Whincup PH et al. Effect of fresh fruit consumption on lung function and wheeze in children. *Thorax* 1997;52:628–633.

84. Gilliland FD, Berhane KT, Li YF, Gauderman WJ, McConnell R, Peters J. Children's lung function and antioxidant vitamin, fruit, juice, and vegetable intake. *Am J Epidemiol* 2003;158:576–584.

85. Nja F, Nystad W, Lodrup Carlsen KC, Hetlevik O, Carlsen KH. Effects of early intake of fruit or vegetables in relation to later asthma and allergic sensitization in school-age children. *Acta Paediatr* 2005;94:147–154.

86. Hogg JC. Pathophysiology of airflow limitation in chronic obstructive pulmonary disease. *Lancet* 2004;364:709–721.
87. Badgett RG, Tanaka DJ, Hunt DK et al. Can moderate chronic obstructive pulmonary disease be diagnosed by historical and physical findings alone? *Am J Med* 1993;94:188–196.
88. Saetta M, Ghezzo H, Kim WD et al. Loss of alveolar attachments in smokers. A morphometric correlate of lung function impairment. *Am Rev Respir Dis* 1985;132:894–900.
89. Hurst JR, Vestbo J, Anzueto A et al. Susceptibility to exacerbation in chronic obstructive pulmonary disease. *N Engl J Med* 2010;363:1128–1138.
90. Mao JT, Goldin JG, Dermand J et al. A pilot study of all-trans-retinoic acid for the treatment of human emphysema. *Am J Respir Crit Care Med* 2002;165:718–723.
91. Mao JT, Tashkin DP, Belloni PN, Baileyhealy I, Baratelli F, Roth MD. All-trans retinoic acid modulates the balance of matrix metalloproteinase-9 and tissue inhibitor of metalloproteinase-1 in patients with emphysema. *Chest* 2003;124:1724–1732.
92. Roth MD, Connett JE, D'Armiento JM et al. Feasibility of retinoids for the treatment of emphysema study. *Chest* 2006;130:1334–1345.
93. O'Sullivan BP, Freedman SD. Cystic fibrosis. *Lancet* 2009;373:1891–1904.
94. Riordan JR, Rommens JM, Kerem B et al. Identification of the cystic fibrosis gene: Cloning and characterization of complementary DNA. *Science* 1989;245:1066–1073.
95. National Institute of Health. Consensus Development Conference Statement. *Genetic Testing for Cystic Fibrosis*, April 14–16, 1997, Bethesda, MD: NIH, 2009.
96. Sommerburg O, Lindner M, Muckenthaler M et al. Initial evaluation of a biochemical cystic fibrosis newborn screening by sequential analysis of immunoreactive trypsinogen and pancreatitis-associated protein (IRT/PAP) as a strategy that does not involve DNA testing in a Northern European population. *J Inherit Metab Dis* 2010;33:S263–S271.
97. Rowntree RK, Harris A. The phenotypic consequences of CFTR mutations. *Ann Hum Genet* 2003;67:471–485.
98. Homnick DN, Cox JH, DeLoof MJ, Ringer TV. Carotenoid levels in normal children and in children with cystic fibrosis. *J Pediatr* 1993;122:703–707.
99. Lagrange-Puget M, Durieu I, Ecochard R et al. Longitudinal study of oxidative status in 312 cystic fibrosis patients in stable state and during bronchial exacerbation. *Pediatr Pulmonol* 2004;38:43–49.
100. Greer RM, Buntain HM, Lewindon PJ et al. Vitamin A levels in patients with CF are influenced by the inflammatory response. *J Cyst Fibros* 2004;3:143–149.
101. Hakim F, Kerem E, Rivlin J et al. Vitamins A and E and pulmonary exacerbations in patients with cystic fibrosis. *J Pediatr Gastroenterol Nutr* 2007;45:347–353.
102. Wood LG, Fitzgerald DA, Lee AK, Garg ML. Improved antioxidant and fatty acid status of patients with cystic fibrosis after antioxidant supplementation is linked to improved lung function. *Am J Clin Nutr* 2003;77:150–159.
103. Back EI, Frindt C, Nohr D et al. Antioxidant deficiency in cystic fibrosis: When is the right time to take action? *Am J Clin Nutr* 2004;80:374–384.
104. Winklhofer-Roob BM. Antioxidant status in cystic fibrosis patients. *Am J Clin Nutr* 1996;63:138–139.
105. Lepage G, Champagne J, Ronco N et al. Supplementation with carotenoids corrects increased lipid peroxidation in children with cystic fibrosis. *Am J Clin Nutr* 1996;64:87–93.
106. Rust P, Eichler I, Renner S, Elmadfa I. Long-term oral beta-carotene supplementation in patients with cystic fibrosis—Effects on antioxidative status and pulmonary function. *Ann Nutr Metab* 2000;44:30–37.
107. Winklhofer-Roob BM, Puhl H, Khoschsorur G, van't Hof MA, Esterbauer H, Shmerling DH. Enhanced resistance to oxidation of low density lipoproteins and decreased lipid peroxide formation during beta-carotene supplementation in cystic fibrosis. *Free Radic Biol Med* 1995;18:849–859.

108. Winklhofer-Roob BM, Schlegel-Haueter SE, Khoschsorur G, van't Hof MA, Suter S, Shmerling DH. Neutrophil elastase/alpha 1-proteinase inhibitor complex levels decrease in plasma of cystic fibrosis patients during long-term oral beta-carotene supplementation. *Pediatr Res* 1996;40:130–134.

109. Renner S, Rath R, Rust P et al. Effects of beta-carotene supplementation for six months on clinical and laboratory parameters in patients with cystic fibrosis. *Thorax* 2001;56:48–52.

110. Sinaasappel M, Stern M, Littlewood J et al. Nutrition in patients with cystic fibrosis: A European Consensus. *J Cyst Fibros* 2002;1:51–75.

111. Borowitz D, Baker RD, Stallings V. Consensus report on nutrition for pediatric patients with cystic fibrosis. *J Pediatr Gastroenterol Nutr* 2002;35:246–259.

112. Maqbool A, Graham-Maar RC, Schall JI, Zemel BS, Stallings VA. Vitamin A intake and elevated serum retinol levels in children and young adults with cystic fibrosis. *J Cyst Fibros* 2008;7:137–141.

113. Graham-Maar RC, Schall JI, Stettler N, Zemel BS, Stallings VA. Elevated vitamin A intake and serum retinol in preadolescent children with cystic fibrosis. *Am J Clin Nutr* 2006;84:174–182.

114. Omenn GS, Goodman GE, Thornquist MD et al. Risk factors for lung cancer and for intervention effects in CARET, the Beta-Carotene and Retinol Efficacy Trial. *J Natl Cancer Inst* 1996;88:1550–1559.

115. Sommerburg O, Langhans CD, Arnhold J et al. Beta-carotene cleavage products after oxidation mediated by hypochlorous acid—A model for neutrophil-derived degradation. *Free Radic Biol Med* 2003;35:1480–1490.

116. Alija AJ, Bresgen N, Sommerburg O, Langhans CD, Siems W, Eckl PM. Cyto- and genotoxic potential of beta-carotene and cleavage products under oxidative stress. *Biofactors* 2005;24:159–163.

117. Hurst JS, Contreras JE, Siems WG, van Kuijk FJ. Oxidation of carotenoids by heat and tobacco smoke. *Biofactors* 2004;20:23–35.

118. Papas KA, Sontag MK, Pardee C et al. A pilot study on the safety and efficacy of a novel antioxidant rich formulation in patients with cystic fibrosis. *J Cyst Fibros* 2008;7:60–67.

119. Sagel SD, Sontag MK, Anthony MM, Emmett P, Papas KA. Effect of an antioxidant-rich multivitamin supplement in cystic fibrosis. *J Cyst Fibros* 2011;10:31–36.

120. Sadowska-Woda I, Rachel M, Pazdan J, Bieszczad-Bedrejczuk E, Pawliszak K. Nutritional supplement attenuates selected oxidative stress markers in pediatric patients with cystic fibrosis. *Nutr Res* 2011;31:509–518.

121. Schupp C, Olano-Martin E, Gerth C, Morrissey BM, Cross CE, Werner JS. Lutein, zeaxanthin, macular pigment, and visual function in adult cystic fibrosis patients. *Am J Clin Nutr* 2004;79:1045–1052.

# 15 Chemopreventive Effects of Astaxanthin on Inflammatory Bowel Disease and Inflammation-Related Colon Carcinogenesis

*Masashi Hosokawa and Yumiko Yasui*

## CONTENTS

## INTRODUCTION

Carotenoids are a family of natural pigments with at least 750 members. They possess a polyisoprenoid structure and are synthesized by plants, seaweeds, and microorganisms, but not by animals. Fruits and vegetables are the major sources of dietary carotenoids [1]. Yellow-orange vegetables and fruits provide $\alpha$- and $\beta$-carotene, orange fruits provide $\beta$-cryptoxanthin, dark green leaves and vegetables

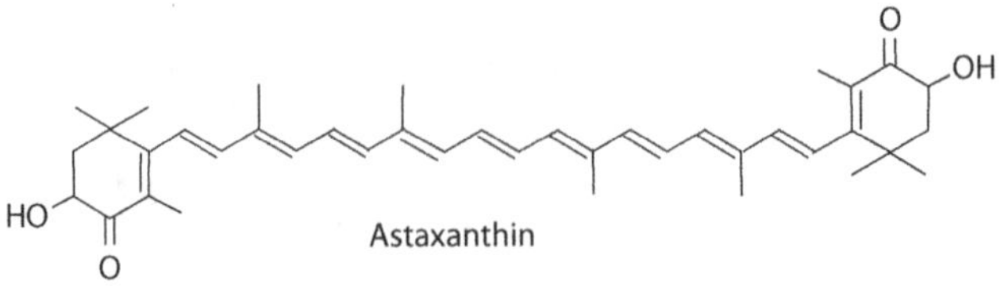

**FIGURE 15.1**   Structure of astaxanthin.

provide lutein and violaxanthin, and tomatoes contain lycopene. On the other hand, marine organisms contain many kinds of carotenoids with unique structures [2,3]. Astaxanthin (3,3′-dihydroxy-β,β-carotene-4,4′-dione) (Figure 15.1), which is a red ketocarotenoid, is contained in salmon and shrimp. Microorganism production of astaxanthin has been performed at an industrial scale. For example, *Haematococcus pluvialis* synthesizes and accumulates astaxanthin ester forms under stressful conditions, such as light and temperature. Other natural sources of astaxanthin used for the industrial production are krill and *Xanthophyllomyces dendrorhous*. Astaxanthin can also be chemically synthesized [4]. It has been reported that astaxanthin has several health beneficial effects, such as antioxidant activity [5], immunomodulation [6], and anticarcinogenesis [7,8]. Furthermore, astaxanthin has recently been used in cosmetics in Japan.

Inflammatory bowel diseases (IBDs) are chronic and relapsing disorders involving uncontrolled inflammation of the gastrointestinal tract. The main types of IBD are Crohn's disease and ulcerative colitis. IBD has been found to have a higher incidence in developed countries, such as North America and Europe than in Asia, Africa, and South America (Table 15.1) [9,10]. In North Europe and South Europe,

**TABLE 15.1**
**Incidence Rates for IBD in Developed Countries and the Asia-Pacific Region**

|  | Incidence Year | (Patients/$10^5$/Year) | |
| --- | --- | --- | --- |
|  |  | Ulcerative Colitis | Crohn's Disease |
| North America | 1993 | 8.3 | 6.9 |
|  | 2004 | 14.3 | 14.6 |
| North Europe | 1993 | 11.8 | 7.0 |
| South Europe | 1993 | 8.7 | 3.9 |
| United Kingdom | 1994 | 13.9 | 8.3 |
| India | 2000 | 6.0 | – |
| New Zealand | 2004 | 16.5 | 7.6 |
| South Korea | 2005 | 3.08 | 1.34 |
| Japan | 1991 | 1.9 | 0.51 |
| Hong Kong | 2001 | 1.2 | 1.0 |

*Source:*   Ahuja, V. and Tandon, R.K., *J. Dig. Dis.*, 11, 134, 2010.

the IBD affects more than 1.8 and 1.2 millions of people each year, respectively [10]. The incidence and the prevalence rates of IBD in the Asia-Pacific region are lower than those of Europe and North America [10]. However, the incidence of IBD has been rising with the development and the economic growth in underdeveloped countries. Environmental influences are thus considered important factors in the emergence of IBD worldwide.

Symptoms of IBD are abdominal pain, vomiting, rectal bleeding, and weight loss. An imbalance of pro-inflammatory cytokines, such as interleukin (IL)-1, IL-6, tumor necrosis factor $\alpha$ (TNF-$\alpha$), and anti-inflammatory cytokines, such as IL-10 and IL-4, is considered to play a pivotal role as the pathogenesis of IBD, although its mechanism remains poorly understood [11]. In addition, excessive expression of cyclooxygenase 2 (COX-2) and inducible nitric oxide synthase (iNOS) to produce prostaglandin $E_2$ and nitric oxide also cause the immune dysregulation observed in IBD [12]. Further, chronic inflammation of the gastrointestinal tract is a risk factor for the colon carcinogenesis [13]. Patients with ulcerative colitis are at an increased risk of colon cancer, the incidence of which varies worldwide [14]. In particular, the risk of colorectal cancer increases with increased duration and extent of ulcerative colitis. Therefore, anti-inflammatory compounds have been considered potential chemopreventive agents for colorectal cancer.

Most of the current therapies for IBD include glucocorticosteroids, sulfasalazine, and 5-aminosalicylic acid. However, the adverse effects after prolonged treatment and a high recurrence rate unavoidably limit their clinical application. Recently, chemoprevention has included the uses of specific natural or synthetic compounds in an attempt to prevent and improve the process of IBD and carcinogenesis [15]. In particular, considerable attention has been focused on the identification of the dietary phytochemicals that have the ability to interfere with inflammation and the carcinogenic process [16–18]. In this chapter, we introduce the chemopreventive effects of astaxanthin on IBD and inflammation-related colon carcinogenesis.

## PREVENTIVE EFFECT OF ASTAXANTHIN ON DEXTRAN SULFATE SODIUM-INDUCED COLITIS

### ANTI-INFLAMMATORY EFFECT OF ASTAXANTHIN

IBD is an immune-mediated intestinal inflammatory condition. Dextran sulfate sodium (DSS) is commonly used to chemically induce intestinal inflammation in rodent models. DSS-induced colitis is characterized by bloody diarrhea, epithelial cell damage, and immune cell infiltration, as well as increased mRNA expression of the pro-inflammatory cytokines observed in patients with ulcerative colitis [19].

ICR mice (male, 5 week old) were allocated in three groups and fed diets containing 0 (group 1), 100 (group 2), and 200 ppm (group 3) of chemically synthesized astaxanthin (free form, >95%) (Figure 15.2A). Four weeks after the start of the astaxanthin feeding, the mice in groups 1 through 3 were given 1.5% DSS in their drinking water for 7 days (Figure 15.2A). Mice in group 4 were served as an untreated control.

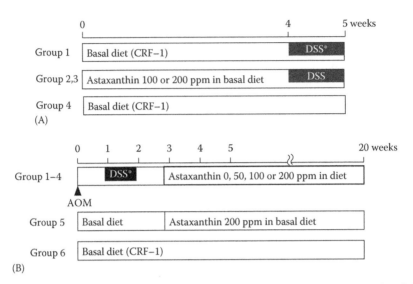

**FIGURE 15.2** Experimental protocols for animal studies. (A) DSS-induced colitis, (B) azoxymethane (AOM)/DSS-induced colitis-associated colon carcinogenesis. *1.5% DSS in tap water.

The diets containing astaxanthin at 100 and 200 ppm did not produce any observable toxicity or any gross change in any examined organ. The mean weights of the body and liver were not significantly different in any of the groups. Histochemical stainings of the colonic mucosa and the scoring of the colonic inflammation were conducted at week 5. DSS-induced colitis was characterized by crypt destruction and loss, glandular architecture distortion, erosion, and mucosal ulceration as shown in group 1 (1.5% DSS alone) (Figure 15.3A) [20]. Further, immune cells such as lymphocytes and macrophages infiltrated into submucosa and lamina propria in the mice in group 1, but not in the control mice (Figure 15.3A and D). In contrast to 1.5% DDS-treated mice, 100 and 200 ppm astaxanthin diets prevented DSS-induced pathological changes of the colonic mucosa (Figure 15.3B and C). Crypts are clearly observed in the mice fed 200 ppm of astaxanthin. The inflammation score of group 1 (1.5% DSS alone) was significantly greater than that of group 4 (untreated mice) (Figure 15.3E). In fact, dietary astaxanthin dose dependently decreased the colonic inflammation score. In particular, 200 ppm astaxanthin diet (group 3) significantly decreased the colonic inflammation score when compared with group 1 ($P < 0.001$).

Astaxanthin is also produced by microalgae. *H. pluvialis* contains 1.5%–3% astaxanthin and recently became commercially available. The chemical form of astaxanthin in *H. pluvialis* is approximately 70% monoester, 25% diester, and 5% free. The ICR mice (male, 5 week old) were fed astaxanthin from *H. pluvialis* for 4 weeks and then given 1.5% DSS in their drinking water for 7 days together with the astaxanthin-containing diet. As shown in Figure 15.4, astaxanthin from *H. pluvialis* also prevented DSS-induced colitis. These results indicate that natural astaxanthin from microalgae also has high potential application as an anti-inflammatory compound.

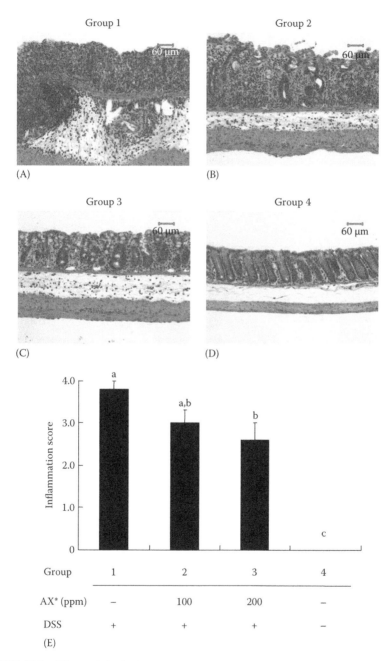

**FIGURE 15.3** Histopathological observation and inflammatory score of the large bowel in ICR mice fed astaxanthin. (A) DSS-alone-treated group (group 1), (B) DSS + 100 ppm astaxanthin (group 2), (C) DSS + 200 ppm astaxanthin (group 3), (D) untreated group (group 4), (E) inflammation scores of the large bowel in mice. Inflammation was graded according to the morphological criteria described by Cooper et al. [36]. Different letters are significantly different from each other at $P < 0.05$. *AX: astaxanthin.

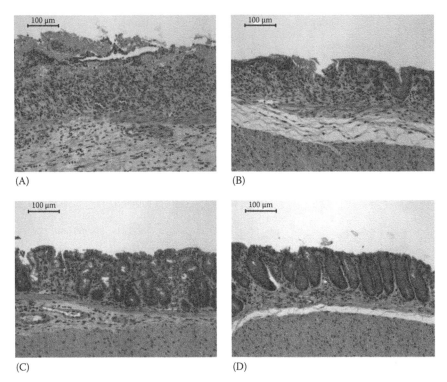

**FIGURE 15.4**  Histopathological observation of the large bowel in ICR mice fed astaxanthin mixture extracted from *H. pluvialis.* (A) DSS-alone-treated group, (B) DSS + 100 ppm astaxanthin group, (C) DSS + 200 ppm astaxanthin group, (D) untreated group.

## Suppression of Pro-Inflammatory Cytokine mRNA Expression in the Colonic Mucosa by Astaxanthin

IBD evidence suggests a disturbed balance between pro-inflammatory and anti-inflammatory cytokines. Increased levels of the pro-inflammatory cytokines such as IL-6, IL-1β, and TNF-α are observed in the bowel. We therefore measured mRNA expression levels of IL-6, IL-1β, and TNF-α in the colon mucosa of the mice fed astaxanthin. The mRNA expression levels of all inflammatory cytokines in the mice treated with DSS (group 1) were increased compared with the untreated mice (group 4) (Figure 15.5). The mRNA expressions of TNF-α, IL-6, and IL-1β were dose dependently suppressed by astaxanthin (Figure 15.5A through C). When astaxanthin was fed at 100 ppm, the expression levels of IL-6 and IL-1β mRNA decreased to 70% and 60%, respectively, in the DSS-treated mice without astaxanthin (group 1). The astaxanthin feeding at 200 ppm caused an 80% reduction in the expression of IL-6 and IL-1β mRNA compared with DSS treatment alone (group 1). In addition, COX-2 mRNA expression level was also decreased to 50% and 80% in group 1 by astaxanthin at 100 and 200 ppm, respectively (Figure 15.5D). Astaxanthin feeding at 100 ppm increased the mRNA expression of iNOS, whereas

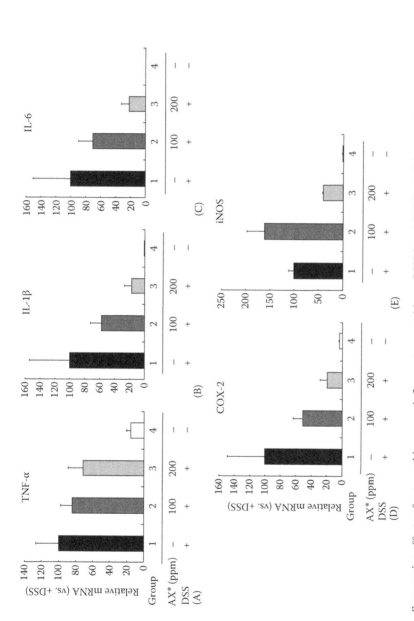

**FIGURE 15.5** Suppressive effects of astaxanthin on pro-inflammatory cytokines, COX-2 and iNOS mRNA expression in the large bowel of DSS-treated mice. Group 1: DSS-alone-treated group, group 2: DSS + 100 ppm astaxanthin, group 3: DSS + 200 ppm astaxanthin, group 4: untreated group. (A) TNF-α, (B) IL-1β, (C) IL-6, (D) COX-2, and (E) iNOS. *AX: astaxanthin.

astaxanthin feeding at 200 ppm decreased it (Figure 15.5E). These data indicate that astaxanthin down-regulated the mRNA expression of pro-inflammatory cytokines and inflammation-related enzymes in the colonic mucosa of the DSS-treated mice. These anti-inflammatory effects by astaxanthin are important in the prevention of DSS-induced colitis.

## INHIBITION OF NUCLEAR TRANSCRIPTION FACTOR KAPPAB ACTIVATION BY ASTAXANTHIN

Aberrant activation and expression of nuclear transcription factor kappaB (NF-κB) are known to be associated with a number of chronic inflammatory diseases [21]. Inflammatory genes, such as IL-6, IL-1β, and TNF-α, are the most common target genes participating in the activation of NF-κB signaling. In the colons of the DSS-treated mice, NF-κB was strongly stained by immunohistochemical analysis (Figure 15.6A). By astaxanthin feeding, the staining level of NF-κB in the colon was attenuated to that of control mice (group 4). Immunohistochemical scoring also showed that astaxanthin feeding dose dependently decreased NF-κB staining, and significant decreases were observed in the mice fed 100 (group 2, $P < 0.05$) and 200 ppm (group 3, $P < 0.001$) of astaxanthin, respectively (Figure 15.6B and C).

DSS ingestion causes macrophage and granulocyte infiltration in the colon, similar to the pathological changes observed in patients with ulcerative colitis. In macrophages and epithelial cells isolated from the inflamed gut of a patient with IBD, a subunit of NF-κB, p65, markedly expresses and regulates mRNA expression of pro-inflammatory cytokines [22]. Astaxanthin inhibited nuclear translocation of phosphorylated p65 of NF-κB in macrophage-like RAW264.7 cells. It has also been reported that astaxanthin inhibits the inhibitory subunit IκB kinase-dependent NF-κB activation in RAW264.7 cells [23,24]. Thus, inhibition of NF-κB activation in macrophages is thought to be one of the molecular mechanisms related with antiulcerative colitis effects by astaxanthin (Figure 15.7). Our finding thus demonstrates that astaxanthin prevents DSS-induced colitis by suppressing the inflammatory events via inhibiting NF-κB activation in the colonic mucosa.

## PREVENTION OF INFLAMMATION-RELATED COLORECTAL CANCER BY ASTAXANTHIN

### INCIDENCE AND MULTIPLICITY OF COLONIC MUCOSAL ULCER AND DYSPLASIA

The risk of colorectal cancer increases with longer duration and greater extent of colitis and degree of inflammation in the bowel. Meta-analysis by Eaden et al. [25] has shown that the risk of colorectal cancer in patients with ulcerative colitis is 2% at 10 years, 8% at 20 years, and 18% at 30 years of disease duration. In addition,

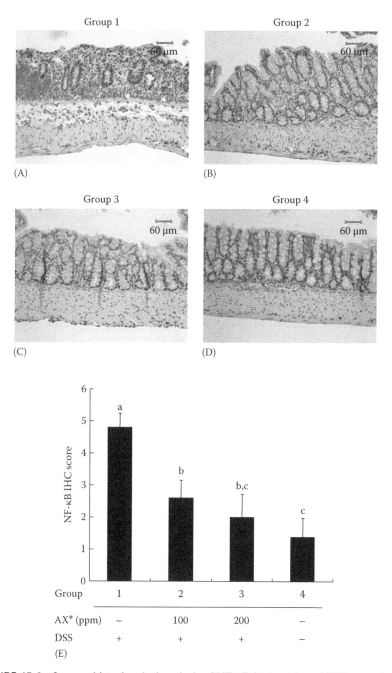

**FIGURE 15.6** Immunohistochemical analysis of NF-κB in the colon of DSS-treated mice fed astaxanthin. (A) DSS-alone-treated group (group 1), (B) DSS + 100 ppm astaxanthin (group 2), (C) DSS + 200 ppm astaxanthin (group 3), (D) untreated group (group 4). (E): Immunohistochemical (IHC) scores of NF-κB expression in the colon of mice. Different letters are significantly different from each other at $P < 0.05$. *AX: astaxanthin.

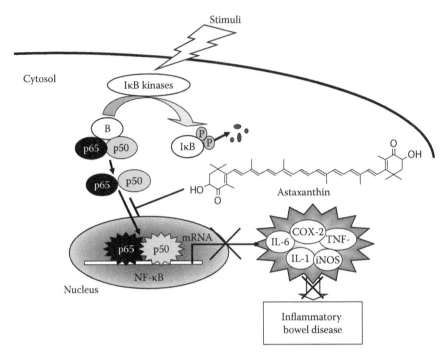

**FIGURE 15.7** Inhibitory mechanism of NF-κB activation by astaxanthin.

Rutter et al. [26] also demonstrated that the degree of colonic inflammation is associated with an increased risk of dysplasia and colorectal cancer. A mouse model was recently established for colitis-related colon carcinogenesis to investigate the pathogenesis and chemoprevention of inflammation-related colorectal cancer [27,28]. In our model, several colon carcinogens, including azoxymethane (AOM), were used as an initiator at a low dose, and a colitis-inducing agent DSS was used as a tumor promoter [28,29].

We investigated the possible inhibitory effect of astaxanthin against the colitis-associated colon carcinogenesis using the AOM/DSS mouse model. The mice in groups 1–4 were given a single intraperitoneal injection of AOM (10 mg/kg body weight) (Figure 15.2B). Starting 1 week after the injection, 1.5% DSS was administered to the mice in the drinking water for 7 days. The mice subsequently received the diets containing 0, 50, 100, and 200 ppm of astaxanthin for 17 weeks. Group 5 was not treated with AOM and DSS and was fed only a diet containing 200 ppm of astaxanthin. Group 6 was served as the untreated control. The mice were sequentially sacrificed at 20 weeks to determine the effects of astaxanthin on colon tumorigenesis.

The incidences and multiplicity of colonic mucosal ulcer and colonic dysplasia at 20 weeks are listed in Table 15.2 [20]. The incidence of mucosal ulcer was decreased with increasing dose of astaxanthin. The incidence of mucosal

**TABLE 15.2**

**Incidence and Multiplicity of Ulcer, Dysplasia, Adenoma, and Adenocarcinoma in Colonic Lesions**

| Group (Treatment) | | Mucosal Ulcer | Dysplasia | Adenoma | Adenocarcinoma |
|---|---|---|---|---|---|
| 1 | AOM/DSS[a] | 13/17, 77% | 15/17, 88% | 7/12, 58% | 11/12, 92% |
| | | $(2.4 \pm 1.77)$ | $(1.8 \pm 1.07)$ | $(1.3 \pm 1.44)$ | $(1.7 \pm 0.98)$ |
| 2 | AOM/DSS/ | 6/12, 50% | 7/12, 58% | 4/12, 33% | 4/12, 33%** |
| | 50 ppm AX[b] | $(0.7 \pm 0.78)$ | $(0.9 \pm 0.90)*$ | $(0.9 \pm 1.62)$ | $(0.6 \pm 1.61)$ |
| 3 | AOM/DSS/ | 5/12, 42% | 6/12, 50%* | 5/12, 42% | 4/12, 33%** |
| | 100 ppm AX | $(0.7 \pm 0.89)$ | $(0.6 \pm 0.69)**$ | $(0.7 \pm 0.98)$ | $(0.7 \pm 1.07)$ |
| 4 | AOM/DSS/ | 4/12, 33%* | 5/12, 42%* | 6/12, 50% | 4/12, 33%** |
| | 200 ppm AX | $(0.4 \pm 0.67)**$ | $(0.5 \pm 0.67)**$ | $(0.9 \pm 1.31)$ | $(0.5 \pm 0.80)*$ |
| 5 | 200 ppm AX | 0/2, 0% | 0/2, 0% | 0/2, 0% | 0/2, 0% |
| 6 | None | 0/2, 0% | 0/2, 0% | 0/2, 0% | 0/2, 0% |

Values are expressed as incidence, % of total mice (multiplicity/in colon of a mouse).

[a] AOM: Azoxymethane, DSS: dextran sulfate sodium.

[b] AX: Astaxanthin.

*$P < 0.05$ versus AOM/DSS.

**$P < 0.01$.

ulcer in the mice fed 200 ppm astaxanthin diet was significantly decreased compared with the control (group 1). Similarly, administration with 100 and 200 ppm astaxanthin significantly decreased the incidence of colonic dysplasia at 20 weeks ($P < 0.05$). In addition, the multiplicity of mucosal ulcer and dysplasia in colonic lesions was significantly reduced by astaxanthin treatment (mucosal ulcer: $P < 0.01$ at 50 and 100 ppm of astaxanthin, $P < 0.001$ at 200 ppm of astaxanthin; dysplasia: $P < 0.05$ at 50 ppm of astaxanthin, and $P < 0.01$ at 100 and 200 ppm of astaxanthin).

## SUPPRESSION OF INCIDENCE AND MULTIPLICITY OF COLONIC ADENOMA AND ADENOCARCINOMA BY ASTAXANTHIN

The incidences and multiplicity of colonic tumor at 20 weeks are shown in Table 15.2 [20]. Group 1 had an incidence of colonic adenocarcinoma of 92%. Treatment with astaxanthin >50 ppm significantly reduced the incidence of adenocarcinoma ($P < 0.01$). Dietary astaxanthin decreased the incidence of colonic adenoma, but not to a significant degree. Dietary administration of astaxanthin also lowered the multiplicities of colonic adenoma and adenocarcinoma (Table 15.2). By the treatment with 200 ppm astaxanthin (group 4), the multiplicity of adenocarcinoma significantly reduced compared with group 1 ($P < 0.05$).

## INHIBITION OF TUMOR CELL PROLIFERATION AND INDUCTION OF APOPTOSIS BY ASTAXANTHIN

Biomarkers of cell proliferation and apoptosis in colonic adenocarcinomas were immunohistochemically analyzed at 20 weeks. The mean proliferating cell nuclear antigen (PCNA)-labeling indices, which are related to tumor cell disordered proliferation, in groups 3 ($P < 0.01$) and 4 ($P < 0.001$) were significantly smaller than that in group 1 (Figure 15.8). The positive indices of survivin, which is highly expressed in cancer cells and inhibits apoptosis, of groups 2 ($P < 0.05$), 3 ($P < 0.001$), and 4 ($P < 0.001$) were significantly lower than that of group 1 (Figure 15.8). These data show that astaxanthin inhibits tumor cell proliferation and induces apoptosis in colonic adenocarcinoma in the AOM/DSS-treated mice.

## INHIBITION OF INFLAMMATORY FACTOR EXPRESSION IN ADENOCARCINOMAS BY ASTAXANTHIN

Inflammation is a critical tumor promoter [28,30] because several pro-inflammatory cytokines such as TNF-$\alpha$ and IL-1$\beta$ can promote tumor growth. The expression levels of TNF-$\alpha$ and IL-1$\beta$ in adenocarcinomas were measured immunohistochemically at 20 weeks. Dietary astaxanthin at 200 ppm suppressed the expression of TNF-$\alpha$ ($P < 0.01$) and IL-1$\beta$ ($P < 0.05$) compared with group 1 as shown in Figure 15.9.

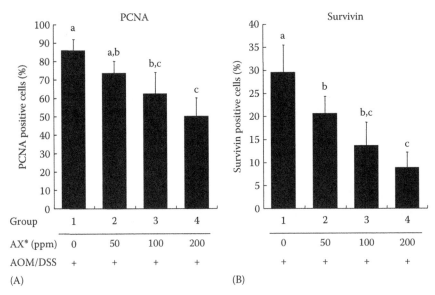

**FIGURE 15.8** Indices of (A) proliferating cell nuclear antigen (PCNA)–label nuclei and (B) survivin in adenocarcinomas developed in the colon of AOM/DSS-treated mice fed astaxanthin. Group 1: azoxymethane (AOM)/DSS-treated group, group 2: AOM/DSS + 50 ppm astaxanthin, group 3: AOM/DSS + 100 ppm astaxanthin, group 4: AOM/DSS + 200 ppm astaxanthin (group 4). Different letters are significantly different from each other at $P < 0.05$. *AX: astaxanthin.

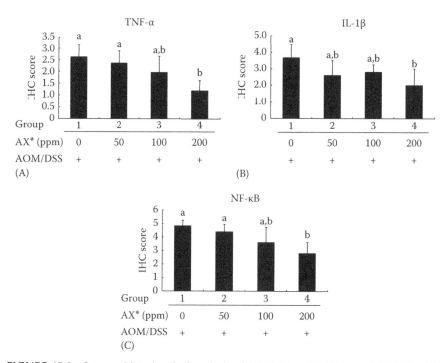

**FIGURE 15.9**   Immunohistochemical analysis of (A) TNF-α, (B) IL-1β, and (C) NF-κB in the colon tumor of DSS/AOM-treated mice fed astaxanthin. Group 1: azoxymethane (AOM)/ DSS-treated group, group 2: AOM/DSS + 50 ppm astaxanthin, group 3: AOM/DSS + 100 ppm astaxanthin, group 4: AOM/DSS + 200 ppm astaxanthin (group 4). Different letters are significantly different from each other at $P < 0.05$. *AX: astaxanthin.

Further, the administration of 200 ppm astaxanthin significantly reduced the expression level of NF-κB ($P < 0.01$) compared with group 1.

NF-κB activation is known to be associated with various types of cancers. Greten et al. [31] revealed the crucial involvement of the IκB kinase B/NF-κB in AOM/DSS treatment induced colon carcinogenesis. The expression of pro-inflammatory cytokines such as TNF-α, IL-1β, and IL-6, and inflammation-related enzymes such as COX-2 and iNOS, is encoded by target genes of the NF-κB signaling pathway [32]. In addition, anti-apoptotic proteins, cell surface adhesion molecules, and matrix metalloproteinase-9, which are involved in tumor initiation, promotion, and metastasis, are also regulated by NF-κB. Infiltrated macrophages surrounding tumors also produce TNF-α through NF-κB activation [33]. Blocking TNF-α has also been reported to reduce colorectal carcinogenesis in mice treated with AOM/DSS [34]. These results indicate that the anti-inflammatory effects of astaxanthin are also important in the prevention of colitis-associated colorectal carcinogenesis.

Our data demonstrate that the dietary administration of astaxanthin ameliorated the AOM/DSS-induced colonic proliferative lesions in mice. The anticarcinogenic effects of astaxanthin seem to be caused by the suppression of inflammatory factors including TNF-α, and IL-1β via inhibition of NF-κB signaling in the colonic mucosa and lesions. Further, astaxanthin suppressed cell proliferation and induced apoptosis by regulating the expression of survivin, a member of the apoptosis inhibitor family, in the malignancies.

Astaxanthin has attracted considerable interest because it has potent health-promoting effects in the prevention of various diseases such as cancer, liver disease, and metabolic syndrome [35]. Our results indicate that dietary astaxanthin inhibits the DSS-induced inflammation and AOM/DSS-induced colitis-associated colon car-cinogenesis in mice by suppressing the expression of pro-inflammatory cytokines via NF-κB activation. Taken together, astaxanthin is suggested to be a candidate chemopreventive agent against IBD and its associated colon cancer.

## REFERENCES

1. Rao AV, Rao LG. Carotenoids and human health. *Pharmacol Res* 2007; 55: 207–216.
2. Hosokawa M, Okada T, Mikami N, Konishi I, Miyashita K. Bio-functions of marine carotenoids. *Food Sci Biotechnol* 2009;18: 1–11.
3. Maoka T. Carotenoids in marine animals. *Mar Drugs* 2011; 9: 278–293.
4. Cooper RD, Davis JB, Leftwick AP, Price C, Weedon BC. Carotenoids and related compounds. Part XXXII. Synthesis of astaxanthin, phoenicoxanthin, hydroxyechine-none, and the corresponding diosphenols. *J Chem Soc, Perkin Trans 1.* 1975; 21: 2195–2204.
5. Pashkow FJ, Watumull DG, Campbell CL. Astaxanthin: A novel potential treatment for oxidative stress and inflammation in cardiovascular disease. *Am J Cardiol* 2008; 101: 58D–68D.
6. Jyonouchi H, Zhang L, Tomita Y. Studies of immunomodulating actions of carotenoids. II. Astaxanthin enhances in vitro antibody production to T-dependent antigens without facilitating polyclonal B-cell. *Nutr Cancer* 1993; 19: 269–280.
7. Tanaka T, Morishita Y, Suzuk M, Kojima T, Okumura A, Mori H. Chemoprevention of mouse urinary bladder carcinogenesis by the naturally occurring carotenoid astaxanthin. *Carcinogenesis* 1994; 15: 15–19.
8. Tanaka T, Kawamori T, Ohnishi M, Makita H, Mori H, Satoh K, Hara A. Suppression of azoxymethane-induced rat colon carcinogenesis by dietary administration of naturally occurring xanthophylls astaxanthin and canthaxanthin during the postinitiation phase. *Carcinogenesis* 1995; 16: 2957–2963.
9. Logan I, Bowlus CL. The geoepidemiology of autoimmune intestinal diseases. *Autoimmun Rev* 2010; 9: A372–A378.
10. Ahuja V, Tandon RK. Inflammatory bowel disease in the Asia-Pacific area: A com-parison with developed countries and regional differences. *J Dig Dis* 2010; 11: 134–147.
11. Rogler G, Andus T. Cytokines in inflammatory bowel disease. *World J Surg* 1998; 22: 382–389.
12. van der Woude CJ, Kleibeuker JH, Jansen PL, Moshage H. Chronic inflammation, apoptosis and (pre-)malignant lesions in the gastro-intestinal tract. *Apoptosis* 2004; 9: 123–130.
13. Kulaylat MN, Dayton MT. Ulcerative colitis and cancer. *J Surg Oncol* 2010; 101: 706–712.
14. Ekbom A, Helmick C, Zack M, Adami HO. Ulcerative colitis and colorectal cancer. A population-based study. *N Engl J Med* 1990; 323: 1228–1233.

15. Yasui Y, Tanaka T. Chemoprevention of colorectal carcinogenesis by natural anti-inflammatory agents. *Anti-Inflamm Anti-Allergy Agents Med Chem* 2010; 9: 150–157.

16. Gullett NP, Ruhul Amin AR, Bayraktar S, Pezzuto JM, Shin DM, Khuri FR, Aggarwal BB, Surh YJ, Kucuk O. Cancer prevention with natural compounds. *Semin Oncol* 2010; 37: 258–281.

17. Surh YJ. Cancer chemoprevention with dietary phytochemicals. *Nat Rev Cancer* 2003; 3: 768–780.

18. Gupta SC, Kim JH, Prasad S, Aggarwal BB. Regulation of survival, proliferation, invasion, angiogenesis, and metastasis of tumor cells through modulation of inflammatory pathways by nutraceuticals. *Cancer Metastasis Rev* 2010; 29: 405–434.

19. Blumberg RS, Saubermann LJ, Strober W. Animal models of mucosal inflammation and their relation to human inflammatory bowel disease. *Curr Opin Immunol* 1999; 11: 648–656.

20. Yasui Y, Hosokawa M, Mikami N, Miyashita K, Tanaka T. Dietary astaxanthin inhibits colitis and colitis-associated colon carcinogenesis in mice via modulation of the inflammatory cytokines. *Chem Biol Interact* 2011; 193: 79–87.

21. Atreya I, Atreya R, Neurath MF. NF-kappaB in inflammatory bowel disease. *J Intern Med* 2008; 263: 591–596.

22. Neurath MF, Pettersson S, Meyerzum Büschenfelde KH, Strober W. Local administration of antisense phosphorothioate oligonucleotides to the p65 subunit of NF-kappaB abrogates established experimental colitis in mice. *Nat Med* 1996; 2: 998–1004.

23. Lee SJ, Bai SK, Lee KS, Namkoong S, Na HJ, Ha KS, Han JA, Yim SV, Chang K, Kwon YG, Lee SK, Kim YM. Astaxanthin inhibits nitric oxide production and inflammatory gene expression by suppressing I(kappa)B kinase-dependent NF-kappaB activation. *Mol Cells* 2003; 16: 97–105.

24. Ohgami K, Shiratori K, Kotake S, Nishida T, Mizuki N, Yazawa K, Ohno S. Effects of astaxanthin on lipopolysaccharide-induced inflammation in vitro and in vivo. *Invest Ophthalmol Vis Sci* 2003; 44: 2694–2701.

25. Eaden JA, Abrams KR, Mayberry JF. The risk of colorectal cancer in ulcerative colitis: A meta-analysis. *Gut* 2001; 48: 526–535.

26. Rutter M, Saunders B, Wilkinson K, Rumbles S, Schofield G, Kamm M, Williams C, Price A, Talbot I, Forbes A. Severity of inflammation is a risk factor for colorectal neoplasia in ulcerative colitis. *Gastroenterology* 2004; 126: 451–459.

27. Tanaka T, Kohno H, Yoshitani S, Takashima S, Okumura A, Murakami A, Hosokawa M. Ligands for peroxisome proliferator-activated receptors alpha and gamma inhibit chemically induced colitis and formation of aberrant crypt foci in rats. *Cancer Res* 2001; 61: 2424–2428.

28. Tanaka T, Kohno H, Suzuki R, Yamada Y, Sugie S, Mori H. A novel inflammation-related mouse colon carcinogenesis model induced by azoxymethane and dextran sodium sulfate. *Cancer Sci* 2003; 94: 965–973.

29. Yasui Y, Suzuki R, Miyamoto S, Tsukamoto T, Sugie S, Kohno H, Tanaka T. Alipophilicstatin, pitavastatin, suppresses inflammation-associated mouse colon carcinogenesis. *Int J Cancer* 2007; 121: 2331–2339.

30. Tanaka T. Colorectal carcinogenesis: Review of human and experimental animal studies. *J Carcinog* 2009; 8: 5.

31. Greten FR, Eckmann L, Greten TF, Park JM, Li ZW, Egan LJ, Kagnoff MF, Karin M. IKKbeta links inflammation and tumorigenesis in a mouse model of colitis-associated cancer. *Cell* 2004; 118: 285–296.

32. Pahl HL. Activators and target genes of Rel/NF-kappaB transcription factors. *Oncogene* 1999; 18: 6853–6866.

33. Kim IW, Myung SJ, Do MY, Ryu YM, Kim MJ, Do EJ, Park S, Yoon SM, Ye BD, Byeon JS, Yang SK, Kim JH. Western-style diets induce macrophage infiltration and contribute to colitis-associated carcinogenesis. *J Gastroenterol Hepatol* 2010; 25: 1785–1794.

34. Popivanova BK, Kitamura K, Wu Y, Kondo T, Kagaya T, Kaneko S, Oshima M, Fujii C, Mukaida N. Blocking TNF-alpha in mice reduces colorectal carcinogens is associated with chronic colitis. *J Clin Invest* 2008; 118: 560–570.

35. Yuan JP, Peng J, Yin K, Wang JH. Potential health-promoting effects of astaxanthin: A high-value carotenoid mostly from microalgae. *Mol Nutr Food Res* 2011; 55: 150–165.

36. Cooper HS, Murthy SN, Shah RS, Sedergran DJ. Clinicopathologic study of dextran sulfate sodium experimental murine colitis. *Lab Invest* 1993; 69: 238–249.

# Part V

Meeting Dietary Supply
of Vitamin A and Carotenoids

# 16 Carotenoids in Laboratory Medicine

*Ingolf Schimke and Michael Vogeser*

## CONTENTS

## INTRODUCTION

Laboratory medicine, a synonym for clinical chemistry, clinical pathology, and clinical biochemistry, is the branch of clinical medicine concerned with developing and carrying out analyses of body fluids and other biological materials using tools of physics, chemistry, biochemistry, immunology, and molecular biology for the risk assessment, diagnosis, treatment monitoring, and follow-up in health and disease of man.*

For the characterization of the body status of vitamin A (the term vitamin A or retinoids are used in general aspects, otherwise the exact terms of the specific retinoids are used) by tools of laboratory medicine, the preferred samples are whole blood, serum, and plasma. Breast milk is another sample type that can be used in the case of solving specific problems, especially concerning the maternal vitamin supply and the supply of the neonates. The use of tissue samples, such as biopsies, for the analysis of vitamin A status is also possible.

For the vitamin A body status, retinol and retinyl esters of saturated long-chain fatty acids of animal origin and carotenoids of plant origin, especially β-carotene, are absorbed in the intestine. The dietary retinoids, together with retinyl esters, originating from β-carotene mucosal cleavage, reduction, and esterification are packed into chylomicrons and transported to the liver for storage. In case of inconsistent uptake, and to guarantee constant retinol supply for organs and tissues, liver-stored retinyl esters are mobilized and secreted in the circulation bound to retinol-binding protein (RBP). The formation of the binary vitamin A/RBP complex is thought to be essential for solubilization of the hydrophobic vitamin A and

---

* Adapted from Richterich R and Colombo JP [1] and supplemented by Müller MM (IFCC past president), personal report.

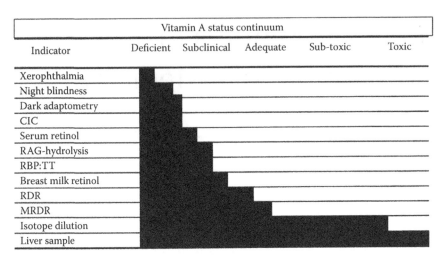

**FIGURE 16.1**  The estimated relationship of vitamin A status indicator to liver reserves of vitamin A. CIC, conjunctival impression cytology; RAG, retinoyl β-glucuronide; RBP:TT, retinol-binding protein to transthyretin ratio; RDR, relative dose response; MRDR, modified relative dose response. (From Tanumihardjo, S.A., *J. Nutr.*, 134, 290S, 2004, reproduced with permission of *The Journal of Nutrition*; License Number: 2838650017296.)

for the protection of the vitamin against nonspecific oxidation. In the circulation, transthyretin (TT), formerly known as prealbumin, is also associated with these molecules, and the resulting 1:1:1 ternary complex avoids the renal loss of low-molecular-weight vitamin A and RBP. Both RBP and TT were, although not exclusively, synthesized in the liver.

The body vitamin A status presents as a continuum ranging from deficiency to toxicity (Figure 16.1, Table 16.1). In industrialized countries, characterization of vitamin A status is mainly requested in patients with suspected fat malabsorption, for example, in cystic fibrosis. In regions with a high prevalence of general malnutrition, analysis of the vitamin A status is primarily done for epidemiological reasons but it would be also desirable for individualized care. Deficiency of vitamin A status is a serious health problem, predominantly in developing countries where nearly 40% of the inhabitants are vitamin A deficient.

Toxic hypervitaminosis and excessive deficiency are clearly associated with typical clinical symptoms and can be diagnosed in this way (Figure 16.1, Table 16.1). In the case of toxic vitamin A status, signs are often nonspecific and present in acute toxicity subjects with symptoms such as nausea, headache, fatigue, loss of appetite, dizziness, dry skin, desquamation, and cerebral edema. In chronic toxicity, symptoms include dry itchy skin, desquamation, loss of appetite, headache, cerebral edema, and bone-and-joint pain. Another indication, dependent on the severity of the vitamin A toxicity, is increased liver enzymes in the blood. Among the clinical symptoms for vitamin A deficiency are those resulting from the relationship between vitamin A and the eye, such as impaired adaptation to the dark, night blindness, altered conjunctiva, and xerophthalmia.

## TABLE 16.1

### Vitamin A Level in Blood Serum/Plasma and Clinical Signs of Deficiency or Toxicity

| Category | Serum Retinol Concentration | Liver Stores | Clinical Signs |
|---|---|---|---|
| Deficient | <0.35 μmol/L (100 μg/L) | Severely depleted (<5 μg/g) | Night blindness, ocular manifestation |
| Marginal[a] | 0.35–0.7 μmol/L (100–200 μg/L) | Severely depleted | None (levels responsive to provision of vitamin A) |
| Adequate | 1.05–3 μmol/L (300–850 μg/L) | ~20–300 μg/g | None |
| Excessive | High normal to >3 μmol/L (>85 μg/L) | High (>300 μg/g) | Not apparent or very mild, elevated liver enzymes in plasma |
| Toxic | Retinyl esters are elevated and might be higher than retinol | Very high in liver and in peripheral tissues | Headache, bone/joint pain, elevated liver enzymes in plasma, clinical signs of liver disease |

*Source:* Adapted from Barua, A. et al., Vitamin A: Sources, absorption, metabolism, functions and assessment of status, in: Herrmann, W. and Obeit, R., eds., *Vitamins in the Prevention of Human Diseases*, Walter de Gruyter, Berlin, Germany, 2011, with permission of Walter de Gruyter.

[a] The range 0.7–1.05 μmol/L is sometimes used to denote marginal status, and <0.7 μmol/L to indicate deficient status; these definitions might be more appropriate for adults, who typically have higher plasma retinol concentration than do children.

Due to clear clinical symptoms for both excessive deficiency and hypervitaminosis, tools of laboratory medicine are necessary for the confirmation of the clinical diagnosis and, more importantly, for the therapy monitoring. However, methods of laboratory medicine for characterizing vitamin A deficiency have become more and more important in areas where adequate staff and equipment are not available to perform the clinical diagnostic tests, especially when whole population groups must be diagnosed, such as in developing countries.

Between the extremes of deficiency and toxicity of vitamin A status, there are regions of the vitamin A continuum that are indicated as marginal (subclinical), adequate, and sub-toxic. However, monitoring the progression of vitamin A status from marginal and sub-toxic, respectively, to clear deficiency and hypervitaminosis are problematic using both clinical signs and the tools of laboratory medicine.

Laboratory medicine offers different strategies for the assessment of the body status of vitamin A (Figure 16.1, Table 16.1). Most frequently, the direct estimation of the serum/plasma retinol concentration is used. The vitamin A forms, including retinol and retinyl esters, can be detected in blood. Additionally, β-carotene (the most well-known provitamin A) can also be measured in the blood.

Unfortunately, blood retinol does not reflect sensitively enough the vitamin A status, especially in the intermediate stages. The main organ responsible for vitamin A storage is the liver, and the blood concentration of retinol does not distinctly change before this store is critically depleted or overfilled.

Information about the vitamin A status can be also obtained from the measurement of RBP and TT. The concentration of RBP is tightly correlated to the retinol concentration in the blood and has consequently been seen as a surrogate marker to assess the vitamin A status.

For better characterization of the liver stores of vitamin A, the "relative dose response" (RDR) test and "modified RDR" (MRDR) test have been developed. To conduct these tests, retinyl ester (RDR) and 3,4-didehydroretinol ester (MRDA) are administered and time dependently monitored in the serum. Another response test (RAG hydrolysis test) is based on the administration of β-glucuronide and subsequent monitoring of the serum retinoic acid.

In general, serum and plasma can be used for the analysis of the vitamin A status and can be sampled in a quantity adequate for routine testing in centralized laboratory medicine. However, some of these methods are adapted for small sample volumes, which is important for diagnosis using samples from neonates and children. With the combination of well-trained staff, state-of-the-art analysis devices (which perform measurements that are highly standardized and automated), and well-designed management programs for the control of procedures and devices, centralized laboratory medicine guarantees economic investigations of the highest quality.

In the case of population-based studies of the vitamin A status (mainly in the developing countries), there are large economic and logistic problems. For this purpose, it has been recommended to use cost-sparing and robust field-friendly techniques [4]. The field techniques should include easy collection of blood samples, for example, the sampling of capillary blood as frequently handled and transported as dried blood spot (DBS) on filter paper. Furthermore, it should be possible to perform the tests directly in the field using devices that are easy to handle, but which can be managed by minimal trained staff. However, for such in laboratory medicine named point-of-care testing, if used for quantitative measurement, the protocol should fulfill all the requirements of accuracy, sensitivity, specificity, both within runs and, even more importantly, in day-to-day variation as guaranteed in centralized testing. In case of not fulfilling these requirements, field-friendly sampled probes (e.g., capillary blood, DBS) should be adequately stored, transported, and analyzed in centralized laboratories with standardized methods adapted to this special material.

For differential diagnosis, there is clear evidence that the serum biomarkers of the vitamin A status can decrease independently of changes in the levels of the liver stores for vitamin A [5]. During acute-phase response to inflammation, RBP and TT (well known as negative acute-phase proteins) as well as retinol decrease with the disease severity. As an explanation for lower retinol levels in the serum, any increased consumption related to the disease was suggested. Additionally, an increased loss of retinol in the urine can be seen, which could be attributed to the impaired tubular reabsorption of the retinol transporter RBP.

In cases of nutritional deficiency, only retinol and RBP are affected, while TT remains unaffected. This fact can be used to discriminate between changes in the vitamin A status related to nutrition or acute-phase response. Furthermore, it has been recommended to also measure acute-phase proteins, such as CRP, to adjust for RBP as a result of infection [5]. Reduced RBP synthesis can also be seen in chronic and acute liver disease, protein malnutrition, and zinc deficiency. Diabetic nephropathy and chronic renal disease can be accompanied with increased RBP levels.

## DIRECT MEASUREMENT OF VITAMIN A

One of the first assays for the determination of vitamin A in blood was published in 1929 [6] and based on the Carr–Price reaction [7] in which vitamin A and carotenes react with antimony trichloride in chloroform to produce measurable blue-colored products. However, multiple steps for the extraction of vitamin A and carotene from blood and subsequent evaporation were necessary. Later and especially directed to the estimation of vitamin A in blood serum or plasma, improvements to the original assay have been performed, mainly by considering sensitivity and specificity aspects (e.g., separate estimation of vitamin A and β-carotene), assay standardization (e.g., improved color stabilization), introduction of standards and controls, as well as unit declaration. This made the assay more practicable and introduced it into laboratory medicine [8,9]. Using such assay for the vitamin A measurement in five serum samples, the story of human vitamin A measurements began. Following continuous modification of the photometric vitamin A measurement, which included direct measurement of its ultraviolet absorption or fluorescence and adaptation to small sample volume, this method was intermittently used until the end of the last century [10–12].

A new era in the laboratory medicine of vitamin A began with the introduction of a chromatographic step before photometric measurement [13]. With the "determination of serum retinol by high-speed liquid chromatography" [14], which requires only 100 μL of serum, the door has been opened for the establishment of high-performance liquid chromatography (HPLC) as a routine tool for vitamin A measurement in laboratory medicine.

### MEASUREMENT OF RETINOL AND β-CAROTENE BY HPLC

The standard procedure for the measurement of retinoids in biological matrices includes sample collection, sample preparation, the actual chromatography, and the detection, followed by quantification with the help of internal standards, calibrators, and controls. For all these analytical steps, a large number of modifications have been tested and introduced [15]. Although the quantification of retinoic acids, retinal, retinyl esters, or other special compounds with vitamin A activity is not a routine method in laboratory medicine, respective assays have been described [16–19]. For laboratory medicine, the measurement of serum retinol using any isocratic HPLC system with UV detector is the most suitable and frequently performed. This technique is simple and robust compared with other HPLC methods. Regardless of this, specialized clinical laboratories that guarantee high-quality standards and perform continuous vitamin A analyses with high flow capacity should be chosen.

In most of the laboratories, commercially available test kits were used for the quantification of serum vitamin A by HPLC. These kits include the analytical column, mobile phases, sample preparation consumables, as well as calibration and quality control samples. In a minority of laboratories, vitamin A is quantified using in-house HPLC methods.

In the majority of assay kits, sample preparation is based on simple protein precipitation. In this step, the complete release of the target analyte from its protein bonds (mainly RBP) is achieved. All commercially available assays include internal standardization. Retinyl esters (C-15, C-17, C-19) are suggested to be used for this purpose [15,17].

The kit solutions claim specificity for retinol while retinal, retinyl esters, and retinoic acids are disclosed by their different chromatographic properties. The majority of the commercially available kits represent combi-assays, which also include the quantification of vitamin E (tocopherol). A representative chromatogram of such combi-assay is shown in Figure 16.2.

Serum reference materials for retinol are available from the U.S. National Institute of Standardization (NIST), allowing traceability of tests on this material. Several schemes of proficiency testing for the serum retinol measurement are implemented (e.g., from UKNEQAS [the United Kingdom] and Instand [Germany]), and quality control materials are available for continuous internal quality assessment. Consequently, the level of standardization for serum retinol measurement is rather high. As a sample material, serum or plasma can be used, and blood should be collected in the morning before breakfast and prior to any medication.

Handling of the samples for vitamin A measurement requires specific arrangements. Based on the polyene structure, retinol and retinoids, in general, are sensitive to light,

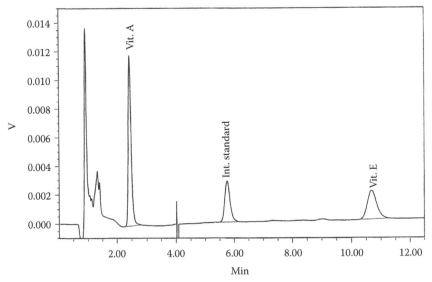

**FIGURE 16.2** Representative chromatogram: Measurement of serum vitamin A by HPLC with UV detection. (Reproduced with permission of Chromsystems, Müchen, Germany.)

oxygen, trace elements, and heat. Consequently, such conditions should be avoided as far as possible during sampling and transport. Sampled blood should be free of hemolysis. Fast centrifugation of blood for serum preparation is favorable, and serum storage should be performed at temperatures below −20°C in darkness. Comparisons of retinol in fresh serum and in serum adequately stored over a number of years have shown no significant difference in concentrations [20,21]. If whole blood is stored for 48 h at 4°C, a loss of nearly 5% for retinol and β-carotene can be expected [21]. This loss increases in cases of room temperature storage. To avoid oxidative alteration of the retinoids and carotenoids, addition of antioxidants to the sample (such as ascorbic acid, butylhydroxytoluene, or related compounds) is suggested.

With respect to a field-friendly procedure for the serum retinol analysis, DBS prepared from capillary blood as sample material has been trialed and validated for retinol measurement by HPLC [22,23]. As a result, blood sampling in the field is followed by preparation of DBS, which is transported humidity controlled in special cooled bags to centralized laboratories for HPLC analysis. In this setting, DBS retinol concentration is comparable with serum retinol.

A still more adequate field-friendly procedure would include the retinol measurement in the field. For this purpose, a technology has been developed that includes the fixing of the RBP/retinol complex inside of blood capillaries by anti-RBP antibodies. After the resting blood has been flushed from the capillary, the retinol fluorescence of the fixed complex can be directly measured using a portable fluorometer [24].

Table 16.1 demonstrates the ranges for the serum retinol concentration that are indicative for differences in the body status of vitamin A. However, as mentioned, the ranges for marginal and deficient status are rather appropriate for adults.

The age-related reference intervals are 0.70–1.40 μmol/L (1–6 years); 0.91–1.71 μmol/L (7–12 years); 0.91–2.51 μmol/L (13–19 years); and 1.05–2.80 μmol/L (adults). Within the reference range, male subjects present with nearly 20% higher values than females [25].

Conversion of units for retinol

$$[\mu mol/L] \times 286.46 = [\mu g/L]; \quad [\mu g/L] \times 0.00349 = [\mu mol/L]$$

The chromatographic determination of β-carotene is also performed on an isocratic HPLC system with UV/Vis detection. Using specially optimized reverse phase-chromatography, α-, cis-β-, and all-trans-β-carotene (as well as some other carotenoids) can be separated and quantified. Serum and plasma can be used. The pre-analytic requirements are comparable with those of the retinol measurement. However, β-carotene seems to be less stable than retinol [21]. Serum reference materials for β-carotene are available from the NIST, allowing traceability of tests on this material. Schemes of proficiency testing for the serum β-carotene measurement are implemented (e.g., from UKNEQAS [the United Kingdom]), and quality control materials are available for continuous internal quality assessment.

The reference interval for serum β-carotene is 0.19–1.58 μmol/L (100–855 μg/L).

Conversion of units for β-carotene

$$[\mu mol/L] \times 536.87 = [\mu g/L]; \quad [\mu g/L] \times 0.00186 = [\mu mol/L]$$

Elevated β-carotene levels were found as a result of disturbed conversion to vitamin A in patients with hypothyroidism, as well as in patients with diabetes mellitus and hyperlipidemia. Furthermore, the blood concentration of β-carotene depended on intestinal fat reabsorption. Consequently, lower concentrations were found in pancreatic insufficiency and malabsorption.

With respect to the antioxidative properties of β-carotene, serum/plasma measurement of the blood antioxidative capacity is frequently performed and thought to be of interest in antiaging medicine [26,27]. For this purpose, an ensemble of assays is offered, but unfortunately, the most of such assays do not fulfill the requirements of laboratory medicine. However, with respect to the benefit of the measurement of antioxidative capacity, this is the subject of strong debate.

With respect to the maternal vitamin A status, and to characterize the neonate supply of vitamin A, breast milk is a further sample material that can be used for the HPLC-based measurement of retinol and β-carotene [28,29]. Its sampling compared with blood is less invasive and therefore often more acceptable in field studies.

## MEASUREMENT OF RETINOL-BINDING PROTEIN AND TRANSTHYRETIN

The RBP secretion from the liver depends on the vitamin A content. In cases of reduced vitamin A supply and depletion of the liver stores, RBP accumulates in the endoplasmic reticulum of the hepatocytes. As soon as the liver vitamin A level returned to normal, RBP were released in the circulation. Therefore, RBP is seen as a surrogate marker for serum retinol. By comparing RBP and retinol measurement, it has been shown that children classified as being vitamin A-deficient using HPLC retinol measurements could also be identified with high sensitivity and specificity by their decreased serum RBP concentration [30,31].

For the measurement of RBP, enzyme-linked immunosorbent assays (ELISA) are the preferred tool for the estimation of RBP. Other tools for the measurement of RBP are radial immunodiffusion and nephelometry. The advantages of the ELISA method compared to the measurement of retinol via HPLC are lower cost, smaller sample volumes, and the greater stability of RBP. The ELISA technique was also adapted to capillary blood and DBS. No differences in the RBP concentrations in venous and capillary blood, as well as between DBS and serum, have been found [24,32,33]. Cutoffs ranging between 0.7 and 1.14 μmol/L were proposed as being comparable with high sensitivity and high specificity to a serum retinol concentration of <0.7 μmol/L [34].

Reference intervals are 0.53–1.63 μmol/L (neonate); 0.86–2.4 μmol/L (6 month old); and 1.44–2.88 μmol/L (adult).

Conversion of units

$$[\mu mol/L] \times 210 = [mg/L]; \quad [mg/L] \times 0.04876 = [\mu mol/L]$$

TT can be measured using ELISA or radial immunodiffusion. A reference interval for TT of 200–400 mg/L adult (20–60 years) was documented [25]. A cutoff <0.36 was defined for RBP/TT ratio as being indicative of vitamin A deficiency, which is comparable to <0.7 µmol/L retinol [35].

## DOSE RESPONSE TESTS, ISOTOPE DILUTION, AND LIVER TISSUE ANALYSIS

Dose response tests are based on the RBP liver accumulation in cases of vitamin A depletion in the liver [36]. The accumulation happens before the serum vitamin A concentration decreases, and this indicates deficiency. Consequently, administrated retinyl ester (relative dose response, RDR) or 3,4-dihydroretinol (modified dose response, MRDR) binds to the accumulated RBP, and the binary complexes are released into the circulation.

In the *RDR*, the levels of retinol are measured before and 5 h after the application of a defined oral dose of retinyl ester. Deficient liver vitamin A is defined as $[(retinol_{5h} - retinol_{0h})/retinol_{5h}] \times 100 < 20$.

The main advantage of the *MRDR* is the need for a single sample collection. To perform this test, retinyl ester 3,4-dihydroretinol, which is nearly absent at baseline, is administered, and the ratio of retinol and 3,4-dihydroretinol is measured in parallel by HPLC in the sample collected 5 h after the treatment. The molar ratio of <0.06 for 3,4-dihydroretinol to retinol was recommended to indicate low liver vitamin A levels that should be treated.

The retinoyl glucuronide (*RAG*) response test based on the finding that RAG is neither absorbed nor hydrolyzed in case of adequate vitamin A status but with depletion of the vitamin A stores administrated RAG is increasingly adsorbed and is hydrolyzed to retinoic acid which can be measured via HPLC. A significant concentration of retinoic acid can be measured in serum only when the vitamin A status is deficient or marginal. In contrast to the measurement of retinol or RBP, results of RAG response are independent of acute-phase reaction [37].

Rather far from the general methodology of laboratory medicine, the *Retinol Isotope Dilution* is a direct and quantitative estimation of the vitamin A body status in relation to the liver-stored vitamin A concentration. In this test, deuterated or $^{13}$C-retinol is applied and, after reaching the equilibrium which is time-consuming, the tracer is estimated by gas chromatography–mass spectrometry. The exact nature of the human body vitamin A status documented with the retinol isotope dilution technique has been demonstrated by parallel measurements of vitamin A by HPLC directly in respective liver tissue samples [38,39]. This tissue vitamin A measurement is adapted for the use of very small liver samples (e.g., to 7 mg tissue performed by needle biopsy technique) [40]. In the same setting, vitamin A could also be estimated in other tissues that are amenable to biopsies. However, information about the tissue vitamin A content using biopsies as samples for measurements is limited. As primarily demonstrated for the liver, the tissue pattern of the vitamin A stores are nonuniform across this tissue [41]. Consequently, a single or only a few biopsies are not representative of the whole organ. But, more importantly, vitamin A measurement in human biopsies for single-patient-oriented diagnostics or treatment monitoring is strongly opposed by the ethics. This means that, to the best of our knowledge, there is no indication for biopsy taking solely for the analysis of vitamin A levels.

## CONCLUSION

The laboratory medicine offers different well-standardized and controlled strategies for the assessment of the body status of vitamin A. Deficiency in the vitamin A status can be indicated by HPLC measurement of retinol and β-carotene. RBP and RBP/TT ratio are accepted as surrogate marker for the classification of the vitamin A status. In dose response tests such as RDR, MRDR, and GAG, which are more sensitive in the intermediate vitamin A status, the metabolite analysis uses the tools of laboratory medicine. Biological variables and pre-analytical variables that affect the markers and that, in this way, can influence interpretation are well known.

Laboratory medicine supplies cost-effective and time-saving strategies for the characterization of the vitamin A status that can be used both in the patient-oriented diagnostics and in the treatment monitoring as well as in population-based field studies.

## REFERENCES

1. Richterich R, Colombo JP. *Clinical Chemistry: Theory, Practice, and Interpretation*; John Wiley & Sons, Chichester, U.K., 1981.
2. Tanumihardjo SA. Assessing vitamin A status: Past present and future. *J Nutr* 2004;134:290S–293S.
3. Barua A, Statcewicz-Sapuntzakis M, Furr HC. Vitamin A: Sources, absorption, metabolism, functions and assessment of status. In: Herrmann W, Obeit R (eds.). *Vitamins in the Prevention of Human Diseases*. Walter de Gruyter, Berlin, Germany, 2011.
4. Garrett DA, Sangha JK, Kothari MT, Boyle D. Field-friendly techniques for assessment of biomarkers of nutrition for development. *Am J Clin Nutr* 2011;94:685S–690S.
5. Schweigert FJ. Inflammation-induced changes in the nutritional biomarkers serum retinol and carotenoids. *Curr Opin Clin Nutr Metab Care* 2001;4:477–481.
6. Rösiö B. The occurrence of vitamin A in blood and blood serum of domestic animals, in cow milk, milk products and in some feeds. *Z Physiol Chem* 1929;182:289–304.
7. Carr FH, Price EA. Colour reactions attributed to vitamin A. *Biochem J* 1926; 20:497–501.
8. Rosenthal E, Erdélyi J. A new colour test for the determination of vitamin A. *Biochem J* 1934;28:41–44.
9. Rosenthal E, Szilárd C. A new method of determining the vitamin A content of blood. *Biochem J* 1935;29:1039–1042.
10. Hoch H. Micro-method for estimating vitamin A by the Carr-Price reaction. *Biochem J* 1943;37:425–429.
11. Koch W, Kaplan D. A simultaneous Carr-Price reaction for the determination of vitamin A. *J Biol Chem* 1948;173:363–369.

12. Biswas AB, Mitra NK, Chakraborty I, Basu S, Kumar S. Evaluation of vitamin A status during pregnancy. *J Indian Med Assoc* 2000;98:525–529.

13. Garry PJ, Pollack JD, Owen GM. Plasma vitamin A assay by fluorometry and use of a silicic acid column technique. *Clin Chem* 1970;16:766–772.

14. De Ruyter MG, De Leenheer AP. Determination of serum retinol (vitamin A) by high-speed liquid chromatography. *Clin Chem* 1976;22:1593–1595.

15. Gundersen TE, Blomhoff R. Qualitative and quantitative liquid chromatographic determination of natural retinoids in biological samples. *J Chromatogr A* 2001;935:13–43.

16. Barua AB, Furr HC, Janick-Buckner D, Olson JA. Simultaneous analysis of individual carotenoids, rentinol, retyl esters, and tocopherols in serum by isocratic non-aqueous reversed-phase HPLC. *Food Chem* 1993;46:419–424.

17. Schmidt CK, Brouwer A, Nau H. Chromatographic analysis of endogenous retinoids in tissues and serum. *Anal Biochem* 2003;315:36–48.

18. Kane MA, Folias AE, Napoli JL. HPLC/UV quantitation of retinal, retinol, and retinyl esters in serum and tissues. *Anal Biochem* 2008;378:71–79.

19. Arnold SL, Amory JK, Walsh TJ, Isoherranen N. A sensitive and specific method for measurement of multiple retinoids in human serum with UHPLC-MS/MS. *J Lipid Res* 2012;53:587–598.

20. Driskell WJ, Lackey AD, Hewett JS, Bashor MM. Stability of vitamin A in frozen sera. *Clin Chem* 1985;31:871–872.

21. Craft NE, Brown ED, Smith JC Jr. Effects of storage and handling conditions on concentrations of individual carotenoids, retinol, and tocopherol in plasma. *Clin Chem* 1988;34:44–48. Erratum in: *Clin Chem* 1988;34:1505.

22. Craft NE, Haitema T, Brindle LK, Yamini S, Humpfrey JH, West KP. Retinol analysis in dried blood spots by HPLC. *J Nutr* 2000;130:882–885.

23. Craft NE, Bulux J, Valdez C, Li Y, Solomons NW. Retinol concentrations in capillary dried blood spots from healthy volunteers: Method validation. *Am J Clin Nutr* 2000;72:450–454.

24. Craft NE. Innovative approaches to vitamin A assessment. *J Nutr* 2001; 131:1626S–1630S.

25. Roberts WL, McMillin GA, Burtis CA. Reference information for the clinical laboratory. In: Burtis CA, Ashwood ER, Brons DE (eds.). *Tietz Textbook of Clinical Chemistry and Molecular Diagnostics*. Elsevier Saunders, St. Louis, MO, 2006.

26. Huang D, Ou B, Prior RL. The chemistry behind antioxidant capacity assays. *J Agric Food Chem* 2005;53:1841–1856.

27. Magalhães LM, Segundo MA, Reis S, Lima JL. Methodological aspects about in vitro evaluation of antioxidant properties. *Anal Chim Acta* 2008;613:1–19.

28. Strobel M, Heinrich F, Biesalski HK. Improved method for rapid determination of vitamin A in small samples of breast milk by high-performance liquid chromatography. *J Chromatogr A* 2000;898:179–183.

29. Tanumihardjo SA, Penniston KL. Simplified methodology to determine breast milk retinol concentrations. *J Lipid Res* 2002;43:350–355.

30. Gamble MV, Ramakrishnan R, Palafox NA, Briand K, Berglund L, Blaner WS. Retinol binding protein as a surrogate measure for serum retinol: Studies in vitamin A-deficient children from the Republic of the Marshall Islands. *Am J Clin Nutr* 2001;73:594–601.

31. Hix J, Rasca P, Morgan J, Denna S, Panagides D, Tam M, Shankar AH. Validation of a rapid enzyme immunoassay for the quantitation of retinol-binding protein to assess vitamin A status within populations. *Eur J Clin Nutr* 2006;60:1299–1303.

32. Gorstein JL, Dary O, Pongtorn, Shell-Duncan B, Quick T, Wasanwisut E. Feasibility of using retinol-binding protein from capillary blood specimens to estimate serum retinol concentrations and the prevalence of vitamin A deficiency in low-resource settings. *Public Health Nutr* 2008;11:513–520.

33. Baingana RK, Matovu DK, Garrett D. Application of retinol-binding protein enzyme immunoassay to dried blood spots to assess vitamin A deficiency in a population-based survey: The Uganda Demographic and Health Survey 2006. *Food Nutr Bull* 2008;29:297–305.

34. de Pee S, Dary O. Biochemical indicators of vitamin A deficiency: Serum retinol and serum retinol binding protein. *J Nutr* 2002;132(9 Suppl):2895S–2901S.

35. Rosales FJ. Effects of storage and handling conditions on concentrations of individual carotenoids, retinol, and tocopherol in plasma. Vitamin a supplementation of vitamin a deficient measles patients lowers the risk of measles-related pneumonia in Zambian children. *J Nutr* 2002;132:3700–3703.

36. Tanumihardjo SA. Vitamin A: Biomarkers of nutrition for development. *Am J Clin Nutr* 2011;94:658S–665S.

37. Sarma PC, Goswami BC, Gogoi K, Bhattacharjee H, Barua AB. A new approach to the assessment of marginal vitamin A deficiency in children in suburban Guwahati, India: Hydrolysis of retinoyl glucuronide to retinoic acid. *Br J Nutr* 2009;101:794–797.

38. Furr HC, Amedee-Manesme O, Clifford AJ, Bergen HR 3rd, Jones AD, Anderson DP, Olson JA. Vitamin A concentrations in liver determined by isotope dilution assay with tetradeuterated vitamin A and by biopsy in generally healthy adult humans. *Am J Clin Nutr* 1989;49:713–716.

39. Haskell MJ, Handelman GJ, Peerson JM, Jones AD, Rabbi MA, Awal MA, Wahed MA, Mahalanabis D, Brown KH. Assessment of vitamin A status by the deuterated-retinol-dilution technique and comparison with hepatic vitamin A concentration in Bangladeshi surgical patients. *Am J Clin Nutr* 1997;66:67–74. Erratum in: *Am J Clin Nutr* 1999;69:576.

40. Amédée-Manesme O, Furr HC, Olson JA. The correlation between liver vitamin A concentrations in micro- (needle biopsy) and macrosamples of human liver specimens obtained at autopsy. *Am J Clin Nutr* 1984;39:315–319.

41. Olson JA, Gunning D, Tilton R. The distribution of vitamin A in human liver. *Am J Clin Nutr* 1979;32:2500–2507.

# 17 Supply of Vitamin A in Developing Countries

*Grace Jui-Ting Tsai and Klaus Kraemer*

## CONTENTS

## INTRODUCTION

Vitamin A deficiency (VAD) is a significant public health concern in the developing world, especially in Africa and Southeast Asia. Estimates from the World Health Organization between 1995 and 2005 identified VAD to be affecting 190 million school children (~32%) and nearly 20 million pregnant women (~10%).[1] Causes for VAD include poverty, infections, and lack of access to traditional foods containing vitamin A.[2] Vitamin A sources in developing countries are primarily consumed in the form of provitamin A carotenoids from plant sources, with animal-derived preformed vitamin A (retinol) comprising little of the daily vitamin A intake.[3] The vitamin plays essential roles in human physiology for normal functioning of the immune system, visual system, and maintenance of cell function for growth, bone development, red blood cell production, and reproduction.[4] Long-term deficiency can cause night blindness and xerophthalmia and is a major contributor to mortality in developing countries. Regular delivery of vitamin A doses have proved effective in reducing child mortality by as much as 24% on average.[5] In addition to supplementation programs, other types of interventions used to improve vitamin A status of populations include dietary diversification, fortification of common food staples, and biofortification.[6]

In this chapter, VAD, the diseases and disorders it contributes to, demographics of those affected, the bioavailabilities of vitamin A sources, supplementation, fortification sources, horticulture-based approaches, and recommendations for the secure supply of vitamin A will be discussed.

## BACKGROUND ON VITAMIN A DEFICIENCY

### Effects of VAD

Clinical consequences associated with suboptimal vitamin A status are collectively known as vitamin A deficiency disorders (VADDs), which include the specific ocular manifestations of xerophthalmia and its blinding sequelae (includes night blindness, Bitot's spot, corneal scarring, phthisis), infectious morbidity and mortality, and impaired growth.[7] Worldwide, approximately 4.4 million preschool children are affected by xerophthalmia and 6 million pregnant women suffer from night blindness.[8] Xerophthalmia is characterized by the excessive dryness of the conjunctiva and cornea and may lead to corneal ulceration and blindness. In the immune system, vitamin A plays an important role in resistance to diseases, enhancing the body's first and second lines of defenses, known respectively as the innate and adaptive components for infectious risks such as measles and corneal blindness. Inadequate nutritional status during pregnancy and lactating stages can negatively influence vitamin A status in the subsequent breast-fed infant by impeding growth, development, and health.[9] In areas of comorbidity where affected communities are vitamin A deficient and exposed to high rates of infectious diseases, a cycle is perpetuated whereby metabolism is increased, leading to excessive excertion that overall exacerbates the existing deficiency.[10] Despite the lower prevalence of these conditions among children than adults, the earlier oneset of life-threatening diseases as a result of vitamin A deficiency would persist throughout their life spans.

### Current Statistics on Vitamin A Deficiency

In the developing world, an estimated 190 million children are vitamin A deficient,[1] with a prevalence of about 30%.[11] VAD is defined by serum retinol level below

0.70 µmol/L (20 µg/dL) and is considered a public health concern if prevalent among 15% of the population. In 1987, WHO estimated VAD to be endemic in 39 countries. In 1995 and 2005, VAD was a public health problem in 45 and 122 countries (Figure 17.1), respectively.[1] These affected regions often share environmental exposures, particularly poverty and infectious diseases, and limited availability of vitamin A–rich food.[12]

Vitamin A nutritional status is a major determinant of morbidity and mortality in preventable childhood blindness, especially in preschool children (0–59 months). In 2004, it was estimated among this age group that VADDs, defined as health disorders attributable to varying degrees of VAD,[13] were responsible for approximately 0.7 million deaths (6.5% total deaths globally).[14] The WHO regions of Africa and Asia have the highest prevalence of VAD, with 91.5 million preschool children residing in South Asia, which comprises the majority of those affected (Figure 17.1).[1] The lowest prevalence of VAD is in South America, Central America, and the Caribbean with a rate of 10% in respective regions.[11] For night blindness, the highest proportion of preschool-aged children affected is in Africa with a rate of 2.0%, which is four times higher than the estimated 0.5% in South Asia.[1] In East Asia, South America, Central America, and the Caribbean, the rates of VAD have been reduced by nearly 50% since 1990. The reduction rate in most of Africa (except North Africa) and south central Asia is lagging behind.[11]

## HISTORICAL DATA ON VAD COVERAGE AND DEVELOPMENT OF PROGRAMS

In the 1990s, limited efforts have been undertaken in the control of VAD, which led to an informal technical consultation on vitamin A in December 1997. The meeting was convened by UNICEF in association with selected representatives of international organizations including the WHO, the Micronutrient Initiative (MI), UNICEF, and Helen Keller International (HKI), Canadian International Development Agency (CIDA), the United States Agency for International Development (USAID), and leading technical experts on VAD (Johns Hopkins University Bloomberg School of Public Health) to exchange knowledge about accelerating progress. Vitamin A was stressed to be essential for child survival, and oral supplementation was deemed a reliable and effective method to combat VAD and mortality in children under the age of 5. The potential effectiveness of food fortification was also recognized as a partial solution. Countries where the rate of under-5 child mortality was greater than 70 deaths/1,000 live births were advised to commence distributing high-dose vitamin A supplements (VASs) immediately, regardless of whether the country's vitamin A status had been evaluated, thus catapulting progress. The Global Alliance for Improved Nutrition has been established at the Special Session of the UN General Assembly on Children in 2002 with the purpose of reducing malnutrition through sustainable strategies and supporting large-scale food fortification programs in coalition with National Fortification Alliances of governments, industries, international organizations, and other stakeholders. Concurrently, the global initiative "VISION 2020: the Right to Sight" was launched in 1999 with goals for the elimination of avoidable blindness by the year 2020 and halting the projected doubling of avoidable visual impairment between 1990 and 2020. Thus far, it has been successful in slowing down the projected rate of blindness attributable to VAD.[15]

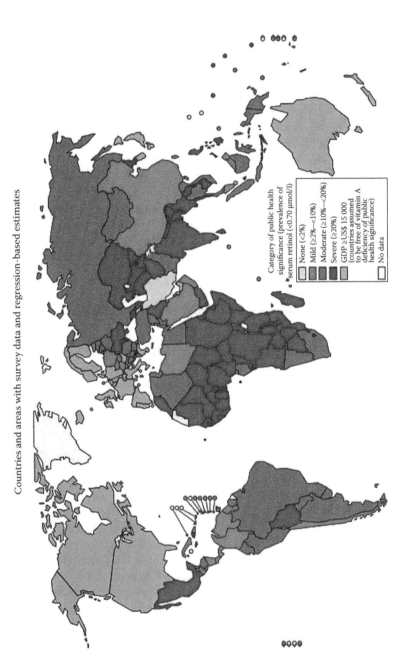

**FIGURE 17.1** Biochemical vitamin A deficiency (retinol) in preschool-aged children as a public health problem by country 1995–2005. (From World Health Organization, Global prevalence of vitamin A deficiency in populations at risk 1995–2005, WHO Global Database on Vitamin A Deficiency, World Health Organization, Geneva, Switzerland, 2009, Available from: http://whqlibdoc.who.int/publications/2009/9789241598019_eng.pdf)

Since the late 1990s, high-dose vitamin A have been regularly delivered in combination with ongoing national immunization days (NIDS), which are campaigns held with periodic frequency in countries for mass vaccination. As NIDS are being phased out in some countries with the gradual success of polio eradication, other alternative distribution systems were devised to effectively deliver VASs. Other distribution methods are at fixed sites through schools, hospitals/clinics, or other comprehensive community infrastructure through which to reach children in remote areas. Along with NIDS, the other most widely used method for VAD control in developing countries is through "Child Health Days," which often includes two or more child care survival interventions, such as deworming, mosquito net distribution, measles vaccinations, and other immunizations. Figure 17.2 summarizes vitamin A supplementation coverage by distribution strategy devised. In controlled trials, all-cause mortality among children aged 6–59 months can be reduced on average by 24% and has contributed to a decrease in VAD-related blindness in many countries,[5] including the maintenance of night blindness, an early indicator of VAD, below the 1% threshold that would be indicative of a public health problem.[16] A significant reduction of 30% of diarrhea-specific deaths in children was also observed. In women, high-dosed oral VASs of 200,000 IU used to be provided to postpartum

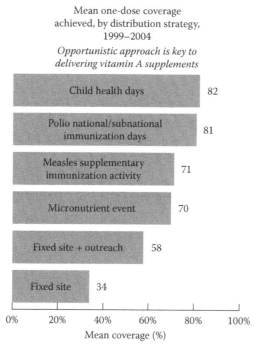

FIGURE 17.2    One-dose vitamin A supplementation coverage by distribution strategy. Note: Estimates derived using coverage data from all vitamin A distributions carried out over the period 1999–2004, broken down by delivery strategy. (From UNICEF, Vitamin A supplementation: A decade of progress, UNICEF, New York, 2007.)

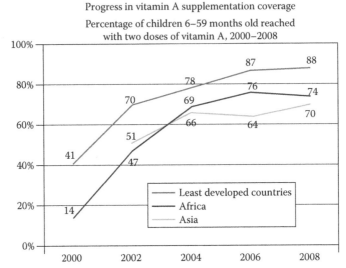

Progress in vitamin A supplementation coverage

Percentage of children 6–59 months old reached with two doses of vitamin A, 2000–2008

**FIGURE 17.3** Percent coverage with two doses of VASs in children 6–59 months old from 2000 to 2008. *Note:* Vitamin A supplementation two-dose (full coverage) trends are based on a subset of 16 African countries and 18 least developed countries with data in even years between 2000 and 2008 and on a subset of 11 Asian countries with data in even years between 2002 and 2008. The trend line for Asia begins in 2002 because of a lack of data for trend analysis prior to that. (From UNICEF, Tracking progress on child and maternal nutrition. UNICEF, New York, 2009, available from www.unicef.org/publications/files/Tracking_Progress_on_Child_and_Maternal_Nutrition_EN_110309.pdf)

mothers within 6 weeks of delivery to increase the retinol status of their breast milk, although in practice coverage remained quite low.[1]

In the past, distribution mechanisms through biannual delivery days have often been inconsistent. The coverage levels varied among administrations, and some countries were unable to distribute the second needed dose. Coverage has been more expansive since then, as seen in many African countries where full coverage increased fivefold between years 2000 and 2008 (Figure 17.3). In Asia, full coverage increased from 51% to 70% in 2002. Overall, the proportion of children covered in developing countries more than doubled within the same period from 41% to 88%.[17] However, despite the increase in coverage in developing countries, there have been minimal effects in reducing the prevalence of low serum retinol. This is mostly due to the fact that most programs work to reduce child mortality rather than to increase serum retinol itself. This is especially true in the Philippines, where there has been an insignificant increase in serum retinol level during a 6 month supplementation period with coverage above 80%. Even when serum retinol levels were increased, it lasted only 2 months to reach pre-supplementation levels. On the other hand, certain Central American countries have maintained low prevalence of low serum retinol due to vitamin A–fortified sugar.[11]

## SOURCES OF VITAMIN A IN DEVELOPING COUNTRIES

Vitamin A exists in two forms: preformed vitamin A (retinol and its esterified form, retinyl ester) and provitamin A carotenoids. Preformed vitamin A is derived from animal sources, which includes fish oil, eggs, dairy products, and meat (particularly liver and kidney). By far, the most abundant and efficient provitamin A carotenoid is β-carotene; other provitamin A carotenoids are α-carotene and β-cryptoxanthin. The body then converts these precursor carotenoids from plants into vitamin A. β-Carotene is primarily present in dark green leafy vegetables, yellow-orange plants, tubers, fruits, and red palm oil.[18] Non-provitamin A carotenoids found in food, such as lycopene, lutein, and zeaxanthin, cannot be converted into vitamin A. Both provitamin A and preformed vitamin A must be metabolized to active retinoid ligand in order to fulfill essential roles in biological development (Table 17.1).[19]

### PROVITAMIN A CAROTENOID SOURCES

Among vegetable leaves, only those that are dark green are considered good sources.[18] β-Carotene and xanthophylls lutein and zeaxanthin account for over 80% of all carotenoids, with α-carotene, γ-carotene, β-cryptoxanthin, and lycopene comprising the majority of remainder.[18]

In addition to dark green vegetables, fruits are good sources of provitamin A carotenoids. The vitamin A activity of fruits is generally lower than that of leafy vegetables.[18] However, the wider acceptance of fruits, especially among children, is

---

### TABLE 17.1
### Common Food Sources of Vitamin A

| Contain Retinol (Highly Bioavailable) | Contain β-Carotene and Other Carotenoids | Fortified Foods That Contain Preformed Vitamin A |
|---|---|---|
| Breast milk | Red palm oil[a] | Sugar |
| Egg | Mango[a] | Cooking fat |
| Liver | Papaya[a] | Edible oil |
| Kidney | Yellow sweet potatoes | Ghee |
| Small fish with liver | Amaranth leaves | Margarine |
| Cod liver oil | Carrot | Monosodium glutamate (MSG) |
| Whole milk | Pumpkin/yellow squash | |
| Fortified skim milk | Collard greens | Corn flour/Wheat flour |
| Butter | Spinach | Noodles |
| Cheese made with whole milk | Apricots | Milk |

*Source:* Adapted from The World Bank, *Nutrition Toolkit Tool #2: Basic Facts: Nuts and Bolts on Nutrition, Annex C: Micronutrients*, Table C-4, 2004. Copyright 2011 by The World Bank.

[a] Among the plant-based sources of vitamin A, these have higher bioavailability.

an advantage for intervention programs.[18] Mango, papaya, and watermelon have the highest concentration of provitamin A carotenoids among fruits. Tomatoes are also one of the provitamin A carotenoid sources, although their lycopene content exceeds the β-carotene content. This has led to research on the development of strains of tomatoes with increased β-carotene concentrations. Another provitamin A source is the pumpkin, in which its main carotenoids are α-carotene and β-carotene. The native buriti fruit (*Mauritia vinifera*) paste is recommended in northeast Brazil, where the staple diet is lacking in preformed vitamin A and provitamin A, as the fruit contains as much β-carotene as that found in red palm oil.[18] Less analysis has been done on roots and tubers than on vegetables and fruits,[18] and most of the variet-ies of roots and tubers examined have low concentrations. However, recent breeding for improved β-carotene content is being emphasized for the pigmented varieties with higher existing carotenoid levels, which include the orange varieties of sweet potatoes and yellowish cassavas. HarvestPlus and its partners have recently released three varieties of high-yielding provitamin A cassava, and multiplication programs are under way to make them widely available in Nigeria, Democratic Republic of Congo, followed by other African countries.

Red palm oil contains the highest concentration of β-carotene among all naturally occurring sources. It contains unsaturated fatty acids and is also rich in vitamin E, ubiquinones, and phytosterols. Red palm oil is a major source of domestic or edible cooking oil in many regions of west and central Africa, but the tree is also cultivated in South America as well as in parts of Asia. Its use in cooking adds a distinctive flavor that may not be well accepted by some people; however, more recent refinery techniques have made red palm oil more readily accepted than in its crude form.[18] The potential for the use of this oil in combating VADD in the future is promising.

Despite the lack of readily available fruits and vegetables in low-income regions, conversely, a sole reliance on plant sources is not reliable because the bioefficacy of carotenoids A is lower than previously believed. Based upon the updated conversion standards from the Institute of Medicine (IOM), the bioconversion ratio was doubled from 6:1 to 12:1.[20] The implication is that it decreases the previously estimated amount of vitamin A in developing countries' plant food supply, with even greater reduction for dark green leafy vegetables. This would require double the food quantity required to sustain the level of usable vitamin A from these sources. As a result, vitamin A from plant sources should be accompanied by adequate quantities of animal or for-tified food sources of preformed vitamin A, since serum retinol concentrations are lower in response to vegetable feeding than supplement or liver feeding.[21] Finding the appropriate food-based interventions in lower-income areas and food diversity is important in ensuring the consumption of moderate amounts of vitamin A–rich foods.

In developing countries, seasonality is a major barrier to obtaining the benefits of fruits and vegetables, many of which are located in the tropical zone. Orange fruits like mangoes and papaya are usually consumed during their short growing seasons and cannot be stored for extended periods. Leafy vegetables have longer, but still limited, seasons. According to Oomen's study, communities often experi-ence an annual lean season, which frequently coincides with the rainy season, when young children and their mothers tend to be affected the most. Chronic floods and droughts may cause a large portion of the total population to have raised levels of night blindness and other deficiencies.[18] It is of importance that vitamin A–rich foods are available and affordable. Seasonal price movements and supply of green leafy vegetables caused consumption patterns to shift threefold in the Philippines.[22] This begs new approaches where programs are encouraged to increase the production of vitamin A–rich food seasonally. Farm-based horticultural approaches can be a partial resolution to this problem. Horticultural program extension networks can be designed to enhance production, distribution, and marketing capacity of products such as vegetables and fruits that are easily perishable. Harvesting and preserving excess foods produced during the peak agricultural season would encourage the production of low-cost vitamin A–rich foods during the off-season periods for local markets.[23] Promoting the preservation and processing of local food systems, in conjunction with both the private sector and producer and consumer groups would better serve the needs of the vulnerable population.

## Preformed Vitamin A

Preformed vitamin A is derived from animal sources. One source is fish liver oil, which is more widely used as food supplements than as dietary ingredient. If available, it could be a valuable source of vitamin A for developing regions of the world. In fact, most mammalian livers, such as those of calves, ox, sheep, and chicken, have high vitamin A concentrations that are comparable to those found in cod liver oil.[18] Beef, mutton, and pork also contain preformed vitamin A. Other animal sources include halibut, cod, shark, herring, and mackerel, from most highly vitamin A content to least. Sources from daily produce include butter, margarine (vitaminized), eggs, milk, and cheese (fatty type), from high concentration to low. Milk, butter, and cheese are all moderate sources. Food coloring also can be sources of β-carotene, as they contain isolated β-carotene or natural extracts from leaves, carrots, tomatoes, algae, and red palm oil. Others may contain β-apo-8′-carotenal and canthaxanthin, which also have vitamin A activity.[18]

## Food Fortification and Biofortification

Fortification programs have moved to the forefront as an effective way to regulate and improve overall vitamin A status in developing countries. The three methods of fortification are commercial fortification (nutritionally enriching food products by adding micronutrients to food during processing), home fortification (adding a micronutrient powder to food prepared at home), and biofortification (selective plant breeding to develop varieties of crops that contain high concentrations of bioavailable micronutrients). However, implementation of these programs still stands to be a barrier to the progress of eliminating VAD. In particular, developing countries may not have sufficient technical capabilities and resources for food fortification, such as food processing facilities, which are often not well structured. Governments may not have the technical capability to perform the required tests and monitor food fortification for quality at the local level. Effective fortified product distribution infrastructure is needed, as well as widespread awareness and demand for fortified foods, or regulatory support from the government.[24] Foods that countries select to fortify are frequently consumed on a regular basis, in adequate and relatively stable amounts, that would allow it to deliver the needed levels of vitamin A to people of different ages and societal levels. Currently, foods that are vitamin A fortified include vegetable oil, cereal staples such as wheat and maize—often consumed as noodles, breads, pasta, and other flour—grain products, sugar, or starch tubers and roots such as potatoes, yams, and cassava. Successful fortification of novel food items includes vegetable oil in Morocco and three high-yielding varieties of provitamin A cassavas in Nigeria.[18]

### Fortified Items and Effectiveness

Fortification of cereal flours, maize meals, and rice with vitamin A is technically feasible, with good nutrient retention in the products.[25] Changes in the process of flour manufacturing are not required, because cereal flour production usually consists of a blend of cultivars, enzymes, and other substances.[25] In 1993, the Venezuelan Government started a program fortifying precooked maize. In the Philippines, intake of vitamin A–fortified wheat flour used in making the commonly consumed bread pandesal significantly improved the plasma retinol and liver stores of school children.[25] Furthermore, the introduction of fortified wheat flour in programs in India reduced the proportion of the population with inadequate vitamin A intakes by 34% and 74% among the two groups that the intervention impacted.[26] More than 17 countries have proposed to fortify wheat and maize flour with vitamin A. Vitamin A fortification can now be more quickly implemented from the planning stages than was the case even in the relatively recent past.

Worldwide production of vegetable oils (canola, corn, cottonseed, coconut, olive, palm, peanut, safflower, soybean, sunflower) is growing, and consumption is also projected to increase, particularly in emerging economies such as India.[25] Vegetable oils can serve as good vehicles for fortification, by providing good stability, if protected from UV light, which allows vitamin A to be readily absorbed by the body.[25] Vegetable oils are widely consumed by most households; for instance, cooking oil was chosen as a vehicle in West Africa because it is a dietary staple that is purchased and consumed in small quantities even by poor families. This has prompted *Tache D'Huile*, a public–private initiative that aims to supply vitamin A–fortified cooking oil, to strengthen its production in eight countries in West Africa. The effectiveness in improving people's access to vitamin A through fortification has been demonstrated in Brazil, where vitamin A added to oil when given along with a rice-based diet was efficacious in raising vitamin A status.[27] In the Philippines, 18 months of sustained vitamin A–fortified cooking oil was provided and was associated with a reduction in the prevalence of VAD.[28]

Many countries also have established a standard for the addition of vitamin A to margarine to make it nutritionally equivalent to the natural vitamin-A complex in butter, which margarine has often replaced. Several countries fortify margarine as either a mandatory or a voluntary practice.[25] In the Philippines, partnerships in public and private sectors with researchers and policy makers have promoted a mandated fortification program that has proven to be very efficacious. The daily consumption of Star margarine fortified with vitamin A comprised 150% of RDI and subsequently decreased the trend of low serum retinol levels in preschool-aged children.[29] In India, a fortified hydrogenated vegetable oil, Vanaspati, is fortified with vitamin A that supplies 0.4%–21% of the RDI.[30] It is replacing ghee, or clarified butter, as the most commonly used fat.

Sugar is suitable as a vehicle for vitamin fortification in countries where it is centrally manufactured in a few plants and is often consumed by a majority of people across the country. In the early 1970s, a program consisting of several Central and South American countries found that fortifying white, refined sugar with vitamin A had a positive impact on reducing VAD.[31] Surveys have shown that there was a significant reduction from 26% to 10% in the number of children with low serum retinol and from 33% to 14% in those with an abnormal relative dose–response.[32] The Institute of Nutrition of Central America and Panama (INCAP) applied and promoted the use of appropriate innovations, encouraged legislation, and introduced national programs in three countries (Guatemala, Honduras, and Costa Rica). Currently, fortified sugar in Guatemala and El Salvador contributes a significant portion of the vitamin A RDI for people older than 3 years of age and about one-third of RDI for young children.[25]

In the 1970s, in the Philippines and Indonesia, a program fortifying monosodium glutamate (MSG) was pursued and was proven to be efficacious in raising serum retinol levels in children and reducing xerophthalmia and mortality.[33] However, it failed to pass product stability and acceptability criteria, which led to the failed implementation of the program at the national scale.

## Acceptance and Effects

One concern that arises with biofortification is the appearance, color, texture, or organoleptic properties of certain foods. While fortified staples may be more easily substituted for unfortified staples, biofortified foods may not be as easily adopted by consumers and may require modification of behavior. Depending on the extent to which the vitamin A pigment is visible in a crop, and the population area it is targeted for, this may influence acceptance. For example, "Golden Rice" is a genetically modified version of the rice *Oryza sativa* that induces a yellow color from β-carotene in the endosperm. Plant breeders have also developed β-carotene-fortified varieties of maize that induces an orange color. Evaluations took place in Zambia to assess the likelihood of acceptance of orange maize, a dietary staple there. In much of Africa, the orange color presents a cultural challenge because of its population's preference for white maize. Nevertheless, orange maize was likely to be accepted by consumers in Zambia and does not translate into unwillingness of consumption.[34,35] In Zambia where children were fed white maize vs. orange maize, results demonstrated a quick adaptation to orange maize, and an optimistic outlook for similar adaptation patterns in other biofortified-maize target countries.[34] Adoption of fortified foods will depend on whether there are any distinguishable differences between the non-fortified and fortified substitutes; therefore, food appearance and taste must be taken into account when selecting foods to fortify. For these foods to achieve significant impact, consumers are assumed to favor biofortified foods over non-biofortified equivalents.

Nutrition education campaigns as part of food fortification programs would help dissipate misconceptions to the targeted populations that their foods have been altered in an unacceptable manner. It would also increase the consumption of fortified food and assist in behavior change from white varieties of food—particularly in the case of maize, potatoes, and rice—to orange varieties of food.

## Biofortification Constraints

Biofortification technology needs to be improved in terms of the retention of vitamin A in crops and delivery efficiency following harvest. Transportation of biofortified foods may take a long time, and by the time it arrives, little of the vitamin may remain. In certain vast developing countries such as India, implications may arise from transporting perishable foods such as animal and animal products. Throughout processing, transport, storage, and cooking, the stability of vitamin A needs to remain at an acceptable level. Another issue that arises with biofortification is the technological constraints: not all food items are suitable for fortification. Furthermore, the bioavailability—the fraction of the dose that is absorbed by the body—differs from that used in supplements.[36,37] Research to study the impact of biofortification on nutrient status needs high priority.

## Home Fortification

The development of home fortification in the 1990s, also known as "point-of-use fortification," is the direct addition of micronutrients to food before consumption. Micronutrient powders (MNPs) (e.g., Sprinkles™, Vitashakti™, Anuka™, MixMe™) containing mixtures of essential vitamins and minerals are sprinkled onto the plate or bowl of food just before eating at home or any other place where meals are consumed (e.g., schools, refugee camps); this would reduce the impact on bioavailability and minimize loss of nutrients through processing and preparation. It can be more effective for certain kinds of populations, due to the intervention's advantage in being able to selectively target groups that are less likely to obtain enough micronutrients from foods fortified for the general population, such as infants and young children, people in isolated areas, or refugees.[36] The use of MNP, even with relatively low and infrequent use, has been proven to be efficacious for reducing VAD among young children in resource-poor settings.[38] Evidence of the effectiveness of large-scale MNP implementation is still scarce. However, it has been suggested that in this type of setting, MNP with an integrated program is feasible. Impact of MNP programs will vary widely, and efforts to assess MNP impact needs to account for context-specific factors to determine the effectiveness of individual programs.

## Multiple Micronutrient Fortification and Safety

High vitamin A consumption is less common in developing countries in comparison to developed countries. The IOM has set 3,000 µg (10,000 IU) as the safe upper limit (UL) for daily vitamin A intake for adults.[39] Adverse effects in response to both acute and chronic overconsumption of vitamin A are a possibility when people consume high amounts of preformed vitamin A. Acute toxicity occurs at doses of about 25,000 IU/kg body weight, and chronic toxicity occurs at doses of more than 50,000 IU/day when taken for months in older children and adults.[40] Because vitamin A is fat soluble, it can be stored in the liver for a long time and can build up to a toxic level. Excessive consumption of vitamin A over years might weaken the bones and contribute to fractures and osteoporosis.[41–44]

In developing countries, inadvertent ingestion of high-dose vitamin A through overly frequent supplementation is the most likely cause of vitamin A toxicity. For serious or sustained side effects to occur, frequent high dosages (50,000–100,000 IU daily) have to be administered persistently for 3–6 months.[45] Nausea, vomiting, abdominal pain, blurred vision, headache, and muscle incoordination are among the characteristics of acute toxicity. These symptoms usually resolve themselves fairly quickly and will not have a long-term negative health impact. Steps must be taken in all supplementation programs to establish a record system to reduce to an absolute minimum that children will receive high-dose supplements too frequently, thereby posing a health risk.

Adverse effects from overconsumption of vitamin A–fortified food sources should also be kept in mind as developing countries are considering alternatives to vitamin A supplementation programs. Recently, control strategies for VAD have grown considerably to include fortification of several foods with multiple micronutrients to achieve wider coverage. In certain countries, more than 20 different food items are fortified, and children will be exposed to multiple vitamin A sources. In South Africa, biscuits for primary schoolchildren are being fortified with iron, iodine, and β-carotene.[46] In Guatemala, fortified foods account for about half the vitamin A foods that poor urban toddlers obtain.[47] Fortification should ensure that a fortified food item consumed by most of the target population should pose no risk of overdosing for those regularly consume the highest quantities of the fortified product.[25] If vitamin A–fortified foods are regulated with dosages not exceeding the recommended dietary allowance and are consumed in normal amounts, they do not usually provide more than 1000–1500 IU/day, and this will carry minimal risk of chronic toxicity. Thus, fortification and supplementation in concordance with existing recommendations would not result in adverse effects[48] and would be beneficial in improving vitamin A status.

With MNP intervention, overdose is unlikely, even for children who receive it along with high-dose vitamin A capsules twice a year. A young child would need to consume many MNP sachets (approximately 20) to reach toxicity levels.[49] Consumption of plants and biofortified foods poses no risk of vitamin A toxicity because the body regulates the amount of β-carotene derived from these sources to be converted to vitamin A.

Quality assurance minimizing overconsumption may be more difficult to achieve for developing countries. Segments of the population that may be exposed to more than one initiative should be monitored by future surveys, more specifically, intakes of vitamin A derived from natural foods, fortified foods, and supplements. Estimates may be useful for assessing levels of intake from these sources in developing countries to avoid risk of overdose while maximizing appropriate nutrient consumption.

## Agriculture/Horticulture/Farm-Based Approach for Food Diversification; Biofortification Adoption

In terms of distribution, food fortification is driven by private sector food industries to reach a broad population; however, household demand would be dependent upon household purchasing power in the commercial market, which means that an alternative option would be to introduce fortified food through public or nonprofit channels and provide goods to some consumers through a local nongovernmental organization (NGO). In developing countries, where underserved and economically deprived populations have restricted market access to nutritious foods, decentralized planning formed at the district and subdistrict levels to ensure high yields of vitamin A–rich foods is required. This can be achieved through farm-based approaches, which helps improve communities' local food production systems by creating year-round gardens with fruits and vegetables of the vitamin A–dense varieties, fish ponds, and small farms for raising poultry, etc. In addition, these community gardens can be developed alongside the school gardens. Groups such as HKI lend support and start-up supplies, such as seeds and chickens to local NGOs, which promotes capacity building with local organizations and farmers for farm-based interventions. For example, HKI's Homestead Food Production (HFP) Program addressed issues within schools by promoting school gardens and providing nutrition education to reduce malnutrition. The gardening program is aimed at encouraging healthy eating and ensuring that school children have long-term access to varied micronutrient-rich fruits and vegetables, especially during the dry season. In addition, the school system can provide vitamin A supplementation to children in schools twice per academic year. Farm-based approaches can also assist farmers and empower the community.[50] Studies have shown that children from a family with developed gardens consume 1.6 times more vegetables and have a lower risk of night blindness than children without such gardens.[50] Projects conducted in Asia and Africa involving gardening and small animal husbandry have been shown to increase dietary diversity and nutritional content of home diets, improve total dietary vitamin A intake, and reduce the risk of VAD-associated xerophthalmia.[50] This intervention is useful in circumstances beyond the household's control—economic upheaval, political unrest, or an increase in agricultural prices—since increased home food production can increase year-round availability and establish continuous access to nutritious foods. Economically, strengthening HFP would ensure steady supply of vitamin A irrespective of the positive or negative effects of the market.[51]

Initiatives to increase vitamin A supply need to be targeted at small-scale producers to grow biofortified crops. The home-based approach with biofortification has an advantage in rural settings because 75% of the impoverished live mostly on subsistence farming and relies on the production of cheaper and more widely available staple crops, such as wheat or rice. This may be more difficult in areas such as Africa, where an agricultural infrastructure is lacking. Local production of biofortified food items provides a feasible means to reach rural areas when VASs or fortified food products are not widely available.[52] If implemented, farm-based interventions are found to have high benefit–cost ratio in terms of economic cost and have the potential to be an effective and sustainable long-term approach to improve vitamin A status to achieve high population coverage in developing countries. This practice should be promoted through education and advocacy. Biofortified crops have yet to be commercialized, and there have been few preliminary assessments on its adoption. The effects of the biofortified traits on the input and yield of the crops are uncertain, and there needs to be adequate cost advantages to ensure access and availability for the vulnerable population. Ultimately, the success of the technology itself is the most important factor. Once biofortified seeds are distributed to subsistence farmers, they are able to reproduce the seeds themselves, which leads to fewer recurrent costs.[52] For example, growers may aim to increase access to agricultural products such as the orange-fleshed sweet potato and biofortified rice.[53] Table 17.2 compares different approaches to combat VAD.

**TABLE 17.2**

**Comparisons between Supplementation, Food Fortification, and the Horticulture Approach to Combat VAD**

|  | Supplementation | Food Fortification | Horticulture Approach |
|---|---|---|---|
| Effectiveness and time frame | Effective strategy usually for short term | Effective medium- to long-term measure | Effective medium- to long-term measure |
| Delivery requirements | An effective health delivery system | A suitable vehicle and organized processing facilities | Access to marginal or homestead land, plus agricultural supplies for plant cultivation |
| Coverage | Reaches only populations receiving the service | Reaches all segments of target population | Reaches families in households, schoolchildren, and populations of the most deprived sectors with gardens |
| Compliance | Requires sustainable motivation of participants | Does not require intensive cooperation and individual compliance of individual | Requires strong motivation of participants for nutrition education and agriculture training |
| Cost of maintenance | Relatively high financial resources needed | Low cost compared to supplementation—to maintain the system self-financing in the end | Minimal capital investment, can produce more without increasing costs—relatively low cost |
| External resources | Foreign currency or external support required for obtaining supplements | Adequate technology that is locally available or can be easily transferred | Provision planting material (tools, seeds, and chemicals) and education promotion |
| Sustainability | Relates to compliance and existing resources | Fortificant compounds may need to be imported | Relates to level of agriculture education and use of local resources for self-reliance |

*Source:* Adapted from Lotfi, M. et al., *Micronutrient Fortification of Foods. Current Practices, Research and Opportunities*, The Micronutrient Initiative, Ottawa, Ontario, Canada, 1996, Table 1-2.

## ROLES OF VITAMIN A DURING PARTICULAR TIMES IN THE LIFE CYCLE: MATERNITY, PREGNANCY, LACTATION, NEONATAL PERIOD, INFANCY, AND CHILDHOOD

### INCREASED BODILY NEEDS

The demand for vitamin A increases during the life stages of infancy, preschool, pregnancy, and lactation, which increases the risk for VAD.[1] When in the womb, the fetus requires vitamin A for growth. In general, infants are born with low vitamin A liver stores, which make them dependent upon breast milk to meet their needs. Considerable interest in vitamin A status in the neonatal period has shifted not only to meet the goal of increasing the body's vitamin A stores, but also to improve survival through vitamin A supplementation to neonates in the few days following birth. The potential benefits within the first month of the newborn period are being investigated; however, there is conflicting evidence with no clear indication of the biochemical mechanisms.

### IMPORTANCE OF BREAST MILK

Breast milk is the main source of vitamin A for infants to optimize their vitamin A levels. Infants of well-nourished mothers can meet their daily requirements from an average of ~725 mL/day breast milk per day.[51] However, in some countries, poor maternal nutritional status may not provide infants the adequate amount of vitamin A from breast milk. Poor maternal vitamin A status results in low breast retinol content, as VAD in the mother increases the risk for early onset of VAD in infants, as is the early cessation of breast-feeding.[54] Correcting VAD at the early stage is imperative, since VAD increases the risk of dying due to infectious diseases and contribute to impaired physical and mental developments. Any amount of breast milk appears to significantly protect children from xerophthalmia through the third year of their life.[51] Vitamin A content in breast milk has been shown to increase for up to 8 months after fortification of sugar in Guatemala[31] and MSG in Indonesia,[33] providing indication of public health potential among women of childbearing age with this intervention.

### EFFECTS ON PREGNANT AND LACTATING WOMEN

In addition to children, 20 million pregnant and lactating women are also affected by VAD. The lower prevalence of VAD at the maternal life stages when compared to preschool-aged children could be partly attributable to a lingering lack of research data.[55] Regardless, this problem is still of high importance due to the global approximation that 15% of pregnant women have VAD and 8% suffer from night blindness.[1] Pregnant women in high-risk areas are most susceptible to VAD during the third trimester of pregnancy due to higher demands of the vitamin from both the mother and the unborn child. Those affected are also predisposed to higher risks for anemia, morbidity, and mortality.[9] Vitamin A supplementation is associated with an insignificant reduction in maternal mortality in pregnancy,[56,57] and given the improvements in global maternal mortality rates, it is unlikely that many additional trials will further explore this relationship. Nonetheless, in this population group, vitamin A supplementation reduced the incidence of night blindness and is associated with fewer low-birth-weight babies in pregnant women who are HIV positive.[56] In addition, the frequent use of bariatric surgery to treat severely obese women who are childbearing is known to alter vitamin A levels, although this type of surgery is not commonly performed in developing countries.[58] A recent study also found that increasing vitamin A status before and during pregnancy may reduce the risk of bacterial vaginosis (BV) in VAD-affected regions. BV in pregnancy is linked to preterm birth, but the associations are not well understood.[59]

## EVIDENCE FOR SUPPLEMENTATION IN NEONATES, INFANTS, AND CHILDREN UNDER 5 YEARS

Although vitamin A supplementation is generally important for decreasing infant mortality, in contrast to what previous study results may have suggested, the supplementation of vitamin A or β-carotene to pregnant women has an insignificant effect on reducing infant death risk in malnourished areas [60–64] within the first year of life or within the first month of life; nor did it reduce illnesses. While three trials of newborn vitamin A dosing conducted in Indonesia, India, and Bangladesh have shown a reduction in mortality during infancy, other trials conducted in Nepal, Zimbabwe, and Guinea-Bissau did not find any overall reduction in infant mortality.[2] A heterogeneous effect was observed between the sexes,[2] but causality for these differences was not established. This may indicate that the effect of vitamin A supplementation is influenced by features that differ between the various locations. These apparent heterogeneous results may be attributable to the research design being conducted in separate settings with varying levels of maternal VAD and infant mortality, as well as the higher HIV prevalence in Zimbabwe. Since mortality in early infancy has shown to be a strong contributor to overall child mortality for children under the age of 5 years in developing countries, it is essential to have sound scientific evidence regarding the effects of vitamin A supplementation in neonates.

There are currently four large-scale trials being conducted in various geographic regions—Pakistan, India, Ghana, and Tanzania—to evaluate the potential of neonatal vitamin A supplementation delivery and the efficacy in reducing mortality.[9] The results will establish a definitive role of this intervention and define future policy making. Furthermore, studies trying to understand the impact of neonatal vitamin A supplementation on immune function regulation and organ maturation, as well as the role of vitamin A in metabolism at birth, are currently under way and would further help shape future recommendation and guidelines.[9] However, West et al. added that notwithstanding the effects on mortality, achieving full vitamin A nutritional status through adequate diet, supplementation, or fortification is still an important public health goal.[57]

## FUTURE DIRECTIONS

Large-scale awareness programs through nutrition campaigns to encourage optimal breast-feeding, complementary feeding practices, and to provide vitamin A supplementation in both the mother and the child are critical for the prevention of VAD for children under 5 years. Due to the low coverage of high-dose vitamin A supplementation to postpartum women,[1] education efforts may prompt mothers to seek increased vitamin A intake during the essential period in which infant's nutritional status is highly dependent on their mothers. Where breast-feeding practices are suboptimal, breast-feeding promotion activities must be an integral part of any plan to prevent VAD. In vitamin A–deficient regions, infants who do not receive breast milk are more likely to have a higher risk of VAD and should receive VASs. It is important to promote a combination of breast-feeding practices with programs that support the maternal diet to ensure adequacy of vitamin A in breast milk.[65]

Regarding neonatal vitamin A supplementation (within 28 days after birth), there is currently no global policy to initiate vitamin A supplementation as a public health intervention for reducing infant mortality and morbidity[9] due to conflicting study results. However, it is still recommended that children under 5 years in areas at risk of VAD and high mortality receive VASs. Future comparative effectiveness trials may guide policymakers craft policies for vitamin A distribution. The precise details of how these programs should be structured and carried out, including the dose, frequency, and time frame of the intervention, are not clear. Operationally, the feasibility of vitamin A supplementation to newborns immediately following birth needs to be evaluated in terms of how to reach most neonates in developing countries within 2 days after birth. Improved indicators to determine vitamin A status in neonates also need to be developed.[2]

## DIFFERENTIAL EFFECT BETWEEN AFRICA AND ASIA

The provision of high-dose VASs to children 6–59 months has shown that by 2009, there was a 71% overall coverage in developing countries and 85% in the least developed countries.[66] There may be a protective effect of vitamin A supplementation in Asia but not Africa or South America.[60] The response to vitamin A supplementation has been shown to differ for newborns between regions, but has not been previously reported for supplementation in children aged 6–59 months.[60] The factors contributing to the differential effect between these regions are not well understood. One deduction could be that the largest proportions of preschool-aged children with biochemical VAD live in South Asia,[1] while Africa ranks second in the highest incidences of biochemical VAD. There are insufficient data to determine whether there is a significant different response to vitamin A supplementation between Asia, Africa, and Latin America in children aged 6–59 months.[60]

## CONCLUSION

To successfully combat VAD, short-term interventions and breast-feeding must be linked to more long-term programs. The basis for lifelong health begins as early as in the womb of the mother, continuing through infancy and childhood. A combination of breast-feeding and vitamin A supplementation, accompanied by long-term food-based solutions, including the promotion of vitamin A–rich foods and fortification, is key to improving vitamin A status.[67] Supplementing vitamin A to children aged 6–59 months can reduce their overall mortality about nearly one-fourth in areas with significant levels of VAD.

To achieve satisfactory vitamin A status in populations of developing countries, it is essential to ensure continued access to sufficient supplies of a variety of vitamin A–rich foods and supplements. Twice-yearly vitamin A supplementation is now a fundamental component in health service packages in many countries. For deficient children, the periodic supply of high-dose vitamin A has reduced overall mortality.[67] However, the challenge is to ensure that the remaining 30%[11] of children not reached by services are covered to ascertain universal supplementation sustainability for child survival. Countries that lag behind in the vitamin A supplementation implementation need to adopt practices including twice-yearly events like Child Health Days to broaden the outreach initiative of the primary health-care system.

This requires additional resources and commitments from governments to mobilize and efficiently use supplementary resources. It is important to strengthen collaboration with NGOs and the government in coordinating vitamin A supplementation in areas of need. For instance, despite public health mandates to provide VASs to target groups, these treatments do not reach all of their intended beneficiaries in rural populations. In India, only a quarter of children receive regular VASs.[68] This may be improved through the approach of government contracting to NGOs for health services delivery. Contracting is based on the renewal of lease contracts that are performance-based, and NGOs are expected to achieve certain health targets, such as covering children in a given population with VASs.[69] As access to vitamin A increases, there is a need for continuous nutrition monitoring, surveillance, and management systems to place effective vitamin A distribution systems in the community for supplement delivery to at-risk groups.

For large populations with extensive deficiencies and low levels of resource availability, fortification may offer a more long-term sustainable solution. In areas in which fortified foods are not consumed and deficiencies are too severe, small-scale, targeted interventions such as supplementation or point-of-use fortification may be required. Supplements can be dispensed at higher dosages to treat more severe deficiencies. At all levels of coverage, supplementation has shown a twofold improvement in population vitamin A status when compared with fortification.[70] The cost of supplementation rises sharply corresponding to increasing coverage level, partly because of the difficulty of reaching remote areas. At 80% coverage, supplementation costs more than fortification per DALY saved, for vitamin A.[36] Fortification is always less costly than supplementation since it does not require contact with a provider. This means that cost-effectiveness of fortification is greater than the cost-effectiveness of supplementation, regardless of the coverage of fortification.

Commercial fortification, when compared with MNP intervention, remains more cost-effective per capita. The provision of MNP is relatively more expensive than high-dose vitamin A supplementation; however a short-term intervention with MNP is cost-effective and provides other micronutrients in addition to vitamin A.[71] Based upon current evidence, MNP scale-up can be feasibly and cost-effectively implemented. MNP is similar to supplementation in that it is likely to be more effective for targeted groups, particularly infants and young children from 6 to 24 months. Based upon a short-term study in a resource-poor setting, MNP has been shown to decrease the prevalence of VAD in preschool-aged children by 7.5%.[38] Comparison of MNP and supplementation for combating VAD in a large-scale setting is rare, because most assessments have been primarily for anemia. Based upon current data, fortification is the preferred option in areas with low resource availability, although investments in both fortification and supplementation would remain cost-effective or very cost-effective when coverage is increased to the highest level.[70]

The control of VAD in developing countries requires the combination of both vitamin A supplementation and also food-based approaches of both naturally occurring sources and fortified foods. The differences in adoption and compliance rates between fortification and supplementation rate also need to be taken into account. Taking over where supplementation leaves off, food fortification can maintain vitamin A status for high-risk groups. For vulnerable populations in developing countries, home gardens can foster a diverse diet and contribute to the development and well-being of rural households.[72]

Furthermore, nutritional education, extension, and training for the benefits of vitamin A are needed at all levels to improve vitamin A status. Innovative strategies need to be developed and applied to improve the knowledge, attitudes, and practices. Coordination of nutrition education policy is currently lacking and needs to be defined. The development in awareness among school-aged children, women, and other consumers has to be strengthened. Women's greater role in food preparation in homes in developing countries makes improving women's knowledge about vitamin A–rich foods critical for ensuring vitamin A security. A comprehensive strategy informs the public about issues of nutritional benefit, safety, and encompasses behavioral measures that would foster a sustained demand for fortified products by the consumer to enhance the impact of fortified foods.

## REFERENCES

1. World Health Organization (WHO). Global prevalence of vitamin A deficiency in populations at risk 1995–2005. WHO Global Database on Vitamin A Deficiency. Geneva, Switzerland: World Health Organization, 2009. Available from: http://whqlibdoc.who.int/publications/2009/9789241598019_eng.pdf (accessed March 2012)
2. World Health Organization (WHO). Report: WHO technical consultation on vitamin A in newborn health: Mechanistic studies. Geneva, Switzerland: World Health Organization, 2012. Available from: http://libdoc.who.int/publications/2012/9789241503167_eng.pdf (accessed March 2012)
3. West KP, Jr., Darnton-Hill I. Vitamin A deficiency. In: Semba RD, Bloem MW, eds. *Nutrition and Health in Developing Countries*. 2nd edn. Totowa, NJ: Humana Press, 2009. pp. 377–433.
4. Sommer A, West KP, Jr. *Vitamin A Deficiency: Health, Survival and Vision*. New York: Oxford University Press, 1996.
5. Beaton GH, Martorell R, Aronson KJ et al. Effectiveness of vitamin A supplementation in the control of young child morbidity and mortality in developing countries. ACC/SCN State of the Art Series, Nutrition. 1993, Policy Discussion Paper: No. 13.
6. World Health Organization (WHO)/Food and Agriculture Organization of the United Nations (FAO). The role of food fortification in the control of micronutrient malnutrition. Geneva, Switzerland: WHO/FAO, 2006. http://www.who.int/entity/nutrition/publications/micronutrients/GFF_Part_1_en.pdf (accessed March 2012)
7. Semba RD, Bloem MW, eds. *Nutrition and Health in Developing Countries*. 2nd edn. Totowa, NJ: Humana Press, 2008.

8. West, KP, Jr. Vitamin A Deficiency disorders in children and women. *Food Nutr Bull.* 2003; 24: S78–S90.

9. World Health Organization (WHO). Guideline: Neonatal vitamin A supplementation. Geneva, Switzerland: World Health Organization, 2011. http://whqlibdoc.who.int/publications/2011/9789241501798_eng.pdf (accessed March 2012)

10. Stephensen CB, Alvarez JO, Kohatsu J et al. Vitamin A is excreted in the urine during acute infection. *Am J Clin Nutr.* 1994; 60: 388–392. http://www.scopus.com/record/display.url?eid=2-s2.0-0027933347&origin=inward&txGid=GUkIZ6xlRfzzrsWrtaz-Xyu%3a4 (accessed April 2012)

11. United Nations System Standing Committee on Nutrition (UNSCN). Sixth report on the world nutrition situation. Geneva, Switzerland: United Nations, 2010. http://www.unscn.org/files/Publications/RWNS6/report/SCN_report.pdf (accessed April 2012)

12. West KP, Jr., McLaren D. The epidemiology of vitamin A deficiency disorders. Johnson GJ, Minassian DC, Weale, RA, West SK, eds. *The Epidemiology of Eye Disease.* 2nd edn. New York: Oxford University Press, 2003.

13. West KP, Jr. Extent of vitamin A deficiency among preschool children and women of reproductive age. *J Nutr.* 2002; 132: S2857–S2866. http://jn.nutrition.org/content/132/9/2857S.full (accessed March 2012)

14. Black RE, Allen LH, Bhutta ZA et al. Maternal and child undernutrition: Global and regional exposures and health consequences. *Lancet.* 2008; 371(9608): 243–260. http://www.thelancet.com/journals/lancet/article/PIIS0140-6736(07)61690-0/fulltext (accessed April 2012)

15. Babalola OE. Addressing residual challenges of VISION 2020: The right to sight. *Middle East Afr J Opthalmol.* 2011; 18: 91–92. http://www.ncbi.nlm.nih.gov/pmc/articles/PMC3119297/ (accessed April 2012)

16. UNICEF. Bangladesh Health and Nutrition: Control of vitamin A deficiency. Available from: http://www.unicef.org/bangladesh/health_nutrition_444.htm (accessed March 2012)

17. UNICEF. Child Info: Monitoring the progress of children and women. Available from: http://www.childinfo.org/vitamina_progress.html (accessed March 2012)

18. McLaren DS, Kraemer K; *Sight and Life Manual on Vitamin A Deficiency Disorders (VADD).* 3rd edn Basel, Switzerland: Task Force Sight and Life 2012.

19. Ross CA. Vitamin A. In: Coates PM, Betz JM, Blackman MR et al. eds. *Encyclopedia of Dietary Supplements.* 2nd edn. London, U.K.: Informa Healthcare, 2010, pp. 778–791.

20. Institute of Medicine. *Dietary References Intakes for Vitamin A, Vitamin K, Arsenic, Boron, Chromium, Copper, Iodine, Iron, Manganese, Molybdenum, Nickel, Silicon, Vanadium, and Zinc.* Washington, DC: National Academy Press, 2001.

21. Haskell MJ, Pandey P, Graham JM et al. Recovery from impaired dark adaptation in nightblind pregnant Nepali women who receive small daily doses of vitamin A as amaranth leaves, carrots, goat liver, vitamin A-fortified rice, or retinyl palmitate. *Am J Clin Nutr.* 2005; 81: 461–471. http://www.ajcn.org/content/81/2/461.long (accessed March 2012)

22. Haddad LJ and Bouis HE. The impact of nutritional status on agricultural productivity: Wage evidence from the Philippines. *Oxford Bull Econ Stat.* 1991; 53: 45–68.

23. FAO: Agricultural production and trade policies. Available from: http://www.fao.org/docrep/X5244E/X5244e05.htm (accessed March 2012)

24. Bishai D, Nalubola R. The history of food fortification in the United States: Its relevance for current fortification efforts in developing countries. *Econ Dev Cultur Change.* 2002; 51: 37–53.

25. Dary O, Mora JO. Food fortification to reduce vitamin A deficiency: International Vitamin A Consultative Group recommendations. *J Nutr.* 2002; 132: 2927S–2933S.

26. Fiedler JL, Babu S, Smitz MF et al. Indian social safety net programs as platforms for introducing wheat flour fortification: A case study of Gujarat, India. *Food Nutr Bull.* 2012; 33: 11–30.

27. Favaro RMD. Evaluation of the effect of heat treatment on the biological value of vitamin A fortified soybean oil. *Nutr Res.* 1992; 12: 1357–1363.

28. Mason JB, Ramirez MA, Fernandez CM et al. Effects on vitamin A deficiency in children of periodic high-dose supplements and of fortified oil promotion in a deficient area of the Philippines. *Int J Vitam Nutr Res.* 2011; 81: 295–305.

29. Solon FS. A case report on the fortification of margarine with vitamin A: The Philippine experience. In: Bowley A, ed *Food Fortification to End Micronutrient Malnutrition. State of the Art.* Ottawa, Ontario, Canada: The Micronutrient Initiative, International Development Research Centre; 1997: pp. 32–36.

30. Sridhar KK. Tackling micronutrient malnutrition: Two case studies in India. In: Bowley A, ed *Food Fortification to End Micronutrient Malnutrition. State of the Art.* Ottawa, Ontario, Canada: The Micronutrient Initiative, International Development Research Centre; 1997: pp. 27–31.

31. Arroyave G. Vitamin A deficiency control in Central America. In: Bauernfeind JC, ed. *Vitamin A Deficiency and Its Control.* Orlando, FL: Academic Press Inc.; 1986: pp. 405–424.

32. Pineda O. Fortification of sugar with vitamin A. *Nutriview* 1993; 2: 6–7.

33. Muhilal, Murdiana A, Azis I et al. Vitamin A-fortified monosodium glutamate and vitamin A status: A controlled field trial. *Am J Clin Nutr.* 1988; 48: 1265–1270. Available from: http://www.ajcn.org/content/48/5/1265.full.pdf (accessed March 2012)

34. Nuss ET, Arscott SA, Bresnahan K et al. Comparative intake of white- versus orange-colored maize by Zambian children in the context of promotion of biofortified maize. *Food Nutr Bull.* 2012; 33: 63–71.

35. Meenakshi JV, Banerji A, Manyong V et al. Using a discrete choice experiment to elicit the demand for a nutritious food: Willingness-to-pay for orange maize in rural Zambia. *J Health Econ.* 2012; 31: 62–71. http://www.sciencedirect.com/science/article/pii/S0167629612000033 (accessed April 2012)

36. Horton S. The economics of food fortification. *J Nutr.* 2006; 136: 1068–1071.

37. Le HT, Brouwer ID, Burema J et al. Efficacy of iron fortification compared to iron supplementation among Vietnamese schoolchildren. *Nutr J.* 2006; 5: 32–40.

38. Suchdev PS, Ruth JL, Woodruff BA et al. Selling sprinkles micronutrient powder reduces anemia, iron deficiency, and vitamin A deficiency in young children in western Kenya: A cluster-randomized controlled trial. *Am J Clin Nutr.* 2012; 95: 1223–1230. http://www.ajcn.org/content/95/5/1223.long (accessed March 2012)

39. National Academy of Sciences, Institute of Medicine, Food and Nutrition Board. Dietary reference intakes: UL for vitamins and elements. http://iom.edu/Activities/Nutrition/SummaryDRIs/~/media/Files/Activity%20Files/Nutrition/DRIs/ULs%20for%20Vitamins%20and%20Elements.pdf (accessed May 2012)

40. Anwar S, Matiur Rahman AKM, Bannerjee M et al. Diplopia and acute sixth nerve palsy in hypervitaminosis A—A case report and a review literature. *Bangladesh J Child Health.* 2011; 35: 36–37.

41. Promislow JH, Goodman-Gruen D, Slymen DJ et al. Retinol intake and bone mineral density in the elderly: The Rancho Bernardo Study. *J Bone Miner Res.* 2002; 17: 1349–1358.

42. Feskanich D, Singh V, Willett WC et al. Vitamin A intake and hip fractures among postmenopausal women. *JAMA.* 2002; 287: 47–54.

43. Melhus H, Michaëlsson K, Kindmark A et al. Excessive dietary intake of vitamin A is associated with reduced bone mineral density and increased risk for hip fracture. *Ann Intern Med.* 1998; 129: 770–778.

44. Michaëlsson K, Lithell H, Vessby B et al. Serum retinol levels and the risk of fracture. *N Engl J Med.* 2003; 348: 287–294.
45. Sommer A. Vitamin A deficiency disorders: Origins of the problem and approaches to its control. The Johns Hopkins University Bloomberg School of Public Health, 2001. Available from: http://www.biotech-info.net/disorders.html (accessed March 2012)
46. van Stuijvenberg ME, Kvalsvig JD, Faber M et al. Effect of iron-, iodine-, and α-carotene-fortified biscuits on the micronutrient status of primary school children: A randomized controlled trial. *Am J Clin Nutr.* 1999; 69: 497–503.
47. Krause VM, Delisle H, Solomons NW. Fortified foods contribute one half of recommended vitamin A intake in poor urban Guatemalan toddlers. *J Nutr.* 1998; 128: 860–864.
48. UNICEF/MI/WHO/CIDA/USAID. Vitamin A global initiative: A strategy for acceleration of progress in combating vitamin A deficiency. Consensus of an Informal Technical Consultation convened by UNICEF in association with MI, WHO, CIDA and USAID, 1998. Available from: http://www.unicef.org/immunization/files/Vit_A_strategy.pdf (accessed March 2012)
49. Sprinkles Global Health Initiative. Micronutrient sprinkles for use in infants and young children: Guidelines on recommendations for use and program monitoring and evaluation. Toronto, Canada: SGHI, 2008. Available from: http://www.sghi.org/resource_centre/GuidelinesGen2008.pdf (accessed April 2012)
50. Helen Keller International. Homestead food production. Available from: http://www.hki.org/reducing-malnutrition/homestead-food-production/(accessed March 2012)
51. West KP Jr., Mehra S. Vitamin A intake and status in populations facing economic stress. *Am J Clin Nutr.* 2010; 140: 201S–207S. Available from: http://jn.nutrition.org/content/140/1/201S.full (accessed March 2012)
52. Gómez-Galera S, Rojas E, Sudhakar D et al. Critical evaluation of strategies for mineral fortification of staple food crops. *Transgenic Res.* 2011; 19: 165–180.
53. Klemm RD, West KP Jr., Palmer AC et al. Vitamin A fortification of wheat flour: Considerations and current recommendations. *Food Nutr Bull.* 2010; 31(1 Suppl): S47–S61.
54. Tarwotjo I, Sommer A, Soegiharto T et al. Dietary practices and xerophthalmia among Indonesian children. *Am J Clin Nutr.* 1982; 35: 574–581.
55. Rice AL, West KP, Black RE. Vitamin A deficiency. In: Ezzati M, Lopez AD, Rogers A, Murray CJL, eds. *Comparative Quantification of Health Risks: Global and Regional Burden of Disease Attributable to Selected Major Risk Factors.* Geneva, Switzerland: World Health Organization, 2003.
56. Thorne-Lyman AL, Fawzi WW. Vitamin A and carotenoids during pregnancy and maternal, neonatal and infant health outcomes: A systematic review and meta-analysis. *Paediatr Perinat Epidemiol.* 2012; 26: 36–54. doi: 10.1111/j.1365-3016.2012.01284.x.
57. West KP, Christian P, Labrique AB et al. Effects of vitamin A or beta carotene supplementation on pregnancy-related mortality and infant mortality in rural Bangladesh cluster randomized trial. *JAMA.* 2011; 305(18): 1986–1995. Available from: http://jama.jamanetwork.com/article.aspx?volume=305&issue=19&page=1986 (accessed March 2012)
58. Sauvant P, Féart C, Atgié C. Vitamin A supply to mothers and children: Challenges and opportunities. *Curr Opin Clin Nutr Metab Care.* 2012; 15: 310–314.
59. Christian P, Labrique AB, Ali H et al. Maternal vitamin A and β-carotene supplementation and risk of bacterial vaginosis: A randomized controlled trial in rural Bangladesh. *Am J Clin Nutr.* 2011; 94: 1643–1649. Available from: http://www.ajcn.org/content/94/6/1643.long (accessed March 2012)
60. Imdad A, Yakoob MY, Sudfeld C et al. Impact of vitamin A supplementation on infant and childhood mortality. *BMC Public Health.* 2011; 11(Suppl 3): S20. Available from: http://www.biomedcentral.com/1471-2458/11/S3/S20 (accessed April 2012)

61. Gogia S, Sachdev HS. Neonatal vitamin A supplementation for prevention of mortality and morbidity in infancy: Systematic review of randomized controlled trials. *BMJ*. 2009; 338: b919. http://www.bmj.com/content/338/bmj.b919?view=long&pmid=19329516 (accessed April 2012)

62. Gogia S, Sachdev HS. Maternal postpartum vitamin A supplementation for the prevention of mortality and morbidity in infancy: A systematic review of randomized controlled trials. *Int J Epidemiol*. 2010; 39: 1217–1226.

63. Gogia S, Sachdev HS. Vitamin A supplementation for the prevention of morbidity and mortality in infants six months of age or less. *Cochrane Database Syst Rev*. 2011: 10 http://www.mrw.interscience.wiley.com/cochrane/clsysrev/articles/CD007480/frame. html (accessed April 2012)

64. Haider BA, Bhutta ZA. Neonatal vitamin A supplementation for the prevention of mortality and morbidity in term neonates in developing countries. *Cochrane Database Syst Rev*. 2011: 10. http://onlinelibrary.wiley.com/doi/10.1002/14651858.CD006980.pub2/ abstract

65. Rice AL, Stoltzfus RJ, de Francisco A, Kjolhede CL. Low breast milk vitamin A concentration reflects an increased risk of low liver vitamin A stores in women. *Adv Exp Med Biol*. 2000; 478: 375–376.

66. United Nations Children's Fund. The state of the world's children: Special Edition. New York: UNICEF, 2009. Available from: http://www.unicef.org/rightsite/sowc/pdfs/ SOWC_Spec%20Ed_CRC_Main%20Report_EN_090409.pdf (accessed March 2012)

67. World Health Organization. Micronutrient deficiencies: Vitamin A deficiency. Geneva, Switzerland: WHO, 2008. Available from: http://www.who.int/nutrition/topics/vad/en (accessed March 2012)

68. International Institution for Population Sciences. *National Family Health Survey (NFHS-3), 2005–06: India: Volume 1 (Survey)*. Mumbai, India: International Institution for Population Sciences (IIPS) and Macro International, 2007.

69. Loevinsohn B. Performance based contracting for health services in developing countries: A toolkit. Washington, DC: World Bank, 2010. Available from: http://siteresources. worldbank.org/INTHSD/Resources/topics/415176-1216235459918/ContractingEbook. pdf (accessed May 2012)

70. The World Health Report 2002: Reducing risks, promoting healthy life (Chapter 5). Geneva, Switzerland: World Health Organization, 2002. Available from: http://www. who.int/whr/2002/chapter5/en/index5.html (accessed March 2012)

71. Sharieff W, Horton, S, Zlotkin S. Economic gains of a home fortification program: Evaluation of "sprinkles" from the provider's perspective. *Can J Public Health*. 2006; 97: 20–23. Available from: http://search.proquest.com/docview/232007214 (accessed April 2012)

72. Food and Agriculture Organization of the United Nations (FAO). Fortification of food with micronutrients and meeting dietary micronutrient requirements: Role and position of FAO. Geneva, Switzerland: WHO/FAO, 2006. Available from: http://www.who.int/ nutrition/topics/vad/en/ (accessed March 2012)

# 18 Vitamin A Intakes Are Below Recommendations in a Significant Proportion of the Population in the Western World

*Hans-Konrad Biesalski, Barbara Trösch, and Petra Weber*

## CONTENT

In the developing world, vitamin A deficiency is a major health problem and responsible for blindness and increased childhood mortality before the age of 5 due to a weakened immune system. The reason for vitamin A deficiency in developing countries is the inadequacy of the diet, while, in contrary, in developed countries, it is often claimed that vitamin A intake is too high and might cause harmful effects. The range of foods available in these countries is wider than ever, but changing lifestyles and dietary habits can significantly impact on intakes of various micronutrients, leading to inadequate intakes even in affluent societies [1]. Obtaining accurate information on dietary intakes of whole populations remains a challenge. However, dietary surveys, despite their well-known limitations, are considered as an established tool to do so. To assess and compare the current vitamin A intakes in industrialized countries, large population-based surveys from Germany [2], the United Kingdom [3], the Netherlands [4], and the United States [5–7] were selected. While the first three are published surveys, the data from the U.S. National Health and Nutrition Examination Survey can be downloaded and analyzed. For this, we employed the Food and Nutrient Database for Dietary Studies (Version 2.0 for NHANES 2003–2004, version 3.0 for 2005–2006, and version 4.0 for 2007–2008) to determine the vitamin content of foods. Intake data from the two 24 h recalls were used to estimate percentiles using the National Cancer Institute method [8].

Comparing the intakes for adults with national recommendations, it becomes apparent that a significant proportion of the population does not achieve adequate intakes for vitamin A (Figure 18.1). Of particular concern is the fact that more

Vitamin A intake adults

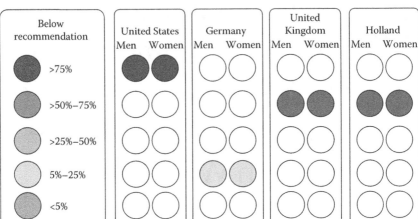

(A)

U.S. elderly (>70 yrs)         U.S. women of childbearing age

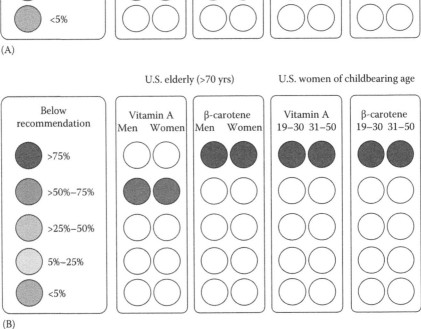

(B)

**FIGURE 18.1** (A) Adults aged 19–50 years in the United States [5–7], Germany [2], and the Netherlands [4] and aged 19–49 years for the United Kingdom [3] and (B) elderly and women of childbearing age in the United States with intakes below the specific recommended reference value for the respective country [9–12]. The level of recommendation covering the needs of 97.5% of the population was used where it existed.

than 75% of women of childbearing age in the NHANES cohort do not achieve the current DRI for vitamin A. This has previously also been shown in pregnant women in a small study in Germany (Schulz et al., 2007). As the most detailed data are available for Germany and the United States, these two countries were investigated more closely (Figure 18.2). Intakes of vitamin A and β-carotene are higher in Germany than in the United States. No difference is apparent in the intake pattern of men and women in either country. In comparison with other European countries [9],

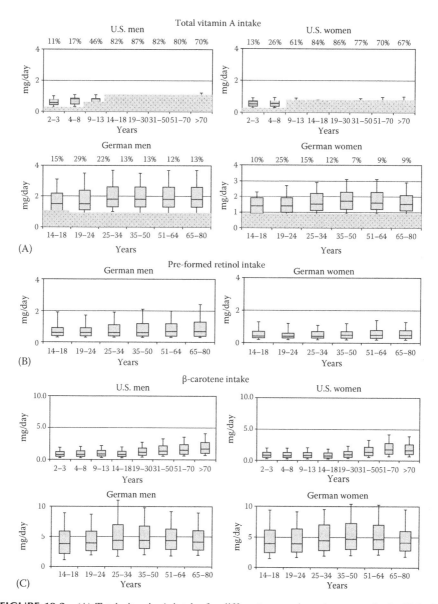

**FIGURE 18.2** (A) Total vitamin A intake for different age and gender groups in the United States and Germany. (B) Intake of preformed retinol for different age and gender groups in Germany. (C) Intake of β-carotene in different age and gender groups in the United States and Germany. Box plots show 10th, 25th, 50th, 75th, and 90th percentiles. Red indicates national recommendations for the respective subpopulation. (From Institute of Medicine, *Dietary Reference Intakes of Vitamin A, Vitamin K, Arsenic, Boron, Chromium, Copper, Iodine, Iron, Manganese, Molybdenum, Nickel, Silicon, Vanadium, and Zinc*, National Academic Press, Washington, DC, 2001; Deutsche Gesellschaft für Ernährung, Österreichische Gesellschaft für Ernährung, Schweizerische Gesellschaft für Ernährungand Schweizerische Vereinigung für Ernährung, *Referenzwerte für die Nährstoffzufuhr*, Umschau Verlag, Frankfurt/Main, Germany, 2008.)

Germany is at the upper and the United States at the lower end of average intakes for both total vitamin A and β-carotene. Even though only mean intakes are given in the European Nutrition and Health Report, they still indicate that inadequate intakes affect a large part of the population [13].

The difference between intakes in Germany and the United States is partially due to the difference in conversion factor used: In the United States, one retinol activity equivalent is defined as 12 units of β-carotene or 24 units of other provitamin A carotenoids [9], while in the Germany, the Netherlands, and the United Kingdom, it is 6 and 12, respectively [2–4]. These conversion factors are derived from differences in the interpretations of the evidence for the relative absorption and conversion of β-carotene [9,10,12]. More recent evidence suggests that the efficiency of conversion also depends on individual factors such as BMI, genotype, and specific metabolic conditions [14–16]. However, β-carotene intakes are also higher in Germany than in the United States, suggesting an actual higher intake rather than an artificial one due to the conversion alone. The main sources of β-carotene in Germany are vegetables, mushrooms, and legumes, followed by soups as well as fruits and nonalcoholic beverages [2]. There are no official recommendations for β-carotene intakes; however, 2–4 mg/day are stated as desired intakes in the German survey [2]. The Institute of Medicine regards the range of 3–6 mg/day, which corresponds to an intake of five or more fruits and vegetables per day, as advisable [17]. However, in the light of the relatively small contribution of preformed retinol to the total vitamin A intake even in affluent countries, it was proposed to increase these recommendations to 7 mg/day [14]. In Germany, around 90% achieve 2 mg/day and nearly 25% of the population 7 mg/day β-carotene (Figure 18.2). In the United States, however, only around 25% achieve the lower (2 mg β-carotene) and considerably less than 10% the higher range (7 mg β-carotene) of these recommendations (Figure 18.2). The situation in the United Kingdom is similar to the United States, with around 70% below 2 mg and 100% below 7 mg [3]. In the Dutch survey, no information on β-carotene intakes is given [4].

Information on preformed retinol intake is available only from the German (Figure 18.2) and the British survey [3] and shows that preformed retinol levels are also higher in the German diet. It was calculated that around one-third of total vitamin A is derived from preformed retinol, half is taken up as β-carotene, and the rest as various sorts of carotenoids [14]. Consequently, risk of excessive intakes of preformed vitamin A and of the resulting adverse health outcomes can be regarded as low [14]. Even the individuals at the high end of intakes in Germany (95th percentile) are below the upper intake of 3 mg preformed vitamin A per day still regarded as safe by the Institute of Medicine [2,9]. A similar conclusion was reached by an evaluation of retinol intakes in adults from food and, where information was available, fortification and supplementation in various European countries [18]. Only a small percentage of postmenopausal women may exceed the upper limit of preformed retinol, if 1.5 mg RAE/day is used for this subpopulation as proposed by the European Safety Authority [19]. Consequently, risk of excessive intakes of preformed vitamin A and of the resulting adverse health outcomes can be regarded as low [14].

Vitamin A intakes of specific population groups such as the elderly or women of childbearing age in the United States also show widespread prevalence of inadequacy (Figure 18.1). The situation was found to be only slightly better in institutionalized

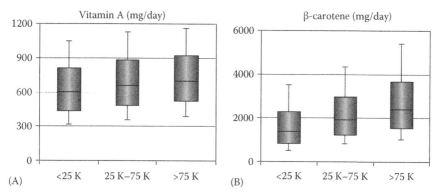

**FIGURE 18.3** Intake of (A) total vitamin A and (B) β-carotene in elderly (>70 years) in the United States stratified by household income (<25 K: less than $25,000, 25 K–75 K: between $25,000 and $75,000, >75 K: more than $75,000 per year). Box plots show 10th, 25th, 50th, 75th, and 90th percentiles. Red indicates national recommendations for the respective subpopulation. (From Institute of Medicine, *Dietary Reference Intakes of Vitamin A, Vitamin K, Arsenic, Boron, Chromium, Copper, Iodine, Iron, Manganese, Molybdenum, Nickel, Silicon, Vanadium, and Zinc,* National Academic Press, Washington, DC, 2001; Deutsche Gesellschaft für Ernährung, Österreichische Gesellschaft für Ernährung, Schweizerische Gesellschaft für Ernährungand Schweizerische Vereinigung für Ernährung, *Referenzwerte für die Nährstoffzufuhr,* Umschau Verlag, Frankfurt/Main, Germany, 2008.)

elderly in Germany, where more than 25% had vitamin A intakes below recommendations [20]. While all the U.S. elderly had rather low vitamin A intakes, a clear trend along income lines was observed for both vitamin A and β-carotene intakes when stratifying the data according to household income (Figure 18.3). Low vitamin intakes in this age group are particularly relevant as dietary factors play an important role in modulating the signaling pathways thought to influence aging and longevity [21] and low vitamin A intake was found to be linked to higher frequency of genome damage [22]. Moreover, while dietary antioxidants such as β-carotene might not directly contribute to longevity, they are likely to provide health benefits with respect to chronic diseases [23]. As described in Chapter 13, low levels of vitamin A even without clinical signs of vitamin A deficiency may promote respiratory tract infections, in particular in children and the elderly. Data from human studies demonstrate that low intakes of vitamin A favors infections, highlighting the importance of vitamin A for the immune system (see Chapter 13). Another group at risk is pregnant women, in particular those with low birth spacing or multiple pregnancy. Recent data from Germany show that even in cases of a normal socioeconomic level, very low plasma levels of the mother and the newborn are the consequence of an inadequate intake [24]. Whether that might have consequences for the child's development has not been established yet. Nevertheless, it is well known that low vitamin A intake results in depleted liver stores, which might be needed in cases of an increased demand as seen typically during infections and tissue repair.

In summary, vitamin A status in Western populations, based on the results of dietary intake surveys relative to current RDA in the respective countries, is critical in a significant part of the population, women of childbearing age, children, and elderly being in particular at risk.

## REFERENCES

1. Troesch, B., B. Hoeft, M. McBurney, M. Eggersdorfer, and P. Weber, Dietary surveys indicate vitamin intakes below recommendations are common in representative Western countries. *British Journal of Nutrition*, in press.
2. Max Rubner-Institut, Nationale Verzehrsstudie II. Ergebnisbericht Teil 2. Die bundesweite Befragung zur Ernährung von Jugendlichen und Erwachsenen. 2008 (http://www.was-esse-ich.de/uploads/media/NVSII_Abschlussbericht_Teil_2.pdf; accessed February 7, 2011): Karlsruhe.
3. Henderson, L., K. Irving, J.F. Gregory, C.J. Bates, A. Prentice, J. Perks, G. Swan, and M. Farron, Volume 3—Vitamin and mineral intake and urinary analytes, in *The National Diet and Nutrition Survey: Adults Aged 19 to 64 Years*. 2003.
4. van Rossum, C.T.M., H.P. Fransen, J. Verkaik-Kloosterman, E.J.M. Buurma-Rethans, and M.C. Ocké, *Dutch National Food Consumption Survey 2007–2010—Diet of Children and Adults Aged 7 to 69 Years*. 2011.
5. Centers for Disease Control and Prevention and National Center for Health Statistics. NHANES 2003–2004. Data, documentation, codebooks, SAS code. Dietary interview. Individual foods, total nutrient intakes first and second day. 2009 [cited August 2010]; Available from http://www.cdc.gov/nchs/nhanes/nhanes2003-2004/diet03_04.htm
6. Centers for Disease Control and Prevention and National Center for Health Statistics. NHANES 2005–2006. Data, documentation, codebooks, SAS code. Dietary interview. Individual foods, total nutrient intakes first and second day. 2009 [cited August 2010]; Available from: http://www.cdc.gov/nchs/nhanes/nhanes2005-2006/diet05_06.htm
7. Centers for Disease Control and Prevention and National Center for Health Statistics. NHANES 2007–2008. Data, documentation, codebooks, SAS code. Dietary interview. Individual foods, total nutrient intakes first and second day. 2009 [cited August 2010]; Available from: http://www.cdc.gov/nchs/nhanes/nhanes2007-2008/diet07_08.htm
8. Tooze, J.A., V. Kipnis, D.W. Buckman, R.J. Carroll, L.S. Freedman, P.M. Guenther, S.M. Krebs-Smith, A.F. Subar, and K.W. Dodd, A mixed-effects model approach for estimating the distribution of usual intake of nutrients: The NCI method. *Statistics in Medicine*, 2010. 29(27): 2857–2868.
9. Institute of Medicine, *Dietary Reference Intakes of Vitamin A, Vitamin K, Arsenic, Boron, Chromium, Copper, Iodine, Iron, Manganese, Molybdenum, Nickel, Silicon, Vanadium, and Zinc*. 2001, Washington, DC: National Academic Press.
10. Deutsche Gesellschaft für Ernährung, Österreichische Gesellschaft für Ernährung, Schweizerische Gesellschaft für Ernährungand Schweizerische Vereinigung für Ernährung, *Referenzwerte für die Nährstoffzufuhr*. 2008, Frankfurt/Main, Germany: Umschau Verlag.
11. Gezondheidsraad, *Naar een adequate inname van vitamine A*. 2008, Den Haag, the Netherlands: Gezondheidsraad.
12. Department of Health, Dietary reference values for food, energy and nutrients for the United Kingdom, No. 41, in *Report on Health and Social Subjects*. 1991: London, U.K.: Department of Health.
13. Elmadfa, I., A. Meyer, V. Nowak, V. Hasenegger, P. Putz, R. Verstraeten, A.M. Remaut-DeWinter et al., *European Nutrition and Health Report 2009*. February 6, 2010 edn. Forum of Nutrition, ed. I. Elmadfa. Vol. 62. 2009, Basel, Switzerland: Karger.
14. Grune, T., G. Lietz, A. Palou, A.C. Ross, W. Stahl, G. Tang, D. Thurnham, S.A. Yin, and H.K. Biesalski, Beta-carotene is an important vitamin A source for humans. *Journal of Nutrition*, 2010. 140(12): 2268S–2285S.

15. Lietz, G., A. Oxley, W. Leung, and J. Hesketh, Single nucleotide polymorphisms upstream from the β-carotene 15,15'-monoxygenase gene influence provitamin A conversion efficiency in female volunteers. *Journal of Nutrition*, 2012. 142: 161S–165S.
16. Leung, W.C., S. Hessel, C. Méplan, J. Flint, V. Oberhauser, F. Tourniaire, J.E. Hesketh, J. von Lintig, and G. Lietz, Two common single nucleotide polymorphisms in the gene encoding β-carotene 15,15'-monoxygenase alter β-carotene metabolism in female volunteers. *FASEB Journal*, 2009. 23(4): 1041–1053.
17. Institute of Medicine, *Dietary Reference Intakes of Vitamin C, Vitamin E, Selenium, and Carotenoids*. 2000, Washington, DC: National Academic Press.
18. Flynn, A., T. Hirvonen, G.B. Mensink, M.C. Ocke, L. Serra-Majem, K. Stos, L. Szponar, I. Tetens, A. Turrini, R. Fletcher, and T. Wildemann, Intake of selected nutrients from foods, from fortification and from supplements in various European countries. *Food and Nutrition Research*, 2009. 53: 1–51.
19. Scientific Committee on Food and N.A.A. Scientific Panel on Dietetic Products, *Tolerable Upper Intake Levels for Vitamins and Minerals*, ed. E.F.S. Authority. 2006.
20. Deutsche Gesellschaft für Ernährung e. V., *Ernährungsbericht 2008*. 2008, Bonn, Germany: Deutsche Gesellschaft für Ernährung e. V.
21. Ostojic, S., N. Pereza, and M. Kapovic, A current genetic and epigenetic view on human aging mechanisms. *Collegium Antropologicum*, 2009. 33(2): 687–699.
22. Fenech, M., Genome health nutrigenomics and nutrigenetics—Diagnosis and nutritional treatment of genome damage on an individual basis. *Food and Chemical Toxicology*, 2008. 46(4): 1365–1370.
23. Chong-Han, K., Dietary lipophilic antioxidants: Implications and significance in the aging process. *Critical Reviews in Food Science and Nutrition*, 2010. 50(10): 931–937.
24. Schulz, C., U. Engel, R. Kreienberg, and H.K. Biesalski, Vitamin A and beta-carotene supply of women with Gemini or short birth intervals: A pilot study. *European Journal of Nutrition*, 2007. 46(1): 12–20.

# 19 Critical Appraisal of Vitamin A Supplementation Program in India

*Dechenla Tshering Bhutia and Saskia de Pee*

## CONTENTS

## INTRODUCTION

The vitamin A supplementation (VAS) program in India is one of the oldest VAS programs in the developing world. It started in the 1970s as a temporary fix to address night blindness while other sustainable solutions could be developed, but it gained importance as an important and cost-effective intervention to increase child survival. However, a program that aims to achieve universal coverage but even after 40 years is not able to cover half of the targeted children is of concern. The reasons for the low coverage of VAS in India should be understood if universal coverage is to be achieved. A strategy for covering children from 6 months to 5 years of age with doses of vitamin A at 6 monthly intervals requires constant commitment both from the health service and from the beneficiaries.

There can never be a perfect program, but a sound program, which is evidence based and feasible to implement, keeping in mind the diversity of states in a country like India, is needed. Understanding the various issues responsible for the low coverage of the VAS program is crucial in order to develop recommendations for improvement.

## PROBLEM OF VITAMIN A DEFICIENCY

Vitamin A deficiency (VAD) is the leading cause of preventable childhood blindness and reduced immunity that results in increased mortality from childhood diseases. It contributes to the vicious cycle of infection and poor nutritional status and increases the risk of death among children [1].

In 2002, West estimated that around 4.4 million children in developing countries suffered from xerophthalmia, out of a total of 127 million preschool children who suffered from biochemical VAD. Around 44% of children with biochemical VAD live in South and South East Asia. India contributes almost 64% (35.3 million) of them and represents around 88% of all cases of xerophthalmia in South and South East Asia region [2]. VAD precipitates the death of around 0.3 million children every year in India [3]. According to WHO in 2009, an estimated 190 million preschool children are suffering from biochemical VAD. The prevalence of biochemical VAD among under-5 children in India is 62% [1], which is one of the highest in the world. Though clinical VAD is less than 1% [4], it has been found to be as high as 1.4% in some states, and the prevalence of biochemical VAD ranges from 52% to 88% among states [5]. The prevalence of VAD is higher than the WHO cutoff for indicating a public health problem (Bitot's spot—0.5%, night blindness in children—1%) [3].

## CASE FOR VITAMIN A SUPPLEMENTATION IN INDIA

Vitamin A supplementation (VAS) is one of the most cost-effective interventions to reduce childhood mortality [6]. Improving vitamin A status is associated with a 24% reduction in all-cause childhood mortality [7]. VAS is considered a key intervention in reducing under-5 mortality rate (U5MR) and achieving the Millennium Development Goal (MDG) 4 target of reducing the U5MR by two-thirds by 2015. Coverage of at least 70% is required with two doses of vitamin A per year to expect a mortality-reducing effect [6]. According to UNICEF, India is identified as priority for VAS with a U5MR of ≥70/1000 live births in 2004 [6]. With slow progress of 34% decline in U5MR since 1990 in India, it still remains high at 63/1000 live births in 2010 [8,9].

At present, the VAS program in India is integrated with the Reproductive and Child Health (RCH) program [10]. The current strategy for VAS among under-5 children in India is the universal coverage with megadoses of vitamin A every 6 months starting from 9 months of age with measles vaccine and continued until 5 years [6]. Under the RCH program, VAS has been incorporated with the Expanded Program on Immunization (EPI) [11]. Though VAS program has been in existence since four decades, the coverage has been bleak. NFHS III, conducted in 2005–2006, reports low VAS coverage of 25% (urban—26.8%, rural—24.2%) among 12–35 month children in the past 6 months [12].

UNICEF's coverage survey in India in 2006 found that 58% of 12–23 month old children had received at least one dose of VAS. Since the child receives the first dose of VAS along with measles vaccine, and measles vaccination coverage was 70.9%; this means that (1) 29% of children are missed because they do not visit the health center for measles immunization and (2) 20% of those that do get measles immunization fail to receive vitamin A [13]. Semba et al. using NFHS III data, reported a coverage of 20.2% with one dose in past 6 months among 6–59 month old children, ranging from 6% to 43% among states [14]. With low coverage and wide interstate variation in VAD, VAS is unlikely to have an impact on the U5MR in India.

Although UNICEF uses a U5MR of 70/1000 live births as cutoff for declaring VAD as a public health problem, the International Vitamin A Consultative Group (IVACG) meeting in 2002 pointed out that even a U5MR of 50/1000 live births indicates very likely existence of VAD [15]. Looking at the current U5MR of 63/1000 live births in India [9], VAD remains a public health problem that needs immediate corrective actions [15].

## DESCRIPTION OF INDIA

India is the second most populous country in the world with a total population of 1.21 billion [16]. With only 2.4% of the world's land area, India is home to around 17% of the world population. The demographic profile of India represents a high birth rate and decreasing death rate (20.97 and 7.48 per 1000 midyear population, respectively). The majority of the population lives in rural areas (68.8%), and the literacy rate is 74.04% but is lower in females (65.46%). India is characterized by diverse religious and ethnic groups with Hindus (80.5%) being the dominant religion followed by Muslims (13.4%), Christians (2.3%), and other religions [16].

The latest census of 2011 estimates that around 13% of the population comprises of 0–6 year old children [16]. With 1.6 million under-5 child deaths annually, India accounts for 21% of all child deaths in the world (under-5 deaths globally in the world is 7.6 million) [9]. Less than 50% of 12–23 month children are fully immunized under the EPI [12]. India represents one of the highest prevalence of VAD in the world. Among states, prevalence ranges from 31% to 57% for preschool children suffering from biochemical VAD and 1%–2% for clinical VAD [4].

India with 28 states and 7 union territories has a healthcare system extending from national to village level. At central level, the Union Ministry of Health and Family Welfare (MOHFW) is responsible for policy making, and assisting and coordinating activities at the state ministries. The state ministry is similar to the central ministry and is mostly independent in matters of healthcare delivery. The district level, which is headed by the chief medical officer (CMO), acts as a link between the state and the periphery and is responsible for implementing health programs laid down at higher levels [17]. In the rural areas, the most peripheral units of administration are the local self-governing bodies called *Panchayat*. These units are a form of decentralized government where every village is responsible for its own developmental affairs. The members belong to the village, and one-third of the membership places are reserved for women [18].

The healthcare delivery system at the periphery in rural areas has a three-tier system. At the subdistrict level, the community health center (CHC) caters to a population of 80,000–120,000 with speciality services in maternal and child health (MCH), surgery, and medicine. The primary health center (PHC) level is established for 20,000–30,000 population and provides promotive, preventive, and basic health services to the population. The subcenter (SC) level represents the peripheral outpost catering to 3000–5000 population with services mostly limited to MCH care, immunization, and family planning. The various levels are connected through a chain of referral system and health management information system [17]. The urban areas are covered by urban health posts, dispensaries, and hospitals [19]. Apart from this, Integrated Child Development Service (ICDS) centers are located in urban and rural areas run by ICDS workers called *Anganwadi* worker (AWW). It caters to preschool children (0–6 years) and mothers with provision of supplementary nutrition, preschool education, immunization, and nutrition education. One ICDS center caters to 400–800 total population [20]. ICDS scheme in India is the largest integrated child program in the world. By 2010, there were 1,241,749 ICDS centers all over India, and the scheme is still expanding [20].

Total Health Expenditure (THE) forms 4.25% of Gross Domestic Product (GDP), but with government spending on health at only 0.84% of the GDP, most is paid by the population itself (71.13% of THE) and most of that on private health care. Social insurance contributes to only 1.13% of the THE [21]. This results in almost one-fourth (24%) of people, most of whom fall below the poverty line, unable to afford medical expenses [22]. According to 2005 estimates, 42% of people in India are living below the poverty line, using a purchasing power parity (PPP) rate of $1.25 per day (Rs. 21.6 in urban areas and Rs. 14.3 in rural areas) [23].

## HOW IS VAS IMPLEMENTED IN INDIA?

VAS program in India is implemented under the MOHFW. It was initiated in 1970 in response to the high prevalence of blindness due to xerophthalmia as a result of VAD, under the name National Program for the Prevention of Nutritional Blindness (NP-PNB) [24]. Clinical VAD was as high as 2% in 1975–1979. According to the program, children in the age group 1–5 years were to be administered 200,000 IU of vitamin A at 6 monthly intervals [25]. In 1992, during the Eighth five year plan (FYP) (1992–1997), the VAS program was integrated with the child survival and safe motherhood (CSSM) program and linked to EPI to improve coverage. The age for receiving VAS was revised to 9 months to 3 years as signs of clinical deficiency were largely restricted to this age group. Dosing was prescribed as 100,000 IU for <1 year and 200,000 IU for 1–3 years to be provided as a total of five megadoses at 6 monthly intervals. Routine immunization can primarily deliver the first two doses (first dose at 9 months along with measles vaccine and second dose at 18 months with booster dose of DPT and OPV) [26]. In 2006, following the recommendations of WHO, UNICEF, and Women and Child Development, which was based on the fact that VAS has been shown to reduce mortality among children under 5, the MOHFW broadened the age range from 9 months to 5 years with nine megadoses of vitamin A to all children as follows [10]:

6–11 months: One dose of 100,000 IU of vitamin A
1–5 years: 200,000 IU of vitamin A at 6 monthly intervals

India provides VAS in the form of syrup with 1 mL equivalent to 100,000 IU of vitamin A. Vitamin A is produced by local manufacturers, and its supply is provided and monitored through Government of India and UNICEF [6,27]. UNICEF provides vitamin A funded by Micronutrient Initiative (MI) in 12 states where the U5MR is more than 50/1000 live births [28]. USAID is also supporting VAS among under-5 children in Uttar Pradesh and Jharkhand states by assisting health departments to identify constraints in supply, distribution, and estimation of vitamin A supply needs [29].

VAS is carried out through the healthcare delivery system of PHC, SC, and ICDS centers. Along with supplementation, the health workers have to educate mothers on VAD and its prevention. The latest recommendations to improve coverage of VAS from the 10th FYP (2002–2007) is to continue VAS with EPI for the first two doses (with measles and DPT booster vaccinations) and provide the subsequent doses at 6 monthly intervals, by AWW, during April–May (pre-summer) and September–October (pre-winter) [30]. The 11th FYP (2007–2012) has continued with the goal of eliminating VAD as a public health problem and reducing biochemical deficiency by 50% [31].

## HISTORIC DATA ON VAS COVERAGE IN INDIA

Since the inception of VAS program in 1970, there has not been high enough coverage by the program. The linkage of VAS with routine immunization has not made much difference. The full immunization coverage among 12–23 month children in India is less than half (43.5%) (2005–2006), very similar to the previous figure of 42% (1998–1999) [12]. NNMB survey in 2005–2006 showed coverage of 25% for two doses of vitamin A among 12–59 month children [32]. The 10th FYP recommendation of biannual administration of subsequent doses of vitamin A to children aged 18–59 months [30] has been tried in Uttar Pradesh with the assistance from UNICEF and has resulted in improved coverage of VAS [26] of 68% with first dose of VAS with measles immunization and 36% coverage for second to fifth dose by 2008 [33].

Though low coverage is seen in surveys, UNICEF State of the World Children (SOWC) reports a higher coverage of 64% in 2005 with two annual doses of vitamin A that increased to 66% (2009) among 6–59 month children [8,34]. NFHS data are from household surveys where denominator is clear, i.e., all children included in the interview; UNICEF reported data are administrative data, i.e., reported by the health sector where it is based on the number of children given VAS and the number of children they would have expected [6]. While recall of VAS receipt may not be perfect, the main issue is with the latter, when the number of children expected is underestimated, coverage is reported higher than it actually is. The coverage level reported for 2005 of 64% for two annual doses [8] is doubtful when coverage in the same year among 12–35 months within past 6 months, as obtained from household surveys, was only 25% [12]. Coverage reported in the latest UNICEF SOWC report of 2012 however is much closer to the data reported by the NFHS: 34% was reported to have received two annual dosages of vitamin A in 2010 [9]. Real coverage is likely somewhere between 25% and 34%, because household surveys may underestimate due to recall bias and administrative data may overestimate due to problems of estimating number of eligible children. A coverage below one-third is much too low, and therefore reasons for the low coverage need to be identified in order to improve VAS programming and reduce VAD. In this chapter, we will refer to coverage as obtained from household surveys, because of greater reliability of the denominator.

Table 19.1 shows U5MR and coverage of VAS per state. Uttar Pradesh with the highest U5MR has the lowest VAS coverage of 5.9% but Kerala with the lowest U5MR still has a low coverage of 33.6%. None of the states exceeds coverage of 43% within past 6 months among 12–59 month old children. Table 19.1 shows coverage for subsequent doses of less than half compared to initial doses. Although the U5MR in India is decreasing, the possibility that this is due to VAS is unlikely because of the low coverage.

## TABLE 19.1
## U5MR and Coverage of VAS by States, India

| States/UT | U5MR (per 1000 Live Births) (2005–2006) [12] | Coverage of One Dose of VAS | | |
|---|---|---|---|---|
| | | Among 12–23 Months (2006) | Within Past 6 Months (12–23 Months) (2006) | Within Past 6 Months (12–59 Months) (2005–2006) |
| | | [13] (%) | | [14] (%) |
| Uttar Pradesh | 96.4 | 36.5 | 23.1 | 5.9 |
| Madhya Pradesh | 94.2 | 66.3 | 30.5 | 16.4 |
| Jharkhand | 93.0 | 59.2 | 37.9 | 18.5 |
| Orissa | 90.6 | 79.0 | 43.5 | 29.7 |
| Chhattisgarh | 90.3 | 57.6 | 26.6 | 12.0 |
| Arunachal Pradesh | 87.7 | 45.6 | 26.0 | 15.2 |
| Rajasthan | 85.4 | 70.5 | 65.5 | 10.5 |
| Assam | 85.0 | 37.6 | 11.3 | 11.6 |
| Bihar | 84.8 | 46.9 | 34.9 | 28.0 |
| Meghalaya | 70.5 | 44.0 | 8.9 | 17.7 |
| Nagaland | 64.7 | 23.3 | 10.9 | 6.7 |
| Andhra Pradesh | 63.2 | 65.0 | 40.0 | 23.4 |
| Gujarat | 60.9 | 46.0 | 18.3 | 15.4 |
| West Bengal | 59.6 | 79.9 | 50.9 | 36.5 |
| Tripura | 59.2 | 46.0 | 17.7 | 31.6 |
| Uttaranchal | 56.8 | 62.6 | 32.4 | 16.2 |
| Karnataka | 54.7 | 62.5 | 42.0 | 14.6 |
| Mizoram | 52.9 | 67.2 | 42.3 | 42.5 |
| Haryana | 52.3 | 60.6 | 44.1 | 11.8 |
| Punjab | 52.0 | 58.5 | 43.7 | 17.5 |
| Jammu and Kashmir | 51.2 | 56.5 | 47.0 | 19.6 |
| Maharashtra | 46.7 | 68.0 | 52.9 | 38.5 |
| Delhi | 46.7 | 30.9 | 18.1 | 15.9 |
| Manipur | 41.9 | 23.8 | 6.9 | 8.2 |
| Himachal Pradesh | 41.5 | 81.7 | 60.3 | 42.9 |
| Sikkim | 40.1 | 78.3 | 55.5 | 17.0 |
| Tamil Nadu | 35.5 | 61.0 | 36.7 | 39.4 |
| Goa | 20.3 | 80.0 | 11.6 | 37.0 |
| Kerala | 16.3 | 72.0 | 25.1 | 33.6 |
| Total | 74.3 | 58 | 37 | 20.2 |

U5MR (2010): 63/1000 live births [9].

## APPROACH TO ASSESSING THE DESIGN AND IMPLEMENTATION OF THE VAS PROGRAM

The Poverty Reduction Strategy Paper (PRSP) framework identifies the determinants of coverage and stages that are key to providing effective services. It allows identification of bottlenecks and evaluates constraining factors at each level of service provision to utilization. It also provides an opportunity to express the interaction between the service and the target population. The framework is ideal to identify and understand factors affecting coverage of VAS program and come up with recommendations to increase the delivery and uptake of VAS among under-5 children.

The determinants are arranged in a hierarchical model with the first four stages (physical accessibility, availability of human and material resources, organizational quality, and social accountability) representing *Potential coverage* and the later three stages (initial contact, timing and continuity, technical quality) corresponding to *Actual coverage* (Figure 19.1).

### SEARCH STRATEGY

Various sources were utilized for searching relevant literature and reports:

- Database such as Pubmed, Cochrane, Sciencedirect, Scopus, and Medline were used to access peer-reviewed journals.
- International organization websites of UNICEF, WHO, and World Bank were accessed for recent and updated data. National databases and surveys like NFHS III, MOHFW, NNMB, and ICMR reports were accessed for country-specific information.

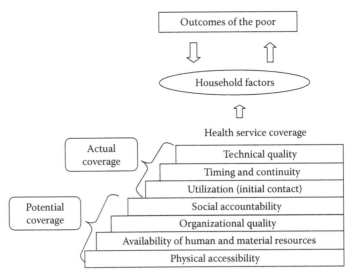

**FIGURE 19.1** PRSP framework: Coverage of vitamin A supplementation program. (From Claeson, M et al., Health, nutrition and population, *A Sourcebook for Poverty Reduction Strategies*, World Bank, Washington, DC, 2002.)

- Google, Google Scholar, and KIT library were accessed.
- Personal communication was established via e-mail with organizations regarding questions relevant to the topic.

## SEARCH DELIMITERS

- The search language used was English because of its widespread use and majority of national and international journals and international organization are found in English language. All relevant national journals are also published in English.
- Articles chosen for review were those that were relevant to the topic and dated from 1991 to 2012. The purpose was to present recent and updated developments and data as well as capture the past trends to get a comprehensive idea about VAS program.

# FINDINGS ABOUT THE PERFORMANCE OF THE VAS PROGRAM

The PRSP framework is used to analyze the factors affecting coverage of VAS program in India.

## PHYSICAL ACCESSIBILITY

Whether services are provided within the reach of the beneficiaries determines the physical accessibility [35].

### Service Supply

PHCs are the basis for provision of health services in rural areas where the majority of the population lives. According to the 10th and 11th FYP, the delivery of VAS should be carried out through the existing system of PHC, SC, and ICDS centers [31].

Table 19.2 shows that the average number of people serviced by different levels of health centers is higher than the required norm. Although this table shows only the population covered, it does not show the distribution of health centers in rural areas. The ground realities may be worse where the dispersed population in hilly and tribal areas and hard-to-reach areas does not access these centers. Moreover, according to National health policy (2002 report), only 24% of villages are equipped with health facilities, whereas 88% of public and private health care facilities are present in urban areas [22].

According to UNICEF's coverage survey in 2006, public health center was the main source of immunization and VAS among children. More than two-thirds of immunization was covered by the public sector either at PHC (23%), SC (24%), government hospitals (17%), outreach sessions (11%), and home/ICDS center/under a tree, etc. (16%) [13]. A program evaluation in 15 states in 2001–2002 reported that outreach activities were rarely carried out for VAS [24].

**TABLE 19.2**
**Distribution of Health Centers in India**

| Type of Health Center | Number of Health Centers | Population per Health Center | Required WHO Norm |
|---|---|---|---|
| SC (subcenter) | 132,000 | 6,311 | 3,000–5,000 |
| PHC (primary health center) | 22,000 | 37,867 | 20,000–30,000 |
| CHC (community health center) | 7,000 | 119,012 | 80,000–120,000 |

*Source:* Census of India, Ministry of Health and Family Welfare, Govt. of India, 2011, http://censusindia.gov.in/2011-prov-results/prov_data_products_india.html, Accessed April 27, 2011; WHO, Country Health System Profile: India, 2007, http://www.searo.who.int/en/Section313/Section1519_10853.htm, Accessed May 10, 2011; MOWCD (Ministry of Women and Child Development), Integrated child development services (ICDS) scheme, 2011, http://wcd.nic.in/, Accessed August 3, 2011; PWC (Price Waterhouse Coopers), Access to healthcare: Challenges and solutions, 2010, http://www.cuts-ccier.org/cohed/pdf/Access_to_Healthcare.pdf, Accessed July 16, 2011; Compiled by authors.

*Note:* Number of ICDS (or 'Angandwadi') centers by 2010: 1,241,749.

This is evident from statements of health providers [24]:

> We are not bothered about those we could not reach because they are not deficient in vitamin A. Otherwise they would have come with night blindness in OPD (Outpatient department). Those who come to us we give them. Clients not coming must be healthy.

Even though 88% of public and private health facilities are located in urban areas [22], the urban poor are equally underserved. The public health facilities in the urban areas, i.e., the government hospitals and ICDS centers, remain sole sources of VAS. In a study in Delhi's slums, almost half of the mothers (46%) said that immunization services were not available [36].

ICDS centers could be a good distribution point for VAS because of their reach in the community. ICDS scheme in India is the largest integrated child program in the world. By 2010, there were 1,241,749 ICDS centers all over India, and the scheme is still expanding [20]. The recommendations of the 11th FYP for administering VAS through ICDS centers, seems a good approach to maximize the accessibility of VAS. However, a study in Pune slums found that even when an ICDS center existed in the area, the coverage of VAS was low [37].

## Distance and Time to Reach Health Facility

Whether services are available within reasonable reach is an important determinant in the access to services. Since the first two doses of vitamin A are included with routine immunization services, access to immunization services also determines access to VAS.

The physical features of India, with lofty mountains to plains to deep valleys and islands [38], are challenging for access and further worsened by poor road conditions especially in rainy seasons [22]. In remote areas, the population is often excluded from the reach of the health center [39]. According to National rural health mission (NRHM) report in 2009, only 44% of villages had access to a PHC within a 5 km range [40]. A survey in 2008 in Jharkhand reported that around 15% of villages had access to a PHC and only 22% to centers catering to child health within the 5 km range [41]. In Delhi, the unreached were the poor living in the outskirts of the city [42].

A study in Agra city slums found that shorter distance to a health center or ICDS center had a positive effect on the immunization coverage. It was seen that children living near the center (within 2 km) were twice more likely to be immunized than those who lived further away [43]. Distance and time are important factors to influence the use of services even if the infrastructure is in place.

## AVAILABILITY OF HUMAN AND MATERIAL RESOURCES

### Human Resources

The performance of VAS program depends upon the availability of human resources to deliver the services. According to 10th FYP, it was recommended that the AWW under supervision of the Auxiliary nurse midwife (ANM) will administer doses of VAS and that once a month VAS and immunization services will be made available at the ICDS center to meet the needs of the target population [30].

The implementation of Integrated Management of Childhood Illnesses (IMCI) in India includes VAS through AWW. In 2009, out of 223 districts implementing IMCI, only 43 districts could train around 80% of their workers. By June 2009, a total of 202,015 health providers (AWW, ANM) were trained, but this number is less than 10% of the healthcare staff needed to serve just the high-burden districts [44]. The NRHM report in 2009 stated that only 48.2% of health centers had adequate staff as per standards [40]. A performance needs assessment (PNA) survey by MOHFW in eight states in 2005 found that lack of sufficient healthcare staff was one of the problems mentioned to affect coverage of immunization services [45]. Similarly, a study in Jharkhand stated that majority of the districts reported major constraints in human resources particularly at district level during rounds of VAS. Planning for vitamin A rounds was inadequate since other RCH activities were given priority [46].

Even when the staffs are available, the workload and multiple responsibilities do not allow giving sufficient priority to VAS. The ANM and AWW are overburdened by other MCH interventions where VAS is only a small part in it, which is often neglected [47]. A study in rural Maharashtra found that the already overburdened AWW was not able to take out time from record keeping and supplementary nutrition activities [48].

NRHM has initiated a new cadre of voluntary workers called the Accredited Social Health Activist (ASHA), who will assist the ANM/AWW in delivering MCH interventions including VAS [45]. This initiative can help reduce the workload of the ANM/AWW.

### Supply and Logistics

Multiple channels of vitamin A supply exist in India. The main channel of supply of vitamin A is through central government to states and districts. Districts also receive supplies from donor agencies mainly UNICEF [24].

Global Alliance for Vitamin A (GAVA) reported that almost 50% of children were not delivered vitamin A due to various constraints including poor supply [37]. A report from Jharkhand showed that only one out of seven districts studied received vitamin A supplies on time [46]. Lahariya et al. also pointed out in their study that one of the reasons for low coverage of VAS was inadequate supply and insufficient focus on provision of VAS [49].

A PNA survey in 2005 by MOHFW found that though stocks of vaccines including vitamin A were adequate, there were problems in vaccine management and transportation of vaccines to outreach sessions [45]. Unavailability of timely supply of vitamin A and poor coordination during routine services contributes to unavailability of vitamin A during supplementation sessions.

## ORGANIZATIONAL QUALITY

"The extent to which services are responsive to consumer concerns and whether these services are delivered in a way that encourages appropriate utilization of relevant interventions" determines the quality of services [35]. It depends on various factors like attitude of health staff, coordination of VAS activities, communication with the beneficiary, operation hours, and modes of payment.

### Coordination of VAS Activities

A program evaluation in 15 states in 2001–2002 showed lack of coordination between different levels. Though providers believed that coordination is required, there was disagreement in the functioning, planning, and financial management among different levels. In some cases, providers had developed their own coordination mechanism by which they conducted monthly meetings and arranged field activities that led to conflicts between different levels. It was found that although program managers were aware of the action plan for the program, district-level workers were not sure of any action plans or responded that they were yet to undergo training for it; some even mentioned that it did not exist, and others mentioned that since it was routine work, no planning is required [24].

### Communication and Attitude of Health Staff

Communication and attitude of health staff reflects responsiveness of service providers to consumer needs. Quality of communication reflects on the quality of health services but also on the user-friendliness of services. An ACC/SCN report had mentioned weak Information, Education and Communication (IEC) components adversely affecting the coverage of VAS in India [50].

A study in Lucknow showed that the behavior and information provided by the staff also determined client satisfaction. Those who did not continue with immunization services responded that the information provided by the staff was not adequate [51]. VAS not only includes administration of vitamin A but also providing information about the next visit and about benefits of VAS. NNMB survey in 2000–2001 found that more than one-third of mothers (34%) were not aware whether VAS was administered to their child and less than 10% of mothers reported being provided with nutrition education on VAD and benefits of VAS [32].

Vaccination cards are an important record of vaccination status and VAS receipt as well as inform the mother about subsequent doses. In a study in Andhra Pradesh, vaccination cards were not received by almost one-third (30%) of mothers even though they had visited the center for immunization [52]. Even among those who received the cards, findings from a survey by CARE in 2006 found that mothers were not aware that VAS was being provided to their child [53].

In 2001, the incident of Assam occurred: High doses of vitamin A were administered to children using a "campaign approach" linked to pulse polio immunization (PPI) and deaths occurred among more than 20 children, which some people believe were related to the VAS campaign. Different hypotheses and opinions exist about this alleged incident, ranging from erroneous administration of too high dose of VA to a conspiracy to discredit the VAS program. The too high-dose hypothesis is based on the fact that UNICEF had introduced 5 mL cups in place of the traditional 2 mL spoons for this campaign, but health providers were not adequately trained or informed about this new introduction nor did they have enough information about side effects of high dose of vitamin A.* However, a link with vitamin A receipt could not be confirmed for any of the deaths (High Court ruling). Whatever was the case, the incident and the attention it got in the media[†] affected the entire public health system as well as the public's confidence in the quality of government health services [54]. It also resulted in the recommendation of the 10th FYP to discontinue administration of VAS with PPI [30]. This episode reflects lack of political commitment and poor organizational quality in delivery of services and impedes acceptability by the community.

Initiatives in Uttar Pradesh with biannual vitamin A months in May and November for administration of VAS encouraged the community to seek VAS for children, and an 18% increase in coverage was observed, between 2000–2001 and 2005, which further motivated the service provider to impart health education on importance of vitamin A and benefits of supplementation [55].

## Modes of Payment and Operation Hours

It includes the financial accessibility as well as the opportunity cost to access facilities for VAS. VAS is provided along with routine immunization in public facilities without any charge. In spite of this, the opportunity cost like transportation, leaving household work or daily labor affects the receipt of VAS by the child. Kar et al. in their study showed that one-fifth of the mothers found the timings of immunization clashing with their household work [56] and the NNMB survey in 2006 also reported that mothers found the timing of services for VAS and immunization inconvenient [32].

UNICEF's survey in 2006 found that almost one-fifth of mothers expressed lack of time to take her child for immunization as reasons for missing doses as well as not receiving any dose [13]. The opportunity cost of leaving work and losing wages was important in deciding whether to bring a child for services [43]. Personally, one

---

* The court judgment in Assam stated that stronger dose of vitamin A was introduced due to replacing 2 mL spoons with 5 mL cups. The health workers were not trained adequately and administered greater doses than many sick children could tolerate. This showed negligence on part of Assam health department on administration of VAS [54].

[†] Assam tribune and BBC news reported the Assam incident and the High Court ruling [57–62].

of the authors has seen that although the opening hours are for the whole day, health centers are often closed after the morning hours.

Although vitamin A is provided free of cost, the financial capability to reach the center and other opportunity costs in combination with inconvenient operating hours might contribute to less use especially by poorer sections of the population.

## SOCIAL ACCOUNTABILITY

When consumers are able to exert their influence on the health system, the health services are more likely to respond to the demands of the consumers [35]. Transparency in the type of service provided and information to the community about services they are entitled to create demand for services and promote accountability of health services.

A program evaluation survey in 15 states found that social mobilization was one of the key elements lacking in the VAS program. There was rarely any involvement of community leaders, NGOs, or other stakeholders in the community during program planning or implementation. Although the health providers stated that they informed the community through various channels about the VAS program, none of the clients had noticed it. Lack of social mobilization efforts was one of the main reasons mentioned by stakeholders in non-utilization of VAS program [24]. A successful approach in Jharkhand state has been the involvement of *Sahiyas* (Community health communicators) under NRHM. The poorly covered areas and left-outs were traced and reached by them, while improving the quality of outreach sessions. This improved the overall immunization coverage among 12–23 months old children in the state from 8% in 2000 to 34.3% in 2005–2006 and is still being carried out, aiming for a further increase [63]. Such initiatives sensitize the public and make them aware of these approaches that encourage them to bring their child for these services.

The village health committee (VHC) initiated by NRHM is an approach for building accountability mechanisms for health and nutrition activities at the community level. VHC is a part of the local self-governing system in the village, called *Panchayats*. The committee constitutes representatives from *panchayat*, NGO, community representatives, and village health workers. VHC is responsible for generating awareness about health services in the community, analyzing priorities and key issues in health and nutrition activities, and providing feedback to the health functionaries. They also oversee the activities of ANM/AWW and the management of SC. In a review of VHC, the gaps identified were lack of community orientation about VHC, community representations, ownership, and non-linkage with government health services [64].

Another initiative by NRHM in improving accountability of health services to the community has been the introduction of the Citizens Charter. Though it does not cater to immunization or VAS specifically, in general, it covers information about services provided at the health center and means through which grievances will be addressed. Under the NRHM, *Rogi Kalyan Samiti*/Hospital Management Committee has been set up. This committee is recommended to form a grievance mechanism system to look into the complaints of the community and provide appropriate feedback [65]. From personal experience, one of the authors has seen that the Citizens Charter apart from displaying the services provided does not mention any grievance contact information, and the grievance mechanism system is limited to paper without any action.

## UTILIZATION (INITIAL CONTACT)

The actual contact between the service provider and the target population is a measure of utilization of services by the beneficiaries. "Utilization is defined as the first use of a service by a consumer in a given year. It is a key indicator of the extent to which the poor express some level of demand for services and come into contact with the health system" [35].

The initial contact with the health service for VAS is when the mother brings her child for measles immunization, because the first dose of vitamin A is administered at this moment. Survey data from both NFHS III and UNICEF (2006) show a higher coverage for measles immunization compared to vitamin A (see "Case for Vitamin A Supplementation in India" section). Also in a study in Ahmedabad, it was seen that out of all children receiving measles vaccine, only 47.8% children received VAS [66]. And the NNMB vitamin A survey that was conducted in eight states in 2000–2001 found that more than half (52%) of the mothers responded that their child did not receive VAS when they visited the health center to get it [32]. Some reasons reported for nonreceipt include absence of ANM or lack of supply [67]. This shows that in addition to relatively poor coverage of immunization, to which VAS is linked, there are further issues with VAS program implementation, such as lack of supplies, inadequate number and training of staff, and limited awareness also among the population.

## TIMING AND CONTINUITY

"It examines whether consumers receive the necessary number of contacts for services that require repeated interventions and whether time sensitive services are delivered in a timely manner. It must be repeated at regular intervals in order to be effective" [35].

Effective coverage is defined as reaching a minimum of 70% of 6–59 month old children with two doses of VAS per year to achieve reduction of U5MR. Since the country indicators used for coverage mentioned one dose in past 6 months as indicators of effective coverage, these data are also being reported [6].

As mentioned earlier, the coverage of VAS program is only 25%–37% (2005) [12,13] for one dose in past 6 months. NNMB survey in 2000–2001 in eight states reported that coverage with at least one dose of VAS among 12–59 month old children in the previous year was 58%. One dose coverage was highest in Orissa (80%) (the higher coverage is due to linking of VAS with PPI) followed by Tamil Nadu (63%) and Karnataka (57%) with the lowest in Kerala (44%). Of these, 36% had received it in the previous 6 months and around 25% received two doses [32].

Semba et al. using NFHS III, 2005–2006 data, reported VAS coverage with one dose in past 6 months of 45.4% among 12–23 months, 16.4% among 36–47 months, and 9.4% among 48–59 month old children. The overall coverage was 20.2% (Table 19.1), and ranged from 6% to 43% among states. It was also seen that most of the states with higher coverage of VAS had lower U5MR [14]. UNICEF's coverage survey results also showed a lower coverage among older children [13].

No state has reached the target of 70% coverage with two annual doses of VAS among 6–59 month old children.

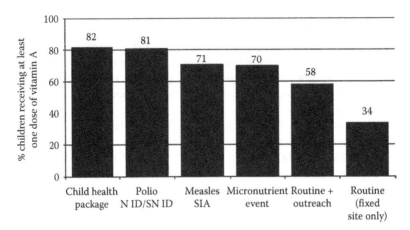

**FIGURE 19.2**   Mean coverage of one dose VA by distribution strategy, 1999–2004. (From UNICEF, Vitamin A supplementation: A decade of progress, 2007, http://www.unicef.org/publications/files/Vitamin_A_Supplementation.pdf, Accessed May 09 2011.)

**Note:** The estimates are derived from all vitamin A distribution coverage data available globally during 1999–2004 and are illustrated according to delivery strategy, from UNICEF VAS database.

UNICEF's analysis of VA coverage achieved with different distribution strategies in different countries found that relying on routine health services, particularly immunization, may not be enough to reach all children aged 6–59 months. Figure 19.2 shows that fixed site administration, such as practiced in India, reached on average 34% of eligible children, whereas interventions like national immunization days (NIDs) and child health days reached on average 82% coverage. Until good coverage of VAS can be sustained through routine services, supplementary activities are necessary to increase VAS coverage [6].

Few strategies have been tried in different states in India. A three year project in 2006 by Maharashtra government was carried out with support from UNICEF, comprising VAS with IEC and capacity building of health workers. By 2008, 73% coverage was reached among 9–59 month old children [68]. The integration of VAS with PPI in Chandigarh slums and Orissa showed a high coverage of more than 90% among 12–59 month children [69,70]. Though these initiatives were continued, lack in continuity brought the coverage levels down to previous levels.

TECHNICAL QUALITY

"The capacity of the sector to provide the appropriate combination of technology and empathy at a given level of utilization is key to ensure that interventions are translated into effective outcomes" [35]. In relation to VAS, it measures whether the doses of vitamin A are administered according to guidelines, health providers are adequately trained, and workload of providers allows them to provide appropriate services [35].

In India, there is no program guideline for VAS or any operations manual. Policy changes have been communicated through workshops or training programs. A program evaluation report in 2001–2002 in 15 states reported that monitoring and supervision was absent and, if present, was limited to review meetings and checking registers. Most of the workers were not aware of any monitoring. Providers did not have clarity on the components of the program, and the majority could not mention the target clients for VAS [24].

The incident in Assam discussed earlier points to a lack of adequate training of health staff in the administration of VAS, inadequate supervision, and negligence. It was also reported that 21.4% of infants not eligible for VAS also received vitamin A [54,71].

From the experience of one of the authors, vitamin A syrup is often given with a spoon without measuring the appropriate quantity. The same spoon is used for all children, and when the child moves or cries, most of the syrup is spilled and sometimes syrup is still left in the spoon. The program evaluation survey also pointed out that mothers did not have problems with VAS but mentioned that they did not like using the same spoon to administer the syrup to all children [24].

As discussed earlier, provider workload also has implications for the quality of services provided. A World Bank report stated that 80% of the time of AWW was spent on supplementary nutrition and preschool education leaving very little room for other activities [4]. A qualitative study in rural Maharashtra in the ICDS centers reported similar findings: out of the multiple responsibilities of the AWW, VAS was not even mentioned as one of the activities that was carried out [48]. This shows that quantity alone is not enough for effective coverage but quality of services is equally important.

## HOUSEHOLD FACTORS

Apart from factors mentioned earlier affecting VAS coverage, there are important determinants such as socioeconomic status, literacy, and cultural factors that have a bearing on the utilization of the program. These factors cannot be tackled by the health care system, but have an important impact on the demand for VAS. Children belonging to households of higher socioeconomic status (SES) were more likely to receive a dose of vitamin A compared to children from lower quintiles [13]. The lower socioeconomic quintiles face more problems with accessibility and opportunity costs. UNICEF's coverage survey in 2006 showed that the mother's literacy status was an important factor in determining whether a child received VAS: 70% of children of literate mothers and 48% of those of illiterate mothers had received VAS [13]. Similarly, a survey by USAID found that important predictors for receipt of routine immunization services are literacy status of mothers, standard of living, and social category. Sex difference is not seen in the uptake of VAS [13,14,32], and there were no cultural barriers affecting the utilization of VAS program [24].

Table 19.3 shows the summary of findings according to PRSP framework.

**TABLE 19.3**

**Summary of Findings from Analysis of PRSP Framework**

| Stage | Factors | Findings |
|---|---|---|
| Physical accessibility | Service supply—Infrastructure | Number of health centers not proportionate to the population, concentrated more in urban areas, rural areas, and urban slums—underserved |
| | Distance and time to reach health facility | Mothers find inconvenience with the timing of health center service provision, underserved areas far from health center, perceived distance is large, nearer the health center, better the utilization |
| Availability of human and material resources | Health care staff to administer VAS—ANM, ICDS workers, Health workers | ANM and ICDS workers are link persons for VAS, distribution of staff not according to population, initiative taken by NRHM to retain staff in rural areas, lack of motivation of health care staff, increased workload affected availability of staff for VAS |
| | Supply of vitamin A syrup, transportation | Multiple channels of supply, irregular supply, shortage and not timely, problems in management and transportation of vitamin A even when supplies were adequate |
| Organizational quality | Coordination of VAS activities | Conflict between different levels of health staff, action plan for the program absent or not known, poor communication by healthcare staff, awareness low among beneficiaries due to lack of information sharing, other than vaccination, no other information imparted to mothers |
| | Attitude of health staff and communication with the beneficiary | |
| | Modes of payment and operation hours | Opportunity cost high Operation hours not favorable to beneficiaries |
| Social accountability | Involvement of the community in management and delivery of services | Noninvolvement of stakeholders in planning and implementation |
| | Mobilization of communities for health promotion activities | Social mobilization efforts and transparency in services missing |
| | Services catering to grievances and suggestions by the consumers | Citizen's charter nonfunctional, absence of grievance committee, does not cater to illiterate population, VHC in infancy stage |

## TABLE 19.3 (continued)
## Summary of Findings from Analysis of PRSP Framework

| Stage | Factors | Findings |
|---|---|---|
| Utilization (Initial contact) | Initial contact of beneficiaries to receive first dose of vitamin A along with measles vaccine | Service provided at initial contact poor, children receiving measles vaccine but not VAS, some beneficiaries not offered the services, disparity in information in vaccination cards and recall of mothers |
| Timing and continuity | Two annual rounds of VAS One dose of vitamin A in the past 6 months | None of the states reached effective coverage of 70% of target children with two annual dose of vitamin A, disparity in coverage data from different sources, coverage of subsequent doses lower, state variation in coverage of VAS |
| Technical quality | Doses of vitamin A administered according to guidelines | No program guidelines, monitoring and supervision absent, improper method of administration of vitamin A, lack of knowledge of VAS among health staff |
| | Training of health providers and Workload of providers | Lack of training of health providers, overburdening of work at ICDS center leaving no time for VAS |
| Household factors | SES, literacy, cultural factors, gender difference | Positive predictors for immunization and VAS are improvement in mother's literacy status, higher SES. No cultural barriers or sex difference was seen |

## WHY HAS VAS BEEN A GREATER SUCCESS IN OTHER COUNTRIES (BANGLADESH, INDONESIA)?

India, Bangladesh, and Indonesia were among the very first countries to start VAS programs, back in the late 1970s, early 1980s. It is therefore interesting to compare the history of the VAS program in India to that in Bangladesh and Indonesia, where programs have achieved high, sustained, coverage.

### BANGLADESH

National program on VAS was initiated in 1973, which was around the same time as in India, with support from UNICEF and CIDA. Since then, various strategies like door-to-door administration of vitamin A capsules (VACs) twice yearly among 12–59 month old children in the 1990s to linking of VAS with EPI and later integration with NID have been tried. The government also initiated the "National vitamin A week" where a week-long administration of vitamin A took place, which later changed to biannual administration in 2003 and continued till 2005 along with NIDs.

Since 2005, the first dose for children aged 6–11 months is provided through EPI, together with measles immunization, and the subsequent doses are delivered through "Child health days" twice a year, also called the "National vitamin A plus Campaign." The campaigns are also used for delivering other nutrition and health interventions like deworming, imparting nutrition messages, salt testing for iodine content, etc. This is done through health workers and community volunteers [72,73]. This has led to increase in the coverage of VAS with one dose in past 6 months from 79% in 1999 to 82% in 2005 and 88% in 2007 [74,8]. In 2010, 100% of 6–59 month old children had received two doses of vitamin A in the past year [9].

Bangladesh was one of the priority countries for VAS, along with India, based on an U5MR of more than 70/1000 live births in 2004. With the present U5MR of 48/1000 live births (2010) [9], it has been able to achieve the MDG target of reducing the U5MR to 48/1000 live births by 2015 [75]. Bangladesh has used different interventions to reach especially those children that are hard to reach. In 2007, an estimated 600,000 children belonging to poor households who were not reached by the VAS program were identified. Coordinated efforts from the government and volunteers tracked the "left-outs" and administered vitamin A. A total of 360,000 unreached children were reached [73]. Strong political commitment and working together with programs like EPI and development partners and community mobilization have achieved universal coverage of VAS in Bangladesh [9].

## INDONESIA

The VAS program in Indonesia dates back nearly three decades. In 1973–1975, a pilot project was conducted in 20 subdistricts of Java among preschool children that included 6 monthly administration of VAC. This was further expanded into a program in the rest of Indonesia in the 1980s. Till mid-1990s, the target children for VAS were 12–59 months old. This was later extended to 6–59 months [76]. Although Indonesia is not included in the priority list of countries for VAS since the U5MR is below 70/1000 live births, the government has been dedicated in its efforts to control VAD, recognizing it as a public health problem [6]. Indonesia has one of the strongest VAS programs [77]. Indonesia's strategy is the administration of biannual VAS through "Vitamin A months" in February and August through health posts (*Posyandu*). Information about VAS is provided by midwives and cadres to under-5 mothers during *Posyandu* days, and mothers are asked to disseminate this information to other mothers. Posters, banners, and local media disseminate information about "Vitamin A months" [78]. Along with VAS, the health posts also provide other interventions like immunization, nutrition education and counseling, and growth monitoring [73,78].

Indonesia is self-sufficient in procuring VAC with 50% financed by central government and 50% by district governments [79]. Transparency of service with dissemination of information through personal as well as mass media and commitment of the health providers is a sign of good quality of service and accountability.

The coverage increased from 64% in 2000, to 76% in 2005, and 80% in 2010 for two doses of VA among 6–59 month old children [8,9].

## HOW SHOULD THE IDENTIFIED ISSUES OF INDIA'S PROGRAM BE TACKLED?

Table 19.1 shows variation in U5MR and coverage of VAS among states, but U5MR is too high in most states and VAS coverage is low in all states. Since the U5MR correlates with VAD [15] and the ultimate objective of VAS is reduction in U5MR, it implies that states with higher U5MR are at higher risk of VAD. Meanwhile, there has not been a significant improvement in U5MR since 1990, which can be explained by lack of progress in states with U5MR as high as 90–96/1000 live births. Therefore, it is necessary to prioritize states for improvement of the VAS program.

Around 10 states have a U5MR higher than UNICEF's cutoff of VAD as a public health problem (Table 19.1). Although other states are somewhat below the cutoff, VAD still exists in these states. With this information, the states in India can be divided into four categories according to the likely magnitude of the VAD problem (Table 19.4).

Table 19.4 shows that 10 states are at highest risk of children dying due to VAD. These states are the ones with U5MR reaching up to 96/1000 live births and very low VAS coverage, not exceeding 25%. It is clear that children in these states are at high risk of mortality with an immediate need to scale up VAS coverage.

The majority of states lie in the range of 50–70/1000 live births (category II). These states are next in line for scaling up VAS since other interventions have not had enough impact to further reduce mortality rates. Within this category, states like Nagaland and Haryana show very low VAS coverage compared to other states. And within states, also among those with U5MR of 20–50/1000 live births, there are

---

**TABLE 19.4**

**Distribution of States in India According to U5MR**

| High-Risk States | U5MR (Per 1000 Live Births) | Number of States/ Union Territory[a] | Comment |
|---|---|---|---|
| I | >70 | 10 | VAD is a public health problem in under-5 children that needs immediate action |
| II | 50–70 | 11 | VAD is a public health problem |
| III | 20–50 | 7 | Problem of VAD likely exists |
| IV | <20 | 1 | VAD may/may not be a problem |

[a] Priority I: Uttar Pradesh, Madhya Pradesh, Jharkhand, Orissa, Chhattisgarh, Arunachal Pradesh, Rajasthan, Assam, Bihar, Meghalaya; Priority II: Nagaland, Andhra Pradesh, Gujarat, West Bengal, Tripura, Uttaranchal, Karnataka, Mizoram, Haryana, Punjab, Jammu and Kashmir; Priority III: Maharashtra, Delhi, Manipur, Himachal Pradesh, Sikkim, Tamil Nadu, Goa; Priority IV: Kerala.

areas and subgroups with higher mortality rates, such as in rural areas, urban slums, and lower socioeconomic quintiles.

A number of factors are responsible for low coverage. It is also clear that deficiency in one component leads to another component being deficient. Although all stages influence the coverage of VAS, there are some factors that come up as most important determinants. The identification of these factors is crucial to enable a change in the other components.

Looking at the framework and at the summary of results in Table 19.3, we see that although availability of human and material resources and providing infrastructure are necessary, this in itself may not be able to markedly improve VAS coverage. The low coverage is to a large extent related to solely relying on routine services for administration of VAS. The decreasing coverage with older age indicates that a major gap lies in not *being able to cover children for subsequent doses through the fixed health facility approach.* Increasing the number and coverage of health facilities and providing resources may improve the coverage of the first two doses with EPI activities, both through increased coverage of EPI and increased coverage of VAS among those who come for EPI, but what about the subsequent doses? With no further activities by the health services in reaching the children, the subsequent doses may not be taken. Therefore, the problem is particularly with the channel through which VAS is being delivered.

Child health days similar to Bangladesh and Indonesia can be a good approach to cover the targeted children and limit dropouts and left-outs. Other health and nutrition interventions can be provided along with VAS. While the campaign approach where VAS was linked with PPI was tried in India, it was discontinued after the Assam incident. The immediate stopping of all campaigns at that time shows the hurriedness of policy makers to just curb the situation rather than going into details of solving the actual problem. Even the recommendation of biannual administration of VAS through ICDS centers is taking very long to implement. With a wide distribution of ICDS centers, it might be a very good channel for delivering VAS, and children in both rural and urban areas can be reached. This also allows beneficiaries from the lowest quintile to access services with minimal opportunity costs.

WHO has suggested that if VAS coverage levels through routine activities are below 80%, it indicates a need for supplementary coverage activities [80]. UNICEF data on coverage with various interventions (Figure 19.2) also show that VAS distribution through routine fixed services has the lowest coverage, 34% on average [6]. This is the current situation in India with coverage with one dose in past six months not exceeding 20.2% among 12–59 months old children [14].

A good strategy for VAS delivery needs program guidelines to standardize the overall operation of the program; however, this is currently lacking. This *lack of program guidelines* directly affects the service quality including technical quality. This is the primary reason for lack of coordination between different implementation levels, absence of training modules for VAS affecting the technical quality, and poor supervision and monitoring. Availability of guidelines itself is a first step to prioritizing the program and streamlining it.

From our viewpoint, the inclusion of the VAS program under the umbrella of the RCH program has masked the VAS program among other popular programs. It remains an insignificant, often sidetracked, part within the routine EPI activities. Furthermore, VAS has not been recognized as an effective intervention to reduce the U5MR in India, especially at the implementation level. Opinions among public health professionals exist that VAD is no longer a public health problem in India, and even if it is, then this is limited to small pockets of the population. Due to this, universal coverage of VAS is not promoted, and it is stated that resources can be directed to other effective interventions, and if VAD is a problem, other sustainable interventions such as dietary modification are encouraged instead. The resistance from public health professionals with opinions in favor of other interventions to reduce U5MR rather than VAS may also have hindered the implementation of VAS, because even though policy statements are in place, program implementation guidelines do not exist.

Given the existing donor support, supplies of vitamin A should not be a problem if the channels of procurement and supply are guided by standard guidelines. With regard to the form of the supplement, India is the only country using syrup; all other countries use capsules. The reason for using syrup is not mentioned in any literature. VAC is easy to administer, hygienic, and children receive appropriate doses in contrast to syrup that needs measurement, which is rarely accurate [76].

Ultimately, the health services are accountable to the community who will utilize them. The *lack of social mobilization efforts* within the program reveals a deficiency in accountability measures. Structures are in place with establishment of VHC and Citizens Charter but are not put into action. The elements of "answerability" of health providers toward the community and "enforceability" of laws in events of non-adherence are missing [81]. This would keep a check on the quality of the program and facilitate ownership of the program. NGOs and community volunteers are capable of acting as important resource persons for generating awareness about the program and tracing left-outs and dropouts. If the community for whom the services are intended does not sense an ownership, the program cannot reach its target.

There is abundant global evidence about successful interventions to increase VAS coverage. Still, India has not been able to adopt effective implementation measures to combat VAD. Examples from Bangladesh and Indonesia suggest that strong political commitment and utilizing the existing health resources in the country can result in better outcomes. These are important reasons for why these countries have been able to reduce their U5MR to achieve MDG goals while U5MR in India is still very high and India has the largest number of children with VAD in the world.

## RECOMMENDATIONS

*At policy level*

1. Advocacy to policy makers to translate policy statements into action, including program guidelines, and follow through with political commitment
2. The VAS program needs revision, which can be implemented in phases (Table 19.5)

**TABLE 19.5**

**Recommended Implementation of VAS Program in Order of Priority**

| Phase (Years) | Strategy |
|---|---|
| I (3–4) | Routine immunization and biannual administration of vitamin A through ICDS centers |
| II (3–4) | Child health days for a week twice yearly including other child health interventions like growth monitoring, nutrition education, and immunization using ICDS centers and temporary posts in inaccessible areas |

3. Formulation of VAS program guidelines with inclusion of
   a. Standard protocols on training
   b. Channels of procurement and supply
   c. Indicators to monitor progress
   d. Format for action plan for each implementation level
4. Vitamin A syrup should be changed to VACs
5. Meanwhile, strengthen the EPI so that coverage of first two doses of VAS not only increases because all children that come for measles and DPT booster vaccination are offered VAS, but also the number of children that come for these immunizations increases.

*At implementation level* (For program managers and healthcare providers)

1. Involvement of
   a. Stakeholders like community representatives, NGOs, women's group in program implementation, and establishment of VHC and grievance committees
   b. Staffs from all levels of health services to coordinate activities for VAS
   c. ASHAs and community volunteers to
      i. Trace left-outs and dropouts
      ii. Organize outreach sessions
      iii. Arrange IEC activities
2. Regular monitoring should be done for
   a. Immunization records
   b. Immunization and outreach sessions
3. Training of health staff to
   a. Update skills and performance
   b. Conduct rapid coverage estimates following VAS distribution to elicit accurate coverage estimates
4. AWW/ANM should be oriented to
   a. Keep record of target children for VAS
   b. Hold VAS session twice a year at ICDS centers.

### RESEARCH

- Operational research in states in each priority stage to assess strengths and weaknesses of different distribution strategies to come up with feasible and effective approach for VAS in India
- Comparison study on coverage estimates from routine administrative data and household survey data

## ABBREVIATIONS

| | |
|---|---|
| ACC/SCN | Administrative Committee on Coordination/Subcommittee on Nutrition |
| ANM | Auxiliary Nurse Midwife |
| ASHA | Accredited Social Health Activist |
| AWW | Anganwadi Worker |
| CHC | Community Health Center |
| CIDA | Canadian international Development Agency |
| CMO | Chief Medical Officer |
| CSSM | Child Survival AND Safe Motherhood |
| EPI | Expanded Program on Immunization |
| FYP | Five Year Plan |
| GAVA | Global Alliance for Vitamin A |
| GDP | Gross Domestic Product |
| ICDS | Integrated Child Development Services |
| ICMR | Indian Council of Medical Research |
| IEC | Information, Education and Communication |
| IMCI | Integrated Management of Childhood Illnesses |
| IVACG | International Vitamin A Consultative Group |
| MCH | Maternal and Child Health |
| MDG | Millennium Development Goals |
| MI | Micronutrient Initiative |
| MOHFW | Ministry OF Health and Family Welfare |
| NFHS | National Family Health Survey |
| NGO | Non Governmental Organization |
| NID | National Immunization Day |
| NNMB | National Nutrition Monitoring Bureau |
| NP-PNB | National Program for the prevention of Nutritional Blindness |
| NRHM | National Rural Health Mission |
| OPD | Outpatient Department |
| PHC | Primary Health Center |
| PNA | Performance Needs Assessment |
| PPI | Pulse Polio Immunization |
| PRSP | Poverty Reduction Strategy Paper |
| RCH | Reproductive and Child Health |
| SC | Subcenter |
| SES | Socioeconomic status |

| SOWC | State of the World Children |
|---|---|
| THE | Total health expenditure |
| UNICEF | United Nations Children's Fund |
| U5MR | Under-5 mortality rate |
| USAID | U.S. Agency for International Development |
| VAC | Vitamin A capsule |
| VAD | Vitamin A deficiency |
| VAS | Vitamin A supplementation |
| VHC | Village Health Committee |
| WHO | World Health Organization |

## GLOSSARY

**Biochemical VAD:** A serum retinol concentration of less than 0.70 μmol/L, which is associated with increased vulnerability to a variety of infectious diseases and, therefore, an increased risk of mortality and morbidity [4].

**Clinical VAD:** A severe form of vitamin A deficiency, resulting in xerophthalmia, symptoms of which include night blindness, Bitot's spot, xerosis, and keratomalacia. If not treated early enough, it can eventually leads to blindness [4]. WHO cutoff for indicating a public health problem [3]
Bitot's spot prevalence among under-5 children > 0.5%
Night blindness prevalence among under five children > 1%

**Dropouts of VAS:** Eligible children under the age of 5 years who received a dose of vitamin A but failed to receive subsequent dose due to any reason (For chapter).

**Effective coverage of VAS program:** A VAS coverage threshold of 70% among 6–59 month children with two annual doses of vitamin A supplements at which countries can expect to observe reductions in child mortality [6].

**Left-outs of VAS:** Eligible children under the age of 5 years who are targeted for VAS but failed to receive the dose due to any reason.

**Missed opportunity:** Any circumstance in which a child under the age of 5 years who is an eligible candidate for the service and who needs the service does not receive this service when they visit a health facility [82].

**National immunization day/Pulse polio immunization:** A day on which simultaneous administration of extra doses of oral polio vaccine to under-5 children irrespective of their previous immunization status is done on a single day [83].

**Outreach session:** The extending of services beyond current or usual limits [84].

**VAD:** WHO defines vitamin A deficiency as tissue concentrations of vitamin A low enough to have adverse health consequences even if there is no evidence of clinical xerophthalmia [85].

## REFERENCES

1. WHO. Global prevalence of vitamin A deficiency in populations at risk 1995–2005: WHO global database on vitamin A deficiency, 2009. http://whqlibdoc.who.int/publications/2009/9789241598019_eng.pdf (Accessed May 9, 2011).

2. West, K. P. Extent of vitamin A deficiency among preschool children and women of reproductive age. *Journal of Nutrition*, 2002; 132(9): 2857S–2866S.

3. WHO. Indicators for assessing vitamin A deficiency and their application in monitoring and evaluating intervention programmes. World Health Organization, Geneva, Switzerland. WHO/NUT/96.10; 1996.

4. World Bank. India's undernourished children: A call for reform and action, 2005. http://sitcrcsources.worldbank.org/SOUTHASIAEXT/Resources/223546-1147272668285/IndiaUndernourishedChildrenFinal.pdf (Accessed March 18, 2011).

5. Arlappa, N., Ravikumar, B. P. Relevance of continuation of universal vitamin A supplementation program in India. *Indian Paediatrics*, 2011; 47: 246–247.

6. UNICEF. Vitamin A supplementation: A decade of progress, 2007. http://www.unicef.org/publications/files/Vitamin_A_Supplementation.pdf (Accessed May 9, 2011).

7. Imdad, A., Herzer, K., Mayo-Wilson, E., Yakoob, M. Y., Bhutta, Z. A. Vitamin A supplementation for preventing morbidity and mortality in children from 6 months to 5 years of age (Review), *Cochrane Library*, Issue 1, 2011. http://onlinelibrary.wiley.com/doi/10.1002/14651858.CD008524.pub2/pdf (Accessed September 1, 2011).

8. UNICEF. The state of the world's children, 2008. http://www.unicef.org/publications/files/The_State_of_the_Worlds_Children_2008.pdf (Accessed May 9, 2011).

9. UNICEF. The state of the world's children, 2012. http://www.unicef.org/sowc2012/statistics.php (Accessed April 7, 2012).

10. MOHFW. Child health division, government of India, 2006. http://motherchildnutrition.org/india/pdf/mcn-vitamin-a-ifa-supplementation.pdf (Accessed April 12, 2011).

11. WHO. India universal immunization programme review, 2010. http://www.whoindia.org/LinkFiles/Routine_Immunization_Acknowledgements_contents.pdf (Accessed May 9, 2011).

12. NFHS (National family health survey) III, Report. Ministry of health and family welfare. India, 2005–2006. http://www.nfhsindia.org/pdf/India.pdf (Accessed February 14, 2011).

13. UNICEF. All India report: Coverage evaluation survey, 2006. http://www.unicef.org/india/1_-_CES_2009_All_India_Report.pdf (Accessed June 11, 2011).

14. Semba, R.D., De Pee, S., Sun. K., Bloem, M. W., Raju, V. K. The role of expanded coverage of the national vitamin A program in preventing morbidity and mortality among preschool children in India. *Journal of Nutrition* (Online) 2010; November (Supplement) 140: 208–212. http://jn.nutrition.org/content/140/1/208S.full.pdf (Accessed May 10, 2011).

15. Schultink, W. Use of under-five mortality rate as an indicator for vitamin A deficiency in a population. *Journal of Nutrition*, 2002; 132: 2881s–2883s.

16. Census of India. Ministry of health and family welfare, Govt. of India. 2011. http://censusindia.gov.in/2011-prov-results/prov_data_products_india.html (Accessed April 27, 2011).

17. WHO. Country health system profile: India, 2007. http://www.searo.who.int/en/Section313/Section1519_10853.htm (Accessed May 10, 2011).

18. CSC (Common services centers) scheme. Department of Information technology, Government of India. Panchayati Raj, 2012. http://www.csc-india.org/Utilities/ContentResource/PanchayatiRaj/tabid/625/language/hi-IN/Default.aspx (Accessed March 27, 2012).

19. Gupta, R. R. Rural healthcare system in India: The challenges and remedies, 2004. http://www.sajpc.org/vol8/vol8_4/ruralhealthcaresystem.htm (Accessed August, 03 2011).

20. MOWCD (Ministry of women and child development). Integrated child development services (ICDS) scheme, 2011. http://wcd.nic.in/ (Accessed August 3, 2011).

21. NHA (National health accounts). Ministry of health and family welfare. India, 2009. http://www.whoindia.org/LinkFiles/Health_Finance_National_Health_Accounts_2004-05.pdf (Accessed May 11, 2011).

22. PWC (Price Waterhouse Coopers). Access to healthcare: Challenges and solutions, 2010. http://www.cuts-ccier.org/cohed/pdf/Access_to_Healthcare.pdf (Accessed July 16, 2011).

23. World bank. New global poverty estimates—What it means for India, 2005. http://www.worldbank.org.in/WBSITE/EXTERNAL/COUNTRIES/SOUTHASIAEXT/INDIAEXTN/0,,contentMDK:21880725~pagePK:141137~piPK:141127~theSitePK:295584,00.html (Accessed August 03, 2011).

24. Devi, R. Program evaluation. What ails "Routine Public Health Programs" in India: Vitamin-A and iron folic acid supplementation—A case study, 2001–2002. http://www.inclentrust.org/research/Reports/Vitamin%20A%20full%20Report.pdf (Accessed July 17, 2011).

25. Kapil, U., Sachdev, H. P. S. Universal vitamin A supplementation programme in India: The need for a re-look. *National Medical Journal of India*, 2010; 23(5): 257–260.

26. NNMB. Vitamin A deficiency, 2006. http://www.wcd.nic.in/research/nti1947/7.11.2%20Vitamin%20A%20Deficiency%20pr%20map.pdf (Accessed June 3, 2011).

27. MN Project. India: Vitamin A epidemiological data, 2002. http://www.tulane.edu/~internut/Countries/India/indiavitamina.html (Accessed June 4, 2011).

28. MI (Micronutrient initiative). Our priorities in India, 2007. http://www.micronutrient.org/english/View.asp?x=603 (Accessed June 6, 2011).

29. USAID. Micronutrient and child blindness project: Vitamin A supplementation, undated. http://www.a2zproject.org/~a2zorg/node/7 (Accessed June 6, 2011).

30. Tenth five year plan. Sectoral policies and programmes. Planning Commission, Government of India, 2002–2007. http://planningcommission.nic.in/plans/planrel/fiveyr/10th/volume2/10th_vol2.pdf (Accessed June 6, 2011).

31. Eleventh five year plan. Report of the working group on integrating nutrition with health. Government of India, Ministry of Women and Child Development, 2007–2012. http://planningcommission.nic.in/aboutus/committee/wrkgrp11/wg11_integt.pdf (Accessed June 6, 2011).

32. NNMB. Prevalence of vitamin A deficiency among preschool children in rural India, 2006. http://www.nnmbindia.org/VAD-REPORT-final-21Feb07.pdf (Accessed June 4, 2011).

33. Singh, G., Vir, S., Narayan, U., Aguayo, V. Child health and nutrition months significantly improves vitamin A supplementation in Uttar Pradesh, India, 2008. http://www.sightandlife.org/micronutForumBejing09/PDFs/Poster%20Presentations/1_Tuesday/2_Vitamin%20A%20Deficiency%20&%20VAS%20Interventions/TU33_Singh.pdf (Accessed August 3, 2011).

34. UNICEF. The state of the world's children. Adolescence: An age of opportunity, 2011. http://www.unicef.org/sowc2011/(Accessed May 16, 2011).

35. Claeson, M., Griffin, C. C., Johnston, T. A., MaLachlan, M., Soucat, A. L. B., Wagstaff, A., Yazbeck, A. S. Health, nutrition and population. In: Klugman, J (ed). *A Sourcebook for Poverty Reduction Strategies*. World Bank, Washington, DC, 2002; 567–574.

36. Khokhar, A., Chitkara, A., Talwar, R., Sachdev, T. R., Rasania, S. K. A study of reasons for partial immunization and nonimmunization among children aged 12–23 months from an urban community of Delhi. *Indian Journal of Preventive and Social Medicine*, 2005; 36(3,4): 83–86.

37. Boy, E., Mannar, V., Pandav, C., Benoist, B. D., Viteri, F., Fontaine, O., Hotz, C. Achievements, challenges, and promising new approaches in vitamin and mineral deficiency control. *Nutrition Review*, 2007; 67(Suppl. 1): 24–30.

38. Maps of India. Geography of India, 2009. http://www.mapsofindia.com/geography/ (Accessed July 15, 2011).

39. Deogaonkar, M. Socio-economic inequality and its effect on healthcare delivery in India: Inequality and healthcare. *Electronic Journal of Sociology* (Online) 2004. http://www.sociology.org/content/vol8.1/deogaonkar.html (Accessed June 10, 2011).

40. Baru, V. R., Bisht, R. Health service inequities as challenge to health security, 2010. http://www.oxfamindia.org/sites/www.oxfamindia.org/files/working_paper_4.pdf (Accessed July 14, 2011).

41. ICSSR (Indian Council of Social Sciences Research). A baseline survey of minority concentration districts of India, 2008. http://www.icssr.org/Gumla%5B1%5D.pdf (Accessed July 12, 2011).

42. India Health Progress. A study on health care accessibility, 2011. http://www.india-healthprogress.in/sites/default/files/Inference%20Report%20-%20Study%20on%20 Healthcare%20Accessibility.pdf (Accessed June 10, 2011).

43. Ghei, K., Agarwal, S., Subramanyam, M. A., Subramanian, S. V. Association between child immunization and availability of health infrastructure in slums in India. *Archives of Pediatrics & Adolescent Medicine*, 2010; 164(3): 243–249.

44. Paul, V. K., Sachdev, H. S., Mavalankar, D., Ramachandran, P., Sankar, M. J., Bhandari, N., Sreenivas, V., Sundararaman, T., Govil, D., Osrin, D., Kirkwood, B. Reproductive health, and child health and nutrition in India: Meeting the challenge. *Lancet Series*, 2011; 377(9762): 332–349.

45. MOHFW. Performance needs assessment of basic health care workers in immunization in India, 2005. http://www.whoindia.org/LinkFiles/Routine_Immunization_ Performance_Needs_Assessment_of_HWs-India_Report-2005.pdf (Accessed July 16, 2011).

46. Kotecha, P. V., Syed, I., Chandranath, M. 2010. Assessing the sustainability of the Jharkhand district vitamin A supplementation program. A2Z: The USAID micronutrient and child blindness project, AED, Washington, DC. http://www.a2zproject.org/pdf/ Vas_Report_Final_091410.pdf (Accessed June 6, 2011).

47. Park, K. Nutrition and health. In *Parks Textbook of Preventive and Social Medicine*, 19th edn. (pp. 754–755). Banarsidas Bhanot, Jabalpur, India, 2007.

48. Dongre, A. R., Deshmukh, P. R., Garg, B.S. Perceived responsibilities of Anganwadi workers and malnutrition in rural Wardha. *Online Journal of Health and Allied Sciences*, (Online) 2008, 7(1). http://www.ojhas.org/issue25/2008-1-3.htm (Accessed June 6, 2011).

49. Lahariya, C., Khanna, R., Nandan, D. Primary health care and child survival in India. *Indian Journal of Pediatrics*, 2010; 77(3): 283–290.

50. Gillespie, S., Mason, J. 1994. ACC/SCN controlling vitamin A deficiency—Nutrition policy discussion paper No. 14, United Nations, 1994. http://www.unscn.org/layout/ modules/resources/files/Policy_paper_No_14.pdf (Accessed May 25, 2011).

51. Nath, B., Singh, J. V., Awasthi, S., Bhushan, V., Singh, S. K., Kumar, V. Client satisfaction with immunization services in urban slums of Lucknow district. *Indian Journal of Pediatrics*, 2009; 76(5): 479–483.

52. Varma, G. R., Kusuma, Y. S. Immunization coverage in tribal and rural areas of Visakhapatnam district of Andhra Pradesh, India. *Journal of Public Health*, 2008; 16: 389–397 (Online). http://www.springerlink.com/content/r555676j6n718x6n/ (Accessed June 6, 2011).

53. CARE. Widening coverage of micronutrient supplements, 2006. http://www.basics.org/ documents/Widening_Coverage_of_Micronutrient_Supplements.pdf (Accessed June 6, 2011).

54. Kapil, U. Update on vitamin A—Related deaths in Assam, India. *American Journal of Clinical Nutrition*, 2004; 80(4): 1082–1083.

55. World Bank. Implementation completion report of World Bank assisted ICDS-III/WCD Project, 2006. http://wcd.nic.in/PBEvalReport.pdf (Accessed June 13, 2011).

56. Kar, M., Reddaiah, V. P., Kant, S. Primary immunization status of children in slum areas of South Delhi—The challenge of reaching the urban poor. *Indian Journal of Community Medicine*, 2009; 26(3): 151–154.

57. *Assam Tribune*. Vitamin A controversy: AGP demands probe by sitting HC Judge, November 16, 2001: 5.
58. *Assam Tribune*. Vitamin-A overdose likely, November 14, 2001.
59. *Assam Tribune*. Mystery shrouds Vitamin A deaths, November 28, 2001.
60. BBC News. UNICEF denies Assam vaccine deaths, Monday, November 19, 2001, 10:55 GMT [Online] http://news.bbc.co.uk/2/hi/health/1664132.stm (Accessed May 10, 2012).
61. BBC News. Child deaths "Assam and UNICEF's fault", Thursday, January 17, 2002, 11:34 GMT [Online] http://news.bbc.co.uk/2/hi/health/1765777.stm (Accessed May 10, 2012).
62. BBC News. India child deaths blamed on UNICEF, Wednesday, September 3, 2003, 23:11 GMT. [Online] http://news.bbc.co.uk/2/hi/south_asia/3079438.stm (Accessed May 10, 2012).
63. USAID. Expanding immunization coverage to protect children in India, 2011. http://www.mchip.net/node/265 (Accessed July 16, 2011).
64. NIPCCD (National Institute of Public Cooperation and Child Development). Role of village health committees in improving health and nutrition outcomes: A review of evidence from India, 2008. http://nipccd.nic.in/mch/er/ervh.pdf (Accessed August 3, 2011).
65. NRHM. Better healthcare service for the poor, 2008. http://www.nrhmcommunityaction.org/media/documents/Community_Entitlement_Book_English.pdf (Accessed July 16, 2011).
66. Kadri, A. M., Singh, A., Jain, S., Mahajan, R. G., Trivedi, A. Study on immunization coverage in urban slums of Ahmedabad city. *Health and Population: Perspectives and Issues*, 2010; 33(1): 50–54.
67. Sharma, S. Immunization coverage in India. Institute of economic growth, New Delhi, 2007. http://www.iegindia.org/workpap/wp283.pdf (Accessed May 20, 2011).
68. Nair, R., Ayoya, M. A., Pandge, G., Dakure, D., Aguayo, V. High and sustained coverage with vitamin A supplements and deworming tablets can and must be achieved for Indian children: Experience and lessons learned in Maharashtra, 2008. http://www.sightandlife.org/micronutForumBejing09/PDFs/Poster%20Presentations/4_Friday/1.%20MN%20Program%20Management,%20Monitoring,%20&%20Sustainability/F20_NAir.pdf (Accessed June 14, 2011).
69. Swami, H. M., Thakur, J. S., Bhatia, S. P. S. Mass supplementation of vitamin A linked to national immunization day. *Indian Journal of Pediatrics*, 2002; 69(8): 675–678.
70. Gorstein, J., Bhaskaram, P., Khanum, S., Hossaini, R., Balakrishna, N., Goodman, T. S., Benoist, B. D., Krishnaswamy, K. Safety and impact of vitamin A supplementation delivered with oral polio vaccine as part of the immunization campaign in Orissa, India. *Food and Nutrition Bulletin* 2003; 24(4): 319–331.
71. Ramachandran, R. A programme gone awry. *India's National Magazine*, 2001; 18(25): 08–21.
72. UNICEF Bangladesh. Control of vitamin A deficiency, 2003. http://www.unicef.org/bangladesh/health_nutrition_444.htm (Accessed June 16, 2011).
73. Horton, S., Begin, F., Greig, A., Lakshman, A. Best practice paper: Micronutrient supplements for child survival (Vitamin A and Zinc), 2008. http://www.vitaminangels.org/micronutrient-library/micronutrient-supplements-child-survival-vitamin-and-zinc-horton-s-begin-f-et- (Accessed April 15, 2011).
74. DHS (Demographic Health Survey) Bangladesh, key findings, 2007. http://www.measuredhs.com/pubs/pdf/SR158/SR158.pdf (Accessed June 16, 2011).
75. UNDP. Millenium development goals. Goal 4: Reduce child mortality, 2009. http://www.undp.org.bd/mdgs/goals/MDG%204.pdf (Accessed June 16, 2011).

76. HKI (Helen Keller International). Vitamin A capsules: Red and blue. What's the difference? *Crisis Bulletin*, 2000; 2(5): 1–4.

77. Berger, S., De Pee, S., Bloem, M. W., Halati, S., Semba, R. D. Malnutrition and morbidity are higher in children who are missed by periodic vitamin A capsule distribution for child survival in rural Indonesia. *Journal of Nutrition*, 2007; 137: 1328–1333.

78. MOH (Ministry of Health) Indonesia. Evaluation of vitamin A supplementation program in three provinces in Indonesia, 2007. http://www.gizi.net/makalah/download/abstract_final_report-suplement-vita.pdf (Accessed June 17, 2011).

79. Soekirman. The revival of food based programmes—Including fortification. *World Nutrition*, 2010; 1(2): 13–18.

80. WHO. Integration of vitamin A supplementation with immunization: Policy and programme implications, 1998. http://www.who.int/vaccines-documents/DocsPDF/www9837.pdf (Accessed June 16, 2011).

81. Murthy, K. R., Klugman, B. Service accountability and community participation in the context of health sector reforms in Asia: Implications for sexual and reproductive health services. *Health Policy and Planning*, 2004; 19(Suppl.1): i78–i86.

82. EPI. The final stages of polio eradication: Columbia faces the challenge. *EPI Newsletter*, 1991; 13(3): 1–8.

83. Parthasarathy, A. Community pediatrics and national child health programs. In IAP Textbook of Pediatrics, 3rd edn. (pp. 165). Jaypee Brothers, New Delhi, India, 2005.

84. Stenhouse, C., Cunningham, M. Guidelines for the introduction of outreach services, 2002. http://www.ics.ac.uk/intensive_care_professional/standards_and_guidelines/guidelines_for_the_introduction_of_outreach_2003 (Accessed July 16, 2011).

85. WHO. *Vitamin and Mineral Requirements in Human Nutrition*, 2nd edn. World Health Organization and Food and Agriculture Organization of the United Nations, Geneva, Switzerland, 2004.

# 20 Consequences of Common Genetic Variations on β-Carotene Cleavage for Vitamin A Supply

*Georg Lietz, Anthony Oxley, and Christine Boesch-Saadatmandi*

## CONTENTS

## FACTORS AFFECTING THE BIOEFFICACY OF PROVITAMIN A CAROTENOIDS

Because humans are unable to synthesize vitamin A *de novo*, they must consume diets with preformed vitamin A, predominantly as retinyl esters, or provitamin A carotenoids, such as β-carotene and carotenoids containing an unsubstituted β-ionone ring [1,2]. β-Carotene constitutes the main provitamin A source with a daily consumption of around 1–3 mg and is the most suitable and important precursor for vitamin A due to its symmetrical structure [3–5]. However, a number of factors such as the food matrix, the nutrient status of the host, and genetic factors can affect the utilization of β-carotene in humans and are summarized by the mnemonic SLAMENGHI (reviewed by Castenmiller and West [6]).

Dietary carotenoids follow the same absorptive pathways as dietary lipids and are transported in the bloodstream exclusively by lipoproteins, with the adipose tissue being their main storage site [1]. At the intestinal level, carotenoid absorption is dependent on three steps: (1) absorption at the enterocyte brush border membrane, (2) enzymatic conversion of a fraction of absorbed provitamin A

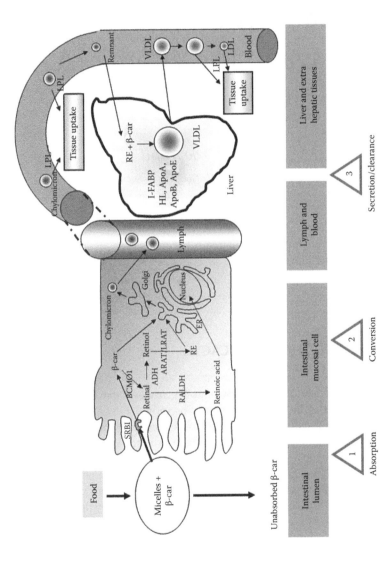

**FIGURE 20.1** Key regulatory proteins influencing provitamin A carotenoid status. This diagram illustrates the influence of key regulatory genes on carotenoid absorption, conversion to vitamin A, incorporation into chylomicrons, and their subsequent clearance. ADH, alcohol dehydrogenases; ARAT, acyl-CoA:retinol acyltransferase; ER, endoplasmic reticulum; LPL, lipoprotein lipase; LRAT, lecithin-retinol acyltransferase; RALDH, retinal-dehyde dehydrogenase; RE, retinyl ester. (From Lietz, G. and Hesketh, J., *Curr. Opin. Lipidol.*, 20, 112, 2009. With permission.)

carotenoids, and (3) secretion of chylomicron particles (Figure 20.1). Although intestinal uptake was thought to be a passive process with simple diffusion over the brush border membrane following a concentration gradient [1], it is now clear that this process is partly mediated by the scavenger receptor class B type I (SR-BI) [7–11]. During et al. [10] showed that absorption of β-carotene is partly dependent on the SR-BI intestinal transporter by demonstrating a 60% decrease in β-carotene absorption through inhibition of SR-BI transport. Expressed at high levels in the intestine, the SR-BI transporter is also involved in the process of cholesterol absorption [12]. However, since intestinal cholesterol absorption is not facilitated by a single transporter protein, this suggests that other proteins such as cluster determinant 36 (CD36), ATP-binding cassette subfamily G member 5 (ABCG5), ATP-binding cassette transporter A1, and Niemann Pick C1-like 1 (NPC1L1) could also be involved in intestinal provitamin A uptake [10,12–14]. Although transporter proteins other than SR-BI affect carotenoid absorption in various tissues (i.e., CD36, ABCG5, and NPC1L1), there is little evidence to date to support their role in carotenoid uptake at the enterocyte. For example, only a near-significant association between plasma lutein and the Q640E variant in ABCG5 was observed in a human intervention study [15], whereas the involvement of CD36 in β-carotene uptake has been confirmed only in mouse adipocytes [16], and no evidence for the role of NPC1L1 in carotenoid uptake exists to date [13]. On the other hand, there are indications that other, not yet identified, epithelial transporters are involved in carotenoid absorption across the brush border membrane, since all-*trans*-β-carotene is preferentially accumulated over 9-*cis* in total plasma and the postprandial triacylglycerol-rich lipoprotein fraction [17–20].

In humans, between 35% and 90% of absorbed all-*trans* β-carotene is oxidatively cleaved by the β,β-carotene 15, 15′-monooxygenase 1 (BCMO1; EC 1.14.99.36) into two molecules of all-*trans* retinal, which subsequently can be oxidized irreversibly to retinoic acid by retinal dehydrogenase or reduced reversibly to retinol by a retinal reductase [1,2]. The other carotenoid cleavage enzyme β,β-carotene-9,10-dioxygenase (BCDO2; (EC 1.14.99.-) cleaves β-carotene at the 9,10 double bond forming β-apo-10-carotenal and β-ionone [21]. The contribution of BCDO2 to vitamin A formation has long been debated and thought to occur via a mechanism similar to β-oxidation [22–27]. However, BCMO1 knockout (KO) mice become vitamin A deficient despite expressing BCDO2 [28]. Furthermore, hepatic BCDO2 expression has been found to be significantly elevated in BCMO1 KO mice compared to wild-type mice, leading to a significant increase in β-apo-12′-carotenal [29] or β-apo-10′-carotenal concentration [30]. Finally, studies in BCMO1 KO mice revealed a large accumulation of β-carotene in tissues (liver, lung, adipose tissue) of animals on a β-carotene-enriched diet [28,30,31]. Thus, in agreement with earlier findings [23], BCMO1 is currently considered the key enzyme for β-carotene conversion into vitamin A.

Unlike initial expectations, β-carotene cleavage is not limited to the digestive tract. In a variety of human tissues, presence of BCMO1 was documented by RNA blotting and immunostaining methods [32,33]. As summarized in von Lintig et al. [34], BCMO1 is expressed in the mucosal and glandular cells of the stomach, small intestine, and colon as well as in hepatocytes and cells comprising the exocrine glands in pancreas, glandular cells in prostate, endometrium, and mammary tissue, kidney cells, and keratinocytes of skin squamous epithelium. Moreover, skeletal muscle cells as well as cells of testis, ovary, and adrenal gland with steroidogenic properties express BCMO1. In addition, BCMO1 expression was found in the retinal pigment epithelium and in the ciliary body pigment epithelia of the eye [33]. The expression/activity of BCMO1 in extra-intestinal tissues has been linked to the capacity of these tissues to directly convert locally stored carotenoids into vitamin A [33]. This has recently been revealed to be of importance in embryonic retinoid metabolism, since β-carotene was shown to serve as an alternative vitamin A source for the *in situ* synthesis of retinoids in developing tissues [35]. Furthermore, local *de novo* synthesis of retinal and/or retinoic acid has shown to regulate liver lipid homeostasis since BCMO1 KO mice develop liver steatosis independent of the vitamin A status of the animal [28]. Tissue-specific β-carotene conversion was also shown to be important for the regulation of body fat, adipose depot mass, and leptin serum concentrations [30].

It is well established that dietary fats increase the intestinal absorption of lipophilic carotenoids through increased incorporation of carotenoids into mixed micelles, with 3–5 g of fat per meal assuring efficient absorption of β-carotene in humans [13,36–40]. Furthermore, the micellarization process of carotenoids was shown to be affected by the length of fatty acyl chains but not their degree of saturation [13,41]. The amount of fat has also shown to influence BCMO1 activity, since increasing fat intake enhanced the post-absorptive conversion of β-carotene in the liver [42]. Likewise, significantly elevated vitamin A concentrations, predominantly as retinyl esters, in several tissues such as liver, lung, kidney, adipose tissue, and testis were observed with increasing fat intake [43]. Intestinal BCMO1 activity was 45% and 58% higher in rats fed a (n-3) PUFA or MUFA diet compared to those fed a (n-6) PUFA diet. The ratios of hepatic retinyl palmitate to hepatic β-carotene also indicated higher conversion rates in rats fed a MUFA and (n-3) PUFA diet compared to rats fed a (n-6) PUFA diet [44]. On the other hand, protein-deficient diets significantly reduce intestinal BCMO1 enzyme activity [45,46], whereas protein-enriched diets revealed significantly higher vitamin A levels in liver (twofold), lung (threefold), and adipose tissue (1.8-fold) [43]. Vitamin A status of the host affects the absorption of β-carotene and is possibly the most important factor that influences BCMO1 activity. It was recently shown that a diet-responsive regulatory network exists at the intestinal level that controls β-carotene absorption and vitamin A production by negative feedback mechanism [47] (Figure 20.2). The intestine-specific homeodomain transcription factor ISX expression is activated by retinoic acid via retinoic acid receptors (RARs) that bind to a specific retinoic acid response element (RARE) within the ISX promoter. Once activated, ISX represses intestinal scavenger receptor class B type 1 (SR-BI) and BCMO1 expression, indicating that intestinal vitamin A uptake and production are under negative feedback control via induction of ISX expression [47].

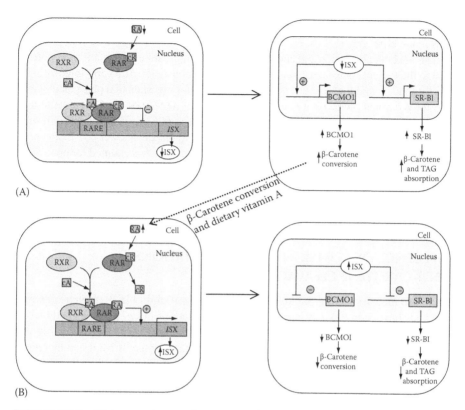

**FIGURE 20.2**   Diet-responsive regulatory network involving the intestine-specific home-odomain transcription factor ISX. (A) Vitamin A deficiency: ISX expression is reduced due to low retinoic acid (RA) and RAR concentrations. SRBI and BCMO1 expression are increased significantly in the small intestine. Enhanced SRBI activity facilitates the absorption of various lipids and carotenoids. (B) Vitamin A sufficiency: RA derived either from β-carotene conversion or from preformed dietary retinoids promotes the binding of RARs, therefore inducing ISX expression. Induction of ISX then leads to the repression of intestinal SR-BI and BCMO1 expression. cR, Co-repressor; cA, co-activator; RARE, retinoic acid response element; RAR, retinoic acid receptor; RXR, retinoid X receptor; ISX, intestine-specific homeodomain transcription factor; SR-BI, intestinal scavenger receptor class B type 1; BCMO1, β, β-carotene 15,15-monooxygenase 1. (From Lietz, G. et al., *Arch. Biochem. Biophys.*, 502, 8, 2010. With permission.)

## LOW RESPONDER/CONVERTER PHENOTYPE

β-Carotene conversion efficiency during absorption varies widely from person to person, even within studies that were conducted in relatively homogeneous groups [13,48–51]. Human intervention studies applying a double-tracer methodology classified 27%–45% of volunteers as poor responders [49–51]. However, there is some confusion about the definition of the low responder phenotype in the literature, as some publications describe the phenotype as a low or nonincrease in plasma β-carotene concentrations after dosing [13,52–54], whereas others use it in relation to provitamin A conversion efficiency [50,55]. Earlier studies concluded that variations in responses between subjects may be caused by differences in absorption and transport of β-carotene rather than by differing conversion efficiency [13,53]. However, more recent studies using stable isotopes concluded that both absorption and conversion of ingested β-carotene to vitamin A contributed to the variable plasma response to ingested β-carotene [50,51,55,56]. Interestingly, low responders and low converters were observed after ingestion of a pharmacological dose of β-carotene with interindividual variations of newly absorbed β-carotene and retinyl palmitate/β-carotene ratios ranging from 53% to 60%, respectively [57]. In support of this finding, low responders were also found to display a lower conversion efficiency [51,58]. Finally, post-intestinal conversion of vitamin A accounted for up to 30% of the total converted retinol over the period of 52 days [51]. Thus, observed variations in cleavage efficiency between individuals could be caused by two aspects: (a) by factors influencing the expression and activity of intestinal BCMO1 and (b) by variations in post-intestinal conversion of provitamin A carotenoids.

## INFLUENCE OF COMMON GENETIC VARIATIONS ON β-CAROTENE CLEAVAGE

There are indications that the high interindividual variability in β-carotene absorption and conversion is partly caused by genetic variations in key enzymes and receptor proteins involved in provitamin A metabolism [2,59,60].

Leung et al. [57] screened the total open reading frame of the BCMO1 coding region for new single nucleotide polymorphisms (SNPs) and identified five SNPs within the coding region, of which two were non-synonymous (R267S: rs12934922; A379V: rs7501331). A subsequent β-carotene supplementation study with healthy female volunteers revealed that carriers of the 379V and 267S/379V variant alleles had 160% and 240% higher fasting β-carotene concentrations combined with 32% and 69% reduced conversion efficiency, respectively (Figure 20.3) [57]. The observation of high SNP frequencies combined with altered β-carotene metabolism provided for the first time a putative explanation for the poor responder phenotype in β-carotene metabolism (Tables 20.1 and 20.2).

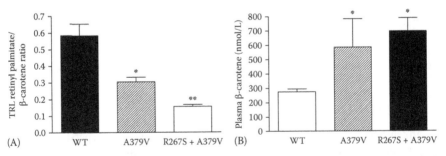

**FIGURE 20.3** Two common non-synonymous SNPs in the BCMO1 gene affect provitamin A conversion efficiency in healthy female volunteers. (A) TRL retinyl palmitate:β-carotene ratio responses after intake of a β-carotene-rich meal (120 mg). (B) Baseline fasting plasma β-carotene concentrations. Data displayed as means ± SEM. *$P < 0.05$, **$P < 0.01$ vs. wild type; one-way ANOVA. Wild type, AA for R267S and CC for A379V (n = 17); A379V, at least one T allele (n = 7); R267S + A379V, at least one T allele in both R267S and A379V (n = 4). (Adapted from Leung, W.C. et al., *FASEB J.*, 23, 1041, 2009.)

**TABLE 20.1**
**Allele Frequencies of rs7501331, rs12934922, rs11645428, rs6420424, rs8044334, and rs6564851 in 11 Different Ethnic Groups[a]**

| Genotype | Genotype Frequency per Population Group (%) | | | | | | | | | | |
|---|---|---|---|---|---|---|---|---|---|---|---|
| | ASW | CEU | CHB | CHD | GIH | JPT | LWK | MEX | MKK | TSI | YRI |
| *rs7501331* | | | | | | | | | | | |
| CC | 89.5 | 45.9 | 67.9 | 72.2 | 58.6 | 71.8 | | 75.4 | 96.8 | 72.5 | 100 |
| CT | 10.5 | 49.5 | 31.4 | 26.9 | 33.3 | 27.3 | | 21.1 | 3.2 | 23.5 | 0.0 |
| TT | 0.0 | 4.5 | 0.7 | 0.9 | 8.1 | 0.9 | | 3.5 | 0.0 | 3.9 | 0.0 |
| *rs12934922* | | | | | | | | | | | |
| AA | | 31.7 | 68.2 | | | 72.1 | | | | | 80.3 |
| AT | | 49.2 | 29.5 | | | 27.9 | | | | | 0.0 |
| TT | | 19 | 2.3 | | | 0.0 | | | | | 1.6 |
| *rs6420424* | | | | | | | | | | | |
| GG | 17.9 | 22.3 | 2.2 | 4.6 | 28.0 | 2.7 | 20.0 | 15.5 | 45.1 | 32.3 | 31.9 |
| AG | 50.0 | 49.1 | 30.7 | 26.6 | 46.0 | 26.5 | 61.8 | 48.3 | 40.5 | 55.6 | 50.7 |
| AA | 32.1 | 28.6 | 67.2 | 68.8 | 26.0 | 70.8 | 18.2 | 36.2 | 14.4 | 12.1 | 17.4 |
| *rs8044334* | | | | | | | | | | | |
| GG | 29.8 | 14.4 | 16.8 | 15.6 | 6.0 | 12.5 | 24.5 | 17.2 | 23.1 | 5.9 | 27.4 |
| GT | 49.1 | 46.8 | 48.9 | 43.1 | 38.0 | 46.4 | 58.2 | 41.4 | 57.7 | 48.0 | 61.0 |
| TT | 21.1 | 38.7 | 34.3 | 41.3 | 56.0 | 41.1 | 17.3 | 41.4 | 19.2 | 46.1 | 11.6 |
| *rs11645428* | | | | | | | | | | | |
| AA | 0.0 | 10.6 | 0.0 | 0.0 | 2.0 | 0.0 | 0.9 | 3.4 | 0.6 | 21.6 | 0.0 |
| AG | 19.3 | 40.7 | 0.0 | 4.6 | 23.8 | 0.0 | 11.1 | 36.2 | 26.3 | 52.9 | 13.6 |
| GG | 80.7 | 48.7 | 100.0 | 95.4 | 74.3 | 100.0 | 88.0 | 60.3 | 73.1 | 25.5 | 86.4 |
| *rs6564851* | | | | | | | | | | | |
| TT | 33.3 | 18.9 | 2.2 | 3.7 | 34.0 | 2.7 | 30.2 | 12.3 | 35.5 | 29.7 | 45.1 |
| GT | 45.6 | 49.5 | 30.9 | 31.5 | 43.0 | 33.9 | 56.6 | 49.1 | 51.0 | 58.4 | 44.4 |
| GG | 21.1 | 31.5 | 66.9 | 64.8 | 23.0 | 63.4 | 13.2 | 38.6 | 13.5 | 11.9 | 10.4 |

*Source:* Adapted from Lietz, G. et al., *J. Nutr.*, 142, 161S, 2011. American Society for Nutrition.

ASW, African ancestry in Southwest United States; CEU, Utah residents with Northern and Western European ancestry from the CEPH collection; CHB, Han Chinese in Beijing, China; CHD, Chinese in Metropolitan Denver, Colorado; GIH, Gujarati Indians in Houston, Texas; JPT, Japanese in Tokyo, Japan; LWK, Luhya in Webuye, Kenya; MEX, Mexican ancestry in Los Angeles, California; MKK, Maasai in Kinyawa, Kenya; TSI, Tuscan in Italy; YRI, Yoruba in Ibadan, Nigeria.

[a] According to HapMap [66].

On the other hand, Lindqvist et al. described a mutation in the BCMO1 gene (T170M) of a subject with hypercarotenemia and mild hypovitaminosis A [61]. In vitro enzyme assays confirmed that the activity of the 170M variant was reduced by 92%, indicating that the amino acid substitution of the small and hydrophilic threonine to the hydrophobic methionine led to the loss of BCMO1 function [61]. Furthermore, this substitution occurred next to a highly conserved histidine at

**TABLE 20.2**

**Single-Nucleotide Polymorphisms in the Coding Region and 5′ Upstream of BCMO1**

| SNP Identification | Polymorphism and Predicted Change | Evidence for Functional Effect |
|---|---|---|
| R267S (rs12934922) | C/T polymorphism that results in Ala → Val substitution | No effect on BCMO1 activity in vitro and in vivo |
| A379V (rs7501331) | A/T polymorphism that results in Arg → Ser substitution | 32% reduced conversion of β-carotene after a pharmacological dose in female volunteers |
| T170M (rs119478057) | C/T polymorphism that resulted in Thr → Met substitution | 90% reduced BCMO1 activity compared to wild type in vitro; causing hypercarotenemia and hypovitaminosis A |
| R267S + A379V | A/T and C/T polymorphisms combined with at least one heterozygote each | 57% reduced BCMO1 activity compared to wild type in vitro 69% reduced conversion of β-carotene after a pharmacological dose in female volunteers |
| 5′ Intron (rs6420424) | A/G polymorphism upstream of BCMO1 | 59% reduced conversion of β-carotene after a pharmacological dose in female volunteers; variant modulates β-carotene status |
| 5′ Intron (rs8044334) | G/T polymorphism upstream of BCMO1 | Variant modulates β-carotene status |
| 5′ Intron (rs11645428) | A/G polymorphism upstream of BCMO1 | 51% reduced conversion of β-carotene after a pharmacological dose in female volunteers; variant modulates β-carotene status |
| 5′ Intron (rs6564851) | G/T polymorphism upstream of BCMO1 | 48% reduced conversion of β-carotene after a pharmacological dose in female volunteers; variant modulates β-carotene and lutein status |

position 172, which was suggested to be one of the four iron-ligating histidines in the active center of carotenoid-cleaving enzymes [62]. The coordination of the iron within the enzyme could be altered due to the amino acid substitution, which might cause structural changes in the active center and reduce the substrate binding and therefore the cleavage activity [61]. However, given the very low frequency of the 170M variation [57], this mutation cannot explain the high interindividual variability in β-carotene absorption and conversion observed in humans.

Screening for SNPs in the human BCMO1 promoter region identified by Gong et al. [63] did not reveal any genetic variations in two different study populations [57,61]. However, a genome-wide association study investigating the effects of common genetic variations on circulating carotenoid levels identified four new SNPs upstream of the BCMO1 gene [64]. The strongest effect on carotenoid status was shown by the rs6564851 SNP, which was associated with increased β-carotene and α-carotene levels and reduced non-provitamin A carotenoid levels, such as lutein, zeaxanthin, and lycopene. Subsequent analysis of these upstream intronic SNPs on β-carotene conversion efficiency revealed that three of the four polymorphisms (rs6420424, rs11645428, and rs6564851) reduce the catalytic activity of BCMO1 in female volunteers by 59%, 51%, and 48%, respectively (Figure 20.4) [60]. More importantly, SNPs negatively affecting provitamin A conversion efficiency occurred

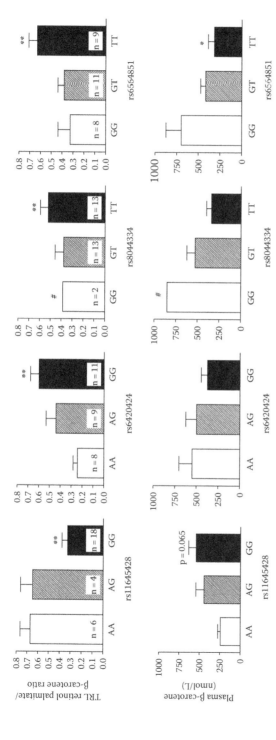

**FIGURE 20.4** SNPs upstream of the BCMO1 gene affect provitamin A conversion efficiency in healthy female volunteers. TRL retinyl palmitate/β-carotene ratios and baseline fasting plasma β-carotene concentrations after intake of a β-carotene-rich meal (120 mg) in 28 female volunteers depending on genotype. Data are displayed as mean ± SEM. * Significantly different from other homozygous allele at P < 0.05; ** significantly different from other homozygous allele at P < 0.01 (ANOVA). ■, reference allele; □, other allele. rs6420424 was analyzed using rs12597639 as the tag SNP. # = no SEM due to small n. (From Lietz, G. et al., *J. Nutr.*, 142, 161S, 2011. American Society for Nutrition.)

at high allele frequencies between 30% and 71% (Table 20.1) [60]. The observed variable response to β-carotene, which has led to the characterization of the poor responder phenotype in up to 45% of volunteers in double-tracer studies [49–51], is therefore most likely caused by a combination of non-synonymous and intronic SNPs within and upstream of BCMO1.

## CONCLUSION

It has been demonstrated that provitamin A conversion is influenced by multiple SNPs and that genetic variability may provide an explanation for the molecular basis of the poor responder phenotype within the population. The importance for using genetic variability to guide local recommendations and policies is further strengthened by the observation of clear differences in SNP frequencies across different ethnic groups. However, no studies to date have investigated how genetic variations can modulate the interindividual response to provitamin A supplementation in a vitamin A–deprived community. Furthermore, there is so far no information available whether genetic variability in key enzymes could precipitate vitamin A deficiency in individuals who mostly rely on dietary provitamin A sources and/or whether other host- or dietary-related factors are more important. Currently, we also do not understand if the genetic basis of reduced BCMO1 enzyme activity could simply be overcome by increasing β-carotene intake, or whether in such instances, only increased preformed vitamin A intake will alleviate vitamin A deficiency. Research to understand the importance of genetic variations of BCMO1 and other important key enzymes and transport proteins in provitamin A metabolism is therefore urgently needed.

## REFERENCES

1. Parker RS. Absorption, metabolism, and transport of carotenoids. *The FASEB Journal* 1996; 10: 542–551.
2. Lietz G, Lange J, Rimbach G. Molecular and dietary regulation of beta,beta-carotene 15,15′-monooxygenase 1 (BCMO1). *Archives of Biochemistry and Biophysics* 2010; 502: 8–16.
3. Granado F, Olmedilla B, Bianco I, Rojashidalgo E. Major fruit and vegetable contributors to the main serum carotenoids in the Spanish diet. *European Journal of Clinical Nutrition* 1996; 50: 246–250.
4. Strobel M, Tinz J, Biesalski HK. The importance of beta-carotene as a source of vitamin A with special regard to pregnant and breastfeeding women. *European Journal of Nutrition* 2007; 46 (Suppl 1): 11–20.
5. Grune T, Lietz G, Palou A et al. Beta-carotene is an important vitamin A source for humans. *The Journal of Nutrition* 2010; 140: 2268S–2285S.
6. Castenmiller JJ, West CE. Bioavailability and bioconversion of carotenoids. *Annual Review of Nutrition* 1998; 18: 19–38.
7. Moussa M, Landrier JF, Reboul E et al. Lycopene absorption in human intestinal cells and in mice involves scavenger receptor class B type I but not Niemann-Pick C1-like 1. *The Journal of Nutrition* 2008; 138: 1432–1436.
8. van Bennekum A, Werder M, Thuahnai ST et al. Class B scavenger receptor-mediated intestinal absorption of dietary beta-carotene and cholesterol. *Biochemistry* 2005; 44: 4517–4525.

9. Yonekura L, Nagao A. Intestinal absorption of dietary carotenoids. *Molecular Nutrition & Food Research* 2007; 51: 107–115.

10. During A, Dawson HD, Harrison EH. Carotenoid transport is decreased and expression of the lipid transporters SR-BI, NPC1L1, and ABCA1 is downregulated in Caco-2 cells treated with ezetimibe. *The Journal of Nutrition* 2005; 135: 2305–2312.

11. During A, Hussain MM, Morel DW, Harrison EH. Carotenoid uptake and secretion by CaCo-2 cells: Beta-carotene isomer selectivity and carotenoid interactions. *Journal of Lipid Research* 2002; 43: 1086–1095.

12. Werder M, Han CH, Wehrli E, Bimmler D, Schulthess G, Hauser H. Role of scavenger receptors SR-BI and CD36 in selective sterol uptake in the small intestine. *Biochemistry* 2001; 40: 11643–11650.

13. Borel P, Grolier P, Mekki N et al. Low and high responders to pharmacological doses of beta-carotene: Proportion in the population, mechanisms involved and consequences on beta-carotene metabolism. *Journal of Lipid Research* 1998; 39: 2250–2260.

14. Kramer W, Girbig F, Corsiero D et al. Intestinal cholesterol absorption: Identification of different binding proteins for cholesterol and cholesterol absorption inhibitors in the enterocyte brush border membrane. *Biochim Biophys Acta* 2003; 1633: 13–26.

15. Herron KL, McGrane MM, Waters D et al. The ABCG5 polymorphism contributes to individual responses to dietary cholesterol and carotenoids in eggs. *The Journal of Nutrition* 2006; 136: 1161–1165.

16. Moussa M, Gouranton E, Gleize B et al. CD36 is involved in lycopene and lutein uptake by adipocytes and adipose tissue cultures. *Molecular Nutrition & Food Research* 2011; 55: 578–584.

17. Gaziano JM, Johnson EJ, Russell RM et al. Discrimination in absorption or transport of beta-carotene isomers after oral supplementation with either all-*trans*-beta-carotene or 9-*cis*-beta-carotene. *American Journal of Clinical Nutrition* 1995; 61: 1248–1252.

18. Johnson EJ, Qin J, Krinsky NI, Russell RM. Concentrations of beta-carotene isomers in serum, breast milk, and buccal mucosa cell after chronic oral doses of all-*trans* (tBC) and 9-*cis* (9cBC) beta-carotene. *FASEB Journal* 1997; 11: 2263.

19. Stahl W, Schwarz W, Vonlaar J, Sies H. All-*trans* beta-carotene preferentially accumulates in human chylomicrons and very-low-density lipoproteins compared with the 9-*cis* geometrical isomer. *The Journal of Nutrition* 1995; 125: 2128–2133.

20. You CS, Parker RS, Goodman KJ, Swanson JE, Corso TN. Evidence of *cis-trans* isomerization of 9-*cis*-beta-carotene during absorption in humans. *American Journal of Clinical Nutrition* 1996; 64: 177–183.

21. Poliakov E, Gentleman S, Chander P et al. Biochemical evidence for the tyrosine involvement in cationic intermediate stabilization in mouse beta-carotene 15, 15′-monooxygenase. *BMC Biochemistry* 2009; 10: 31.

22. Barua AB, Olson JA. Beta-carotene is converted primarily to retinoids in rats in vivo. *The Journal of Nutrition* 2000; 130: 1996–2001.

23. Olson JA. Provitamin A function of carotenoids: The conversion of beta-carotene into vitamin A. *The Journal of Nutrition* 1989; 119: 105–108.

24. Wang XD, Krinsky NI, Tang GW, Russell RM. Retinoic acid can be produced from excentric cleavage of beta-carotene in human intestinal mucosa. *Archives of Biochemistry and Biophysics* 1992; 293: 298–304.

25. Wang XD, Russell RM, Liu C, Stickel F, Smith DE, Krinsky NI. Beta-oxidation in rabbit liver in vitro and in the perfused ferret liver contributes to retinoic acid biosynthesis from beta-apocarotenoic acids. *The Journal of Biological Chemistry* 1996; 271: 26490–26498.

26. Wolf G, Phil D. The enzymatic cleavage of beta-carotene—Still controversial. *Nutrition Reviews* 1995; 53: 134–137.

27. Wyss A. Carotene oxygenases: A new family of double bond cleavage enzymes. *The Journal of Nutrition* 2004; 134: 246S–250S.

28. Hessel S, Eichinger A, Isken A et al. CMO1 deficiency abolishes vitamin A production from beta-carotene and alters lipid metabolism in mice. *The Journal of Biological Chemistry* 2007; 282: 33553–33561.

29. Shmarakov I, Fleshman MK, D'Ambrosio DN et al. Hepatic stellate cells are an important cellular site for beta-carotene conversion to retinoid. *Archives of Biochemistry and Biophysics* 2010; 504: 3–10.

30. Amengual J, Gouranton E, van Helden YG et al. Beta-carotene reduces body adiposity of mice via BCMO1. *PLoS One* 2011; 6: e20644.

31. Lindshield BL, King JL, Wyss A et al. Lycopene biodistribution is altered in 15,15′-carotenoid monooxygenase knockout mice. *The Journal of Nutrition* 2008; 138: 2367–2371.

32. Lindqvist A, Andersson S. Biochemical properties of purified recombinant human beta-carotene 15,15′-monooxygenase. *The Journal of Biological Chemistry* 2002; 277: 23942–23948.

33. Lindqvist A, Andersson S. Cell type-specific expression of beta-carotene 15,15′-monooxygenase in human tissues. *Journal of Histochemistry and Cytochemistry* 2004; 52: 491–499.

34. von Lintig J. Colors with functions: Elucidating the biochemical and molecular basis of carotenoid metabolism. *Annual Review of Nutrition* 2010; 30: 35–56.

35. Kim YK, Wassef L, Chung S et al. Beta-carotene and its cleavage enzyme beta-carotene-15,15′-oxygenase (CMOI) affect retinoid metabolism in developing tissues. *The FASEB Journal* 2011; 25: 1641–1652.

36. Dimitrov NV, Meyer C, Ullrey DE et al. Bioavailability of beta-carotene in humans. *The American Journal of Clinical Nutrition* 1988; 48: 298–304.

37. Jalal F, Nesheim MC, Agus Z, Sanjur D, Habicht JP. Serum retinol concentrations in children are affected by food sources of beta-carotene, fat intake, and anthelmintic drug treatment. *The American Journal of Clinical Nutrition* 1998; 68: 623–629.

38. Roodenburg AJ, Leenen R, van het Hof KH, Weststrate JA, Tijburg LB. Amount of fat in the diet affects bioavailability of lutein esters but not of alpha-carotene, beta-carotene, and vitamin E in humans. *The American Journal of Clinical Nutrition* 2000; 71: 1187–1193.

39. van het Hof KH, West CE, Weststrate JA, Hautvast JG. Dietary factors that affect the bioavailability of carotenoids. *The Journal of Nutrition* 2000; 130: 503–506.

40. Hollander D, Ruble PE Jr. Beta-carotene intestinal absorption: Bile, fatty acid, pH, and flow rate effects on transport. *The American Journal of Physiology* 1978; 235: E686–E691.

41. Huo T, Ferruzzi MG, Schwartz SJ, Failla ML. Impact of fatty acyl composition and quantity of triglycerides on bioaccessibility of dietary carotenoids. *Journal of Agricultural and Food Chemistry* 2007; 55: 8950–8957.

42. Deming DM, Boileau AC, Lee CM, Erdman JW Jr. Amount of dietary fat and type of soluble fiber independently modulate postabsorptive conversion of beta-carotene to vitamin A in Mongolian gerbils. *The Journal of Nutrition* 2000; 130: 2789–2796.

43. Lakshman MR, Liu QH, Sapp R, Somanchi M, Sundaresan PR. The effects of dietary taurocholate, fat, protein, and carbohydrate on the distribution and fate of dietary beta-carotene in ferrets. *Nutrition and Cancer* 1996; 26: 49–61.

44. Raju M, Lakshminarayana R, Krishnakantha TP, Baskaran V. Micellar oleic and eicosapentaenoic acid but not linoleic acid influences the beta-carotene uptake and its cleavage into retinol in rats. *Molecular and Cellular Biochemistry* 2006; 288: 7–15.

45. Parvin SG, Sivakumar B. Nutritional status affects intestinal carotene cleavage activity and carotene conversion to vitamin A in rats. *The Journal of Nutrition* 2000; 130: 573–577.

46. Hosotani K, Kitagawa M. Effects of dietary protein, fat and beta-carotene levels on beta-carotene absorption in rats. *International Journal for Vitamin and Nutrition Research* 2005; 75: 274–280.

47. Lobo GP, Hessel S, Eichinger A et al. ISX is a retinoic acid-sensitive gatekeeper that controls intestinal {beta},{beta}-carotene absorption and vitamin A production. *The FASEB Journal* 2010; 24(6): 1656–1666.

48. Edwards AJ, You CS, Swanson JE, Parker RS. A novel extrinsic reference method for assessing the vitamin A value of plant foods. *The American Journal of Clinical Nutrition* 2001; 74: 348–355.

49. Hickenbottom SJ, Follett JR, Lin Y et al. Variability in conversion of beta-carotene to vitamin A in men as measured by using a double-tracer study design. *The American Journal of Clinical Nutrition* 2002; 75: 900–907.

50. Lin Y, Dueker SR, Burri BJ, Neidlinger TR, Clifford AJ. Variability of the conversion of beta-carotene to vitamin A in women measured by using a double-tracer study design. *The American Journal of Clinical Nutrition* 2000; 71: 1545–1554.

51. Wang Z, Yin S, Zhao X, Russell RM, Tang G. Beta-carotene-vitamin A equivalence in Chinese adults assessed by an isotope dilution technique. *British Journal of Nutrition* 2004; 91: 121–131.

52. Bowen PE, Garg V, Stacewicz-Sapuntzakis M, Yelton L, Schreiner RS. Variability of serum carotenoids in response to controlled diets containing six servings of fruits and vegetables per day. *Annals of the New York Academy of Sciences* 1993; 691: 241–243.

53. Brown ED, Micozzi MS, Craft NE et al. Plasma carotenoids in normal men after a single ingestion of vegetables or purified beta-carotene. *The American Journal of Clinical Nutrition* 1989; 49: 1258–1265.

54. Furr HC, Clark RM. Intestinal absorption and tissue distribution of carotenoids. *Journal of Nutritional Biochemistry* 1997; 8: 364–377.

55. Tang G, Qin J, Dolnikowski GG, Russell RM. Short-term (intestinal) and long-term (postintestinal) conversion of beta-carotene to retinol in adults as assessed by a stable-isotope reference method. *The American Journal of Clinical Nutrition* 2003; 78: 259–266.

56. Hickenbottom SJ, Lemke SL, Dueker SR et al. Dual isotope test for assessing beta-carotene cleavage to vitamin A in humans. *The European Journal of Nutrition* 2002; 41: 141–147.

57. Leung WC, Hessel S, Meplan C et al. Two common single nucleotide polymorphisms in the gene encoding beta-carotene 15,15′-monoxygenase alter beta-carotene metabolism in female volunteers. *The FASEB Journal* 2009; 23: 1041–1053.

58. Tourniaire F, Gouranton E, von Lintig J et al. Beta-carotene conversion products and their effects on adipose tissue. *Genes and Nutrition* 2009; 4: 179–187.

59. Borel P. Genetic variations involved in interindividual variability in carotenoid status. *Molecular Nutrition & Food Research* 2011; 56(2): 228–240.

60. Lietz G, Oxley A, Leung W, Hesketh J. Single nucleotide polymorphisms upstream from the beta-carotene 15,15′-monoxygenase gene influence provitamin A conversion efficiency in female volunteers. *The Journal of Nutrition* 2011; November 23. [Epub ahead of print] DOI 10.3945/jn.111.140756.

61. Lindqvist A, Sharvill J, Sharvill DE, Andersson S. Loss-of-function mutation in carotenoid 15,15′-monooxygenase identified in a patient with hypercarotenemia and hypovitaminosis A. *The Journal Nutrition* 2007; 137: 2346–2350.

62. Kloer DP, Schulz GE. Structural and biological aspects of carotenoid cleavage. *Cellular and Molecular Life Sciences: CMLS* 2006; 63: 2291–2303.

63. Gong X, Tsai SW, Yan B, Rubin LP. Cooperation between MEF2 and PPARgamma in human intestinal beta,beta-carotene 15,15′-monooxygenase gene expression. *BMC Molecular Biology* 2006; 7: 7.

64. Ferrucci L, Perry JR, Matteini A et al. Common variation in the beta-carotene 15,15′-monooxygenase 1 gene affects circulating levels of carotenoids: A genome-wide association study. *American Journal of Human Genetics* 2009; 84: 123–133.

65. Lietz G, Hesketh J. A network approach to micronutrient genetics: Interactions with lipid metabolism. *Current Opinion in Lipidology* 2009; 20: 112–120.

66. HapMap. The International HapMap Project. 2003. http://hapmap.nebi.nlm.nih.gov

# Index

## A

Acitretin and etretinate, 65–66
Acute otitis media (AOM)
  pathogenesis
    epithelial integrity, 141
    physicochemical injuries, 142
  and vitamin A
    cell-mediated immunity, 144
    immunomodulatory functions, 143
    lymphocyte proliferation, 144–145
    medical treatment, 143
    nutritional blindness and morbidity, 144
    predisposition, nasopharyngeal
      colonization, 142–143
Acute suppurative otitis media (ASOM), 143
Adenine nucleotide translocator (ANT), 235
Adipocyte differentiation-related protein
  (ADRP), 15
Adipogenesis
  cell cycle-controlling mechanisms, 27
  endocrine and/or dietary stimuli, 28
  multipotent mesenchymal stem cells, 26
  nutrients and hormones, 27
  preadipocytes, 27
Adipose triglyceride lipase (ATGL), 7, 28
ADRP, *see* Adipocyte differentiation-related
  protein (ADRP)
Age-related macular degeneration (AMD)
  PUFA, 78
  RPE, 78
  vision disorder, 77
  zeaxanthin stereoisomers, 79
Alitretinoin, 66–67
AMD, *see* Age-related macular
  degeneration (AMD)
ANM, *see* Auxiliary nurse midwife (ANM)
ANT, *see* Adenine nucleotide translocator (ANT)
Antioxidative and pro-oxidative mechanisms
  β-carotene and α-tocopherol, 236
  carotenoids, 233–234
  hypothetic scheme, 236
  retinoids, 235
  retinol, 237
AOM, *see* Acute otitis media (AOM)
ASOM, *see* Acute Suppurative Otitis Media
  (ASOM)
Astaxanthin
  diets, 292
  DSS, 291

experimental protocols, animal studies,
  291–292
ICR mice, 293–294
microalgae, 292
NF-κB, 296–298
pro-inflammatory cytokine mRNA
  expression, 294–296
structure, 290
Asthma
  M2R, 276–277
  serum retinol levels, 278
  vitamin A deficiency, 277
  vitamin C and E, 276
ATGL, *see* Adipose triglyceride lipase
  (ATGL)
Auxiliary nurse midwife (ANM), 361

## B

BAT, *see* Brown adipose tissue (BAT)
β-Carotene (BC)
  androgen receptor, 217
  apoptosis, 216
  cell cultures, 205
  cell invasion assay, 210, 211
  cell proliferation, 205
  cytotoxicity assay, 205
  and DHT, 208
  dietary factors, 217
  genes coding, proteins, 210
  growth factors and steroid hormones,
    220–221
  LNCaP, 210, 211
  microarray analysis, 213, 214
  MMP9 transcripts, 217, 218
  nuclear receptor, 219–220
  prostate cancer, 213, 219
  retinol dehydrogenase, 209
  RNA extraction, cDNA synthesis and
    RT-PCR analysis, 205–206
  RQ-PCR, 205–208
  steroid hormones, 217
  uptake, 205
Beta-carotene and retinol efficacy trial
  (CARET), 228
Bexarotene, 67–68
β-Lactoglobulin (β-LG), 29
BMI, *see* Body mass index (BMI)
Body mass index (BMI), 23
BPD, *see* Bronchopulmonary dysplasia (BPD)

**397**